AF771929

Quadratic Formula

If $ax^2 + bx + c = 0$, where $a \neq 0$, then

$$x = \frac{-b \pm \sqrt{b^2 - 4ac}}{2a}.$$

Straight Lines

$m = \dfrac{y_2 - y_1}{x_2 - x_1}$	(slope formula)
$y - y_1 = m(x - x_1)$	(point-slope form)
$y = mx + b$	(slope-intercept form)
$x = \text{constant}$	(vertical line)
$y = \text{constant}$	(horizontal line)

Inequalities

If $a < b$, then $a + c < b + c$.
If $a < b$ and $c > 0$, then $ac < bc$.
If $a < b$ and $c > 0$, then $a(-c) > b(-c)$.

Logarithms

$\log_b x = y$ where $x = b^y$

$\log_b (mn) = \log_b m + \log_b n$

$\log_b \dfrac{m}{n} = \log_b m - \log_b n$

$\log_b m^r = r \log_b m$

$\log_b 1 = 0$

$\log_b b = 1$

$\log_b b^r = r$

$b^{\log_b m} = m$

$\log_b m = \dfrac{\log_a m}{\log_a b}$

Counting

$${}_nP_r = \frac{n!}{(n-r)!}$$

$${}_nC_r = \frac{n!}{r!(n-r)!}$$

Greek Alphabet

alpha	Α	α	nu	Ν	ν
beta	Β	β	xi	Ξ	ξ
gamma	Γ	γ	omicron	Ο	o
delta	Δ	δ	pi	Π	π
epsilon	Ε	ϵ	rho	Ρ	ρ
zeta	Ζ	ζ	sigma	Σ	σ
eta	Η	η	tau	Τ	τ
theta	Θ	θ	upsilon	Υ	υ
iota	Ι	ι	phi	Φ	ϕ, φ
kappa	Κ	κ	chi	Χ	χ
lambda	Λ	λ	psi	Ψ	ψ
mu	Μ	μ	omega	Ω	ω

Mathematics for Applications

Ernest F. Haeussler, Jr.
Richard S. Paul
Larry J. Goldstein
David I. Schneider
Martha J. Siegel

From **Introductory Mathematical Analysis for Business, Economics, and the Life and Social Sciences, Ninth Edition**
by Ernest F. Haeussler, Jr. and Richard S. Paul

Finite Mathematics & Its Applications, Sixth Edition
by Larry J. Goldstein, David I. Schneider and Martha J. Siegel

Prentice Hall
Pearson Custom Publishing

Excerpts taken from:

Introductory Mathematical Analysis for Business, Economics, and the Life and Social Sciences, Ninth Edition,
by Ernest F. Haeussler, Jr. and Richard S. Paul

A Pearson Education Company
Upper Saddle River, New Jersey 07458

Finite Mathematics and Its Applications, Sixth Edition,
by Larry J. Goldstein, David I. Schneider, and Martha J. Siegel

This special edition published in cooperation with Pearson Custom Publishing.

Printed in the United States of America

10 9 8 7 6 5 4 3 2 1

Please visit our web site at www.pearsoncustom.com

ISBN 0-536-62007-5

BA 992680

PEARSON CUSTOM PUBLISHING
75 Arlington Street, Suite 300, Boston, MA 02116
A Pearson Education Company

Contents

Chapter 1: Algebra Refresher 1

Chapter 2: Equations 31

Chapter 3: Applications of
Equations and Inequalities 55

Chapter 4: Functions and Graphs 81

Chapter 5: Lines, Parabolas, and Systems 119

Answers to Odd-Numbered Problems 171

Chapter 6: Linear Equations and Straight Lines 185

Chapter 7: Matrices 235

Chapter 8: Linear Programming,
A Geometric Approach 291

Chapter 9: The Mathematics of Finance 321

Tables 1-6 353

Answers to Exercises 361

Chapter 1
Algebra Refresher

0.1 Purpose

This chapter is designed to give you a brief review of some terms and methods of manipulative mathematics. No doubt you have been exposed to much of this material before. However, because these topics are important in handling the mathematics that comes later, perhaps an immediate second exposure to them would be beneficial. Devote whatever time is necessary to the sections in which you need review.

OBJECTIVE

To become familiar with sets, the classification of real numbers, and the real-number line.

0.2 Sets and Real Numbers

In simplest terms, a *set* is a collection of objects. For example, we can speak of the set of even numbers between 5 and 11, namely, 6, 8, and 10. An object in a set is called an *element* or *member* of that set.

One way to specify a set is by listing its elements, in any order, inside braces. For example, the previous set is $\{6, 8, 10\}$, which we can denote by a letter such as A. A set A is said to be a subset of a set B if and only if every element of A is also an element of B. For example, if $A = \{6, 8, 10\}$ and $B = \{6, 8, 10, 12\}$, then A is a subset of B.

Certain sets of numbers have special names. The numbers 1, 2, 3, and so on form the set of **positive integers** (or **natural numbers**):

$$\begin{matrix}\textit{set of} \\ \textit{positive integers}\end{matrix} = \{1, 2, 3, \ldots\}.$$

The three dots mean that the listing of elements is unending, although we know what the elements are.

The positive integers, together with 0 and the **negative integers** $-1, -2, -3, \ldots$, form the set of **integers:**

$$\begin{matrix}\textit{set of} \\ \textit{integers}\end{matrix} = \{\ldots, -3, -2, -1, 0, 1, 2, 3, \ldots\}.$$

The reason for $q \neq 0$ is that we cannot divide by zero.

The set of **rational numbers** consists of numbers, such as $\frac{1}{2}$ and $\frac{5}{3}$ that can be written as a ratio (quotient) of two integers. That is, a rational number is a number that can be written as p/q, where p and q are integers and $q \neq 0$. (The symbol "$\neq$" is read "is not equal to.") For example, the numbers $\frac{19}{20}$, $\frac{-2}{7}$, and $\frac{-6}{-2}$, are rational. We remark that $\frac{2}{4}$, $\frac{1}{2}$, $\frac{3}{6}$, $\frac{-4}{-8}$, and 0.5 all represent the same rational number. The integer 2 is rational, since $2 = \frac{2}{1}$. In fact, every integer is rational.

Every integer is a rational number.

All rational numbers can be represented by decimal numbers that *terminate,* such as $\frac{3}{4} = 0.75$ and $\frac{3}{2} = 1.5$, or by *nonterminating repeating decimal numbers* (composed of a group of digits that repeats without end), such as $\frac{2}{3} = 0.666\ldots$, $\frac{-4}{11} = -0.3636\ldots$, and $\frac{2}{15} = 0.1333\ldots$. Numbers represented by *nonterminating nonrepeating* decimals are called **irrational numbers.** An irrational number cannot be written as an integer divided by an integer. The numbers π (pi) and $\sqrt{2}$ are irrational.

The real numbers consist of all decimal numbers.

Together, the rational numbers and irrational numbers form the set of **real numbers.** Real numbers can be represented by points on a line. First we choose a point on the line to represent zero. This point is called the *origin.* (See Fig. 0.1.) Then a standard measure of distance, called a "unit distance," is chosen and is successively marked off both to the right and to the left of the origin. With each point on the line we associate a directed distance, or *signed number,* which depends on the position of the point with respect to the origin. Positions to the right of the origin are considered positive (+) and positions to the left are negative (−). For example, with the point $\frac{1}{2}$ unit to the right of the origin there corresponds the signed number $\frac{1}{2}$, which is called the **coordinate** of that point. Similarly, the coordinate of the point 1.5 units to the left of the origin is −1.5. In Figure 0.1, the coordinates of some points are marked. The arrowhead indicates that the direction to the right along the line is considered the positive direction.

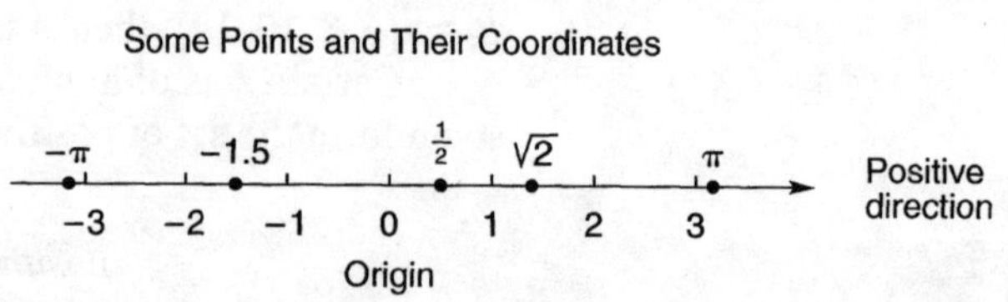

FIGURE 0.1 The real-number line.

To each point on the line there corresponds a unique real number, and to each real number there corresponds a unique point on the line. For this reason, we say that there is a *one-to-one correspondence* between points on the line and real numbers. We call this line a **coordinate line** or the **real-number line.** We feel free to treat real numbers as points on a real-number line and vice versa.

■ Exercise 0.2

In Problems 1–12, classify the statement as either true or false. If false, give a reason.

1. -7 is an integer.

2. $\frac{1}{6}$ is rational.

3. -3 is a natural number.

4. 0 is not rational.

5. 5 is rational.

6. $\frac{7}{0}$ is a rational number.

7. $\frac{4}{2}$ is not a positive integer.

8. π is a real number.

9. $\frac{0}{6}$ is rational.

10. 0 is a natural number.

11. -3 is to the right of -4 on the real number line.

12. Every integer is positive or negative.

OBJECTIVE

To state and illustrate the following properties of real numbers: transitive, commutative, associative, inverse, and distributive. To define subtraction and division in terms of addition and multiplication, respectively.

0.3 Some Properties of Real Numbers

We now state a few important properties of the real numbers. Let a, b, and c be real numbers.

1. The Transitive Property of Equality

If $a = b$ and $b = c$, then $a = c$.

Thus, two numbers that are both equal to a third number are equal to each other. For example, if $x = y$ and $y = 7$, then $x = 7$.

2. The Commutative Properties of Addition and Multiplication

$$a + b = b + a \quad \text{and} \quad ab = ba.$$

This means that two numbers can be added or multiplied in any order. For example, $3 + 4 = 4 + 3$ and $7(-4) = (-4)(7)$.

3. The Associative Properties of Addition and Multiplication

$$a + (b + c) = (a + b) + c \quad \text{and} \quad a(bc) = (ab)c.$$

This means that in addition or multiplication, numbers can be grouped in any order. For example, $2 + (3 + 4) = (2 + 3) + 4$; in both cases, the sum is 9. Similarly, $2x + (x + y) = (2x + x) + y$ and $6(\frac{1}{3} \cdot 5) = (6 \cdot \frac{1}{3}) \cdot 5$.

4. The Inverse Properties

For each real number a, there is a unique real number denoted $-a$ such that

$$a + (-a) = 0.$$

The number $-a$ is called the **additive inverse,** or **negative,** of a.

For example, since $6 + (-6) = 0$, the additive inverse of 6 is -6. The additive inverse of a number is not necessarily a negative number. For example, the additive inverse of -6 is 6, since $(-6) + (6) = 0$. That is, the negative of -6 is 6, so we can write $-(-6) = 6$.

For each real number a, except 0, there is a unique real number denoted a^{-1} such that

$$a \cdot a^{-1} = 1.$$

The number a^{-1} is called the **multiplicative inverse** of a.

Zero does not have a multiplicative inverse because there is no number that, when multiplied by 0, gives 1.

Thus, all numbers except 0 have a multiplicative inverse. You may recall that a^{-1} can be written $\frac{1}{a}$ and is also called the *reciprocal* of a. For example, the multiplicative inverse of 3 is $\frac{1}{3}$, since $3(\frac{1}{3}) = 1$. Hence, $\frac{1}{3}$ is the reciprocal of 3. The reciprocal of $\frac{1}{3}$, is 3, since $(\frac{1}{3})(3) = 1$. **The reciprocal of 0 is not defined.**

5. The Distributive Properties

$$a(b + c) = ab + ac \quad \text{and} \quad (b + c)a = ba + ca.$$

For example, although $2(3 + 4) = 2(7) = 14$, we can write

$$2(3 + 4) = 2(3) + 2(4) = 6 + 8 = 14.$$

Similarly,

$$(2 + 3)(4) = 2(4) + 3(4) = 8 + 12 = 20,$$

$$\text{and} \quad x(z + 4) = x(z) + x(4) = xz + 4x.$$

The distributive property can be extended to the form

$$a(b + c + d) = ab + ac + ad.$$

In fact, it can be extended to sums involving any number of terms.

Subtraction is defined in terms of addition:

$$a - b \quad \text{means} \quad a + (-b),$$

where $-b$ is the additive inverse of b. Thus, $6 - 8$ means $6 + (-8)$.

In a similar way, we define **division** in terms of multiplication. If $b \neq 0$, then $a \div b$, or $\frac{a}{b}$, is defined by

$$\frac{a}{b} = a(b^{-1}).$$

Since $b^{-1} = \frac{1}{b}$,

$$\frac{a}{b} = a(b^{-1}) = a\left(\frac{1}{b}\right).$$

$\frac{a}{b}$ means a times the reciprocal of b.

Thus, $\frac{3}{5}$ means 3 times $\frac{1}{5}$, where $\frac{1}{5}$ is the multiplicative inverse of 5. Sometimes we refer to $a \div b$ or $\frac{a}{b}$ as the *ratio* of a to b. We remark that since 0 does not have a multiplicative inverse, **division by 0 is not defined.**

The following examples show some manipulations involving the preceding properties.

EXAMPLE 1 Applying Properties of Real Numbers

a. $x(y - 3z + 2w) = (y - 3z + 2w)x$, by the commutative property of multiplication.

b. By the associative property of multiplication, $3(4 \cdot 5) = (3 \cdot 4)5$. Thus, the result of multiplying 3 by the product of 4 and 5 is the same as the result of multiplying the product of 3 and 4 by 5. In either case, the result is 60. ■

EXAMPLE 2 Applying Properties of Real Numbers

a. *Show that* $2 - \sqrt{2} = -\sqrt{2} + 2$.
Solution: By the definition of subtraction, $2 - \sqrt{2} = 2 + (-\sqrt{2})$.

However, by the commutative property of addition, $2 + (-\sqrt{2}) = -\sqrt{2} + 2$. Hence, by the transitive property, $2 - \sqrt{2} = -\sqrt{2} + 2$. More concisely, we omit intermediate steps and directly write

$$2 - \sqrt{2} = -\sqrt{2} + 2.$$

b. *Show that* $(8 + x) - y = 8 + (x - y)$.

Solution: Beginning with the left side, we have

$$\begin{aligned} (8 + x) - y &= (8 + x) + (-) && \text{(definition of subtraction)} \\ &= 8 + [x + (-y)] && \text{(associative property)} \\ &= 8 + (x - y) && \text{(definition of subtraction).} \end{aligned}$$

Hence, by the transitive property,

$$(8 + x) - y = 8 + (x - y).$$

c. *Show that* $3(4x + 2y + 8) = 12x + 6y + 24$.

Solution: By the distributive property,

$$3(4x + 2y + 8) = 3(4x) + 3(2y) + 3(8).$$

But by the associative property of multiplication,

$$3(4x) = (3 \cdot 4)x = 12x \quad \text{and similarly} \quad 3(2y) = 6y.$$

Thus, $3(4x + 2y + 8) = 12x + 6y + 24$. ■

EXAMPLE 3 Applying Properties of Real Numbers

a. *Show that* $\frac{ab}{c} = a\left(\frac{b}{c}\right)$ *for* $c \neq 0$.

Solution: By the definition of division,

$$\frac{ab}{c} = (ab) \cdot \frac{1}{c} \text{ for } c \neq 0.$$

But by the associative property,

$$(ab) \cdot \frac{1}{c} = a\left(b \cdot \frac{1}{c}\right).$$

However, by the definition of division, $b \cdot \frac{1}{c} = \frac{b}{c}$. Thus,

$$\frac{ab}{c} = a\left(\frac{b}{c}\right).$$

We can also show that $\frac{ab}{c} = \left(\frac{a}{c}\right)b$.

b. *Show that* $\frac{a+b}{c} = \frac{a}{c} + \frac{b}{c}$ for $c \neq 0$.

Solution: By the definition of division and the distributive property,

$$\frac{a+b}{c} = (a+b)\frac{1}{c} = a \cdot \frac{1}{c} + b \cdot \frac{1}{c}.$$

However,

$$a \cdot \frac{1}{c} + b \cdot \frac{1}{c} = \frac{a}{c} + \frac{b}{c}.$$

Hence,

$$\frac{a+b}{c} = \frac{a}{c} + \frac{b}{c}.$$

We remark that $\frac{a}{b+c} \neq \frac{a}{b} + \frac{a}{c}$. For example,

$$\frac{3}{2+1} \neq \frac{3}{2} + \frac{3}{1}.$$

■

Finding the product of several numbers can be done only by considering products of numbers taken two at a time. For example, to find the product of x, y, and z, we could first multiply x by y and then multiply that product by z; that is, we find $(xy)z$. Or alternatively, we could multiply x by the product of y and z; that is, we find $x(yz)$. The associative property of multiplication guarantees that both results are identical, regardless of how the numbers are grouped. Thus, it is not ambiguous to write xyz. This concept can be extended to more than three numbers and applies equally well to addition.

One final comment is appropriate before we end this section: Not only should you be aware of the manipulative aspects of the properties of the real numbers; you should also be aware of, and familiar with, the terminology involved.

■ Exercise 0.3

In Problems 1–10, classify the statements as either true or false.

1. Every real number has a reciprocal.

2. The reciprocal of $\frac{2}{5}$ is $\frac{5}{2}$.

3. The additive inverse of 5 is $\frac{1}{5}$.

4. $2(3 \cdot 4) = (2 \cdot 3)(2 \cdot 4)$.

5. $-x + y = y - x$.

6. $(x + 2)(4) = 4x + 8$.

7. $\frac{x+2}{2} = \frac{x}{2} + 1$.

8. $3\left(\frac{x}{4}\right) = \frac{3x}{4}$.

9. $x + (y + 5) = (x + y) + (x + 5)$.

10. $8(9x) = 72x$.

In Problems 11–20, state which properties of the real numbers are being used.

11. $2(x + y) = 2x + 2y$.

12. $(x + 5) + y = y + (x + 5)$.

13. $2(3y) = (2 \cdot 3)y$.

14. $\frac{6}{7} = 6 \cdot \frac{1}{7}$.

15. $2(x - y) = (x - y)(2)$.

16. $x + (x + y) = (x + x) + y$.

17. $8 - y = 8 + (-y)$.

18. $5(4 + 7) = 5(7 + 4)$.

19. $(7 + x)y = 7y + xy$.

20. $(-1)[-3 + 4] = (-1)(-3) + (-1)(4)$.

In Problems 21–26, show that the statements are true by using properties of the real numbers.

21. $5a(x + 3) = 5ax + 15a$.

22. $(2 - x) + y = 2 + (y - x)$.

23. $(x + y)(2) = 2x + 2y$.

24. $2[27 + (x + y)] = 2[(y + 27) + x]$.

25. $x[(2y + 1) + 3] = 2xy + 4x$.

26. $(x + 1)(y + 1) = xy + x + y + 1$.

27. Show that $a(b + c + d) = ab + ac + ad$.
[*Hint:* $b + c + d = (b + c) + d$.]

OBJECTIVE

To list and illustrate the most common properties of real numbers.

0.4 Operations with Real Numbers

The following list states important properties of real numbers that you should study thoroughly. Being able to manipulate real numbers is essential to your success in mathematics. A numerical example follows each property. All denominators are different from zero. A knowledge of addition and subtraction of real numbers is assumed.

Property	*Example*
1. $a - b = a + (-b)$.	$2 - 7 = 2 + (-7) = -5$.
2. $a - (-b) = a + b$.	$2 - (-7) = 2 + 7 = 9$.
3. $-a = (-1)(a)$.	$-7 = (-1)(7)$.
4. $a(b + c) = ab + ac$.	$6(7 + 2) = 6 \cdot 7 + 6 \cdot 2 = 54$.
5. $a(b - c) = ab - ac$.	$6(7 - 2) = 6 \cdot 7 - 6 \cdot 2 = 30$.
6. $-(a + b) = -a - b$.	$-(7 + 2) = -7 - 2 = -9$.
7. $-(a - b) = -a + b$.	$-(2 - 7) = -2 + 7 = 5$.
8. $-(-a) = a$.	$-(-2) = 2$.
9. $a(0) = 0$.	$2(0) = 0$.
10. $(-a)(b) = -(ab) = a(-b)$.	$(-2)(7) = -(2 \cdot 7) = 2(-7) = -14$.
11. $(-a)(-b) = ab$.	$(-2)(-7) = 2 \cdot 7 = 14$.
12. $\frac{a}{1} = a$.	$\frac{7}{1} = 7,\ \frac{-2}{1} = -2$.
13. $\frac{a}{b} = a\left(\frac{1}{b}\right)$.	$\frac{2}{7} = 2\left(\frac{1}{7}\right)$.
14. $\frac{a}{-b} = -\frac{a}{b} = \frac{-a}{b}$.	$\frac{2}{-7} = -\frac{2}{7} = \frac{-2}{7}$.
15. $\frac{-a}{-b} = \frac{a}{b}$.	$\frac{-2}{-7} = \frac{2}{7}$.
16. $\frac{0}{a} = 0$ when $a \neq 0$.	$\frac{0}{7} = 0$.
17. $\frac{a}{a} = 1$ when $a \neq 0$.	$\frac{2}{2} = 1,\ \frac{-5}{-5} = 1$.
18. $a\left(\frac{b}{a}\right) = b$.	$2\left(\frac{7}{2}\right) = 7$.
19. $a \cdot \frac{1}{a} = 1$ when $a \neq 0$.	$2 \cdot \frac{1}{2} = 1$.
20. $\frac{a}{b} \cdot \frac{c}{d} = \frac{ac}{bd}$.	$\frac{2}{3} \cdot \frac{4}{5} = \frac{2 \cdot 4}{3 \cdot 5} = \frac{8}{15}$.
21. $\frac{ab}{c} = \left(\frac{a}{c}\right)b = a\left(\frac{b}{c}\right)$	$\frac{2 \cdot 7}{3} = \frac{2}{3} \cdot 7 = 2 \cdot \frac{7}{3}$.

Property	*Example*
22. $\frac{a}{bc} = \left(\frac{a}{b}\right)\left(\frac{1}{c}\right) = \left(\frac{1}{b}\right)\left(\frac{a}{c}\right)$	$\frac{2}{3 \cdot 7} = \frac{2}{3} \cdot \frac{1}{7} = \frac{1}{3} \cdot \frac{2}{7}$.
23. $\frac{a}{b} = \left(\frac{a}{b}\right)\left(\frac{c}{c}\right) = \frac{ac}{bc}$ when $c \neq 0$.	$\frac{2}{7} = \left(\frac{2}{7}\right)\left(\frac{5}{5}\right) = \frac{2 \cdot 5}{7 \cdot 5}$.
24. $\frac{a}{b(-c)} = \frac{a}{(-b)(c)} = \frac{-a}{bc} = \frac{-a}{(-b)(-c)} = -\frac{a}{bc}$.	$\frac{2}{3(-5)} = \frac{2}{(-3)(5)} = \frac{-2}{3(5)} = \frac{-2}{(-3)(-5)} = -\frac{2}{3(5)} = -\frac{2}{15}$.
25. $\frac{a(-b)}{c} = \frac{(-a)b}{c} = \frac{ab}{-c} = \frac{(-a)(-b)}{-c} = -\frac{ab}{c}$.	$\frac{2(-3)}{5} = \frac{(-2)(3)}{5} = \frac{2(3)}{-5} = \frac{(-2)(-3)}{-5} = -\frac{2(3)}{5} = -\frac{6}{5}$.
26. $\frac{a}{c} + \frac{b}{c} = \frac{a+b}{c}$.	$\frac{2}{9} + \frac{3}{9} = \frac{2+3}{9} = \frac{5}{9}$.
27. $\frac{a}{c} - \frac{b}{c} = \frac{a-b}{c}$.	$\frac{2}{9} - \frac{3}{9} = \frac{2-3}{9} = \frac{-1}{9}$.
28. $\frac{a}{b} + \frac{c}{d} = \frac{ad+bc}{bd}$.	$\frac{4}{5} + \frac{2}{3} = \frac{4 \cdot 3 + 5 \cdot 2}{5 \cdot 3} = \frac{22}{15}$.
29. $\frac{a}{b} - \frac{c}{d} = \frac{ad-bc}{bd}$.	$\frac{4}{5} - \frac{2}{3} = \frac{4 \cdot 3 - 5 \cdot 2}{5 \cdot 3} = \frac{2}{15}$.
30. $\frac{\frac{a}{b}}{\frac{c}{d}} = \frac{a}{b} \div \frac{c}{d} = \frac{a}{b} \cdot \frac{d}{c} = \frac{ad}{bc}$.	$\frac{\frac{2}{3}}{\frac{7}{5}} = \frac{2}{3} \div \frac{7}{5} = \frac{2}{3} \cdot \frac{5}{7} = \frac{10}{21}$.
31. $\frac{a}{\frac{b}{c}} = a \div \frac{b}{c} = a \cdot \frac{c}{b} = \frac{ac}{b}$.	$\frac{2}{\frac{3}{5}} = 2 \div \frac{3}{5} = 2 \cdot \frac{5}{3} = \frac{2 \cdot 5}{3} = \frac{10}{3}$.
32. $\frac{\frac{a}{b}}{c} = \frac{a}{b} \div c = \frac{a}{b} \cdot \frac{1}{c} = \frac{a}{bc}$.	$\frac{\frac{2}{3}}{5} = \frac{2}{3} \div 5 = \frac{2}{3} \cdot \frac{1}{5} = \frac{2}{3 \cdot 5} = \frac{2}{15}$.

Property 23 is essentially the **fundamental principle of fractions,** which states that *multiplying or dividing both the numerator and denominator of a fraction by the same number, except* 0, *results in a fraction that is equivalent to (that is, has the same value as) the original fraction.* Thus,

$$\frac{7}{\frac{1}{8}} = \frac{7 \cdot 8}{\frac{1}{8} \cdot 8} = \frac{56}{1} = 56.$$

By Properties 28 and 23, we have

$$\frac{2}{5} + \frac{4}{15} = \frac{2 \cdot 15 + 5 \cdot 4}{5 \cdot 15} = \frac{50}{75} = \frac{2 \cdot 25}{3 \cdot 25} = \frac{2}{3}.$$

We can also do this problem by converting $\frac{2}{5}$ and $\frac{4}{15}$ into equivalent fractions that have the same denominators and then using Property 26. The fractions $\frac{2}{5}$ and $\frac{4}{15}$ can be written with a common denominator of $5 \cdot 15$

$$\frac{2}{5} = \frac{2 \cdot 15}{5 \cdot 15} \quad \text{and} \quad \frac{4}{15} = \frac{4 \cdot 5}{15 \cdot 5}.$$

However, 15 is the *least* such common denominator and is called the *least common denominator* (L.C.D.) of $\frac{2}{5}$ and $\frac{4}{15}$. Thus,

$$\frac{2}{5}+\frac{4}{15}=\frac{2\cdot 3}{5\cdot 3}+\frac{4}{15}=\frac{6}{15}+\frac{4}{15}=\frac{6+4}{15}=\frac{10}{15}=\frac{2}{3}.$$

Similarly,

$$\frac{3}{8}-\frac{5}{12}=\frac{3\cdot 3}{8\cdot 3}-\frac{5\cdot 2}{12\cdot 2} \qquad \text{(L.C.D. = 24)}$$
$$=\frac{9}{24}-\frac{10}{24}=\frac{9-10}{24}$$
$$=-\frac{1}{24}.$$

■ Exercise 0.4

Simplify each of the following if possible.

1. $-2+(-4)$.
2. $-6+2$.
3. $6+(-4)$.
4. $7-2$.
5. $7-(-4)$.
6. $-7-(-4)$.
7. $-8-(-6)$.
8. $(-2)(9)$.
9. $7(-9)$.
10. $(-2)(-12)$.
11. $(-1)6$.
12. $-(-9)$.
13. $-(-6+x)$.
14. $-7(x)$.
15. $-12(x-y)$.
16. $-[-6+(-y)]$.
17. $-2\div 6$.
18. $-2\div(-4)$.
19. $4\div(-2)$.
20. $2(-6+2)$.
21. $3[-2(3)+6(2)]$.
22. $(-2)(-4)(-1)$.
23. $(-5)(-5)$.
24. $x(0)$
25. $3(x-4)$.
26. $4(5+x)$.
27. $-(x-2)$.
28. $0(-x)$.
29. $8\left(\frac{1}{11}\right)$.
30. $\frac{7}{1}$.
31. $\frac{-5x}{7y}$.
32. $\frac{3}{-2x}$.
33. $\frac{2}{3}\cdot\frac{1}{x}$.
34. $\frac{x}{y}(2z)$.
35. $(2x)\left(\frac{3}{2x}\right)$.
36. $\frac{-15x}{-3y}$.
37. $\frac{7}{y}\cdot\frac{1}{x}$.
38. $\frac{2}{x}\cdot\frac{5}{y}$.
39. $\frac{1}{2}+\frac{1}{3}$.
40. $\frac{5}{12}+\frac{3}{4}$.
41. $\frac{3}{10}-\frac{7}{15}$.
42. $\frac{2}{3}+\frac{7}{3}$.
43. $\frac{x}{9}-\frac{y}{9}$.
44. $\frac{3}{2}-\frac{1}{4}+\frac{1}{6}$.
45. $\frac{2}{3}-\frac{5}{8}$.
46. $\frac{6}{\frac{x}{y}}$.
47. $\frac{\frac{x}{6}}{y}$.
48. $\frac{\frac{-7}{2}}{\frac{5}{8}}$.
49. $\frac{7}{0}$.
50. $\frac{0}{7}$.
51. $\frac{0}{0}$.
52. $0\cdot 0$.

OBJECTIVE

To review positive integral exponents, the zero exponent, negative integral exponents, rational exponents, principal roots, radicals, and the procedure of rationalizing the denominator.

0.5 Exponents and Radicals

The product $x\cdot x\cdot x$ is abbreviated x^3. In general, for n a positive integer, x^n is the abbreviation for the product of n x's. The letter n in x^n is called the *exponent*, and x is called the *base*. More specifically, if n is a positive integer, we have:

1. $x^n = \underbrace{x \cdot x \cdot x \cdot \ldots \cdot x}_{n \text{ factors}}$.

2. $x^{-n} = \dfrac{1}{x^n} = \dfrac{1}{\underbrace{x \cdot x \cdot x \cdot \ldots \cdot x}_{n \text{ factors}}}$.

3. $\dfrac{1}{x^{-n}} = x^n$.

4. $x^0 = 1$ if $x \neq 0$. 0^0 is not defined.

EXAMPLE 1 Exponents

a. $\left(\dfrac{1}{2}\right)^4 = \left(\dfrac{1}{2}\right)\left(\dfrac{1}{2}\right)\left(\dfrac{1}{2}\right)\left(\dfrac{1}{2}\right) = \dfrac{1}{16}$.

b. $3^{-5} = \dfrac{1}{3^5} = \dfrac{1}{3 \cdot 3 \cdot 3 \cdot 3 \cdot 3} = \dfrac{1}{243}$.

c. $\dfrac{1}{3^{-5}} = 3^5 = 243$.

d. $2^0 = 1$, $\pi^0 = 1$, $(-5)^0 = 1$.

e. $x^1 = x$. ■

If $r^n = x$, where n is a positive integer, then r is an nth *root* of x. For example, $3^2 = 9$, so 3 is a second root (usually called a *square root*) of 9. Since $(-3)^2 = 9$, -3 is also a square root of 9. Similarly, -2 is a *cube root* of -8, since $(-2)^3 = -8$.

Some numbers do not have an nth root that is a real number. For example, since the square of any real number is nonnegative, there is no real number that is a square root of -4.

The **principal *n*th root** of x is the nth root of x that is positive if x is positive and is negative if x is negative and n is odd. We denote the principal nth root of x by $\sqrt[n]{x}$. Thus,

$$\sqrt[n]{x} \text{ is } \begin{cases} \text{positive if } x \text{ is positive,} \\ \text{negative if } x \text{ is negative and } n \text{ is odd.} \end{cases}$$

For example, $\sqrt[2]{9} = 3$, $\sqrt[3]{-8} = -2$, and $\sqrt[3]{\frac{1}{27}} = \frac{1}{3}$. We define $\sqrt[n]{0} = 0$.

The symbol $\sqrt[n]{x}$ is called a **radical.** Here n is the *index*, x is the *radicand*, and $\sqrt{}$ is the *radical sign.* With principal square roots, we usually omit the index and write $\sqrt{x}$ instead of $\sqrt[2]{x}$. Thus, $\sqrt{9} = 3$.

Pitfall ▼ Although 2 and -2 are square roots of 4, the **principal** square root of 4 is 2, not -2. Hence, $\sqrt{4} = 2$.

If x is positive, the expression $x^{p/q}$, where p and q are integers and q is positive, is defined to be $\sqrt[q]{x^p}$. Hence,

$$x^{3/4} = \sqrt[4]{x^3}; \quad 8^{2/3} = \sqrt[3]{8^2} = \sqrt[3]{64} = 4;$$

$$4^{-1/2} = \sqrt[2]{4^{-1}} = \sqrt{\tfrac{1}{4}} = \tfrac{1}{2}.$$

Here are the basic laws of exponents and radicals:[1]

Law	*Example*
1. $x^m \cdot x^n = x^{m+n}$.	$2^3 \cdot 2^5 = 2^8 = 256$; $x^2 \cdot x^3 = x^5$.
2. $x^0 = 1$ if $x \neq 0$.	$2^0 = 1$.
3. $x^{-n} = \dfrac{1}{x^n}$.	$2^{-3} = \dfrac{1}{2^3} = \dfrac{1}{8}$.
4. $\dfrac{1}{x^{-n}} = x^n$.	$\dfrac{1}{2^{-3}} = 2^3 = 8$; $\dfrac{1}{x^{-5}} = x^5$.
5. $\dfrac{x^m}{x^n} = x^{m-n} = \dfrac{1}{x^{n-m}}$.	$\dfrac{2^{12}}{2^8} = 2^4 = 16$; $\dfrac{x^8}{x^{12}} = \dfrac{1}{x^4}$.
6. $\dfrac{x^m}{x^m} = 1$.	$\dfrac{2^4}{2^4} = 1$.
7. $(x^m)^n = x^{mn}$.	$(2^3)^5 = 2^{15}$; $(x^2)^3 = x^6$.
8. $(xy)^n = x^n y^n$.	$(2 \cdot 4)^3 = 2^3 \cdot 4^3 = 8 \cdot 64 = 512$.
9. $\left(\dfrac{x}{y}\right)^n = \dfrac{x^n}{y^n}$.	$\left(\dfrac{2}{3}\right)^3 = \dfrac{2^3}{3^3}$; $\left(\dfrac{1}{3}\right)^5 = \dfrac{1^5}{3^5} = \dfrac{1}{3^5} = 3^{-5}$.
10. $\left(\dfrac{x}{y}\right)^{-n} = \left(\dfrac{y}{x}\right)^n$.	$\left(\dfrac{3}{4}\right)^{-2} = \left(\dfrac{4}{3}\right)^2 = \dfrac{16}{9}$.
11. $x^{1/n} = \sqrt[n]{x}$.	$3^{1/5} = \sqrt[5]{3}$.
12. $x^{-1/n} = \dfrac{1}{x^{1/n}} = \dfrac{1}{\sqrt[n]{x}}$.	$4^{-1/2} = \dfrac{1}{4^{1/2}} = \dfrac{1}{\sqrt{4}} = \dfrac{1}{2}$.
13. $\sqrt[n]{x}\sqrt[n]{y} = \sqrt[n]{xy}$.	$\sqrt[3]{9}\,\sqrt[3]{2} = \sqrt[3]{18}$.
14. $\dfrac{\sqrt[n]{x}}{\sqrt[n]{y}} = \sqrt[n]{\dfrac{x}{y}}$.	$\dfrac{\sqrt[3]{90}}{\sqrt[3]{10}} = \sqrt[3]{\dfrac{90}{10}} = \sqrt[3]{9}$.
15. $\sqrt[m]{\sqrt[n]{x}} = \sqrt[mn]{x}$.	$\sqrt[3]{\sqrt[4]{2}} = \sqrt[12]{2}$.
16. $x^{m/n} = \sqrt[n]{x^m} = (\sqrt[n]{x})^m$.	$8^{2/3} = \sqrt[3]{8^2} = (\sqrt[3]{8})^2 = 2^2 = 4$.
17. $(\sqrt[m]{x})^m = x$.	$(\sqrt[8]{7})^8 = 7$.

When computing $x^{m/n}$, it is often easier to first find $\sqrt[n]{x}$ and then raise the result to the mth power. Thus, $(-27)^{4/3} = (\sqrt[3]{-27})^4 = (-3)^4 = 81$.

EXAMPLE 2 Exponents and Radicals

a. By Law 1,

$$x^6x^8 = x^{6+8} = x^{14},$$
$$a^3b^2a^5b = a^3a^5b^2b^1 = a^8b^3,$$
$$x^{11}x^{-5} = x^{11-5} = x^6,$$
$$z^{2/5}z^{3/5} = z^1 = z,$$
$$xx^{1/2} = x^1x^{1/2} = x^{3/2}.$$

b. By Law 16,

$$\left(\frac{1}{4}\right)^{3/2} = \left(\sqrt{\frac{1}{4}}\right)^3 = \left(\frac{1}{2}\right)^3 = \frac{1}{8}.$$

c. $$\left(-\frac{8}{27}\right)^{4/3} = \left(\sqrt[3]{\frac{-8}{27}}\right)^4 = \left(\frac{\sqrt[3]{-8}}{\sqrt[3]{27}}\right)^4 \qquad \text{(Laws 16 and 14)}$$
$$= \left(\frac{-2}{3}\right)^4$$
$$= \frac{(-2)^4}{3^4} = \frac{16}{81} \qquad \text{(Law 9).}$$

[1]Although some laws involve restrictions, they are not vital to our discussion.

d. $(64a^3)^{2/3} = 64^{2/3}(a^3)^{2/3}$ (Law 8)

$= (\sqrt[3]{64})^2 a^2$ (Laws 16 and 7)

$= (4)^2 a^2 = 16a^2.$ ∎

Rationalizing the denominator of a fraction is a procedure in which a fraction having a radical in its denominator is expressed as an equivalent fraction without a radical in its denominator. We use the fundamental principle of fractions, as Example 3 shows.

EXAMPLE 3 Rationalizing Denominators

a. $$\frac{2}{\sqrt{5}} = \frac{2}{5^{1/2}} = \frac{2 \cdot 5^{1/2}}{5^{1/2} \cdot 5^{1/2}} = \frac{2 \cdot 5^{1/2}}{5^1} = \frac{2\sqrt{5}}{5}.$$

b. $$\frac{2}{\sqrt[6]{3x^5}} = \frac{2}{\sqrt[6]{3} \cdot \sqrt[6]{x^5}} = \frac{2}{3^{1/6}x^{5/6}} = \frac{2 \cdot 3^{5/6}x^{1/6}}{3^{1/6}x^{5/6} \cdot 3^{5/6}x^{1/6}}$$

$$= \frac{2(3^5x)^{1/6}}{3x} = \frac{2\sqrt[6]{3^5x}}{3x}.$$ ∎

The following examples illustrate various applications of the laws of exponents and radicals.

EXAMPLE 4 Exponents

a. *Eliminate negative exponents in* $\frac{x^{-2}y^3}{z^{-2}}$.

Solution:

$$\frac{x^{-2}y^3}{z^{-2}} = x^{-2} \cdot y^3 \cdot \frac{1}{z^{-2}} = \frac{1}{x^2} \cdot y^3 \cdot z^2 = \frac{y^3z^2}{x^2}.$$

By comparing our answer with the original expression, we conclude that we can bring a factor of the numerator down to the denominator, and vice versa, by changing the sign of the exponent.

b. *Simplify* $\frac{x^2y^7}{x^3y^5}$.

Solution:

$$\frac{x^2y^7}{x^3y^5} = \frac{y^{7-5}}{x^{3-2}} = \frac{y^2}{x}.$$

c. *Simplify* $(x^5y^8)^5$.

Solution:

$$(x^5y^8)^5 = (x^5)^5(y^8)^5 = x^{25}y^{40}.$$

d. *Simplify* $(x^{5/9}y^{4/3})^{18}$.

Solution:

$$(x^{5/9}y^{4/3})^{18} = (x^{5/9})^{18}(y^{4/3})^{18} = x^{10}y^{24}.$$

e. *Simplify* $\left(\frac{x^{1/5}y^{6/5}}{z^{2/5}}\right)^5$.

Solution:

$$\left(\frac{x^{1/5}y^{6/5}}{z^{2/5}}\right)^5 = \frac{(x^{1/5}y^{6/5})^5}{(z^{2/5})^5} = \frac{xy^6}{z^2}.$$

f. *Simplify* $\frac{x^3}{y^2} \div \frac{x^6}{y^5}$.

Solution:

$$\frac{x^3}{y^2} \div \frac{x^6}{y^5} = \frac{x^3}{y^2} \cdot \frac{y^5}{x^6} = \frac{y^3}{x^3}.$$

■

EXAMPLE 5 Exponents

a. *Eliminate negative exponents in* $x^{-1} + y^{-1}$ *and simplify.*

Solution:

$$x^{-1} + y^{-1} = \frac{1}{x} + \frac{1}{y} = \frac{y + x}{xy}. \quad \left(Note\text{: } x^{-1} + y^{-1} \neq \frac{1}{x + y}.\right)$$

b. *Simplify* $x^{3/2} - x^{1/2}$ *by using the distributive law.*

Solution:

$$x^{3/2} - x^{1/2} = x^{1/2}(x - 1).$$

c. *Eliminate negative exponents in* $7x^{-2} + (7x)^{-2}$.

Solution:

$$7x^{-2} + (7x)^{-2} = \frac{7}{x^2} + \frac{1}{(7x)^2} = \frac{7}{x^2} + \frac{1}{49x^2}.$$

d. *Eliminate negative exponents in* $(x^{-1} - y^{-1})^{-2}$.

Solution:

$$(x^{-1} - y^{-1})^{-2} = \left(\frac{1}{x} - \frac{1}{y}\right)^{-2} = \left(\frac{y - x}{xy}\right)^{-2}$$

$$= \left(\frac{xy}{y - x}\right)^2 = \frac{x^2y^2}{(y - x)^2}.$$

e. *Apply the distributive law to* $x^{2/5}(y^{1/2} + 2x^{6/5})$.

Solution:

$$x^{2/5}(y^{1/2} + 2x^{6/5}) = x^{2/5}y^{1/2} + 2x^{8/5}.$$

■

EXAMPLE 6 Radicals

a. *Simplify* $\sqrt[4]{48}$

Solution:

$$\sqrt[4]{48} = \sqrt[4]{16 \cdot 3} = \sqrt[4]{16}\sqrt[4]{3} = 2\sqrt[4]{3}.$$

b. *Rewrite* $\sqrt{2 + 5x}$ *without using a radical sign.*

Solution:

$$\sqrt{2 + 5x} = (2 + 5x)^{1/2}.$$

c. *Rationalize the denominator of* $\frac{\sqrt[5]{2}}{\sqrt[3]{6}}$ *and simplify.*

Solution:

$$\frac{\sqrt[5]{2}}{\sqrt[3]{6}} = \frac{2^{1/5} \cdot 6^{2/3}}{6^{1/3} \cdot 6^{2/3}} = \frac{2^{3/15}6^{10/15}}{6} = \frac{(2^3 6^{10})^{1/15}}{6} = \frac{\sqrt[15]{2^3 6^{10}}}{6}.$$

d. *Simplify* $\frac{\sqrt{20}}{\sqrt{5}}$.

Solution:

$$\frac{\sqrt{20}}{\sqrt{5}} = \sqrt{\frac{20}{5}} = \sqrt{4} = 2.$$

■

EXAMPLE 7 Radicals

a. *Simplify* $\sqrt[3]{x^6 y^4}$

Solution:

$$\sqrt[3]{x^6 y^4} = \sqrt[3]{(x^2)^3 y^3 y} = \sqrt[3]{(x^2)^3} \cdot \sqrt[3]{y^3} \cdot \sqrt[3]{y}$$
$$= x^2 y \sqrt[3]{y} \quad \text{(Law 17)}.$$

b. *Simplify* $\sqrt{\frac{2}{7}}$.

Solution:

$$\sqrt{\frac{2}{7}} = \sqrt{\frac{2 \cdot 7}{7 \cdot 7}} = \sqrt{\frac{14}{7^2}} = \frac{\sqrt{14}}{\sqrt{7^2}} = \frac{\sqrt{14}}{7}.$$

c. *Simplify* $\sqrt{250} - \sqrt{50} + 15\sqrt{2}$.

Solution:

$$\sqrt{250} - \sqrt{50} + 15\sqrt{2} = \sqrt{25 \cdot 10} - \sqrt{25 \cdot 2} + 15\sqrt{2}$$
$$= 5\sqrt{10} - 5\sqrt{2} + 15\sqrt{2}$$
$$= 5\sqrt{10} + 10\sqrt{2}.$$

d. *If x is any real number, simplify* $\sqrt{x^2}$.

Solution:

Note: $\sqrt{x^2} \neq x$.

$$\sqrt{x^2} = \begin{cases} x, & \text{if } x \text{ is positive,} \\ -x, & \text{if } x \text{ is negative,} \\ 0, & \text{if } x = 0. \end{cases}$$

Thus, $\sqrt{2^2} = 2$ and $\sqrt{(-3)^2} = -(-3) = 3$.

■

■ Exercise 0.5

In Problems 1–14, simplify and express all answers in terms of positive exponents.

1. $(2^3)(2^2)$. **2.** $x^6 x^9$. **3.** $w^4 w^8$. **4.** $x^6\, x^4 x^3$.

5. $\dfrac{x^2x^6}{y^7y^{10}}$.

6. $(x^{12})^4$.

7. $\dfrac{(x^2)^5}{(y^5)^{10}}$.

8. $\left(\dfrac{x^2}{y^3}\right)^5$.

9. $(2x^2y^3)^3$.

10. $\left(\dfrac{w^2s^3}{y^2}\right)^2$.

11. $\dfrac{x^8}{x^2}$.

12. $\left(\dfrac{2x^2}{4x^4}\right)^3$.

13. $\dfrac{(x^3)^6}{x(x^3)}$.

14. $\dfrac{(x^2)^3(x^3)^2}{(x^3)^4}$.

In Problems 15–28, evaluate the expressions.

15. $\sqrt{25}$.

16. $\sqrt[3]{64}$.

17. $\sqrt[5]{-32}$.

18. $\sqrt{0.04}$.

19. $\sqrt[4]{\frac{1}{16}}$.

20. $\sqrt[3]{-\frac{8}{27}}$.

21. $(100)^{1/2}$.

22. $(64)^{1/3}$.

23. $4^{3/2}$.

24. $(25)^{-3/2}$.

25. $(32)^{-2/5}$.

26. $(0.09)^{-1/2}$.

27. $\left(\dfrac{1}{16}\right)^{5/4}$.

28. $\left(-\dfrac{27}{64}\right)^{2/3}$.

In Problems 29–40, simplify the expressions.

29. $\sqrt{32}$.

30. $\sqrt[3]{24}$.

31. $\sqrt[3]{2x^3}$.

32. $\sqrt{4x}$.

33. $\sqrt{16x^4}$.

34. $\sqrt[4]{\dfrac{x}{16}}$.

35. $2\sqrt{75} - 4\sqrt{27} + \sqrt[3]{128}$.

36. $\sqrt{\frac{5}{11}}$.

37. $(9z^4)^{1/2}$.

38. $(16y^8)^{3/4}$.

39. $\left(\dfrac{27t^3}{8}\right)^{2/3}$.

40. $\left(\dfrac{1000}{a^9}\right)^{-2/3}$.

In Problems 41–52, write the expressions in terms of positive exponents only. Avoid all radicals in the final form. For example,

$$y^{-1}\sqrt{x} = \frac{x^{1/2}}{y}.$$

41. $\dfrac{x^3y^{-2}}{z^2}$.

42. $\sqrt[5]{x^2y^3z^{-10}}$.

43. $2x^{-1}x^{-3}$.

44. $x + y^{-1}$.

45. $(3t)^{-2}$.

46. $(3 - z)^{-4}$.

47. $\sqrt[3]{7s^2}$.

48. $(x^{-2}y^2)^{-2}$.

49. $\sqrt{x} - \sqrt{y}$.

50. $\dfrac{x^{-2}y^{-6}z^2}{xy^{-1}}$.

51. $x^2\sqrt[4]{xy^{-2}z^3}$.

52. $(\sqrt[5]{xy^{-3}})x^{-1}y^{-2}$.

In Problems 53–58, write the exponential forms in equivalent forms involving radicals.

53. $(8x - y)^{4/5}$.

54. $(ab^2c^3)^{3/4}$.

55. $x^{-4/5}$.

56. $2x^{1/2} - (2y)^{1/2}$.

57. $2x^{-2/5} - (2x)^{-2/5}$.

58. $[(x^{-4})^{1/5}]^{1/6}$.

In Problems 59–68, rationalize the denominators.

59. $\dfrac{3}{\sqrt{7}}$.

60. $\dfrac{5}{\sqrt{11}}$.

61. $\dfrac{4}{\sqrt{2x}}$.

62. $\dfrac{y}{\sqrt{2y}}$.

63. $\dfrac{1}{\sqrt[3]{3x}}$.

64. $\dfrac{4}{3\sqrt[3]{x^2}}$.

65. $\dfrac{\sqrt{32}}{\sqrt{2}}$.

66. $\dfrac{\sqrt{18}}{\sqrt{2}}$.

67. $\dfrac{\sqrt[5]{2}}{\sqrt[4]{xy^2}}$.

68. $\dfrac{\sqrt{2}}{\sqrt[3]{3}}$.

In Problems 69–90, simplify. Express all answers in terms of positive exponents. Rationalize the denominator where necessary to avoid fractional exponents in the denominator.

69. $2x^2y^{-3}x^4$.

70. $\dfrac{2}{x^{3/2}y^{1/3}}$.

71. $\sqrt{\sqrt[3]{t^4}}$.

72. $\{[(2x^2)^3]^{-4}\}^{-1}$.

73. $\dfrac{2^0}{(2^{-2}x^{1/2}y^{-2})^3}$.

74. $\dfrac{\sqrt{s^5}}{\sqrt[3]{s^2}}$.

75. $\sqrt[3]{x^2yz^3}\,\sqrt[3]{xy^2}$.

76. $(\sqrt[5]{2})^{10}$.

77. $3^2(27)^{-4/3}$.

78. $(\sqrt[5]{x^2y})^{2/5}$.

79. $(2x^{-1}y^2)^2$.

80. $\dfrac{3}{\sqrt[3]{y}\,\sqrt[4]{x}}$.

81. $\sqrt{x}\sqrt{x^2y^3}\sqrt{xy^2}$.

82. $\sqrt{75k^4}$.

83. $\dfrac{(x^2y^{-1}z)^{-2}}{(xy^2)^{-4}}$.

84. $\sqrt{6(6)}$.

85. $\dfrac{(x^2)^3}{x^4} \div \left[\dfrac{x^3}{(x^3)^2}\right]^2$.

86. $\sqrt{(-6)(-6)}$.

87. $-\dfrac{8s^{-2}}{2s^3}$.

88. $(x^{-1}y^{-2}\sqrt{z})^4$.

89. $(2x^2y \div 3y^3z^{-2})^2$.

90. $\dfrac{1}{\left(\dfrac{\sqrt{2}x^{-2}}{\sqrt{16}x^3}\right)^2}$.

OBJECTIVE

To add, subtract, multiply, and divide algebraic expressions. To define a polynomial, to use special products, and to use long division to divide polynomials.

0.6 Operations with Algebraic Expressions

If numbers, represented by symbols, are combined by any or all of the operations of addition, subtraction, multiplication, division, and extraction of roots, then the resulting expression is called an *algebraic expression.*

EXAMPLE 1 Algebraic Expressions

a. $\sqrt[3]{\dfrac{3x^3 - 5x - 2}{10 - x}}$ is an algebraic expression in the variable x.

b. $10 - 3\sqrt{y} + \dfrac{5}{7 + y^2}$ is an algebraic expression in the variable y.

c. $\dfrac{(x + y)^3 - xy}{y} + 2$ is an algebraic expression in the variables x and y. ■

The algebraic expression $5ax^3 - 2bx + 3$ consists of three *terms:* $+5ax^3$, $-2bx$, and $+3$. Some of the *factors* of the first term, $5ax^3$, are $5, a, x, x^2, x^3, 5ax$, and ax^2. Also, $5a$ is the *coefficient* of x^3, and 5 is the *numerical coefficient* of ax^3. If a and b represent fixed numbers throughout a discussion, then a and b are called *constants.*

Algebraic expressions with exactly one term are called *monomials.* Those having exactly two terms are *binomials,* and those with exactly three terms are *trinomials.* Algebraic expressions with more than one term are called *multinomials.* Thus, the multinomial $2x - 5$ is a binomial; the multinomial $3\sqrt{y} + 2y - 4y^2$ is a trinomial.

A *polynomial in x* is an algebraic expression of the form[2]

$$c_nx^n + c_{n-1}x^{n-1} + \cdots + c_1x + c_0,$$

where n is a nonnegative integer and the coefficients $c_0, c_1, \ldots, c_n$ are constants with $c_n \neq 0$. We call n the *degree* of the polynomial. Hence, $4x^3 - 5x^2 + x - 2$ is a polynomial in x of degree 3, and $y^5 - 2$ is a polynomial in y of degree 5. A nonzero constant is a polynomial of degree zero; thus, 5 is a polynomial of degree zero. The constant 0 is considered to be a polynomial; however, no degree is assigned to it.

The words "polynomial" and "multinomial" should not be used interchangeably. For example, $\sqrt{x} + 2$ is a multinomial, but not a polynomial. On the other hand, $x + 2$ is a multinomial and a polynomial.

In the following examples, we illustrate operations with algebraic expressions.

[2]The three dots indicate all other terms that are understood to be included in the sum.

EXAMPLE 2 Adding Algebraic Expressions

Simplify $(3x^2y - 2x + 1) + (4x^2y + 6x - 3)$.

Solution: We first remove the parentheses. Next, using the commutative property of addition, we gather all similar terms together. *Similar terms* are terms that differ only by their numerical coefficients. In this example, $3x^2y$ and $4x^2y$ are similar, as are the pairs $-2x$ and $6x$, and 1 and -3. Thus,

$$\begin{aligned} &(3x^2y - 2x + 1) + (4x^2y + 6x - 3) \\ &= 3x^2y - 2x + 1 + 4x^2y + 6x - 3 \\ &= 3x^2y + 4x^2y - 2x + 6x + 1 - 3. \end{aligned}$$

By the distributive property,

$$3x^2y + 4x^2y = (3 + 4)x^2y = 7x^2y$$

$$\text{and} \quad -2x + 6x = (-2 + 6)x = 4x.$$

Hence, $\quad (3x^2y - 2x + 1) + (4x^2y + 6x - 3) = 7x^2y + 4x - 2.$ ■

EXAMPLE 3 Subtracting Algebraic Expressions

Simplify $(3x^2y - 2x + 1) - (4x^2y + 6x - 3)$.

Solution: Here we apply the definition of subtraction and the distributive property:

$$\begin{aligned} &(3x^2y - 2x + 1) - (4x^2y + 6x - 3) \\ &= (3x^2y - 2x + 1) + (-1)(4x^2y + 6x - 3) \\ &= (3x^2y - 2x + 1) + (-4x^2y - 6x + 3) \\ &= 3x^2y - 2x + 1 - 4x^2y - 6x + 3 \\ &= 3x^2y - 4x^2y - 2x - 6x + 1 + 3 \\ &= (3 - 4)x^2y + (-2 - 6)x + 1 + 3 \\ &= -x^2y - 8x + 4. \end{aligned}$$ ■

EXAMPLE 4 Removing Grouping Symbols

Simplify $3\{2x[2x + 3] + 5[4x^2 - (3 - 4x)]\}$.

Solution: We first eliminate the innermost grouping symbols (the parentheses). Then we repeat the process until all grouping symbols are removed—combining similar terms whenever possible. We have

$$\begin{aligned} &3\{2x[2x + 3] + 5[4x^2 - (3 - 4x)]\} \\ &= 3\{2x[2x + 3] + 5[4x^2 - 3 + 4x]\} \\ &= 3\{4x^2 + 6x + 20x^2 - 15 + 20x\} \\ &= 3\{24x^2 + 26x - 15\} \\ &= 72x^2 + 78x - 45. \end{aligned}$$ ■

The distributive property is the key tool in multiplying expressions. For example, to multiply $ax + c$ by $bx + d$ we can consider $ax + c$ to be a single number and then use the distributive property:

$$(ax + c)(bx + d) = (ax + c)bx + (ax + c)d.$$

Using the distributive property again, we have

$$(ax + c)bx + (ax + c)d = abx^2 + cbx + adx + cd$$
$$= abx^2 + (ad + cb)x + cd.$$

Thus, $(ax + c)(bx + d) = abx^2 + (ad + cb)x + cd$. In particular, if $a = 2$, $b = 1$, $c = 3$, and $d = -2$, then

$$(2x + 3)(x - 2) = 2(1)x^2 + [2(-2) + 3(1)]x + 3(-2)$$
$$= 2x^2 - x - 6.$$

We now give a list of special products that may be obtained from the distributive property and are useful in multiplying algebraic expressions.

Special Products

1. $x(y + z) = xy + xz$ (distributive property).
2. $(x + a)(x + b) = x^2 + (a + b)x + ab$.
3. $(ax + c)(bx + d) = abx^2 + (ad + cb)x + cd$.
4. $(x + a)^2 = x^2 + 2ax + a^2$ (square of a binomial).
5. $(x - a)^2 = x^2 - 2ax + a^2$ (square of a binomial).
6. $(x + a)(x - a) = x^2 - a^2$ (product of sum and difference).
7. $(x + a)^3 = x^3 + 3ax^2 + 3a^2x + a^3$ (cube of a binomial).
8. $(x - a)^3 = x^3 - 3ax^2 + 3a^2x - a^3$ (cube of a binomial).

EXAMPLE 5 Special Products

a. By Rule 2,

$$(x + 2)(x - 5) = [x + 2][x + (-5)]$$
$$= x^2 + (2 - 5)x + 2(-5)$$
$$= x^2 - 3x - 10.$$

b. By Rule 3,

$$(3z + 5)(7z + 4) = 3 \cdot 7z^2 + (3 \cdot 4 + 5 \cdot 7)z + 5 \cdot 4$$
$$= 21z^2 + 47z + 20.$$

c. By Rule 5,

$$(x - 4)^2 = x^2 - 2(4)x + 4^2$$
$$= x^2 - 8x + 16.$$

d. By Rule 6,

$$(\sqrt{y^2 + 1} + 3)(\sqrt{y^2 + 1} - 3) = (\sqrt{y^2 + 1})^2 - 3^2$$
$$= (y^2 + 1) - 9$$
$$= y^2 - 8.$$

e. By Rule 7,

$$(3x + 2)^3 = (3x)^3 + 3(2)(3x)^2 + 3(2)^2(3x) + (2)^3$$
$$= 27x^3 + 54x^2 + 36x + 8.$$

■

EXAMPLE 6 Multiplying Multinomials

Find the product $(2t - 3)(5t^2 + 3t - 1)$.

Solution: We treat $2t - 3$ as a single number and apply the distributive property twice:

$$\begin{aligned}(2t - 3)(5t^2 + 3t - 1) &= (2t - 3)5t^2 + (2t - 3)3t - (2t - 3)1 \\ &= 10t^3 - 15t^2 + 6t^2 - 9t - 2t + 3 \\ &= 10t^3 - 9t^2 - 11t + 3.\end{aligned}$$

In Example 3(b) of Sec. 0.3, we showed that $\frac{a+b}{c} = \frac{a}{c} + \frac{b}{c}$. Similarly, $\frac{a-b}{c} = \frac{a}{c} - \frac{b}{c}$. Using these results, we can divide a multinomial by a monomial by dividing each term in the multinomial by the monomial.

EXAMPLE 7 Dividing a Multinomial by a Monomial

a. $\frac{x^3 + 3x}{x} = \frac{x^3}{x} + \frac{3x}{x} = x^2 + 3.$

b.
$$\begin{aligned}\frac{4z^3 - 8z^2 + 3z - 6}{2z} &= \frac{4z^3}{2z} - \frac{8z^2}{2z} + \frac{3z}{2z} - \frac{6}{2z} \\ &= 2z^2 - 4z + \frac{3}{2} - \frac{3}{z}.\end{aligned}$$

Long Division

To divide a polynomial by a polynomial, we use so-called long division when the degree of the divisor is less than or equal to the degree of the dividend, as the next example shows.

EXAMPLE 8 Long Division

Divide $2x^3 - 14x - 5$ *by* $x - 3$.

Solution: Here $2x^3 - 14x - 5$ is the *dividend* and $x - 3$ is the *divisor.* To avoid errors, it is best to write the dividend as $2x^3 + 0x^2 - 14x - 5$. Note that the powers of x are in decreasing order. We have

$$\begin{array}{rl} & 2x^2 + 6x + 4 \leftarrow \text{quotient} \\ x - 3 \,\big)\!\!\! & \overline{2x^3 + 0x^2 - 14x - 5} \\ & \underline{2x^3 - 6x^2} \\ & \quad 6x^2 - 14x \\ & \quad \underline{6x^2 - 18x} \\ & \qquad\quad 4x - 5 \\ & \qquad\quad \underline{4x - 12} \\ & \qquad\qquad\quad 7 \leftarrow \text{remainder.} \end{array}$$

Note that we divided x (the first term of the divisor) into $2x^3$ and got $2x^2$. Then we multiplied $2x^2$ by $x - 3$, getting $2x^3 - 6x^2$. After subtracting $2x^3 - 6x^2$ from $2x^3 + 0x^2$, we obtained $6x^2$ and then "brought down" the term $-14x$. This process is continued until we arrive at 7, the *remainder.* We always stop when the remainder is 0 or is a polynomial whose degree is less than the degree of the divisor. Our answer may be written as

$$2x^2 + 6x + 4 + \frac{7}{x - 3}.$$

That is, the answer has the form

$$\text{quotient} + \frac{\text{remainder}}{\text{divisor}}.$$

A way of checking a division is to verify that

$$(\text{quotient})(\text{divisor}) + \text{remainder} = \text{dividend}.$$

By using this equation, you should be able to verify the result of the example. ■

■ Exercise 0.6

Perform the indicated operations and simplify.

1. $(8x - 4y + 2) + (3x + 2y - 5)$.

2. $(6x^2 - 10xy + 2) + (2z - xy + 4)$.

3. $(8t^2 - 6s^2) + (4s^2 - 2t^2 + 6)$.

4. $(\sqrt{x} + 2\sqrt{x}) + (\sqrt{x} + 3\sqrt{x})$.

5. $(\sqrt{x} + \sqrt{2y}) + (\sqrt{x} + \sqrt{3z})$.

6. $(3x + 2y - 5) - (8x - 4y + 2)$.

7. $(6x^2 - 10xy + \sqrt{2}) - (2z - xy + 4)$.

8. $(\sqrt{x} + 2\sqrt{x}) - (\sqrt{x} + 3\sqrt{x})$.

9. $(\sqrt{x} + \sqrt{2y}) - (\sqrt{x} + \sqrt{3z})$.

10. $4(2z - w) - 3(w - 2z)$.

11. $3(3x + 2y - 5) - 2(8x - 4y + 2)$.

12. $(2s + t) - 3(s - 6) + 4(1 - t)$.

13. $3(x^2 + y^2) - x(y + 2x) + 2y(x + 3y)$.

14. $2 - [3 + 4(s - 3)]$.

15. $2\{3[3(x^2 + 2) - 2(x^2 - 5)]\}$.

16. $4\{3(t + 5) - t[1 - (t + 1)]\}$.

17. $-3\{4x(x + 2) - 2[x^2 - (3 - x)]\}$.

18. $-\{-2[2a + 3b - 1] + 4[a - 2b] - a[2(b - 3)]\}$.

19. $(x + 4)(x + 5)$.

20. $(x + 3)(x + 2)$.

21. $(x + 3)(x - 2)$.

22. $(z - 7)(z - 3)$.

23. $(2x + 3)(5x + 2)$.

24. $(y - 4)(2y + 3)$.

25. $(x + 3)^2$.

26. $(2x - 1)^2$.

27. $(x - 5)^2$.

28. $(\sqrt{x} - 1)(2\sqrt{x} + 5)$.

29. $(\sqrt{2y} + 3)^2$.

30. $(y - 3)(y + 3)$.

31. $(2s - 1)(2s + 1)$.

32. $(z^2 - 3w)(z^2 + 3w)$.

33. $(x^2 - 3)(x + 4)$.

34. $(x + 1)(x^2 + x + 3)$.

35. $(x^2 - 1)(2x^2 + 2x - 3)$.

36. $(2x - 1)(3x^3 + 7x^2 - 5)$.

37. $x\{3(x - 1)(x - 2) + 2[x(x + 7)]\}$.

38. $[(2z + 1)(2z - 1)](4z^2 + 1)$.

39. $(x + y + 2)(3x + 2y - 4)$.

40. $(x^2 + x + 1)^2$.

41. $(x + 5)^3$.

42. $(x - 2)^3$.

43. $(2x - 3)^3$.

44. $(x + 2y)^3$.

45. $\dfrac{z^2 - 4z}{z}$.

46. $\dfrac{2x^3 - 7x + 4}{x}$.

47. $\dfrac{6x^5 + 4x^3 - 1}{2x^2}$.

48. $\dfrac{(3x - 4) - (x + 8)}{4x}$.

49. $(x^2 + 3x - 1) \div (x + 3)$.

50. $(x^2 - 5x + 4) \div (x - 4)$.

51. $(3x^3 - 2x^2 + x - 3) \div (x + 2)$.

52. $(x^4 + 2x^2 + 1) \div (x - 1)$.

53. $t^2 \div (t - 8)$.

54. $(4x^2 + 6x + 1) \div (2x - 1)$.

55. $(3x^2 - 4x + 3) \div (3x + 2)$.

56. $(z^3 + z^2 + z) \div (z^2 - z + 1)$.

OBJECTIVE

To state the basic rules for factoring and apply them to factoring expressions.

0.7 Factoring

If two or more expressions are multiplied together, the expressions are called *factors* of the product. Thus, if $c = ab$ then a and b are both factors of the product c. The process by which an expression is written as a product of its factors is called *factoring.*

Listed next are rules for factoring expressions, most of which arise from the special products discussed in Sec. 0.6. The right side of each identity is the factored form of the left side.

Rules for Factoring

1. $xy + xz = x(y + z)$ (common factor).
2. $x^2 + (a + b)x + ab = (x + a)(x + b)$.
3. $abx^2 + (ad + cb)x + cd = (ax + c)(bx + d)$.
4. $x^2 + 2ax + a^2 = (x + a)^2$ (perfect-square trinomial).
5. $x^2 - 2ax + a^2 = (x - a)^2$ (perfect-square trinomial).
6. $x^2 - a^2 = (x + a)(x - a)$ (difference of two squares).
7. $x^3 + a^3 = (x + a)(x^2 - ax + a^2)$ (sum of two cubes).
8. $x^3 - a^3 = (x - a)(x^2 + ax + a^2)$ (difference of two cubes).

When factoring a polynomial, we usually choose factors that themselves are polynomials. For example, $x^2 - 4 = (x + 2)(x - 2)$. We shall not write $x - 4$ as $(\sqrt{x} + 2)(\sqrt{x} - 2)$.

Always factor completely. For example,

$$2x^2 - 8 = 2(x^2 - 4) = 2(x + 2)(x - 2).$$

EXAMPLE 1 Common Factors

a. *Factor* $3k^2x^2 + 9k^3x$ *completely.*

Solution: Since $3k^2x^2 = (3k^2x)(x)$ and $9k^3x = (3k^2x)(3k)$, each term of the original expression contains the common factor $3k^2x$. Thus, by Rule 1,

$$3k^2x^2 + 9k^3x = 3k^2x(x + 3k).$$

Note that although $3k^2x^2 + 9k^3x = 3(k^2x^2 + 3k^3x)$, we do not say that the expression is completely factored, since $k^2x^2 + 3k^3x$ can yet be factored.

b. *Factor* $8a^5x^2y^3 - 6a^2b^3yz - 2a^4b^4xy^2z^2$ *completely.*

Solution:

$$\begin{aligned} &8a^5x^2y^3 - 6a^2b^3yz - 2a^4b^4xy^2z^2 \\ &= 2a^2y(4a^3x^2y^2 - 3b^3z - a^2b^4xyz^2). \end{aligned}$$

■

EXAMPLE 2 Factoring Trinomials

a. *Factor* $3x^2 + 6x + 3$ *completely.*

Solution: First we remove a common factor. Then we factor the resulting expression completely. Thus, we have

$$3x^2 + 6x + 3 = 3(x^2 + 2x + 1)$$
$$= 3(x + 1)^2 \qquad \text{(Rule 4).}$$

b. *Factor $x^2 - x - 6$ completely.*

Solution: If this trinomial factors into the form $(x + a)(x + b)$, which is a product of two binomials, then we must determine the values of a and b. Since $(x + a)(x + b) = x^2 + (a + b)x + ab$, it follows that

$$x^2 + (-1)x + (-6) = x^2 + (a + b)x + ab.$$

By equating corresponding coefficients, we want

$$a + b = -1 \qquad \text{and} \qquad ab = -6.$$

If $a = -3$ and $b = 2$, then both conditions are met, and hence,

$$x^2 - x - 6 = (x - 3)(x + 2).$$

As a check, it is wise to multiply the right side to see if it agrees with the left side.

c. *Factor $x^2 - 7x + 12$ completely.*

Solution:

$$x^2 - 7x + 12 = (x - 3)(x - 4).$$

■

EXAMPLE 3 Factoring

The following is an assortment of expressions that are completely factored. The numbers in parentheses refer to the rules used.

a. $x^2 + 8x + 16 = (x + 4)^2$ (4).

b. $9x^2 + 9x + 2 = (3x + 1)(3x + 2)$ (3).

c. $6y^3 + 3y^2 - 18y = 3y(2y^2 + y - 6)$ (1)

$= 3y(2y - 3)(y + 2)$ (3).

d. $x^2 - 6x + 9 = (x - 3)^2$ (5).

e. $z^{1/4} + z^{5/4} = z^{1/4}(1 + z)$ (1).

f. $x^4 - 1 = (x^2 + 1)(x^2 - 1)$ (6)

$= (x^2 + 1)(x + 1)(x - 1)$ (6).

g. $x^{2/3} - 5x^{1/3} + 4 = (x^{1/3} - 1)(x^{1/3} - 4)$ (2).

h. $ax^2 - ay^2 + bx^2 - by^2 = (ax^2 - ay^2) + (bx^2 - by^2)$

$= a(x^2 - y^2) + b(x^2 - y^2)$ (1)

$= (x^2 - y^2)(a + b)$ (1)

$= (x + y)(x - y)(a + b)$ (6).

i. $8 - x^3 = (2)^3 - (x)^3 = (2 - x)(4 + 2x + x^2)$ (8).

j. $x^6 - y^6 = (x^3)^2 - (y^3)^2 = (x^3 + y^3)(x^3 - y^3)$ (6)

$= (x + y)(x^2 - xy + y^2)(x - y)(x^2 + xy + y^2)$ (7), (8).

■

Note in Example 3(f) that $x^2 - 1$ is factorable, but $x^2 + 1$ is not. In Example 3(h), we factored by making use of grouping.

■ Exercise 0.7

Factor the following expressions completely.

1. $6x + 4$.

2. $6y^2 - 4y$.

3. $10xy + 5xz$.

4. $3x^2y - 9x^3y^3$.

5. $8a^3bc - 12ab^3cd + 4b^4c^2d^2$.

6. $6z^2t^3 + 3zst^4 - 12z^2t^3$.

7. $x^2 - 25$.

8. $x^2 + 3x - 4$.

9. $p^2 + 4p + 3$.

10. $s^2 - 6s + 8$.

11. $16x^2 - 9$.

12. $x^2 + 5x - 24$.

13. $z^2 + 6z + 8$.

14. $4t^2 - 9s^2$.

15. $x^2 + 6x + 9$.

16. $y^2 - 15y + 50$.

17. $2x^2 + 12x + 16$.

18. $2x^2 + 7x - 15$.

19. $3x^2 - 3$.

20. $4y^2 - 8y + 3$.

21. $6y^2 + 13y + 2$.

22. $4x^2 - x - 3$.

23. $12s^3 + 10s^2 - 8s$.

24. $9z^2 + 24z + 16$.

25. $x^{2/3}y - 4x^{8/3}y^3$.

26. $9x^{4/7} - 1$.

27. $2x^3 + 2x^2 - 12x$.

28. $x^2y^2 - 4xy + 4$.

29. $(4x + 2)^2$.

30. $3s^2(3s - 9s^2)^2$.

31. $x^3y^2 - 10x^2y + 25x$.

32. $(3x^2 + x) + (6x + 2)$.

33. $(x^3 - 4x) + (8 - 2x^2)$.

34. $(x^2 - 1) + (x^2 - x - 2)$.

35. $(y^4 + 8y^3 + 16y^2) - (y^2 + 8y + 16)$.

36. $x^3y - xy + z^2x^2 - z^2$.

37. $x^3 + 8$.

38. $x^3 - 1$.

39. $x^6 - 1$.

40. $27 + 8x^3$.

41. $(x + 3)^3(x - 1) + (x + 3)^2(x - 1)^2$.

42. $(x + 5)^2(x + 1)^3 + (x + 5)^3(x + 1)^2$.

43. $P(1 + r) + P(1 + r)r$.

44. $(x - 3)(2x + 3) - (2x + 3)(x + 5)$.

45. $x^4 - 16$.

46. $81x^4 - y^4$.

47. $y^8 - 1$.

48. $t^4 - 4$.

49. $x^4 + x^2 - 2$.

50. $x^4 - 5x^2 + 4$.

51. $x^5 - 2x^3 + x$.

52. $4x^3 - 6x^2 - 4x$.

OBJECTIVE

To simplify fractions and to add, subtract, multiply, and divide fractions. To rationalize the denominator of a fraction.

0.8 Fractions

Simplifying Fractions

By using the fundamental principle of fractions (Sec. 0.4), we may be able to simplify fractions. That principle allows us to multiply or divide both the numerator and the denominator of a fraction by the same nonzero quantity. The resulting fraction will be equivalent to the original one. The fractions that we shall consider are assumed to have nonzero denominators.

EXAMPLE 1 Simplifying Fractions

a. *Simplify* $\dfrac{x^2 - x - 6}{x^2 - 7x + 12}$.

Solution: First, we completely factor the numerator and denominator:

$$\frac{x^2 - x - 6}{x^2 - 7x + 12} = \frac{(x - 3)(x + 2)}{(x - 3)(x - 4)}.$$

Dividing both numerator and denominator by the common factor $x - 3$ we have

$$\frac{(x-3)(x+2)}{(x-3)(x-4)} = \frac{1(x+2)}{1(x-4)} = \frac{x+2}{x-4}.$$

Usually, we just write

$$\frac{x^2 - x - 6}{x^2 - 7x + 12} = \frac{\overset{1}{\cancel{(x-3)}}(x+2)}{\underset{1}{\cancel{(x-3)}}(x-4)} = \frac{x+2}{x-4}$$

or

$$\frac{x^2 - x - 6}{x^2 - 7x + 12} = \frac{\cancel{(x-3)}(x+2)}{\cancel{(x-3)}(x-4)} = \frac{x+2}{x-4}.$$

The process of eliminating the common factor $x - 3$ is commonly referred to as "cancellation."

b. *Simplify* $\dfrac{2x^2 + 6x - 8}{8 - 4x - 4x^2}$.

Solution:

$$\frac{2x^2 + 6x - 8}{8 - 4x - 4x^2} = \frac{2(x^2 + 3x - 4)}{4(2 - x - x^2)} = \frac{2(x-1)(x+4)}{4(1-x)(2+x)}$$

Note how $1 - x$ is written as $(-1)(x - 1)$ to allow cancellation.

$$= \frac{2(x-1)(x+4)}{2(2)[(-1)(x-1)](2+x)}$$

$$= \frac{x+4}{-2(2+x)} = -\frac{x+4}{2(x+2)}.$$ ■

Multiplication and Division of Fractions

The rule for multiplying $\frac{a}{b}$ by $\frac{c}{d}$ is

$$\frac{a}{b} \cdot \frac{c}{d} = \frac{ac}{bd}.$$

EXAMPLE 2 Multiplying Fractions

a. $\dfrac{x}{x+2} \cdot \dfrac{x+3}{x-5} = \dfrac{x(x+3)}{(x+2)(x-5)}$.

b. $\dfrac{x^2 - 4x + 4}{x^2 + 2x - 3} \cdot \dfrac{6x^2 - 6}{x^2 + 2x - 8} = \dfrac{[(x-2)^2][6(x+1)(x-1)]}{[(x+3)(x-1)][(x+4)(x-2)]}$

$$= \frac{6(x-2)(x+1)}{(x+3)(x+4)}.$$ ■

To divide $\frac{a}{b}$ by $\frac{c}{d}$, where $c \neq 0$, we have

In short, we invert the divisor and multiply.

$$\frac{a}{b} \div \frac{c}{d} = \frac{\frac{a}{b}}{\frac{c}{d}} = \frac{a}{b} \cdot \frac{d}{c}.$$

EXAMPLE 3 Dividing Fractions

a. $\dfrac{x}{x+2} \div \dfrac{x+3}{x-5} = \dfrac{x}{x+2} \cdot \dfrac{x-5}{x+3} = \dfrac{x(x-5)}{(x+2)(x+3)}.$

b. $\dfrac{\dfrac{x-5}{x-3}}{2x} = \dfrac{\dfrac{x-5}{x-3}}{\dfrac{2x}{1}} = \dfrac{x-5}{x-3} \cdot \dfrac{1}{2x} = \dfrac{x-5}{2x(x-3)}.$

c. $\dfrac{\dfrac{4x}{x^2-1}}{\dfrac{2x^2+8x}{x-1}} = \dfrac{4x}{x^2-1} \cdot \dfrac{x-1}{2x^2+8x} = \dfrac{4x(x-1)}{[(x+1)(x-1)][2x(x+4)]}$

$$= \frac{2}{(x+1)(x+4)}.$$ ■

Rationalizing the Denominator

Sometimes the denominator of a fraction has two terms and involves square roots, such as $2 - \sqrt{3}$ or $\sqrt{5} + \sqrt{2}$. The denominator may then be rationalized by multiplying by an expression that makes the denominator a difference of two squares. For example,

$$\frac{4}{\sqrt{5}+\sqrt{2}} = \frac{4}{\sqrt{5}+\sqrt{2}} \cdot \frac{\sqrt{5}-\sqrt{2}}{\sqrt{5}-\sqrt{2}}$$

$$= \frac{4(\sqrt{5}-\sqrt{2})}{(\sqrt{5})^2-(\sqrt{2})^2} = \frac{4(\sqrt{5}-\sqrt{2})}{5-2}$$

$$= \frac{4(\sqrt{5}-\sqrt{2})}{3}.$$

Rationalizing the *numerator* is a similar procedure.

EXAMPLE 4 Rationalizing Denominators

a. $\dfrac{x}{\sqrt{2}-6} = \dfrac{x}{\sqrt{2}-6} \cdot \dfrac{\sqrt{2}+6}{\sqrt{2}+6} = \dfrac{x(\sqrt{2}+6)}{(\sqrt{2})^2-6^2}$

$$= \frac{x(\sqrt{2}+6)}{2-36} = -\frac{x(\sqrt{2}+6)}{34}.$$

b. $\dfrac{\sqrt{5}-\sqrt{2}}{\sqrt{5}+\sqrt{2}} = \dfrac{\sqrt{5}-\sqrt{2}}{\sqrt{5}+\sqrt{2}} \cdot \dfrac{\sqrt{5}-\sqrt{2}}{\sqrt{5}-\sqrt{2}}$

$$= \frac{(\sqrt{5}-\sqrt{2})^2}{5-2} = \frac{5-2\sqrt{5}\sqrt{2}+2}{3} = \frac{7-2\sqrt{10}}{3}.$$ ■

Addition and Subtraction of Fractions

In Example 3(b) of Sec 0.3, it was shown that $\dfrac{a}{c} + \dfrac{b}{c} = \dfrac{a+b}{c}$. That is, if we add two fractions having a common denominator, then the result is a fraction whose denominator is the common denominator. The numerator is the sum of the numerators of the original fractions. Similarly, $\dfrac{a}{c} - \dfrac{b}{c} = \dfrac{a-b}{c}$.

EXAMPLE 5 Adding and Subtracting Fractions

a. $$\frac{p^2-5}{p-2}+\frac{3p+2}{p-2}=\frac{(p^2-5)+(3p+2)}{p-2}$$
$$=\frac{p^2+3p-3}{p-2}.$$

b. $$\frac{x^2-5x+4}{x^2+2x-3}-\frac{x^2+2x}{x^2+5x+6}=\frac{(x-1)(x-4)}{(x-1)(x+3)}-\frac{x(x+2)}{(x+2)(x+3)}$$
$$=\frac{x-4}{x+3}-\frac{x}{x+3}=\frac{(x-4)-x}{x+3}=-\frac{4}{x+3}.$$

c. $$\frac{x^2+x-5}{x-7}-\frac{x^2-2}{x-7}+\frac{-4x+8}{x^2-9x+14}$$
$$=\frac{x^2+x-5}{x-7}-\frac{x^2-2}{x-7}+\frac{-4(\cancel{x-2})}{(\cancel{x-2})(x-7)}$$
$$=\frac{(x^2+x-5)-(x^2-2)+(-4)}{x-7}$$
$$=\frac{x-7}{x-7}=1.$$ ■

To add (or subtract) two fractions with different denominators, use the fundamental principle of fractions to rewrite the fractions as equivalent fractions that have the same denominator. Then proceed with the addition (or subtraction) by the method just described.

For example, to find

$$\frac{2}{x^3(x-3)}+\frac{3}{x(x-3)^2},$$

we can convert the first fraction into an equivalent fraction by multiplying the numerator and denominator by $x-3$:

$$\frac{2(x-3)}{x^3(x-3)^2};$$

we can convert the second fraction by multiplying the numerator and denominator by x^2:

$$\frac{3x^2}{x^3(x-3)^2}.$$

These fractions have the same denominator. Hence,

$$\frac{2}{x^3(x-3)}+\frac{3}{x(x-3)^2}=\frac{2(x-3)}{x^3(x-3)^2}+\frac{3x^2}{x^3(x-3)^2}$$
$$=\frac{3x^2+2x-6}{x^3(x-3)^2}.$$

We could have converted the original fractions into equivalent fractions with *any* common denominator. However, we chose to convert them into fractions with the denominator $x^3(x-3)^2$. This is the **least common denominator (L.C.D.)** of the fractions $2/[x^3(x-3)]$ and $3/[x(x-3)^2]$.

In general, to find the L.C.D. of two or more fractions, first factor each denominator completely. *The L.C.D. is the product of each of the distinct factors appearing in the denominators, each raised to the highest power to which it occurs in any single denominator.*

EXAMPLE 6 Adding and Subtracting Fractions

a. *Subtract:* $\frac{t}{3t+2} - \frac{4}{t-1}$.

Solution: The L.C.D. is $(3t+2)(t-1)$. Thus, we have

$$\frac{t}{(3t+2)} - \frac{4}{t-1} = \frac{t(t-1)}{(3t+2)(t-1)} - \frac{4(3t+2)}{(3t+2)(t-1)}$$
$$= \frac{t(t-1) - 4(3t+2)}{(3t+2)(t-1)}$$
$$= \frac{t^2 - t - 12t - 8}{(3t+2)(t-1)} = \frac{t^2 - 13t - 8}{(3t+2)(t-1)}.$$

b. *Add:* $\frac{4}{q-1} + 3$.

Solution: The L.C.D. is $q - 1$.

$$\frac{4}{q-1} + 3 = \frac{4}{q-1} + \frac{3(q-1)}{q-1}$$
$$= \frac{4 + 3(q-1)}{q-1} = \frac{3q+1}{q-1}.$$

■

EXAMPLE 7 Subtracting Fractions

$$\frac{x-2}{x^2+6x+9} - \frac{x+2}{2(x^2-9)}$$
$$= \frac{x-2}{(x+3)^2} - \frac{x+2}{2(x+3)(x-3)} \qquad [\text{L.C.D.} = 2(x+3)^2(x-3)]$$
$$= \frac{(x-2)(2)(x-3)}{(x+3)^2(2)(x-3)} - \frac{(x+2)(x+3)}{2(x+3)(x-3)(x+3)}$$
$$= \frac{(x-2)(2)(x-3) - (x+2)(x+3)}{2(x+3)^2(x-3)}$$
$$= \frac{2(x^2-5x+6) - (x^2+5x+6)}{2(x+3)^2(x-3)}$$
$$= \frac{2x^2 - 10x + 12 - x^2 - 5x - 6}{2(x+3)^2(x-3)}$$
$$= \frac{x^2 - 15x + 6}{2(x+3)^2(x-3)}.$$

■

Example 8 shows two methods of simplifying a "complex" fraction.

EXAMPLE 8 Combined Operations with Fractions

Simplify $\dfrac{\dfrac{1}{x+h}-\dfrac{1}{x}}{h}$.

Solution: First we combine the fractions in the numerator and obtain

$$\frac{\dfrac{1}{x+h}-\dfrac{1}{x}}{h}=\frac{\dfrac{x}{x(x+h)}-\dfrac{x+h}{x(x+h)}}{h}=\frac{\dfrac{x-(x+h)}{x(x+h)}}{h}$$

$$=\frac{\dfrac{-h}{x(x+h)}}{\dfrac{h}{1}}=\frac{-h}{x(x+h)h}=-\frac{1}{x(x+h)}.$$

The original fraction can also be simplified by multiplying the numerator and denominator by the L.C.D. of the fractions involved in the numerator (and denominator), namely, $x(x+h)$:

$$\frac{\dfrac{1}{x+h}-\dfrac{1}{x}}{h}=\frac{\left[\dfrac{1}{x+h}-\dfrac{1}{x}\right]x(x+h)}{h[x(x+h)]}$$

$$=\frac{x-(x+h)}{x(x+h)h}=\frac{-h}{x(x+h)h}=-\frac{1}{x(x+h)}.$$ ∎

▪ Exercise 0.8

In Problems 1–6, simplify.

1. $\dfrac{x^2-4}{x^2-2x}$.
2. $\dfrac{x^2-5x-6}{x^2-2x-3}$.
3. $\dfrac{x^2-9x+20}{x^2+x-20}$.
4. $\dfrac{3x^2-27x+24}{2x^3-16x^2+14x}$.
5. $\dfrac{6x^2+x-2}{2x^2+3x-2}$.
6. $\dfrac{12x^2-19x+4}{6x^2-17x+12}$.

In Problems 7–48, perform the operations and simplify as much as possible.

7. $\dfrac{y^2}{y-3}\cdot\dfrac{-1}{y+2}$.
8. $\dfrac{z^2-4}{z^2+2z}\cdot\dfrac{z^2}{z-2}$.
9. $\dfrac{2x-3}{x-2}\cdot\dfrac{2-x}{2x+3}$.
10. $\dfrac{x^2-y^2}{x+y}\cdot\dfrac{x^2+2xy+y^2}{y-x}$.
11. $\dfrac{2x-2}{x^2-2x-8}\div\dfrac{x^2-1}{x^2+5x+4}$.
12. $\dfrac{x^2+2x}{3x^2-18x+24}\div\dfrac{x^2-x-6}{x^2-4x+4}$.
13. $\dfrac{\frac{x^2}{6}}{\frac{x}{3}}$.
14. $\dfrac{\frac{4x^3}{9x}}{\frac{x}{18}}$.
15. $\dfrac{\frac{2m}{n^3}}{\frac{4m}{n^2}}$.
16. $\dfrac{\frac{c+d}{c}}{\frac{c-d}{2c}}$.
17. $\dfrac{\frac{4x}{3}}{2x}$.
18. $\dfrac{4x}{\frac{3}{2x}}$.
19. $\dfrac{-9x^3}{\frac{x}{3}}$.
20. $\dfrac{\frac{-9x^3}{x}}{3}$.

21. $\dfrac{\dfrac{x-5}{x^2-7x+10}}{x-2}$.

22. $\dfrac{\dfrac{x^2+6x+9}{x}}{x+3}$.

23. $\dfrac{\dfrac{10x^3}{x^2-1}}{\dfrac{5x}{x+1}}$.

24. $\dfrac{\dfrac{x^2-4}{x^2+2x-3}}{\dfrac{x^2-x-6}{x^2-9}}$.

25. $\dfrac{\dfrac{x^2+7x+10}{x^2-2x-8}}{\dfrac{x^2+6x+5}{x^2-3x-4}}$.

26. $\dfrac{\dfrac{(x+2)^2}{3x-2}}{\dfrac{9x+18}{4-9x^2}}$.

27. $\dfrac{\dfrac{4x^2-9}{x^2+3x-4}}{\dfrac{2x-3}{1-x^2}}$.

28. $\dfrac{\dfrac{6x^2y+7xy-3y}{xy-x+5y-5}}{\dfrac{x^3y+4x^2y}{xy-x+4y-4}}$.

29. $\dfrac{x^2}{x+3}+\dfrac{5x+6}{x+3}$.

30. $\dfrac{2}{x+2}+\dfrac{x}{x+2}$.

31. $\dfrac{1}{t}+\dfrac{2}{3t}$.

32. $\dfrac{4}{x^2}-\dfrac{1}{x}$.

33. $1-\dfrac{p^2}{p^2-1}$.

34. $\dfrac{4}{s+4}+s$.

35. $\dfrac{4}{2x-1}+\dfrac{x}{x+3}$.

36. $\dfrac{x+1}{x-1}-\dfrac{x-1}{x+1}$.

37. $\dfrac{1}{x^2-x-2}+\dfrac{1}{x^2-1}$.

38. $\dfrac{y}{3y^2-5y-2}-\dfrac{2}{3y^2-7y+2}$.

39. $\dfrac{4}{x-1}-3+\dfrac{-3x^2}{5-4x-x^2}$.

40. $\dfrac{2x-3}{2x^2+11x-6}-\dfrac{3x+1}{3x^2+16x-12}+\dfrac{1}{3x-2}$.

41. $(1+x^{-1})^2$.

42. $(x^{-1}+y^{-1})^2$.

43. $(x^{-1}-y)^{-1}$.

44. $(x-y^{-1})^2$.

45. $\dfrac{1+\dfrac{1}{x}}{3}$.

46. $\dfrac{\dfrac{x+3}{x}}{x-\dfrac{9}{x}}$.

47. $\dfrac{3-\dfrac{1}{2x}}{x+\dfrac{x}{x+2}}$.

48. $\dfrac{\dfrac{x-1}{x^2+5x+6}-\dfrac{1}{x+2}}{3+\dfrac{x-7}{3}}$.

In Problems 49 and 50, perform the indicated operations, but do not rationalize the denominators.

49. $\dfrac{2}{\sqrt{x+h}}-\dfrac{2}{\sqrt{x}}$.

50. $\dfrac{x\sqrt{x}}{\sqrt{1+x}}+\dfrac{1}{\sqrt{x}}$.

In Problems 51–60, simplify, and express your answer in a form that is free of radicals in the denominator.

51. $\dfrac{1}{2+\sqrt{3}}$.

52. $\dfrac{1}{1-\sqrt{2}}$.

53. $\dfrac{\sqrt{2}}{\sqrt{3}-\sqrt{6}}$.

54. $\dfrac{5}{\sqrt{6}+\sqrt{7}}$.

55. $\dfrac{2\sqrt{2}}{\sqrt{2}-\sqrt{3}}$.

56. $\dfrac{2\sqrt{3}}{\sqrt{5}-\sqrt{2}}$.

57. $\dfrac{1}{x+\sqrt{5}}$.

58. $\dfrac{x-3}{\sqrt{x}-1}+\dfrac{4}{\sqrt{x}-1}$.

59. $\dfrac{5}{1+\sqrt{3}}-\dfrac{4}{2-\sqrt{2}}$.

60. $\dfrac{4}{\sqrt{x}+2}\cdot\dfrac{x^2}{3}$.

Chapter 2
Equations

Even beginning students in many areas of study are soon faced with solving elementary equations. In this chapter, we shall develop techniques to accomplish this task. These methods will be applied in the next chapter to some practical situations.

OBJECTIVE

To discuss equivalent equations and to develop techniques for solving linear equations, including literal equations.

1.1 LINEAR EQUATIONS

Equations

An **equation** is a statement that two expressions are equal. The two expressions that make up an equation are called its **sides** or **members.** They are separated by the **equality sign, "=."**

Principles in Practice 1
Examples of Equations

You are packaging fencing material for a rectangular garden in which the length is 2 feet more than the width. Write an equation that represents the linear feet P needed for a garden with width w.

EXAMPLE 1 Examples of Equations

a. $x + 2 = 3.$

b. $x^2 + 3x + 2 = 0.$

c. $\frac{y}{y-4} = 6.$

d. $w = 7 - z.$ ■

In Example 1, each equation contains at least one variable. A **variable** is a symbol that can be replaced by any one of a set of different numbers. The most popular symbols for variables are letters from the latter part of the alphabet, such as *x, y, z, w,* and *t.* Hence, equations (a) and (c) are said to be in the variables x and y, respectively. Equation (d) is in the variables w and z. In the equation $x + 2 = 3$, the numbers 2 and 3 are called *constants.* They are fixed numbers.

Here we discuss restrictions on variables.

We *never* allow a variable in an equation to have a value for which any expression in that equation is undefined. Thus, in

$$\frac{y}{y-4} = 6,$$

y cannot be 4, because this would make the denominator zero. (We cannot divide by zero.) In some equations, the allowable values of a variable are restricted for physical reasons. For example, if the variable t represents time, negative values of t may not make sense. Hence, we should assume that $t \geq 0$.

To *solve* an equation means to find all values of its variables for which the equation is true. These values are called *solutions* of the equation and are said to *satisfy* the equation. When only one variable is involved, a solution is also called a **root.** The set of all solutions is called the **solution set** of the equation. Sometimes a letter representing an unknown quantity in an equation is simply called an *unknown.* Let us illustrate these terms.

EXAMPLE 2 Terminology for Equations

a. In the equation $x + 2 = 3$, the variable x is the unknown. The only value of x that satisfies the equation is obviously 1. Hence, 1 is a root and the solution set is $\{1\}$.

b. -2 is a root of $x^2 + 3x + 2 = 0$ because substituting -2 for x makes the equation true: $(-2)^2 + 3(-2) + 2 = 0$.

c. $w = 7 - z$ is an equation in two unknowns. One solution is the pair of values $w = 4$ and $z = 3$. However, there are infinitely many solutions. Can you think of another? ■

Equivalent Equations

Solving an equation may involve performing operations on it. We prefer that any such operation result in another equation having exactly the same solutions as the given equation. When this occurs, the equations are said to be **equivalent.** There are three operations that guarantee equivalence:

1. Adding (subtracting) the same polynomial[1] to (from) both sides of an equation, where the polynomial is in the same variable as that occurring in the equation.

For example, if $-5x = 5 - 6x$, then adding $6x$ to both sides gives the equivalent equation $-5x + 6x = 5 - 6x + 6x$, or $x = 5$.

Equivalence is not guaranteed if both sides are multiplied or divided by an expression involving a variable.

2. Multiplying (dividing) both sides of an equation by the same constant, except by zero.

For example, if $10x = 5$, then dividing both sides by 10 gives the equivalent equation $\frac{10x}{10} = \frac{5}{10}$, or $x = \frac{1}{2}$.

3. Replacing either side of an equation by an equal expression.

For example, if $x(x + 2) = 3$, then replacing the left side by the equal expression $x^2 + 2x$ gives the equivalent equation $x^2 + 2x = 3$.

We repeat: Applying Operations 1–3 guarantees that the resulting equation is equivalent to the given one. However, sometimes in solving an equation we have to apply operations other than 1–3. These operations may *not* necessarily result in equivalent equations. They include the following.

[1]See Sec. 0.6 for a definition of a polynomial.

Operations That May Not Produce Equivalent Equations

4. Multiplying both sides of an equation by an expression involving the variable;

5. Dividing both sides of an equation by an expression involving the variable;

6. Raising both sides of an equation to equal powers.

Operation 6 includes taking roots of both sides.

Let us illustrate the last three operations. For example, by inspection, the only root of $x - 1 = 0$ is 1. Multiplying each side by x (Operation 4) gives $x^2 - x = 0$, which is satisfied if x is 0 or 1. (Check this by substitution.) But 0 *does not* satisfy the *original* equation. Thus, the equations are not equivalent.

Continuing, you may check that the equation $(x - 4)(x - 3) = 0$ is satisfied when x is 4 or 3. Dividing both sides by $x - 4$ (Operation 5) gives $x - 3 = 0$, whose only root is 3. Again, we do not have equivalence, since in this case a root has been "lost." Note that when x is 4, division by $x - 4$ implies division by 0, an invalid operation.

Finally, squaring each side of the equation $x = 2$ (Operation 6) gives $x^2 = 4$, which is true if $x = 2$ or -2. But -2 is not a root of the given equation.

From our discussion, it is clear that when Operations 4–6 are performed, we must be careful about drawing conclusions concerning the roots of a given equation. Operations 4 and 6 *can* produce an equation with more roots. Thus, you should check whether or not each "solution" obtained by the these operations satisfies the *original* equation. Operation 5 *can* produce an equation with fewer roots. In this case, any "lost" root may never be determined. Thus, avoid Operation 5 whenever possible.

In summary, an equation may be thought of as a set of restrictions on any variable in the equation. Operations 4–6 may increase or decrease the number of restrictions, giving solutions different from those of the original equation. However, Operations 1–3 never affect the restrictions.

TECHNOLOGY

A graphics calculator can be used to test for a root. For example, suppose we want to determine whether 3/2 is a root of the equation

$$2x^3 + 7x^2 = 19x + 60.$$

First, we rewrite the equation so that one side is 0. Subtracting $19x + 60$ from both sides gives the equivalent equation

$$2x^3 + 7x^2 - 19x - 60 = 0.$$

For a TI-82 graphics calculator, we enter the expression $2x^3 + 7x^2 - 19x - 60$ as Y_1 and then evaluate Y_1 at $x = 3/2$. Figure 1.1 shows that the result is -66, which is not 0. Thus, 3/2 is not a root. However, Y_1 evaluated at $x = -5/2$ *is* 0. So $-5/2$ is a root of the original equation.

It is worth noting that if the original equation had been in terms of the variable t, that is,

$$2t^3 + 7t^2 = 19t + 60,$$

then we must replace t by x because the calculator evaluates Y_1 at a specified value of x, not t.

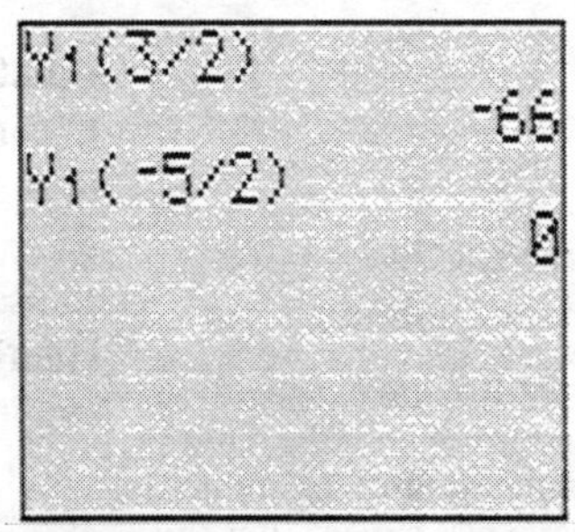

FIGURE 1.1 For $2x^3 + 7x^2 - 19x - 60 = 0$, $^3/_2$ is not a root, but $^{-5}/_2$ is a root.

Linear Equations

The principles presented so far will now be demonstrated in the solution of a *linear equation.*

DEFINITION

*A **linear equation** in the variable x is an equation that can be written in the form*

$$ax + b = 0, \tag{1}$$

where a and b are constants and $a \neq 0$.

A linear equation is also called a first-degree equation or an equation of degree one, since the highest power of the variable that occurs in Eq. (1) is the first.

To solve a linear equation, we perform operations on it until we have an equivalent equation whose solutions are *obvious.* This means an equation in which the variable is isolated on one side, as the following examples show.

Principles in Practice 2
Solving a Linear Equation

A trendy coffeehouse's total daily revenue from the sale of x specialty coffees is given by $r = 2.25x$, and its total daily costs are given by $c = 0.75x + 300$. How many specialty coffees need to be sold each day to break even? In other words, when will revenue equal costs?

EXAMPLE 3 Solving a Linear Equation

Solve $5x - 6 = 3x$.

Solution: We begin by getting the terms involving x on one side and the constant on the other. Then we solve for x by the appropriate mathematical operation. We have

$$5x - 6 = 3x,$$

$$5x - 6 + (-3x) = 3x + (-3x) \quad \text{(adding } -3x \text{ to both sides),}$$

$$2x - 6 = 0 \quad \text{(simplifying, that is, Operation 3),}$$

$$2x - 6 + 6 = 0 + 6 \quad \text{(adding 6 to both sides),}$$

$$2x = 6 \quad \text{(simplifying),}$$

$$\frac{2x}{2} = \frac{6}{2} \quad \text{(dividing both sides by 2)}$$

$$x = 3.$$

Clearly, 3 is the only root of the last equation. Since each equation is equivalent to the one before it, we conclude that 3 must be the only root of $5x - 6 = 3x$. That is, the solution set is $\{3\}$. We can describe the first step in the solution as moving a term from one side of an equation to the other while changing its sign; this is commonly called *transposing.* Note that since the original equation can be put in the form $2x + (-6) = 0$, it is a linear equation.

■

Principles in Practice 3
Solving a Linear Equation

Monica and Pedro have agreed to pool their savings when they have saved the same amount of money. Monica can save \$40 a week, but she must use her first \$125 to pay her credit card debt. Pedro has been saving \$35 a week for three weeks already. When will they pool their savings? How much will they each have saved?

EXAMPLE 4 Solving a Linear Equation

Solve $2(p + 4) = 7p + 2$.

Solution: First, we remove parentheses. Then we collect similar terms and solve. We have

$$
\begin{aligned}
2(p+4) &= 7p+2, \\
2p+8 &= 7p+2 && \text{(distributive property)}, \\
2p &= 7p-6 && \text{(subtracting 8 from both sides)}, \\
-5p &= -6 && \text{(subtracting } 7p \text{ from both sides)}, \\
p &= \frac{-6}{-5} && \text{(dividing both sides by } -5), \\
p &= \frac{6}{5}.
\end{aligned}
$$

■

EXAMPLE 5 Solving a Linear Equation

Solve $\frac{7x+3}{2} - \frac{9x-8}{4} = 6.$

Solution: We first clear the equation of fractions by multiplying *both* sides by the least common denominator (L.C.D.),[2] which is 4. Then we use various algebraic operations to obtain a solution. Thus,

$$
4\left(\frac{7x+3}{2} - \frac{9x-8}{4}\right) = 4(6),
$$

The distributive property requires that *both* terms within the parentheses be multiplied by 4.

$$
\begin{aligned}
4 \cdot \frac{7x+3}{2} - 4 \cdot \frac{9x-8}{4} &= 24 && \text{(distributive property)}, \\
2(7x+3) - (9x-8) &= 24 && \text{(simplifying)}, \\
14x + 6 - 9x + 8 &= 24 && \text{(distributive property)}, \\
5x + 14 &= 24 && \text{(simplifying)}, \\
5x &= 10 && \text{(subtracting 14 from both sides)}, \\
x &= 2 && \text{(dividing both sides by 5)}.
\end{aligned}
$$

■

Every linear equation has exactly one root.

Each equation in Examples 3–5 has one and only one root. This is true of every linear equation in one variable.

Literal Equations

Equations in which some of the constants are not specified, but are represented by letters, such as $a, b, c,$ or d, are called **literal equations,** and the letters are called **literal constants** or **arbitrary constants.** For example, in the literal equation $x + a = 4b$, we may consider a and b to be arbitrary constants. Formulas, such as $I = Prt$, that express a relationship between certain quantities may be regarded as literal equations. If we want to express a particular letter in a formula in terms of the others, this letter is considered the unknown.

Principles in Practice 4
Solving a Literal Equation

The formula $d = rt$ gives the distance d an object travels going a rate r for a time t. What is the rate r for a train that travels d miles in t hours?

EXAMPLE 6 Solving Literal Equations

a. *The equation $I = Prt$ is the formula for the simple interest I on a principal of P dollars at the annual interest rate of r for a period of t years. Express r in terms of I, P, and t.*

[2] The least common denominator of two or more fractions is the smallest number with all the denominators as factors. That is, the L.C.D. is the least common multiple of all the denominators.

Solution: Here we consider r to be the unknown. To isolate r, we divide both sides by Pt. We have

$$I = Prt,$$

$$\frac{I}{Pt} = \frac{Prt}{Pt},$$

$$\frac{I}{Pt} = r \qquad \text{or} \qquad r = \frac{I}{Pt}.$$

When we divided both sides by Pt, we assumed that $Pt \neq 0$, since we cannot divide by 0. Similar assumptions will be made in solving other literal equations.

b. *The equation $S = P + Prt$ is the formula for the value S of an investment of a principal of P dollars at a simple annual interest rate of r for a period of t years. Solve for P.*

Solution:

$$S = P + Prt,$$

$$S = P(1 + rt) \qquad \text{(factoring)},$$

$$\frac{S}{1 + rt} = P \qquad \text{(dividing both sides by } 1 + rt\text{)}.$$

■

Principles in Practice 5
Solving a Literal Equation

The formula $S = 4\pi\left(\frac{d}{2}\right)^2$ gives the surface area S of a sphere with diameter d. What is the length of the side of the smallest box that will hold a ball with surface area S?

EXAMPLE 7 Solving a Literal Equation

Solve $(a + c)x + x^2 = (x + a)^2$ for x.

Solution: We first simplify the equation and then get all terms involving x on one side:

$$(a + c)x + x^2 = (x + a)^2,$$

$$ax + cx + x^2 = x^2 + 2ax + a^2,$$

$$ax + cx = 2ax + a^2,$$

$$cx - ax = a^2,$$

$$x(c - a) = a^2,$$

$$x = \frac{a^2}{c - a}.$$

■

■ Exercise 1.1

In Problems 1–6, determine by substitution which of the given numbers, if any, satisfy the given equation.

1. $9x - x^2 = 0$; $1, 0$.

2. $20 - 9x = -x^2$; $5, 4$.

3. $y + 2(y - 3) = 4$; $\frac{10}{3}, 1$.

4. $2x + x^2 - 8 = 0$; $2, -4$.

5. $x(7 + x) - 2(x + 1) - 3x = -2$; $-3, 0$.

6. $x(x + 1)^2(x + 2) = 0$; $0, -1, 2$.

In Problems 7–16, determine what operations were applied to the first equation to obtain the second. State whether or not the operations guarantee that the equations are equivalent. Do not solve the equations.

7. $x - 5 = 4x + 10$; $x = 4x + 15$.

8. $8x - 4 = 16$; $x - \frac{1}{2} = 2$.

9. $x = 4$; $x^2 = 16$.

10. $\frac{1}{2}x^2 + 3 = x - 9$; $x^2 + 6 = 2x - 18$.

11. $x^2 - 2x = 0$; $x - 2 = 0$.

12. $\dfrac{2}{x - 2} + x = x^2$; $2 + x(x - 2) = x^2(x - 2)$.

13. $\dfrac{x^2 - 1}{x - 1} = 3$; $x^2 - 1 = 3(x - 1)$.

14. $x(x + 5)(x + 9) = x(x + 1)$; $(x + 5)(x + 9) = x + 1$.

15. $\dfrac{x(x + 1)}{x - 5} = x(x + 9)$; $x + 1 = (x + 9)(x - 5)$.

16. $2x^2 - 9 = x$; $x^2 - \frac{1}{2}x = \frac{9}{2}$.

In Problems 17–46, solve the equations.

17. $4x = 10$.

18. $0.2x = 5$.

19. $3y = 0$.

20. $2x - 4x = -5$.

21. $-5x = 10 - 15$.

22. $3 - 2x = 4$.

23. $5x - 3 = 9$.

24. $\sqrt{2}x + 3 = 8$.

25. $7x + 7 = 2(x + 1)$.

26. $6z + 5z - 3 = 41$.

27. $2(p - 1) - 3(p - 4) = 4p$.

28. $t = 2 - 2[2t - 3(1 - t)]$.

29. $\dfrac{x}{5} = 2x - 6$.

30. $\dfrac{5y}{7} - \dfrac{6}{7} = 2 - 4y$.

31. $5 + \dfrac{4x}{9} = \dfrac{x}{2}$.

32. $\dfrac{x}{3} - 4 = \dfrac{x}{5}$.

33. $q = \dfrac{3}{2}q - 4$.

34. $\dfrac{x}{2} + \dfrac{x}{3} = 7$.

35. $3x + \dfrac{x}{5} - 5 = \dfrac{1}{5} + 5x$.

36. $y - \dfrac{y}{2} + \dfrac{y}{3} - \dfrac{y}{4} = \dfrac{y}{5}$.

37. $\dfrac{2y - 3}{4} = \dfrac{6y + 7}{3}$.

38. $\dfrac{p}{3} + \dfrac{3}{4}p = \dfrac{9}{2}(p - 1)$.

39. $w + \dfrac{w}{2} - \dfrac{w}{3} + \dfrac{w}{4} = 5$.

40. $\dfrac{7 + 2(x + 1)}{3} = \dfrac{8x}{5}$.

41. $\dfrac{x + 2}{3} - \dfrac{2 - x}{6} = x - 2$.

42. $\dfrac{x}{5} + \dfrac{2(x - 4)}{10} = 7$.

43. $\dfrac{9}{5}(3 - x) = \dfrac{3}{4}(x - 3)$.

44. $\dfrac{2y - 7}{3} + \dfrac{8y - 9}{14} = \dfrac{3y - 5}{21}$.

45. $\dfrac{3}{2}(4x - 3) = 2[x - (4x - 3)]$.

46. $(3x - 1)^2 - (5x - 3)^2 = -(4x - 2)^2$.

In Problems 47–54, express the indicated symbol in terms of the remaining symbols.

47. $I = Prt$; P.

48. $ax + b = 0$; x.

49. $p = 8q - 1$; q.

50. $p = -3q + 6$; q.

51. $S = P(1 + rt)$; r.

52. $r = \dfrac{2mI}{B(n + 1)}$; m.

53. $S = \dfrac{n}{2}(a_1 + a_n)$; a_1.

54. $S = \dfrac{R[(1 + i)^n - 1]}{i}$; R.

55. Geometry Use the formula $P = 2l + 2w$ to find the width w of a rectangle whose perimeter P is 960 m and whose length l is 360 m.

56. Geometry Use the formula $A = \frac{1}{2}bh$ to find the height h of a triangle whose area A is 75 cm² and whose base b is 15 cm.

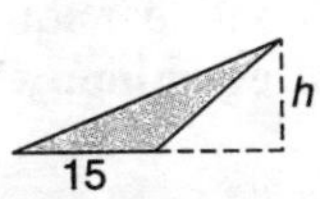

57. Sales Tax A salesperson needs to calculate the cost of an item with a sales tax of 8.25%. Write an equation that represents the total cost c of an item costing x dollars.

58. Revenue A day care center's total monthly revenue from the care of x toddlers is given by $r = 650x$, and its total monthly costs are given by $c = 400x + 2000$. How many toddlers need to be enrolled each month to break even? In other words, when will revenue equal costs?

59. Straight-line Depreciation If you purchase an item for business use, in preparing your income tax you may be able to spread out its expense over the life of the item. This is called *depreciation.* One method of depreciation is *straight-line depreciation,* in which the annual depreciation is computed by dividing the cost of the item, less its estimated salvage value, by its useful life. Suppose the cost is C dollars, the useful life is N years, and there is no salvage value. Then the value V (in dollars) of the item at the end of n years is given by

$$V = C\left(1 - \frac{n}{N}\right).$$

If new office furniture is purchased for \$1600, has a useful life of 8 years, and has no salvage value, after how many years will it have a value of \$1000?

60. Radar Beam When radar is used on a highway to determine the speed of a car, a radar beam is sent out and reflected from the moving car. The difference F (in cycles per second) in frequency between the original and reflected beams is given by

$$F = \frac{vf}{334.8},$$

where v is the speed of the car in miles per hour and f is the frequency of the original beam (in megacycles per second).

Suppose you are driving along a highway with a speed limit of 55 mi/h. A police officer aims a radar

beam with a frequency of 2450 megacycles per second at your car, and the officer observes the difference in frequency to be 420 cycles per second. Can the officer claim that you were speeding?

61. Savings Paula and Sam want to buy a house, so they have decided to save one-fourth of each of their salaries. Paula earns \$22.00 per hour and receives an extra \$8.00 a week because she declined company benefits, and Sam earns \$26.00 per hour plus benefits. They want to save at least \$350.00 each week. How many hours must they each work each week?

62. Gravity The equation $h = -4.9t^2 + m$ is the formula for the height h of an object in meters t seconds after is it dropped from a starting point of m meters. What is the time t an object has been falling if it has fallen from a height m and is now at a height h?

63. Linear Expansion When solid objects are heated, they expand in length—which is why expansion joints are placed in bridges and pavements. Generally, when the temperature of a solid body of length I_0 is increased from T_0 to T, the body's length I is given by

$$I = I_0[1 + \alpha(T - T_0)],$$

where α (the Greek letter *alpha*) is called the *coefficient of linear expansion.* Suppose a metal rod 1 m long at 0°C expands 0.001 m when it is heated from 0°C to 100°C. Find the coefficient of linear expansion.

64. Predator-Prey Relation To study a predator–prey relationship, an experiment[3] was conducted in which a blindfolded subject, the "predator," stood in front of a 3-ft-square table on which uniform sandpaper discs, the "prey," were placed. For 1 minute the "predator" searched for the discs by tapping with a finger. Whenever a disc was found, it was removed and searching resumed. The experiment was repeated for various disc densities (number of discs per 9 ft²). It was estimated that if y is the number of discs picked up in 1 minute when x discs are on the table, then

$$y = a(1 - by)x,$$

where a and b are constants. Solve this equation for y.

In Problems 65–68, use a graphics calculator to determine which of the given numbers, if any, are roots of the given equation.

65. $112x^2 = 6x + 1;\quad \frac{1}{8}, -\frac{2}{5}, -\frac{1}{14}.$

66. $8x^3 + 11x + 21 = 58x^2;\quad 7, -\frac{1}{2}, \frac{4}{3}.$

67. $\dfrac{3.1t - 7}{4.8t - 2} = 7;\quad \sqrt{6}, -\dfrac{47}{52}, \dfrac{14}{61}.$

68. $\left(\dfrac{v}{v + 3}\right)^2 = v;\quad 0, \dfrac{27}{4}, \dfrac{13}{3}.$

OBJECTIVE

To solve fractional and radical equations that lead to linear equations.

1.2 EQUATIONS LEADING TO LINEAR EQUATIONS

Fractional Equations

In this section, we illustrate that solving a nonlinear equation may lead to a linear equation. We begin with a **fractional equation,** which is an equation in which an unknown is in a denominator.

Principles in Practice 1
Solving a Fractional Equation

A boat with a speed r travels 10 miles downstream in a 2-mile-per-hour current in the same time a boat going the same speed travels 6 miles upstream in the current. Write an equation describing this situation, and find the speed of the boats.

EXAMPLE 1 Solving a Fractional Equation

Solve $\dfrac{5}{x - 4} = \dfrac{6}{x - 3}.$

[3] C. S. Holling, "Some Characteristics of Simple Types of Predation and Parasitism," *The Canadian Entomologist,* XCI, no. 7 (1959), 385–98.

Solution:

Strategy We first write the equation in a form that is free of fractions. Then we use standard algebraic techniques to solve the resulting linear equation.

Multiplying both sides by the L.C.D., $(x-4)(x-3)$, we have

$$(x-4)(x-3)\left(\frac{5}{x-4}\right)=(x-4)(x-3)\left(\frac{6}{x-3}\right),$$
$$5(x-3)=6(x-4) \qquad \text{(linear equation)},$$
$$5x-15=6x-24,$$
$$9=x.$$

An alternative solution that avoids multiplying both sides by the L.C.D. is as follows:

$$\frac{5}{x-4}-\frac{6}{x-3}=0.$$

Assuming that x is neither 3 nor 4 and combining fractions gives

$$\frac{9-x}{(x-4)(x-3)}=0.$$

A fraction can be 0 only when its numerator is 0 and its denominator is not. Hence, $x=9$.

In the first step, we multiplied each side by an expression involving the *variable x*. As we mentioned in Sec. 1.1, this means that we are not guaranteed that the last equation is equivalent to the *original* equation. Thus, we must check whether or not 9 satisfies the *original* equation. Substituting 9 for x in that equation, we get

$$\frac{5}{9-4}=\frac{6}{9-3},$$
$$1=1,$$

which is a true statement. Hence, 9 is a root. ■

Some equations that are not linear do not have any solutions. In that case, we say that the solution set is the **empty set** or **null set,** which we denote by { } or ∅. Example 2 will illustrate.

EXAMPLE 2 Solving Fractional Equations

a. *Solve* $\dfrac{3x+4}{x+2}-\dfrac{3x-5}{x-4}=\dfrac{12}{x^2-2x-8}$.

Solution: Observing the denominators and noting that

$$x^2-2x-8=(x+2)(x-4),$$

we conclude that the L.C.D. is $(x+2)(x-4)$. Multiplying both sides by the L.C.D., we have

$$(x+2)(x-4)\left(\frac{3x+4}{x+2}-\frac{3x-5}{x-4}\right)=(x+2)(x-4)\cdot\frac{12}{(x+2)(x-4)},$$
$$(x-4)(3x+4)-(x+2)(3x-5)=12,$$
$$3x^2-8x-16-(3x^2+x-10)=12,$$
$$3x^2-8x-16-3x^2-x+10=12,$$
$$-9x-6=12,$$
$$-9x=18,$$
$$x=-2. \qquad (1)$$

However, the *original* equation is not defined for $x=-2$ (we cannot divide by zero), so there are no roots. Thus, the solution set is ∅. Although

-2 is a solution of Eq. (1), it is not a solution of the *original* equation and is called an **extraneous solution** of that equation.

b. *Solve* $\dfrac{4}{x-5} = 0$.

Solution: The only way a fraction can equal zero is if the numerator is 0 but its denominator is not. Since the numerator, 4, is never 0, the solution set is $\varnothing$. ■

Principles in Practice 2
Literal Equation

The time it takes an airplane to travel a given distance with a tailwind ("with the wind") can be calculated by dividing the distance by the sum of the speed of the airplane and the speed of the wind. Write an equation that calculates the time t it takes an airplane traveling at speed r with a wind w to cover a distance d. Solve your equation for w.

EXAMPLE 3 Literal Equation

If $s = \dfrac{u}{au + v}$, *express u in terms of the remaining letters; that is, solve for u.*

Solution:

Strategy: Since the unknown, u, is in the denominator, we first clear fractions and then solve for u.

$$s = \frac{u}{au + v},$$

$$s(au + v) = u \qquad \text{(multiplying both sides by } au + v\text{)},$$

$$sau + sv = u,$$

$$sau - u = -sv,$$

$$u(sa - 1) = -sv,$$

$$u = \frac{-sv}{sa - 1} = \frac{sv}{1 - sa}.$$

■

Radical Equations

A **radical equation** is one in which an unknown occurs in a radicand. The next two examples illustrate the techniques employed to solve such equations.

Principles in Practice 3
Solving a Radical Equation

The difference between the length of a ramp and the length of the horizontal distance it covers is 2 feet. The square of the vertical distance the ramp covers is 16 square feet. Write an equation for the difference and solve. What is the length of the ramp?

EXAMPLE 4 Solving a Radical Equation

Solve $\sqrt{x^2 + 33} - x = 3$.

Solution: To solve this radical equation, we raise both sides to the same power to eliminate the radical. This operation does *not* guarantee equivalence, so we must check any resulting "solutions." We begin by isolating the radical on one side. Then we square both sides and solve using standard techniques. Thus,

$$\sqrt{x^2 + 33} = x + 3,$$

$$x^2 + 33 = (x + 3)^2 \qquad \text{(squaring both sides)},$$

$$x^2 + 33 = x^2 + 6x + 9,$$

$$24 = 6x,$$

$$4 = x.$$

You should show by substitution that 4 is indeed a root. ■

With some radical equations, you may have to raise both sides to the same power more than once, as Example 5 shows.

EXAMPLE 5 Solving a Radical Equation

Solve $\sqrt{y-3} - \sqrt{y} = -3$.

Solution: When an equation has two terms involving radicals, first write the equation so that one radical is on each side, if possible. Then square and solve. We have

The reason why we desire one radical on each side is to eliminate squaring a binomial with two different radicals.

$$\sqrt{y-3} = \sqrt{y} - 3,$$
$$y - 3 = y - 6\sqrt{y} + 9 \qquad \text{(squaring both sides)},$$
$$6\sqrt{y} = 12,$$
$$\sqrt{y} = 2,$$
$$y = 4 \qquad \text{(squaring both sides)}.$$

Substituting 4 into the left side of the *original* equation gives $\sqrt{1} - \sqrt{4}$, which is -1. Since this does not equal the right side, -3, there is no solution. That is, the solution set is $\varnothing$. Here 4 is an extraneous solution. ■

■ Exercise 1.2

In Problems 1–34, solve the equations.

1. $\dfrac{5}{x} = 25.$

2. $\dfrac{4}{x-1} = 2.$

3. $\dfrac{3}{7-x} = 0.$

4. $\dfrac{5x-2}{x+1} = 0.$

5. $\dfrac{4}{8-x} = \dfrac{3}{4}.$

6. $\dfrac{x+3}{x} = \dfrac{2}{5}.$

7. $\dfrac{q}{3q-4} = 3.$

8. $\dfrac{4p}{7-p} = 1.$

9. $\dfrac{1}{p-1} = \dfrac{2}{p-2}.$

10. $\dfrac{2x-3}{4x-5} = 6.$

11. $\dfrac{1}{x} + \dfrac{1}{5} = \dfrac{4}{5}.$

12. $\dfrac{4}{t-3} = \dfrac{3}{t-4}.$

13. $\dfrac{3x-2}{2x+3} = \dfrac{3x-1}{2x+1}.$

14. $\dfrac{x+2}{x-1} + \dfrac{x+1}{2-x} = 0.$

15. $\dfrac{y-6}{y} - \dfrac{6}{y} = \dfrac{y+6}{y-6}.$

16. $\dfrac{y-3}{y+3} = \dfrac{y-3}{y+2}.$

17. $\dfrac{-4}{x-1} = \dfrac{7}{2-x} + \dfrac{3}{x+1}.$

18. $\dfrac{1}{x-3} - \dfrac{3}{x-2} = \dfrac{4}{1-2x}.$

19. $\dfrac{9}{x-3} = \dfrac{3x}{x-3}.$

20. $\dfrac{x}{x+3} - \dfrac{x}{x-3} = \dfrac{3x-4}{x^2-9}.$

21. $\sqrt{x+6} = 3.$

22. $\sqrt{z-2} = 3.$

23. $\sqrt{5x-6} - 16 = 0.$

24. $6 - \sqrt{2x+5} = 0.$

25. $\sqrt{\dfrac{x}{2} + 1} = \dfrac{2}{3}.$

26. $(x+6)^{1/2} = 7.$

27. $\sqrt{4x-6} = \sqrt{x}.$

28. $\sqrt{7-2x} = \sqrt{x-1}.$

29. $(x-3)^{3/2} = 8.$

30. $\sqrt{y^2-9} = 9 - y.$

31. $\sqrt{y} + \sqrt{y+2} = 3.$

32. $\sqrt{x} - \sqrt{x+1} = 1.$

33. $\sqrt{z^2+2z} = 3 + z.$

34. $\sqrt{\dfrac{1}{w}} - \sqrt{\dfrac{2}{5w-2}} = 0.$

In Problems 35–38, express the indicated letter in terms of the remaining letters.

35. $r = \dfrac{d}{1-dt}$; d.

36. $\dfrac{x-a}{b-x} = \dfrac{x-b}{a-x}$; x.

37. $r = \dfrac{2ml}{B(n+1)}$; n.

38. $\dfrac{1}{p} + \dfrac{1}{q} = \dfrac{1}{f}$; q.

39. **Prey Density** In a certain wildlife preserve, the number y of prey consumed by an individual predator over a given period of time is given by

$$y = \frac{10x}{1 + 0.1x},$$

where x is the *prey density* (the number of prey per unit of area). What prey density would allow a predator to survive if it needs to consume 50 prey over the given period?

40. **Store Hours** Suppose the ratio of the number of hours a video store is open to the number of daily customers is constant. When the store is open 8 hours, the number of customers is 92 less than the maximum number of customers. When the store is open 10 hours, the number of customers is 46 less than the maximum number of customers. Write an equation describing this situation, and find the maximum number of daily customers.

41. **Travel Time** The time it takes a boat to travel a given distance upstream (against the current) can be calculated by dividing the distance by the difference of the speed of the boat and the speed of the current. Write an equation that calculates the time t it takes a boat moving at a speed r against a current c to travel a distance d. Solve your equation for r.

42. **Ramp Length** The difference between the length of a ramp and the length of the horizontal distance it covers is 5 feet. The square of the vertical distance the ramp covers is 45 square feet. Write an equation for the difference and solve. What is the length of the ramp?

43. **Sight Distance** The number of miles a person can see to the horizon from a point above the surface of the Earth is 0.85 of the square root of the person's distance in feet above the surface. Jack is 85 feet higher and sees 4.25 miles farther than Jill. How high are Jack and Jill above the surface?

44. **Automobile Skidding** Police have used the formula $s = \sqrt{30\,fd}$ to estimate the speed s (in miles per hour) of a car if it skidded d feet when stopping. The literal number f is the coefficient of friction, determined by the kind of road (such as concrete, asphalt, gravel, or tar) and whether the road is wet or dry. Some values of f are given in Table 1.1. At 40 mi/h, about how many feet will a car skid on a dry concrete road? Give your answer to the nearest foot.

TABLE 1.1

	Concrete	Tar
Wet	0.4	0.5
Dry	0.8	1.0

OBJECTIVE

To solve quadratic equations by factoring or by using the quadratic formula.

1.3 Quadratic Equations

To learn how to solve more complicated problems, we turn to methods of solving *quadratic equations.*

DEFINITION

A ***quadratic equation*** *in the variable x is an equation that can be written in the form*

$$ax^2 + bx + c = 0, \qquad (1)$$

where a, b, and c are constants and $a \neq 0$.

A quadratic equation is also called a *second-degree equation* or an *equation of degree two,* since the highest power of the variable that occurs is the second. Whereas a linear equation has only one root, a quadratic equation may have two different roots.

Solution by Factoring

A useful method of solving quadratic equations is based on factoring, as the following example shows.

Principles in Practice 1
Solving a Quadratic Equation by Factoring

A number squared is 30 more than the number. What is the number?

EXAMPLE 1 Solving a Quadratic Equation by Factoring

a. *Solve $x^2 + x - 12 = 0$.*

Solution: The left side factors easily:

$$(x - 3)(x + 4) = 0.$$

Think of this as two quantities, $x - 3$ and $x + 4$, whose product is zero. **Whenever the product of two or more quantities is *zero,* at least one of the quantities *must* be zero.** This means that either

$$x - 3 = 0 \qquad \text{or} \qquad x + 4 = 0.$$

Solving these gives $x = 3$ and $x = -4$. Thus, the roots of the original equation are 3 and -4, and the solution set is $\{3, -4\}$.

b. *Solve* $6w^2 = 5w$.

We do not divide both sides by w (a variable) since equivalence is not guaranteed and we may 'lose' a root.

Solution: We write the equation as

$$6w^2 - 5w = 0,$$

so that one side is 0. Factoring gives

$$w(6w - 5) = 0.$$

Setting each factor equal to 0, we have

$$w = 0 \qquad \text{or} \qquad 6w - 5 = 0,$$
$$6w = 5.$$

Thus, the roots are $w = 0$ and $w = \frac{5}{6}$. Note that if we had divided both sides of $6w^2 = 5w$ by w and obtained $6w = 5$, our only solution would be $w = \frac{5}{6}$. That is, we would lose the root $w = 0$. This confirms our discussion of Operation 5 in Sec. 1.1. ■

Principles in Practice 2
Solving a Quadratic Equation by Factoring

The area of a rectangular mural that has a width 10 feet less than its length is 3000 square feet. What are the dimensions of the mural?

EXAMPLE 2 Solving a Quadratic Equation by Factoring

Solve $(3x - 4)(x + 1) = -2$.

Pitfall ▼ You should approach a problem like this with caution. If the product of two quantities is equal to -2, it is not true that at least one of the quantities must be -2. Why? You should **not** set each factor equal to -2; doing so will not provide solutions to the given equation.

Solution: We first multiply the factors on the left side:

$$3x^2 - x - 4 = -2.$$

Rewriting this equation so that 0 appears on one side, we have

$$3x^2 - x - 2 = 0,$$
$$(3x + 2)(x - 1) = 0,$$
$$x = -\frac{2}{3}, 1.$$ ■

Some equations that are not quadratic may be solved by factoring, as Example 3 shows.

EXAMPLE 3 Solving a Higher Degree Equation by Factoring

a. *Solve* $4x - 4x^3 = 0$.

Solution: This is called a *third-degree equation.* We proceed to solve it as follows:

$$4x - 4x^3 = 0,$$

$$4x(1 - x^2) = 0 \quad \text{(factoring)},$$

$$4x(1 - x)(1 + x) = 0 \quad \text{(factoring)}.$$

Do not neglect the fact that the factor x gives rise to a root.

Setting each factor equal to 0 gives $4 = 0$ (impossible), $x = 0$, $1 - x = 0$, or $1 + x = 0$. Thus,

$$x = 0, 1, -1,$$

which we can write as $x = 0, \pm 1$.

Principles in Practice 3
Solving a Higher Degree Equation by Factoring

The volume of a rectangular prism with a square base and height that is 5 times as long as its width is 5 times its width. What are the dimensions of the rectangular prism?

b. *Solve* $x(x + 2)^2(x + 5) + x(x + 2)^3 = 0$.

Solution: Factoring $x(x + 2)^2$ from both terms on the left side, we have

$$x(x + 2)^2[(x + 5) + (x + 2)] = 0,$$

$$x(x + 2)^2(2x + 7) = 0.$$

Hence, $x = 0$, $x + 2 = 0$, or $2x + 7 = 0$, from which it follows that $x = 0, -2, -\frac{7}{2}$. ■

EXAMPLE 4 A Fractional Equation Leading to a Quadratic Equation

Solve

$$\frac{y + 1}{y + 3} + \frac{y + 5}{y - 2} = \frac{7(2y + 1)}{y^2 + y - 6}. \quad (2)$$

Solution: Multiplying both sides by the L.C.D., $(y + 3)(y - 2)$, we get

$$(y - 2)(y + 1) + (y + 3)(y + 5) = 7(2y + 1). \quad (3)$$

Since Eq. (2) was multiplied by an expression involving the variable y, remember (from Sec. 1.1) that Eq. (3) is not necessarily equivalent to Eq. (2). After simplifying Eq. (3), we have

$$2y^2 - 7y + 6 = 0 \quad \text{(quadratic equation)},$$

$$(2y - 3)(y - 2) = 0 \quad \text{(factoring)}.$$

Thus, $\frac{3}{2}$ and 2 are *possible* roots of the given equation. But 2 cannot be a root of Eq. (2), since substitution leads to a denominator of 0. However, you should check that $\frac{3}{2}$ does indeed satisfy the *original* equation. Hence, its only root is $\frac{3}{2}$. ■

Do not hastily conclude that the solution of $x^2 = 3$ consists of $x = \sqrt{3}$ only.

EXAMPLE 5 Solution by Factoring

Solve $x^2 = 3$.

Solution:

$$x^2 = 3,$$

$$x^2 - 3 = 0.$$

Principles in Practice 4
Solution by Factoring

If you earned \$225 for selling x items for x dollars each, how many items did you sell, and at what price did you sell them?

Factoring, we obtain

$$(x - \sqrt{3})(x + \sqrt{3}) = 0.$$

Thus $x - \sqrt{3} = 0$ or $x + \sqrt{3} = 0$, so $x = \pm\sqrt{3}$.

A more general form of the equation $x^2 = 3$ is $u^2 = k$. In the same manner as the preceding we can show that

If $u^2 = k$, then $u = \pm\sqrt{k}$. (4)

Quadratic Formula

Solving quadratic equations by factoring can be quite difficult, as is evident by trying that method on $0.7x^2 - \sqrt{2}x - 8\sqrt{5} = 0$. However, there is a formula called the *quadratic formula*[4] that gives the roots of any quadratic equation.

Quadratic Formula

The roots of the quadratic equation $ax^2 + bx + c = 0$, where a, b, and c are constants and $a \neq 0$, are given by

$$x = \frac{-b \pm \sqrt{b^2 - 4ac}}{2a}.$$

Pitfall ▼ Be certain that you use the quadratic formula correctly.

$$x \neq -b \pm \frac{\sqrt{b^2 - 4ac}}{2a}.$$

Principles in Practice 5

A Quadratic Equation with Two Real Roots

Suppose the height h in feet of fireworks fired straight upward from the ground is given by $h = 160t - 16t^2$, where t is in seconds. When will the fireworks be 300 feet off the ground?

EXAMPLE 6 A Quadratic Equation with Two Real Roots

Solve $4x^2 - 17x + 15 = 0$ *by the quadratic formula.*

Solution: Here $a = 4$, $b = -17$, and $c = 15$. Thus,

$$x = \frac{-b \pm \sqrt{b^2 - 4ac}}{2a} = \frac{-(-17) \pm \sqrt{(-17)^2 - 4(4)(15)}}{2(4)}$$

$$= \frac{17 \pm \sqrt{49}}{8} = \frac{17 \pm 7}{8}.$$

The roots are $\frac{17 + 7}{8} = \frac{24}{8} = 3$ and $\frac{17 - 7}{8} = \frac{10}{8} = \frac{5}{4}$. ■

Principles in Practice 6

A Quadratic Equation with One Real Root

Suppose the weekly revenue r of a company is given by the equation $r = -2p^2 + 400p$, where p is the price of the company's product. What is the price of the product if the weekly revenue is \$20,000?

EXAMPLE 7 A Quadratic Equation with One Real Root

Solve $2 + 6\sqrt{2}y + 9y^2 = 0$ *by the quadratic formula.*

Solution: Look at the arrangement of the terms. Here $a = 9$, $b = 6\sqrt{2}$, and $c = 2$. Hence,

$$y = \frac{-b \pm \sqrt{b^2 - 4ac}}{2a} = \frac{-6\sqrt{2} \pm \sqrt{0}}{2(9)}.$$

Thus,

$$y = \frac{-6\sqrt{2} + 0}{18} = -\frac{\sqrt{2}}{3} \quad \text{or} \quad y = \frac{-6\sqrt{2} - 0}{18} = -\frac{\sqrt{2}}{3}.$$

Therefore, the only root is $-\frac{\sqrt{2}}{3}$. ■

[4]A derivation of the quadratic formula appears in Sec. 1.4 as a supplement.

EXAMPLE 8 A Quadratic Equation with No Real Solution

Solve $z^2 + z + 1 = 0$ by the quadratic formula.

Solution: Here $a = 1, b = 1$, and $c = 1$. The roots are

$$\frac{-b \pm \sqrt{b^2 - 4ac}}{2a} = \frac{-1 \pm \sqrt{-3}}{2}.$$

Now, $\sqrt{-3}$ denotes a number whose square is -3. However, no such real number exists, since the square of any real number is nonnegative. Thus, the equation has no real roots.[5] ■

This describes the nature of the roots of a quadratic equation.

From Examples 6–8, you can see that a quadratic equation has either two different real roots, exactly one real root, or no real roots, depending on whether $b^2 - 4ac > 0$, $= 0$, or < 0, respectively.

TECHNOLOGY

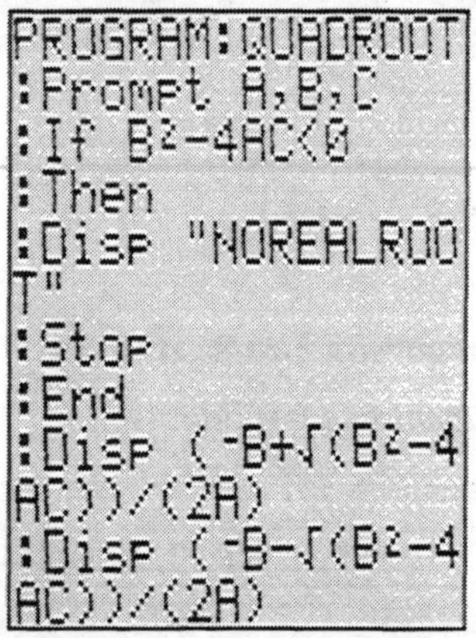

FIGURE 1.2 Program to find real roots of $Ax^2 + Bx + C = 0$.

Using the program feature of a graphics calculator, one can create a program that gives the real roots of the quadratic equation $Ax^2 + Bx^2 + C = 0$. Figure 1.2 shows such a program for the TI-82 graphics calculator. To execute it for

$$20x^2 - 33x + 10 = 0,$$

you are prompted to enter the values of A, B, and C. (See Fig. 1.3.) The resulting roots are $x = 1.25$ and $x = 0.4$.

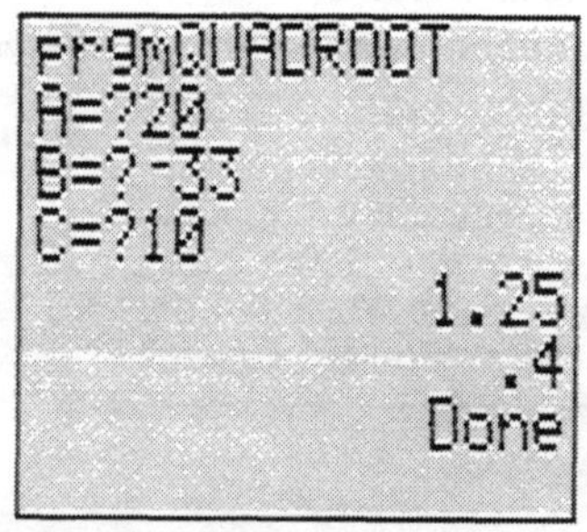

FIGURE 1.3 Roots of $20x^2 - 33x + 10 = 0$.

Quadratic Forms

Sometimes an equation that is not quadratic can be transformed into a quadratic equation by an appropriate substitution. In this case, the given equation is said to have **quadratic form.** The next example will illustrate.

Principles in Practice 7

A Quadratic Equation With No Real Solution

Suppose the height h in feet of fireworks fired straight upward from the ground is given by $h = 160t - 16t^2$, where t is in seconds. When will the fireworks be 500 feet off the ground?

EXAMPLE 9 Solving a Quadratic-Form Equation

Solve $\frac{1}{x^6} + \frac{9}{x^3} + 8 = 0.$

Solution: This equation can be written as

$$\left(\frac{1}{x^3}\right)^2 + 9\left(\frac{1}{x^3}\right) + 8 = 0,$$

[5] $\frac{-1 \pm \sqrt{-3}}{2}$ can be expressed as $\frac{-1 \pm i\sqrt{3}}{2}$, where $i = \sqrt{-1}$ is called the *imaginary unit.*

so it is quadratic in $1/x^3$ and hence has quadratic form. Substituting the variable w for $1/x^3$ gives a quadratic equation in the variable w, which we can then solve:

$$w^2 + 9w + 8 = 0,$$
$$(w + 8)(w + 1) = 0,$$
$$w = -8 \quad \text{or} \quad w = -1.$$

Do not assume that -8 and -1 are solutions of the *original* equation.

Returning to the variable x, we have

$$\frac{1}{x^3} = -8 \quad \text{or} \quad \frac{1}{x^3} = -1.$$

Thus,

$$x^3 = -\frac{1}{8} \quad \text{or} \quad x^3 = -1,$$

from which it follows that

$$x = -\frac{1}{2}, \quad -1.$$

Checking, we find that these values of x satisfy the original equation. ■

■ Exercise 1.3

In Problems 1–30, solve by factoring.

1. $x^2 - 4x + 4 = 0.$

2. $t^2 + 3t + 2 = 0.$

3. $y^2 - 7y + 12 = 0.$

4. $x^2 + x - 12 = 0.$

5. $x^2 - 2x - 3 = 0.$

6. $x^2 - 16 = 0.$

7. $x^2 - 12x = -36.$

8. $3w^2 - 12w + 12 = 0.$

9. $x^2 - 4 = 0.$

10. $2x^2 + 4x = 0.$

11. $z^2 - 8z = 0.$

12. $x^2 + 9x = -14.$

13. $4x^2 + 1 = 4x.$

14. $2z^2 + 7z = 4.$

15. $y(2y + 3) = 5.$

16. $8 + 2x - 3x^2 = 0.$

17. $-x^2 + 3x + 10 = 0.$

18. $\frac{1}{7}y^2 = \frac{3}{7}y.$

19. $2p^2 = 3p.$

20. $-r^2 - r + 12 = 0.$

21. $x(x - 1)(x + 2) = 0.$

22. $(x - 2)^2(x + 1)^2 = 0.$

23. $x^3 - 64x = 0.$

24. $x^3 - 4x^2 - 5x = 0.$

25. $6x^3 + 5x^2 - 4x = 0.$

26. $(x + 1)^2 - 5x + 1 = 0.$

27. $(x + 3)(x^2 - x - 2) = 0.$

28. $3(x^2 + 2x - 8)(x - 5) = 0.$

29. $p(p - 3)^2 - 4(p - 3)^3 = 0.$

30. $x^4 - 3x^2 + 2 = 0.$

In Problems 31–44, find all real roots by using the quadratic formula.

31. $x^2 + 2x - 24 = 0.$

32. $x^2 - 2x - 15 = 0.$

33. $4x^2 - 12x + 9 = 0.$

34. $p^2 + 2p = 0.$

35. $p^2 - 5p + 3 = 0.$

36. $2 - 2x + x^2 = 0.$

37. $4 - 2n + n^2 = 0.$

38. $2x^2 + x = 5.$

39. $6x^2 + 7x - 5 = 0.$

40. $w^2 - 2\sqrt{2}w + 2 = 0.$

41. $2x^2 - 3x = 20.$

42. $0.01x^2 + 0.2x - 0.6 = 0.$

43. $2x^2 + 4x = 5.$

44. $-2x^2 - 6x + 5 = 0.$

In Problems 45–54, solve the given quadratic-form equation.

45. $x^4 - 5x^2 + 6 = 0.$

46. $x^4 - 3x^2 - 10 = 0.$

47. $\frac{2}{x^2} + \frac{3}{x} - 2 = 0.$

48. $x^{-2} + x^{-1} - 12 = 0.$

49. $x^{-4} - 9x^{-2} + 14 = 0.$

50. $\frac{1}{x^4} - \frac{9}{x^2} + 8 = 0.$

51. $(x - 3)^2 + 9(x - 3) + 14 = 0.$

52. $(x + 5)^2 - 8(x + 5) = 0.$

53. $\frac{1}{(x - 2)^2} - \frac{12}{x - 2} + 35 = 0.$

54. $\frac{2}{(x + 4)^2} + \frac{7}{x + 4} + 3 = 0.$

In Problems 55–76, solve by any method.

55. $x^2 = \frac{x + 3}{2}.$

56. $\frac{x}{3} = \frac{6}{x} - 1.$

57. $\frac{3}{x - 4} + \frac{x - 3}{x} = 2.$

58. $\frac{2}{x - 1} - \frac{6}{2x + 1} = 5.$

59. $\frac{6x + 7}{2x + 1} - \frac{6x + 1}{2x} = 1.$

60. $\frac{6(w + 1)}{2 - w} + \frac{w}{w - 1} = 3.$

61. $\frac{2}{r - 2} - \frac{r + 1}{r + 4} = 0.$

62. $\frac{2x - 3}{2x + 5} + \frac{2x}{3x + 1} = 1.$

63. $\frac{y + 1}{y + 3} + \frac{y + 5}{y - 2} = \frac{14y + 7}{y^2 + y - 6}.$

64. $\frac{3}{t + 1} + \frac{4}{t} = \frac{12}{t + 2}.$

65. $\frac{2}{x^2 - 1} - \frac{1}{x(x - 1)} = \frac{2}{x^2}.$

66. $5 - \frac{3(x + 3)}{x^2 + 3x} = \frac{1 - x}{x}.$

67. $\sqrt{x + 2} = x - 4.$

68. $3\sqrt{x + 4} = x - 6.$

69. $q + 2 = 2\sqrt{4q - 7}.$

70. $x + \sqrt{x} - 2 = 0.$

71. $\sqrt{x + 7} - \sqrt{2x} - 1 = 0.$

72. $\sqrt{x} - \sqrt{2x - 8} - 2 = 0.$

73. $\sqrt{x} - \sqrt{2x + 1} + 1 = 0.$

74. $\sqrt{y - 2} + 2 = \sqrt{2y + 3}.$

75. $\sqrt{x + 5} + 1 = 2\sqrt{x}.$

76. $\sqrt{\sqrt{x} + 2} = \sqrt{2x - 4}.$

In Problems 77 and 78, find the roots, rounded to two decimal places.

77. $0.02x^2 - 4.7x + 8.6 = 0.$

78. $0.01x^2 + 0.2x - 0.6 = 0.$

79. Geometry The area of a rectangular picture with a width 2 inches less than its length is 48 square inches. What are the dimensions of the picture?

80. Temperature The temperature has been rising X degrees per day for X days. X days ago it was 15 degrees. Today it is 51 degrees. How much has the temperature been rising each day? How many days has it been rising?

81. Economics One root of the economics equation

$$\overline{M} = \frac{Q(Q + 10)}{44}$$

is $-5 + \sqrt{25 + 44\overline{M}}$. Verify this by using the quadratic formula to solve for Q in terms of $\overline{M}$. Here Q is real income and $\overline{M}$ is the level of money supply.

82. Diet for Rats A group of biologists studied the nutritional effects on rats that were fed a diet containing 10% protein.[6] The protein was made up of yeast and corn flour. By changing the percentage P (expressed as a decimal) of yeast in the protein mix, the group estimated that the average weight gain g (in grams) of a rat over a period of time was given by

$$g = -200P^2 + 200P + 20.$$

What percentage of yeast gave an average weight gain of 70 grams?

83. Drug Dosage There are several rules for determining doses of medicine for children when the adult dose has been specified. Such rules may be based on weight,

[6] Adapted from R. Bressani, "The Use of Yeast in Human Foods," in R. I. Mateles and S. R. Tannenbaum (eds.), *Single-Cell Protein* (Cambridge, MA: MIT Press, 1968).

height, and so on. If A is the age of the child, d is the adult dose, and c is the child's dose, then here are two rules:

$$\text{Young's rule:} \quad c = \frac{A}{A+12}d.$$

$$\text{Cowling's rule:} \quad c = \frac{A+1}{24}d.$$

At what age are the children's doses the same under both rules? Round your answer to the nearest year.

84. Delivered Price of a Good In a discussion of the delivered price of a good from a mill to a customer, DeCanio[7] arrives at and solves the two quadratic equations

$$(2n-1)v^2 - 2nv + 1 = 0,$$

and

$$nv^2 - (2n+1)v + 1 = 0,$$

where $n \geq 1$.

a. Solve the first equation for v.

b. Solve the second equation for v if $v < 1$.

85. Optics An object is 120 cm from a wall. In order to focus the image of the object on the wall, a converging lens with a focal length of 24 cm is used. The lens is placed between the object and the wall at a distance of p centimeters from the object, where

$$\frac{1}{p} + \frac{1}{120-p} = \frac{1}{24}.$$

Find p, rounded to one decimal place.

86. Physics A platinum resistance thermometer of certain specifications operates according to the equation

$$R = 10{,}000 + (4.124 \times 10^{-2})T - (1.779 \times 10^{-5})T^2,$$

where R is the resistance of the thermometer (in ohms) at temperature T (in degrees Celsius). If $R = 13.946$, find the corresponding value of T. Round your answer to the nearest degree. Assume that such a thermometer is usable only if $T < 600°C$.

87. Motion Suppose the height h of an object thrown straight upward from the ground is given by

$$h = 44.1t - 4.9t^2,$$

where h is in meters and t is the elapsed time in seconds.

a. After how many seconds does the object strike the ground?

b. When is the object at a height of 88.2 m?

■ ***In Problems 88–93, use a program to determine any real roots of the equation. Round answers to three decimal places. For Problems 88 and 89, confirm your result algebraically.***

88. $2x^2 - 3x - 27 = 0.$

89. $8x^2 - 18x + 9 = 0.$

90. $14x^2 + 7x - 3 = 0.$

91. $23x^2 - \frac{11}{3}x + 5 = 0.$

92. $\frac{9}{2}z^2 - 6.3 = \frac{z}{3}(1.1 - 7z).$

93. $(\pi t - 4)^2 = 4.1t - 3.$

[7] S. J. DeCanio, "Delivered Pricing and Multiple Basing Point Equilibria: A Revolution," *Quarterly Journal of Economics*, XCIX, no. 2 (1984), 329–49.

1.4 Supplement

Following is a derivation of the quadratic formula. Suppose $ax^2 + bx + c = 0$ is a quadratic equation. Since $a \neq 0$, we can divide both sides by a:

$$x^2 + \frac{b}{a}x + \frac{c}{a} = 0,$$

$$x^2 + \frac{b}{a}x = -\frac{c}{a}.$$

If we add $\left(\frac{b}{2a}\right)^2$ to both sides, then the left side factors as the square of a binomial:

$$x^2 + \frac{b}{a}x + \left(\frac{b}{2a}\right)^2 = \left(\frac{b}{2a}\right)^2 - \frac{c}{a},$$

$$\left(x + \frac{b}{2a}\right)^2 = \frac{b^2 - 4ac}{4a^2}.$$

This equation has the form $u^2 = k$, so, from Eq. (4) of Sec. 1.3,

$$x + \frac{b}{2a} = \pm\sqrt{\frac{b^2 - 4ac}{4a^2}} = \pm\frac{\sqrt{b^2 - 4ac}}{2a}.$$

Solving for x gives

$$x = -\frac{b}{2a} \pm \frac{\sqrt{b^2 - 4ac}}{2a} = \frac{-b \pm \sqrt{b^2 - 4ac}}{2a}.$$

In summary, the roots of the quadratic equation $ax^2 + bx + c = 0$ are given by the **quadratic formula:**

$$x = \frac{-b \pm \sqrt{b^2 - 4ac}}{2a}.$$

1.5 REVIEW

IMPORTANT TERMS AND SYMBOLS

Section 1.1 equation; side (member) of equation; variable; root of equation; solution set; equivalent equations; linear (first-degree) equation; literal equation; arbitrary constant

Section 1.2 fractional equation; empty set, ∅; extraneous solution; radical equation

Section 1.3 quadratic (second-degree) equation; quadratic formula

SUMMARY

When solving an equation, we may apply rules to it that give equivalent equations—that is, equations with exactly the same solutions as the given equation. These rules include adding (or subtracting) the same polynomial to (from) both sides, as well as multiplying (or dividing) both sides by the same constant, except 0.

A linear equation (in x) is of the first degree and has the form $ax + b = 0$, where $a \neq 0$. Every linear equation has exactly one root. To solve a linear equation, we apply operations on it until we obtain an equivalent equation in which the unknown is isolated on one side.

A quadratic equation (in x) is of the second degree and has the form $ax^2 + bx + c = 0$, where $a \neq 0$. Every quadratic equation has either two different real roots, exactly one real root, or no real roots. A quadratic equation may be solved either by factoring or by the quadratic formula:

$$x = \frac{-b \pm \sqrt{b^2 - 4ac}}{2a}.$$

When solving a fractional equation or a radical equation, one often applies operations to it that do not guarantee that the resulting equation is equivalent to the given equation. These operations include multiplying both sides by an expression containing the variable and raising both sides to the same power. In these cases, all solutions obtained at the end of such procedures must be checked by substituting them into the given equation. In this way, extraneous solutions may be found.

REVIEW PROBLEMS

In Problems 1–44, solve the equations.

1. $4 - 3x = 2 + 5x.$

2. $\frac{5}{7}x - \frac{2}{3}x = \frac{3}{21}x.$

3. $3[2 - 4(1 + x)] = 5 - 3(3 - x).$

4. $3(x + 4)^2 + 6x = 3x^2 + 7.$

5. $2 - w = 3 + w.$

6. $x = 2x.$

7. $x = 2x - (7 + x).$

8. $3x - 8 = 4(x - 2).$

9. $2(4 - \frac{3}{5}p) = 5.$

10. $\frac{5}{7}x - \frac{2}{3}x = \frac{3}{21}.$

11. $\dfrac{3x - 1}{x + 4} = 0.$

12. $\dfrac{5}{p + 3} - \dfrac{2}{p + 3} = 0.$

13. $\dfrac{2x}{x - 3} - \dfrac{x + 1}{x + 2} = 1.$

14. $\dfrac{t + 3t + 4}{7 - t} = 14.$

15. $3x^2 + 2x - 5 = 0.$

16. $x^2 - 2x - 2 = 0.$

17. $5q^2 = 7q.$

18. $2x^2 - x = 0.$

19. $x^2 - 10x + 25 = 0.$

20. $r^2 + 10r - 25 = 0.$

21. $3x^2 - 5 = 0.$

22. $x(x - 9) = 0.$

23. $(8t - 5)(2t + 6) = 0.$

24. $2(x^2 - 1) + 2x = x^2 - 6x + 1.$

25. $-3x^2 + 5x - 1 = 0.$

26. $y^2 = 6.$

27. $x(x^2 - 9) = 4(x^2 - 9).$

28. $4x^2(x - 5) - 9(x - 5) = 0.$

29. $\dfrac{6w + 7}{2w + 1} - \dfrac{6w + 1}{2w} = 1.$

30. $\dfrac{3}{x + 1} + \dfrac{4}{x} - \dfrac{12}{x + 2} = 0.$

31. $\dfrac{2}{x^2 - 9} - \dfrac{3x}{x + 3} = \dfrac{1}{x - 3}.$

32. $\dfrac{3}{x^2 - 4} + \dfrac{2}{x^2 + 4x + 4} - \dfrac{4}{x + 2} = 0.$

33. $\sqrt{2x + 5} = 5.$

34. $\sqrt{3x - 4} = \sqrt{2x + 5}.$

35. $\sqrt[3]{11x + 9} = 4.$

36. $\sqrt{x^2 + 5x + 25} = x + 4.$

37. $\sqrt{y} + 6 = 5.$

38. $\sqrt{z^2 + 9} = 5.$

39. $\sqrt{x - 1} + \sqrt{x + 6} = 7.$

40. $\sqrt{2x + 1} = x - 7.$

41. $x + 2 = 2\sqrt{4x - 7}.$

42. $\sqrt{3z} - \sqrt{5z + 1} + 1 = 0.$

43. $y^{2/3} + y^{1/3} - 2 = 0.$

44. $2y^{-2/3} - 5y^{-1/3} - 3 = 0.$

In Problems 45–52, solve the given equation for the indicated letter.

45. $E = 4\pi k\dfrac{Q}{A};\ Q.$

46. $E_1 = i_2R_1 + i_3R_1 + i_2R_2;\ i_3.$

47. $n - 1 = C + \dfrac{C'}{\lambda^2};\ C'.$

48. $\sigma = \dfrac{n_0 - n_e}{\lambda}L;\ n_0.$

49. $T^2 = 4\pi^2\left(\dfrac{L}{g}\right);\ T.$

50. $s = \dfrac{1}{2}at^2;\ t.$

51. $mgh = \dfrac{1}{2}mv^2 + \dfrac{1}{2}I\,\omega^2;\ \omega.$

52. $P = \dfrac{E^2}{R + r} - \dfrac{E^2r}{(R + r)^2};\ E.$

53. Electricity In studies of electrical networks, the following equation occurs:

$$S^2 + \frac{R}{L}S + \frac{1}{LC} = 0.$$

Show that

$$S = -\frac{R}{2L} \pm \sqrt{\left(\frac{R}{2L}\right)^2 - \frac{1}{LC}}.$$

54. Electricity In an electrical circuit, resonance occurs when

$$2\pi f_r L = \frac{1}{2\pi f_r C},$$

where f_r is a resonant frequency, L is inductance, and C is capacitance. Solve for f_r if $f_r > 0$.

In Problems 55 and 56, use a graphics calculator to determine which of the given numbers, if any, are roots of the given equation.

55. $12x^3 + 61x = 83x^2 - 30;\ 4, 6, \frac{5}{4}.$

56. $\sqrt{t^2 + 4} = t + 1;\ \frac{2}{3}, \frac{14}{3}, \frac{3}{2}.$

In Problems 57 and 58, use a program to determine any real roots of the equation. Round answers to three decimal places.

57. $4.6 - 7.2x - 19.3x^2 = 0.$

58. $(9x - 3)^2 - \dfrac{7}{6}(x - 2) = 15.$

MATHEMATICAL *SNAPSHOT*

REAL GROWTH OF AN INVESTMENT[8]

When we speak of the real growth of an investment, we are referring to the growth in its purchasing power, that is, the growth of the quantity of goods that the investment can buy. The real growth depends on the influence of both interest and inflation. Interest raises the value of the investment, whereas inflation slows down the growth of the investment by increasing prices and, hence, decreasing its purchasing power. The rate of real growth does not usually equal the difference between the rate of interest and the rate of inflation but is described by a different formula called "Fisher's effect."

You can understand Fisher's effect by carefully examining the question posed by Edward P. Foldessy in the *Wall Street Journal* (27 May 1986, 37):

> For the year ending with May 1986, the annual rate of interest was 11% and the annual rate of inflation was 3.6%. Under these conditions, what was the annual rate of real growth of an investment?

You might think that the answer is simply obtained by subtracting percentages: $11\% - 3.6\% = 7.4\%$. However, 7.4% is not the answer.

Suppose you analyze the situation in more specific terms. Consider strawberries selling at \$1.00 per pound and assume, because of inflation, that this price increases at the rate of 3.6% a year. From June 1985 to June 1986, the price per pound of \$1.00 rose to

$$\$1.00 + (3.6\% \text{ of } \$1.00) = \$1.036.$$

On the other hand, consider \$100 invested in June 1985 at an 11% annual interest rate. In June 1986 the interest earned is \$100(0.11), so the accumulated amount is

$$\$100 + \$100(0.11) = \$111.$$

Now, compare the purchasing power of \$100 in June 1985 to that of \$111 in June 1986. In 1985 the \$100 bought 100 lb of strawberries at \$1.00 per pound. In 1986 strawberries were \$1.036 per pound, so the accumulated amount of \$111 bought $111/1.036 \approx 107.14$ lb of strawberries (the symbol $\approx$ means *approximately equals*).

What change occurred in the purchasing power of the investment? It increased from 100 lb to 107.14 lb, an increase of 7.14%. That is,

$$\frac{\text{new quantity} - \text{initial quantity}}{\text{initial quantity}} = \frac{107.14 - 100}{100}$$

$$= 0.0714 = 7.14\%.$$

Thus 7.14% is the real rate of growth, which is less than the difference $11\% - 3.6\% = 7.4\%$. Actually, this difference has no meaning, because the three percentages refer to three different quantities: (a) interest (a fraction of the investment—11% of \$100), (b) inflation (a fraction of the unit price of goods—3.6% of \$1.00), and (c) the rate of real growth (a percentage of the purchasing power—7.14% of the initial quantity of strawberries).

To derive a formula for the rate of real growth, g, let y be the annual rate of interest (the yield) and let i

[8] Adapted from Yves Nievergelt, "Fisher's Effect: Real Growth Is Not Interest Less Inflation," *Mathematics Teacher,* 81 (October 1988), 546–47. By permission of the National Council of Teachers of Mathematics.

be the annual rate of inflation. In one year an investment of P dollars (the principal) earns interest of $y \cdot P$ dollars, so it produces an accumulated amount (in dollars) of

$$P + y \cdot P = P(1 + y) \qquad \text{(factoring)}.$$

In one year the price of goods, say p dollars per unit, rises by $i \cdot p$ dollars to a new price of

$$p + i \cdot p = p(1 + i)$$

dollars per unit. The initial purchasing power represents the initial quantity of goods:

$$\text{initial quantity} = \frac{\text{principal}}{\text{initial price}} = \frac{P}{p}.$$

One year later, the new quantity of goods that the accumulated amount of the investment will buy at the new price is given by

$$\text{new quantity} = \frac{\text{new balance}}{\text{new price}} = \frac{P(1 + y)}{p(1 + i)}.$$

Consequently, the rate of growth, or relative change, of the purchasing power is given by

$$\begin{aligned} g &= \frac{\text{new quantity} - \text{initial quantity}}{\text{initial quantity}} \\ &= \frac{\dfrac{P(1 + y)}{p(1 + i)} - \dfrac{P}{p}}{\dfrac{P}{p}}. \end{aligned}$$

Multiplying the numerator and denominator by p/P gives

$$\begin{aligned} g &= \frac{1 + y}{1 + i} - 1 \\ &= \frac{(1 + y) - (1 + i)}{1 + i} = \frac{y - i}{1 + i}. \end{aligned}$$

Thus the rate of real growth is given by the literal equation

$$g = \frac{y - i}{1 + i}. \qquad (1)$$

The relationship in Eq. (1) is Fisher's effect.[9] To illustrate its use, apply it to the previous example, where $y = 11\%$ and $i = 3.6\%$. Fisher's formula gives

$$g = \frac{0.11 - 0.036}{1 + 0.036} \approx 0.0714 = 7.14\%.$$

■ Exercises

1. According to Foldessy (1986), during 1980 the rate of interest averaged 11.4% at a time when inflation advanced by 13.5%.

a. Compute the accumulated amount of a \$100 investment after one year at 11.4%.

b. If a pound of apples cost \$1 in January 1980, how much did it cost a year later?

c. If a pound of applies cost \$1 in January 1980, what quantity of apples did \$100 buy in 1980?

d. What quantity of apples did the accumulated amount [see part (a)] buy a year later?

e. Use the results of parts (c) and (d) to calculate the rate of real growth by means of the equation

$$g = \frac{\text{new quantity} - \text{initial quantity}}{\text{initial quantity}}.$$

f. Verify your answer to part (e) by means of Fisher's formula.

2. Find the rate of real growth, given an interest rate of 10% and an inflation rate of 5%.

---

[9] Irving Fisher, "Appreciation and Interest," *Publications of the American Economic Association*, Third Series, 11 (August 1986), 331–442.

Chapter 3
Applications of Equations and Inequalities

OBJECTIVE

To model situations described by linear or quadratic equations.

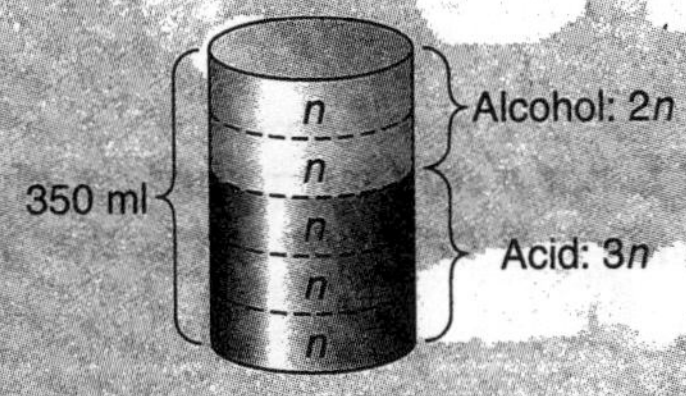

FIGURE 2.1 Chemical solution (Example 1).

Note that the solution to an equation is not necessarily the solution to the problem posed.

2.1 APPLICATIONS OF EQUATIONS

In most cases, to solve practical problems you must translate the relationships stated in the problems into mathematical symbols. This is called *modeling.* The following examples illustrate basic techniques and concepts. Examine each of them carefully before going to the exercises.

EXAMPLE 1 Mixture

A chemist must prepare 350 ml *of a chemical solution made up of two parts alcohol and three parts acid. How much of each should be used?*

Solution: Let n be the number of milliliters in each part. Figure 2.1 shows the situation. From the diagram, we have

$$2n + 3n = 350,$$

$$5n = 350,$$

$$n = \frac{350}{5} = 70.$$

But $n = 70$ is *not* the answer to the original problem. Each *part* has 70 ml. The amount of alcohol is $2n = 2(70) = 140$, and the amount of acid is $3n = 3(70) = 210$. Thus, the chemist should use 140 ml of alcohol and 210 ml of acid. This example shows how helpful a diagram can be in setting up a word problem. ■

EXAMPLE 2 Observation Deck

A rectangular observation deck overlooking a scenic valley is to be built. [*See Fig.* 2.2(a).] *The deck is to have dimensions* 6 m *by* 12 m. *A rectangular shelter of area* 40 m^2 *is to be centered over the deck. The uncovered part of the deck is to be a walkway of uniform width. How wide should this walkway be?*

Solution: A diagram of the deck is shown in Fig. 2.2(b). Let w be the width (in meters) of the walkway. Then the part of the deck for the shelter has dimensions $12 - 2w$ by $6 - 2w$. Since its area must be 40 m^2, where area = (length)(width), we have

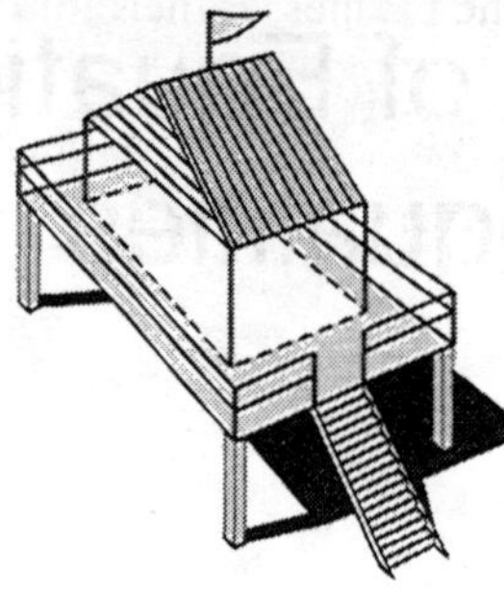

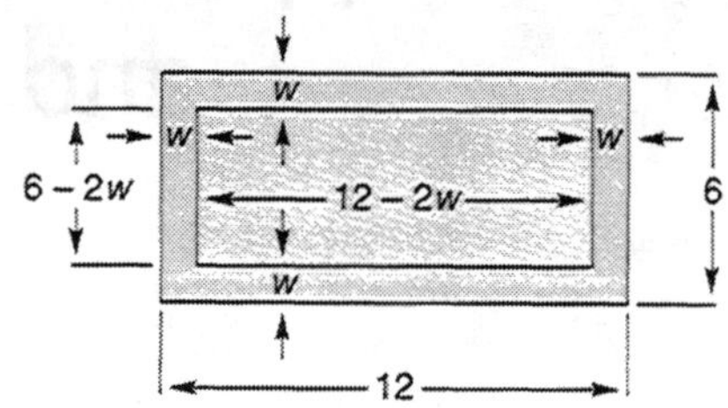

FIGURE 2.2 Deck walkway (Example 2).

$$
\begin{aligned}
(12 - 2w)(6 - 2w) &= 40, \\
72 - 36w + 4w^2 &= 40 \qquad \text{(multiplying)}, \\
4w^2 - 36w + 32 &= 0, \\
w^2 - 9w + 8 &= 0 \qquad \text{(dividing both sides by 4)}, \\
(w - 8)(w - 1) &= 0, \\
w &= 8, 1.
\end{aligned}
$$

Although 8 is a solution to the equation, it is *not* a solution to our problem, because one of the dimensions of the deck itself is only 6 m. Thus, the only possible solution is that the walkway be 1 m wide. ■

The key words introduced here are *fixed cost, variable cost, total cost, total revenue,* and *profit.* This is the time for you to gain familiarity with these terms because they recur throughout the book.

In the next example, we shall refer to some business terms relative to a manufacturing firm. **Fixed cost** (or *overhead*) is the sum of all costs that are independent of the level of production, such as rent, insurance, and so on. This cost must be paid whether or not output is produced. **Variable cost** is the sum of all costs that are dependent on the level of output, such as labor and material. **Total cost** is the sum of variable cost and fixed cost:

$$\text{total cost} = \text{variable cost} + \text{fixed cost}.$$

Total revenue is the money that the manufacturer receives for selling the output:

$$\text{total revenue} = (\text{price per unit})(\text{number of units sold}).$$

Profit is total revenue minus total cost:

$$\text{profit} = \text{total revenue} - \text{total cost}.$$

EXAMPLE 3 Profit

The Anderson Company produces a product for which the variable cost per unit is \$6 *and fixed cost is* \$80,000. *Each unit has a selling price of* \$10. *Determine the number of units that must be sold for the company to earn a profit of* \$60,000.

Solution: Let q be the number of units that must be sold. (In many business problems, q represents quantity.) Then the variable cost (in dollars) is $6q$. The *total* cost for the business is therefore $6q + 80{,}000$. The total revenue from the sale of q units is $10q$. Since

$$\text{profit} = \text{total revenue} - \text{total cost},$$

our model for this problem is

$$60{,}000 = 10q - (6q + 80{,}000).$$

Solving gives

$$\begin{aligned} 60{,}000 &= 10q - 6q - 80{,}000, \\ 140{,}000 &= 4q, \\ 35{,}000 &= q. \end{aligned}$$

Thus, 35,000 units must be sold to earn a profit of \$60,000. ■

EXAMPLE 4 Pricing

Sportcraft manufactures women's sportswear and is planning to sell its new line of slacks to retail outlets. The cost to the retailer will be \$33 per pair of slacks. As a convenience to the retailer, Sportcraft will attach a price tag to each pair. What amount should be marked on the price tag so that the retailer may reduce this price by 20% during a sale and still make a profit of 15% on the cost?

Solution: Here we use the fact that

Note that *price = cost + profit*.

$$\text{selling price} = \text{cost per pair} + \text{profit per pair}.$$

Let p be the tag price per pair, in dollars. During the sale, the retailer actually receives $p - 0.2p$. This must equal the cost, 33, plus the profit, $(0.15)(33)$. Hence,

$$\begin{aligned} \text{selling price} &= \text{cost} + \text{profit}, \\ p - 0.2p &= 33 + (0.15)(33), \\ 0.8p &= 37.95, \\ p &= 47.4375. \end{aligned}$$

From a practical point of view, Sportcraft should mark the price tag at \$47.44. ■

EXAMPLE 5 Investment

A total of \$10,000 was invested in two business ventures, A and B. At the end of the first year, A and B yielded returns of 6% and $5\frac{3}{4}$%, respectively, on the original investments. How was the original amount allocated if the total amount earned was \$588.75?

Solution: Let x be the amount (in dollars) invested at 6%. Then $10{,}000 - x$ was invested at $5\frac{3}{4}$%. The interest earned in A was $(0.06)(x)$, and in B it was $(0.0575)(10{,}000 - x)$, which total 588.75. Hence,

$$\begin{aligned} (0.06)x + (0.0575)(10{,}000 - x) &= 588.75, \\ 0.06x + 575 - 0.0575x &= 588.75, \\ 0.0025x &= 13.75, \\ x &= 5500. \end{aligned}$$

Thus, \$5500 was invested at 6%, and \$10,000 − \$5500 = \$4500 was invested at $5\frac{3}{4}$%. ■

EXAMPLE 6 Bond Redemption

The board of directors of Maven Corporation agrees to redeem some of its bonds in two years. At that time, \$1,102,500 *will be required. Suppose the firm presently sets aside* \$1,000,000. *At what annual rate of interest, compounded annually, will this money have to be invested in order that its future value be sufficient to redeem the bonds?*

Solution: Let r be the required annual rate of interest. At the end of the first year, the accumulated amount will be \$1,000,000 plus the interest, $1{,}000{,}000r$, for a total of

$$1{,}000{,}000 + 1{,}000{,}000r = 1{,}000{,}000(1 + r).$$

Under compound interest, at the end of the second year the accumulated amount will be $1{,}000{,}000(1 + r)$ plus the interest on this, which is $1{,}000{,}000(1 + r)r$. Thus, the total value at the end of the second year will be

$$1{,}000{,}000(1 + r) + 1{,}000{,}000(1 + r)r.$$

This must equal \$1,102,500:

$$1{,}000{,}000(1 + r) + 1{,}000{,}000(1 + r)r = 1{,}102{,}500. \tag{1}$$

Since $1{,}000{,}000(1 + r)$ is a common factor of both terms on the left side, we have

$$1{,}000{,}000(1 + r)(1 + r) = 1{,}102{,}500,$$

$$1{,}000{,}000(1 + r)^2 = 1{,}102{,}500,$$

$$(1 + r)^2 = \frac{1{,}102{,}500}{1{,}000{,}000} = \frac{11{,}025}{10{,}000} = \frac{441}{400},$$

$$1 + r = \pm\sqrt{\frac{441}{400}} = \pm\frac{21}{20},$$

$$r = -1 \pm \frac{21}{20}.$$

Thus, $r = -1 + (21/20) = 0.05$, or $r = -1 - (21/20) = -2.05$. Although 0.05 and −2.05 are roots of Eq. (1), we reject −2.05, since we want r to be positive. Hence, $r = 0.05$, so the desired rate is 5%. ■

At times there may be more than one way to model a word problem, as Example 7 shows.

EXAMPLE 7 Apartment Rent

A real-estate firm owns the Parklane Garden Apartments, which consist of 90 *apartments. At* \$350 *per month, every apartment can be rented. However, for each* \$10 *per month increase, there will be two vacancies with no possibility of filling them. The firm wants to receive* \$31,980 *per month from rents. What rent should be charged for each apartment?*

Solution:

Method I. Suppose r is the rent (in dollars) to be charged per apartment. Then the increase over the \$350 level is $r - 350$. Thus, the number of \$10 increases is $\frac{r-350}{10}$. Because each \$10 increase results in two vacancies, the total number of vacancies will be $2\left(\frac{r-350}{10}\right)$. Hence, the total number of apartments rented will be $90-2\left(\frac{r-350}{10}\right)$. Since

$$\text{total rent} = (\text{rent per apartment})(\text{number of apartments rented}),$$

we have

$$\begin{aligned} 31{,}980 &= r\left[90 - \frac{2(r-350)}{10}\right], \\ 31{,}980 &= r\left[90 - \frac{r-350}{5}\right], \\ 31{,}980 &= r\left[\frac{450 - r + 350}{5}\right], \\ 159{,}900 &= r[800 - r]. \end{aligned}$$

Thus,

$$r^2 - 800r + 159{,}900 = 0.$$

By the quadratic formula,

$$\begin{aligned} r &= \frac{800 \pm \sqrt{(-800)^2 - 4(1)(159{,}900)}}{2(1)} \\ &= \frac{800 \pm \sqrt{400}}{2} = \frac{800 \pm 20}{2} = 400 \pm 10. \end{aligned}$$

Hence, the rent for each apartment should be either \$410 or \$390.

Method II. Suppose n is the number of \$10 increases. Then the increase in rent per apartment will be $10n$ and there will be $2n$ vacancies. Since

$$\text{total rent} = (\text{rent per apartment})(\text{number of apartments rented}),$$

we have

$$\begin{aligned} 31{,}980 &= (350 + 10n)(90 - 2n), \\ 31{,}980 &= 31{,}500 + 200n - 20n^2, \\ 20n^2 - 200n + 480 &= 0, \\ n^2 - 10n + 24 &= 0. \\ (n-6)(n-4) &= 0. \end{aligned}$$

Thus, $n = 6$ or $n = 4$. The rent charged should be either $350 + 10(6) = \$410$ or $350 + 10(4) = \$390$. ■

■ Exercise 2.1

1. **Fencing** A fence is to be placed around a rectangular plot so that the enclosed area is 800 ft^3 and the length of the plot is twice the width. How many feet of fencing must be used?

2. **Geometry** The perimeter of a rectangle is 200 ft, and the length of the rectangle is three times the width. Find the dimensions of the rectangle.

3. Gypsy Moth One of the most important defoliating insects is the gypsy moth caterpillar, which feeds on foliage of shade, forest, and fruit trees. A homeowner lives in an area in which the gypsy moth has become a problem. She wishes to spray the trees on her property before more defoliation occurs. She needs 128 oz of a solution made up of 3 parts of insecticide A and 5 parts of insecticide B. The solution is then mixed with water. How many ounces of each insecticide should be used?

4. Concrete Mix A builder makes a certain type of concrete by mixing together 1 part cement, 3 parts sand, and 5 parts stone (by volume). If 585 ft^3 of concrete are needed, how many cubic feet of each ingredient does he need?

5. Furniture Finish According to *The Consumer's Handbook* [Paul Fargis, ed. (New York: Hawthorn, 1974)], a good oiled furniture finish contains two parts boiled linseed oil and one part turpentine. If you need a pint (16 fluid oz) of this furniture finish, how many fluid ounces of turpentine are needed?

6. Forest Management A lumber company owns a forest that is of rectangular shape, 1 mi by 2 mi. If the company cuts a uniform strip of trees along the outer edges of this forest, how wide should the strip be if $\frac{3}{4}$ sq mi of forest is to remain?

7. Garden Pavement A rectangular plot, 4 m by 8 m, is to be used for a garden. It is decided to put a pavement inside the entire border so that 12 m^2 of the plot is left for flowers. How wide should the pavement be?

8. Ventilating Duct The diameter of a circular ventilating duct is 140 mm. This duct is joined to a square duct system as shown in Fig. 2.3. To ensure smooth airflow, the areas of the circle and square sections must be equal. To the nearest millimeter, what should the length x of a side of the square section be?

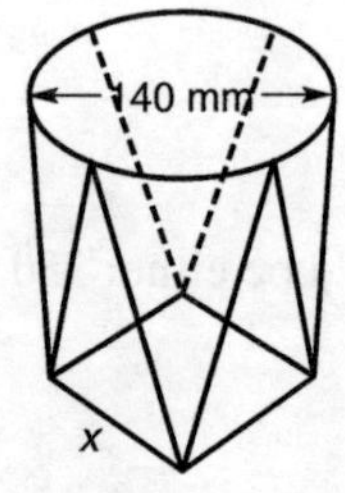

FIGURE 2.3 Ventilating duct (Problem 8).

9. Profit The Geometric Products Company produces a product at a variable cost per unit of \$2.20. If fixed costs are \$95,000 and each unit sells for \$3, how many units must be sold for the company to have a profit of \$50,000?

10. Sales The Clark Company management would like to know the total sales units that are required for the company to earn a profit of \$100,000. The following data are available: unit selling price of \$20; variable cost per unit of \$15; total fixed cost of \$600,000. From these data, determine the required sales units.

11. Investment A person wishes to invest \$20,000 in two enterprises so that the total income per year will be \$1440. One enterprise pays 6% annually; the other has more risk and pays $7\frac{1}{2}$% annually. How much must be invested in each?

12. Investment A person invested \$20,000, part at an interest rate of 6% annually and the remainder at 7% annually. The total interest at the end of 1 year was equivalent to an annual $6\frac{3}{4}$% rate on the entire \$20,000. How much was invested at each rate?

13. Pricing The cost of a product to a retailer is \$3.40. If the retailer wishes to make a profit of 20% on the selling price, at what price should the product be sold?

14. Bond Retirement In two years, a company will require \$1,123,600 in order to retire some bonds. If the company now invests \$1,000,000 for this purpose, what annual rate of interest, compounded annually, must it receive on that amount in order to retire the bonds?

15. Expansion Program In two years, a company will begin an expansion program. It has decided to invest \$2,000,000 now so that in two years the total value of the investment will be \$2,163,200, the amount required for the expansion. What is the annual rate of interest, compounded annually, that the company must receive to achieve its purpose?

16. Business A company finds that if it produces and sells q units of a product, its total sales revenue in dollars is $100\sqrt{q}$. If the variable cost per unit is \$2 and the fixed cost is \$1200, find the values of q for which

$$\text{total sales revenue} = \text{variable cost} + \text{fixed cost}.$$

(That is, profit is zero.)

17. Dormitory Housing A college dormitory houses 210 students. This fall, rooms are available for 76 freshmen. On the average, 95% of those freshmen who request room applications actually reserve a room. How many room applications should the college send out if it wants to receive 76 reservations?

18. Poll A group of people were polled, and 20%, or 700, of them favored a new product over the best-selling brand. How many people were polled?

19. Prison Guard Salary It was reported that in a certain women's jail, female prison guards, called matrons, received 30% (or \$200) a month less than their male counterparts, deputy sheriffs. Find the yearly salary of a deputy sheriff. Give your answer to the nearest dollar.

20. Striking Drivers A few years ago, cement drivers were on strike for 46 days. Before the strike, these drivers earned \$7.50 per hour and worked 260 eight-hour days a year. What percentage increase is needed in yearly income to make up for the lost time within 1 year?

21. Break Even A manufacturer of video-game cartridges sells each cartridge for \$19.95. The manufacturing cost of each cartridge is \$14.95. Monthly fixed costs are \$8000. During the first month of sales of a new game, how many cartridges must be sold in order for the manufacturer to break even (that is, in order that total revenue equal total cost)?

22. Investment Club An investment club bought a bond of an oil corporation for \$5000. The bond yields 8% per year. The club now wants to buy shares of stock in a hospital supply company. The stock sells at \$20 per share and earns a dividend of \$0.50 per share per year. How many shares should the club buy so that its total investment in stocks and bonds yields 5% per year?

23. Vision Care As a fringe benefit for its employees, a company established a vision-care plan. Under this plan, each year the company will pay the first \$35 of an employee's vision-care expenses and 80% of all additional vision-care expenses, up to a maximum *total* benefit payment of \$100. For an employee, find the total annual vision-care expenses covered by this program.

24. Quality Control Over a period of time, the manufacturer of a caramel-center candy bar found that 2% of the bars were rejected for imperfections.

a. If c candy bars are made in a year, how many would the manufacturer expect to be rejected?

b. This year, annual consumption of the candy is projected to be 2,000,000 bars. Approximately how many bars will have to be made if rejections are taken into consideration?

25. Business Suppose that consumers will purchase q units of a product when the price is $(80 - q)/4$ dollars *each.* How many units must be sold in order that sales revenue be \$400?

26. Investment How long would it take to double an investment at simple interest with a rate of 5% per year? [*Hint:* See Example 6(a) of Sec. 1.1, and express 5% as 0.05.]

27. Business Alternatives The inventor of a new toy offers the Kiddy Toy Company exclusive rights to manufacture and sell the toy for a lump-sum payment of \$25,000. After estimating that future sales possibilities beyond one year are nonexistent, the Company management is reviewing an alternative proposal to give a lump-sum payment of \$2000 plus a royalty of \$0.50 for each unit sold. How many units must be sold the first year to make this alternative as economically attractive to the inventor as the original request? [*Hint:* Determine when the incomes under both proposals are the same.]

28. Parking Lot A company parking lot is 120 ft long and 80 ft wide. Due to an increase in personnel, it is decided to double the area of the lot by adding strips of equal width to one end and one side. Find the width of one such strip.

29. Rentals You are the chief financial advisor to a corporation that owns an office complex consisting of 50 units. At \$400 per month, every unit can be rented. However, for each \$20 per month increase, there will be two vacancies with no possibility of filling them. The corporation wants to receive a total of \$20,240 per month from rents in the complex. You are asked to determine the rent that should be charged for each unit. What is your reply?

30. Investment Six months ago, an investment company had a \$3,000,000 portfolio consisting of blue-chip and glamor stocks. Since then, the value of the blue-chip investment increased by $\frac{1}{10}$, whereas the value of the glamor stocks decreased by $\frac{1}{10}$. The current value of the portfolio is \$3,140,000. What is the *current* value of the blue-chip investment?

31. Revenue The monthly revenue of a certain company is given by $R = 800p - 7p^2$, where p is the price in dollars of the product the company manufactures. At what price will the revenue be \$10,000 if the price must be greater than \$50?

32. Price–earnings Ratio The *price–earnings ratio* (P/E) of a company is the ratio of the market value of one share of the company's outstanding common stock to the earnings per share. If P/E increases by 10% and the earnings per share increase by 20%, determine the percentage increase in the market value per share of the common stock.

33. **Market Equilibrium** When the price of a product is p dollars each, suppose that a manufacturer will supply $2p - 8$ units of the product to the market and that consumers will demand to buy $300 - 2p$ units. At the value of p for which supply equals demand, the market is said to be in equilibrium. Find this value of p.

34. **Market Equilibrium** Repeat Problem 33 for the following conditions: At a price of p dollars each, the supply is $3p^2 - 4p$ and the demand is $24 - p^2$.

35. **Security Fence** For security reasons, a company will enclose a rectangular area of 11,200 ft^2 in the rear of its plant. One side will be bounded by the building and the other three sides by fencing. (See Fig. 2.4.) If 300 ft of fencing will be used, what will be the dimensions of the rectangular area?

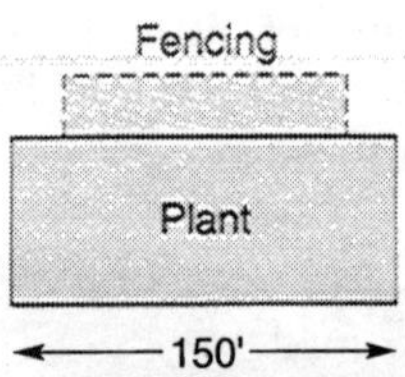

FIGURE 2.4 Security fencing (Problem 35).

36. **Package Design** A company is designing a package for its product. One part of the package is to be an open box made from a square piece of aluminum by cutting out a 3-in square from each corner and folding up the sides. (See Fig. 2.5.) The box is to contain 75 in^3. What are the dimensions of the square piece of aluminum that must be used?

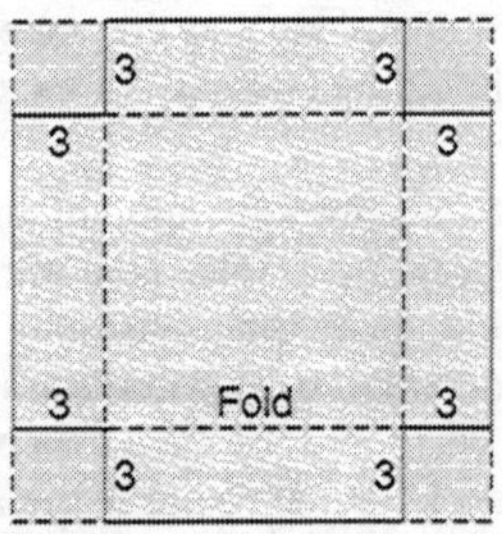

FIGURE 2.5 Box construction (Problem 36).

37. **Product Design** A candy company makes the popular Dandy Bar. The rectangular-shaped bar is 10 centimeters (cm) long, 5 cm wide, and 2 cm thick. (See Fig. 2.6.) Because of increasing costs, the company has decided to cut the volume of the bar by a drastic 28%. The thickness will be the same, but the length and width will be reduced by equal amounts. What will be the length and width of the new bar?

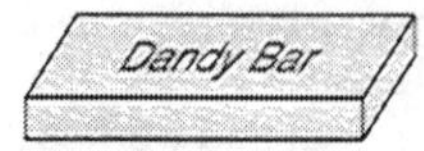

FIGURE 2.6 Candy bar (Problem 37).

38. **Product Design** A candy company makes a washer-shaped candy (a candy with a hole in it); see Fig. 2.7. Because of increasing costs, the company will cut the volume of candy in each piece by 20%. To do this, the firm will keep the same thickness and outer radius, but will make the inner radius larger. At present the thickness is 2 millimeters (mm), the inner radius is 2 mm, and the outer radius is 7 mm. Find the inner radius of the new-style candy. [*Hint:* The volume V of a solid disk is $\pi r^2 h$, where r is the radius and h is the thickness of the disk.)

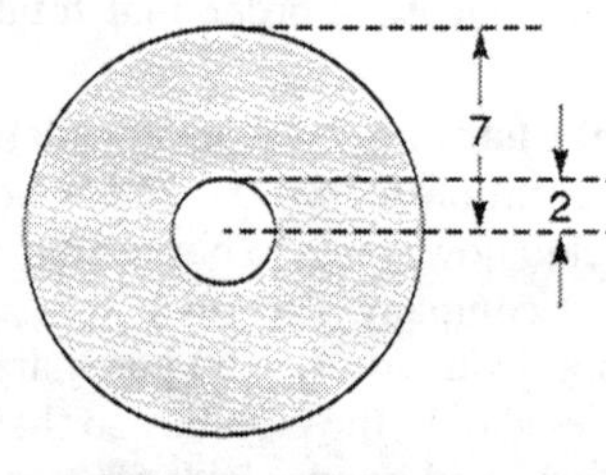

FIGURE 2.7 Washer-shaped candy (Problem 38).

39. **Compensating Balance** *Compensating balance* refers to that practice wherein a bank requires a borrower to maintain on deposit a certain portion of a loan during the term of the loan. For example, if a firm takes out a \$100,000 loan that requires a compensating balance of 20%, it would have to leave \$20,000 on deposit and would have the use of \$80,000. To meet the expenses of retooling, the Victor Manufacturing Company must borrow \$95,000. The Third National Bank, with whom the firm has had no prior association, requires a compensating balance of 15%. To the nearest thousand dollars, what amount of the loan is required to obtain the needed funds?

40. Incentive Plan A machine company has an incentive plan for its salespeople. For each machine that a salesperson sells, the commission is \$40. The commission for *every* machine sold will increase by \$0.04 for each machine sold over 600. For example, the commission on each of 602 machines sold is \$40.08. How many machines must a salesperson sell in order to earn \$30,800?

41. Real Estate A land investment company purchased a parcel of land for \$7200. After having sold all but 20 acres at a profit of \$30 per acre over the original cost per acre, the company regained the entire cost of the parcel. How many acres were sold?

42. Margin of Profit The *margin of profit* of a company is the net income divided by the total sales. A company's margin of profit increased by 0.02 from last year. Last year the company sold its product at \$3.00 each and had a net income of \$4500. This year it increased the price of its product by \$0.50 each, sold 2000 more, and had a net income of \$7140. The company never has had a margin of profit greater than 0.15. How many of its product were sold last year and how many were sold this year?

43. Business A company manufactures products A and B. The cost of producing each unit of A is \$2 more than that of B. The costs of production of A and B are \$1500 and \$1000, respectively, and 25 more units of A are produced than of B. How many of each are produced?

OBJECTIVE

To solve linear inequalities in one variable and to introduce interval notation.

2.2 Linear Inequalities

Suppose a and b are two points on the real-number line. Then either a and b coincide, or a lies to the left of b, or a lies to the right of b. (See Fig. 2.8.)

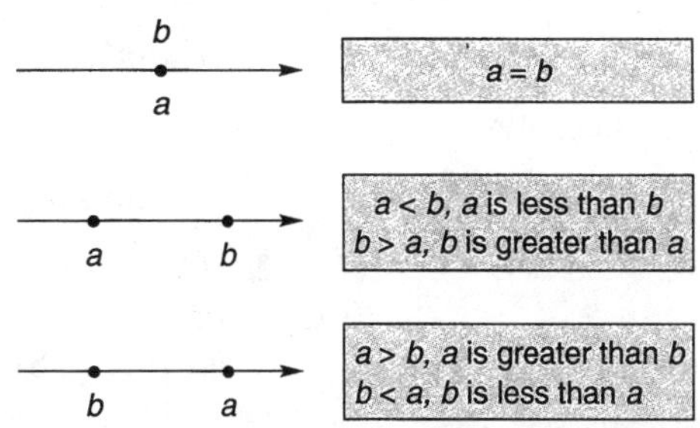

FIGURE 2.8 Relative position of two points.

If a and b coincide, then $a = b$. If a lies to the left of b, we say that a is less than b and write $a < b$, where the *inequality symbol* "$<$" is read "is less than." On the other hand, if a lies to the right of b, we say that a is greater than b, written $a > b$. The statements $a > b$ and $b < a$ are equivalent.

Another inequality symbol "$\leq$" is read "is less than or equal to" and is defined as follows: $a \leq b$ if and only if $a < b$ or $a = b$. Similarly, the symbol "$\geq$" is defined as follows: $a \geq b$ if and only if $a > b$ or $a = b$. In this case, we say that a is greater than or equal to b.

We shall use the words *real numbers* and *points* interchangeably, since there is a one-to-one correspondence between real numbers and points on a line. Thus, we can speak of the points $-5, -2, 0, 7$, and 9 and can write $7 < 9$, $-2 > -5$, $7 \leq 7$, and $7 \geq 0$. (See Fig. 2.9.) Clearly, if $a > 0$, then a is positive; if $a < 0$, then a is negative.

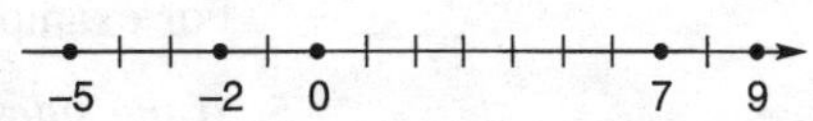

FIGURE 2.9 Points on number line.

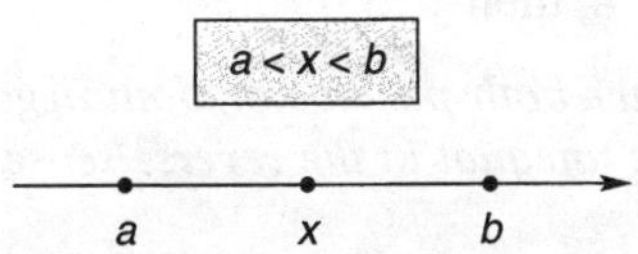

Figure 2.10 $a < x$ and $x < b$.

Suppose that $a < b$ and x is between a and b. (See Fig. 2.10.) Then not only is $a < x$, but also, $x < b$. We indicate this by writing $a < x < b$, which can be thought of as a dual inequality. For example, $0 < 7 < 9$. (Refer back to Fig. 2.9.)

In defining an inequality next, we shall use the less-than relation (<), but the others (>, ≤, ≥) would also apply.

DEFINITION

*An **inequality** is a statement that one number is less than another number.*

Of course, we represent inequalities by means of inequality symbols. If two inequalities have their inequality symbols pointing in the same direction, then the inequalities are said to have the *same sense.* If not, they are said to be *opposite in sense,* or one is said to have the *reverse sense* of the other. Hence, $a < b$ and $c < d$ have the same sense, but $a < b$ has the reverse sense of $c > d$.

Solving an inequality, such as $2(x - 3) < 4$, means finding all values of the variable for which the inequality is true. This involves the application of certain rules, which we now state.

Rules for Inequalities

Keep in mind that the rules also apply to ≤, >, and ≥.

1. *If the same number is added to or subtracted from both sides of an inequality, the resulting inequality has the same sense as the original inequality.* Symbolically,

$$\text{if } a < b, \text{ then } a + c < b + c \text{ and } a - c < b - c.$$

For example, $7 < 10$, so $7 + 3 < 10 + 3$.

2. *If both sides of an inequality are multiplied or divided by the same **positive** number, the resulting inequality has the same sense as the original inequality.* Symbolically,

$$\text{if } a < b \text{ and } c > 0, \text{ then } ac < bc \text{ and } \frac{a}{c} < \frac{b}{c}.$$

For example, $3 < 7$ and $2 > 0$, so $3(2) < 7(2)$ and $\frac{3}{2} < \frac{7}{2}$.

The sense of an inequality must be reversed when multiplying or dividing both sides by a negative number.

3. *If both sides of an inequality are multiplied or divided by the same **negative** number, then the resulting inequality has the **reverse** sense of the original inequality.* Symbolically,

$$\text{if } a < b \text{ and } c > 0, \text{ then } a(-c) > b(-c) \text{ and } \frac{a}{-c} > \frac{b}{-c}.$$

For example, $4 < 7$ but $4(-2) > 7(-2)$ and $\frac{4}{-2} > \frac{7}{-2}$.

4. *Any side of an inequality can be replaced by an expression equal to it.* Symbolically,

$$\text{if } a < b \text{ and } a = c, \text{ then } c < b.$$

For example, if $x < 2$ and $x = y + 4$, then $y + 4 < 2$.

5. *If the sides of an inequality are either both positive or both negative, then their respective reciprocals[1] are unequal in the **reverse** sense.* For example, $2 < 4$, but $\frac{1}{2} > \frac{1}{4}$.

[1]The *reciprocal* of a nonzero number a is defined to be $\frac{1}{a}$.

6. *If both sides of an inequality are positive and we raise each side to the same positive power, then the resulting inequality has the same sense as the original inequality.* Thus, if $0 < a < b$ and $n > 0$, then

$$a^n < b^n \qquad \text{and} \qquad \sqrt[n]{a} < \sqrt[n]{b},$$

where we assume that n is a positive integer in the latter inequality. For example, $4 < 9$, so $4^2 < 9^2$ and $\sqrt{4} < \sqrt{9}$.

The result of applying Rules 1–4 to an inequality is called an *equivalent inequality.* This is an inequality whose solution is exactly the same as that of the original inequality. We shall apply these rules to a *linear inequality.*

DEFINITION

*A **linear inequality** in the variable x is an inequality that can be written in the form*

$$ax + b < 0,$$

where a and b are constants and $a \neq 0$.

The definition also applies to $\leq$, $>$, and $\geq$.

Principles in Practice 1

Solving a Linear Inequality

A salesman has a monthly income given by $I = 200 + 0.8S$, where S is the number of products sold in a month. How many products must he sell to make at least $4500 a month?

EXAMPLE 1 Solving a Linear Inequality

Solve $2(x - 3) < 4$.

Solution:

Strategy: We shall replace the given inequality by equivalent inequalities until the solution is evident.

$$\begin{aligned} 2(x-3) &< 4, & \\ 2x - 6 &< 4 & \text{(Rule 4)}, \\ 2x - 6 + 6 &< 4 + 6 & \text{(Rule 1)}, \\ 2x &< 10 & \text{(Rule 4)}, \\ \frac{2x}{2} &< \frac{10}{2} & \text{(Rule 2)}, \\ x &< 5. & \end{aligned}$$

All of the foregoing inequalities are equivalent. Thus, the original inequality is true for *all* real numbers x such that $x < 5$. For example, the inequality is true for $x = -10, -0.1, 0, \frac{1}{2}$, and 4.9. We may write our solution simply as $x < 5$ and can geometrically represent it by the bold line segment in Fig. 2.11. The parenthesis indicates that 5 is *not included* in the solution. ■

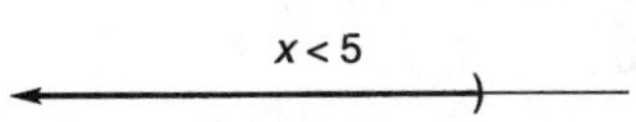

Figure 2.11 All real numbers less than 5.

In Example 1 the solution consisted of a set of numbers, namely, all numbers less than 5. In general, it is common to use the term **interval** to describe such a set. In the case of Example 1, the set of all x such that $x < 5$ can be denoted by the *interval notation* $(-\infty, 5)$. The symbol $-\infty$ is not a number, but is merely a convenience for indicating that the interval extends indefinitely to the left.

There are other types of intervals. For example, the set of all numbers x for which $a \leq x \leq b$ is called a **closed interval** and includes the numbers a and b, which are called *endpoints* of the interval. This interval is denoted by $[a, b]$ and is shown in Fig. 2.12(a). The square brackets indicated that a and b *are included* in the interval. On the other hand, the set of all x for which

Closed interval $[a, b]$
(a)

Open interval (a, b)
(b)

FIGURE 2.12 Closed and open intervals.

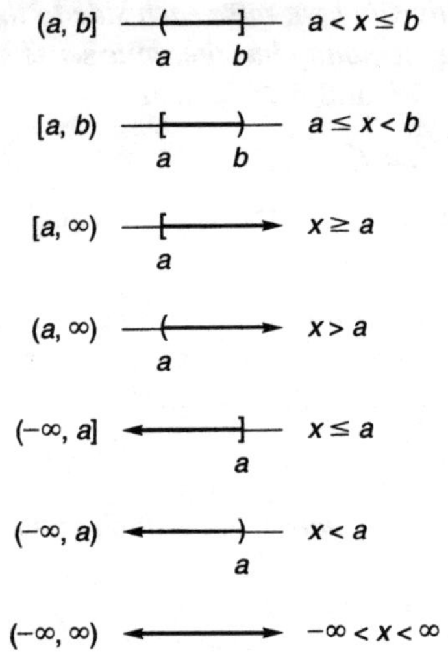

FIGURE 2.13 Intervals.

$a < x < b$ is called an **open interval** and is denoted by (a, b). The endpoints are *not* part of this set. [See Fig. 2.12(b).] Extending these concepts, we have the intervals shown in Fig. 2.13.

EXAMPLE 2 Solving a Linear Inequality

Solve $3 - 2x \leq 6$.

Solution:

$$3 - 2x \leq 6,$$

$$-2x \leq 3 \qquad \text{(Rule 1)},$$

$$x \geq -\frac{3}{2} \qquad \text{(Rule 3)}.$$

Dividing both sides by -2 reverses the sense of the inequality.

The solution is $x \geq -\frac{3}{2}$, or, in interval notation, $[-\frac{3}{2}, \infty)$. This is represented geometrically in Fig. 2.14. ■

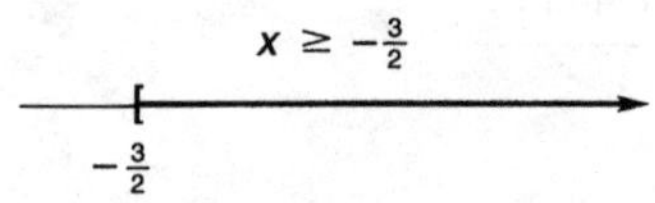

FIGURE 2.14 The interval $[-\frac{3}{2}, \infty)$.

Principles in Practice 2
Solving a Linear Inequality

A zoo veterinarian can purchase four different animal foods with various nutrient values for the zoo's grazing animals. Let x_1 represent the number of bags of food 1, x_2 represent the number of bags of food 2, and so on. The number of bags of each food needed can be described by the following equations:

$$x_1 = 150 - x_4$$

$$x_2 = 3x_4 - 210$$

$$x_3 = x_4 + 60$$

Develop four inequalities from these equations, assuming that each variable must be nonnegative.

EXAMPLE 3 Solving a Linear Inequality

Solve $\frac{3}{2}(s - 2) + 1 > -2(s - 4)$.

Solution:

$$\tfrac{3}{2}(s - 2) + 1 > -2(s - 4),$$

$$2[\tfrac{3}{2}(s - 2) + 1] > 2[-2(s - 4)] \qquad \text{(Rule 2)},$$

$$3(s - 2) + 2 > -4(s - 4),$$

$$3s - 4 > -4s + 16,$$

$$7s > 20 \qquad \text{(Rule 1)},$$

$$s > \frac{20}{7} \qquad \text{(Rule 2)}.$$

The solution is $(\frac{20}{7}, \infty)$; see Fig. 2.15. ■

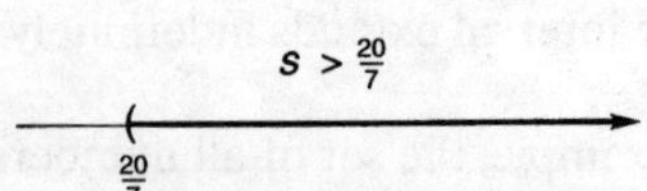

FIGURE 2.15 The interval $(\frac{20}{7}, \infty)$.

EXAMPLE 4 Solving Linear Inequalities

a. *Solve* $2(x - 4) - 3 > 2x - 1$.

Solution:

$$2(x - 4) - 3 > 2x - 1,$$
$$2x - 8 - 3 > 2x - 1,$$
$$-11 > -1.$$

Since it is never true that $-11 > -1$, there is no solution, and the solution set is $\varnothing$.

b. *Solve* $2(x - 4) - 3 < 2x - 1$.

Solution: Proceeding as in part (a), we obtain $-11 < -1$. This is true for all real numbers x, so the solution is $(-\infty, \infty)$; see Fig. 2.16. ■

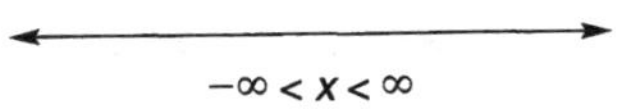

FIGURE 2.16 The interval $(-\infty, \infty)$.

■ Exercise 2.2

In Problems 1–34, solve the inequalities. Give your answer in interval notation, and indicate the answer geometrically on the real-number line.

1. $3x > 12$.

2. $4x < -2$.

3. $4x - 13 \le 7$.

4. $3x \ge 0$.

5. $-4x \ge 2$.

6. $2y + 1 > 0$.

7. $3 - 5s > 5$.

8. $4s - 1 < -5$.

9. $3 < 2y + 3$.

10. $6 \le 5 - 3y$.

11. $2x - 3 \le 4 + 7x$.

12. $-3 \ge 8(2 - x)$.

13. $3(2 - 3x) > 4(1 - 4x)$.

14. $8(x + 1) + 1 < 3(2x) + 1$.

15. $2(3x - 2) > 3(2x - 1)$.

16. $3 - 2(x - 1) \le 2(4 + x)$.

17. $x + 2 < \sqrt{3} - x$.

18. $\sqrt{2}(x + 2) > \sqrt{8}(3 - x)$.

19. $\frac{5}{3}x < 10$.

20. $-\frac{1}{2}x > 6$.

21. $\frac{9y + 1}{4} \le 2y - 1$.

22. $\frac{4y - 3}{2} \ge \frac{1}{3}$.

23. $4x - 1 \ge 4(x - 2) + 7$.

24. $0x \le 0$.

25. $\frac{1 - t}{2} < \frac{3t - 7}{3}$.

26. $\frac{3(2t - 2)}{2} > \frac{6t - 3}{5} + \frac{t}{10}$.

27. $2x + 3 \ge \frac{1}{2}x - 4$.

28. $4x - \frac{1}{2} \le \frac{3}{2}x$.

29. $\frac{2}{3}r < \frac{5}{6}r$.

30. $\frac{7}{4}t > -\frac{2}{3}t$.

31. $\frac{y}{2} + \frac{y}{3} > y + \frac{y}{5}$.

32. $9 - 0.1x \le \frac{2 - 0.01x}{0.2}$.

33. $0.1(0.03x + 4) \ge 0.02x + 0.434$.

34. $\frac{5y - 1}{-3} < \frac{7(y + 1)}{-2}$.

---

35. Earnings Each month last year, a company had earnings that were greater than \$37,000, but less than \$53,000. If S represents the total earnings for the year, describe S by using inequalities.

36. Using inequalities, symbolize the following statement: The number of labor hours x to produce a product is not less than $2\frac{1}{2}$ nor more than 4.

37. Geometry In a right triangle, one of the acute angles x is less than 3 times the other acute angle plus 10 degrees. Solve for x.

38. Spending A student has \$330 to spend on a stereo system and some compact disks. If she buys a stereo that costs \$198 and the disks are \$9.50 each, find the greatest number of disks she can buy.

OBJECTIVE

To model situations in terms of inequalities.

2.3 Applications of Inequalities

Solving word problems may sometimes involve inequalities, as the following examples illustrate.

EXAMPLE 1 Profit

For a company that manufactures thermostats, the combined cost for labor and material is \$5 per thermostat. Fixed costs (costs incurred in a given period, regardless of output) are \$60,000. If the selling price of a thermostat is \$7, how many must be sold for the company to earn a profit?

Solution:

Strategy: Recall that

$$\text{profit} = \text{total revenue} - \text{total cost}.$$

We shall find total revenue and total cost and then determine when their difference is positive.

Let q be the number of thermostats that must be sold. Then their cost is $5q$. The total cost for the company is therefore $5q + 60{,}000$. The total revenue from the sale of q thermostats will be $7q$. Now,

$$\text{profit} = \text{total revenue} - \text{total cost},$$

and we want profit > 0. Thus,

$$\text{total revenue} - \text{total cost} > 0.$$

$$7q - (5q + 60{,}000) > 0,$$

$$2q > 60{,}000,$$

$$q > 30{,}000.$$

Therefore, at least 30,001 thermostats must be sold for the company to earn a profit. ■

EXAMPLE 2 Renting vs. Purchasing

A builder must decide whether to rent or buy an excavating machine. If he were to rent the machine, the rental fee would be \$600 per month (on a yearly basis), and the daily cost (gas, oil, and driver) would be \$60 for each day the machine is used. If he were to buy it, his fixed annual cost would be \$4000, and daily operating and maintenance costs would be \$80 for each day the machine is used. What is the least number of days each year that the builder would have to use the machine to justify renting it rather than buying it?

Solution:

Strategy: We shall determine expressions for the annual cost of renting and the annual cost of purchasing. We then find when the cost of renting is less than that of purchasing.

Let d be the number of days each year that the machine is used. If the machine is rented, the total yearly cost consists of rental fees, which are (12)(600), and daily charges of $60d$. If the machine is purchased, the cost per year is $4000 + 80d$. We want

$$\begin{aligned} \text{cost}_{\text{rent}} &< \text{cost}_{\text{purchase}}, \\ 12(600) + 60d &< 4000 + 80d, \\ 7200 + 60d &< 4000 + 80d, \\ 3200 &< 20d, \\ 160 &< d. \end{aligned}$$

Thus, the builder must use the machine at least 161 days to justify renting it. ■

EXAMPLE 3 Current Ratio

The *current ratio* of a business is the ratio of its current assets (such as cash, merchandise inventory, and accounts receivable) to its current liabilities (such as short-term loans and taxes payable).

After consulting with the comptroller, the president of the Ace Sports Equipment Company decides to make a short-term loan to build up inventory. The company has current assets of \$350,000 *and current liabilities of* \$80,000. *How much can the company borrow if the current ratio is to be no less than* 2.5? *(Note: The funds received are considered as current assets and the loan as a current liability.)*

Solution: Let x denote the amount the company can borrow. Then current assets will be $350{,}000 + x$, and current liabilities will be $80{,}000 + x$. Thus,

$$\text{current ratio} = \frac{\text{current assets}}{\text{current liabilities}} = \frac{350{,}000 + x}{80{,}000 + x}.$$

We want

$$\frac{350{,}000 + x}{80{,}000 + x} \geq 2.5.$$

Although the inequality that must be solved is not linear, it leads to a linear inequality.

Since x is positive, so is $80{,}000 + x$. Hence, we can multiply both sides of the inequality by $80{,}000 + x$ and the sense of the inequality will remain the same. We have

$$\begin{aligned} 350{,}000 + x &\geq 2.5(80{,}000 + x), \\ 150{,}000 &\geq 1.5x, \\ 100{,}000 &\geq x. \end{aligned}$$

Consequently, the company may borrow as much as \$100,000 and yet maintain a current ratio of no less than 2.5. ■

EXAMPLE 4 Publishing

A publishing company finds that the cost of publishing each copy of a certain magazine is \$1.50. *The revenue from dealers is* \$1.40 *per copy. The advertising revenue is* 10% *of the revenue received from dealers for all copies sold beyond* 10,000. *What is the least number of copies that must be sold so as to have a profit for the company?*

Solution:

Strategy: We have profit = total revenue − total cost, so we find an expression for profit and then set it greater than 0.

Let q be the number of copies that are sold. The revenue from dealers is $1.40q$, and the revenue from advertising is $(0.10)[(1.40)(q - 10{,}000)]$. The total cost of publication is $1.50q$. Thus,

$$\begin{aligned} \text{total revenue} - \text{total cost} &> 0. \\ 1.40q + (0.10)[(1.40)(q - 10{,}000)] - 1.50q &> 0, \\ 1.4q + 0.14q - 1400 - 1.5q &> 0, \\ 0.04q - 1400 &> 0, \\ 0.04q &> 1400, \\ q &> 35{,}000. \end{aligned}$$

Therefore, the total number of copies must be greater than 35,000. That is, at least 35,001 copies must be sold to guarantee a profit. ■

■ Exercise 2.3

1. **Profit** The Davis Company manufactures a product that has a unit selling price of \$20 and a unit cost of \$15. If fixed costs are \$600,000, determine the least number of units that must be sold for the company to have a profit.
2. **Profit** To produce 1 unit of a new product, a company determines that the cost for material is \$2.50 and the cost of labor is \$4. The constant overhead, regardless of sales volume, is \$5000. If the cost to a wholesaler is \$7.40 per unit, determine the least number of units that must be sold by the company to realize a profit.
3. **Renting vs. Purchasing** A businesswoman wants to determine the difference between the costs of owning and renting an automobile. She can rent a car for \$400 per month (on an annual basis). Under this plan, the cost per mile (gas and oil) is \$0.10. If she were to purchase the car, the fixed annual expense would be \$3000, and other costs would amount to \$0.18 per mile. What is the least number of miles she would have to drive per year to make renting no more expensive than purchasing?
4. **Shirt Manufacturer** A T-shirt manufacturer produces N shirts at a total labor cost (in dollars) of $1.2N$ and a total material cost of $0.3N$. The constant overhead for

the plant is \$6000. If each shirt sells for \$3, how many must be sold by the company to realize a profit?
5. **Publishing** The cost of publication of each copy of a magazine is \$0.65. It is sold to dealers for \$0.60 each, and the amount received for advertising is 10% of the amount received for all magazines issued beyond 10,000. Find the least number of magazines that can be published without loss—that is, such that profit ≥ 0. (Assume that all issues will be sold.)
6. **Production Allocation** A company produces alarm clocks. During the regular workweek, the labor cost for producing one clock is \$2.00. However, if a clock is produced on overtime, the labor cost is \$3.00. Management has decided to spend no more than a total of \$25,000 per week for labor. The company must produce 11,000 clocks this week. What is the minimum number of clocks that must be produced during the regular workweek?
7. **Investment** A company invests a total of \$30,000 of surplus funds at two annual rates of interest: 5% and $6\frac{3}{4}$%. It wishes an annual yield of no less than $6\frac{1}{2}$%. What is the least amount of money that the company must invest at the $6\frac{3}{4}$% rate?
8. **Current Ratio** The current ratio of Precision Machine Products is 3.8. If the firm's current assets are \$570,000, what are its current liabilities? To raise additional funds, what is the maximum amount the company can borrow on a short-term basis if the current ratio is to be no less than 2.6? (See Example 3 for an explanation of current ratio.)
9. **Sales Allocation** At present, a manufacturer has 2500 units of product in stock. The product is now selling at \$4 per unit. Next month the unit price will increase by \$0.50. The manufacturer wants the total revenue received from the sale of the 2500 units to be no less than \$10,750. What is the maximum number of units that can be sold this month?
10. **Revenue** Suppose consumers will purchase q units of a product at a price of $\frac{100}{q} + 1$ dollars per unit. What is the minimum number of units that must be sold in order that sales revenue be greater than \$5000?

11. Hourly Rate Painters are often paid either by the hour or on a per-job basis. The rate they receive can affect their working speed. For example, suppose they can work either for \$8.50 per hour or for \$300 plus \$3 for each hour less than 40 if they complete the job in less than 40 hours. Suppose the job will take t hours. If $t \geq 40$, clearly the hourly rate is better. If $t < 40$, for what values of t is the hourly rate the better pay scale?

12. Compensation Suppose a company offers you a sales position with your choice of two methods of determining your yearly salary. One method pays \$12,600 plus a bonus of 2% of your yearly sales. The other method pays a straight 8% commission on your sales. For what yearly sales amount is it better to choose the first method?

13. Campus Concert The dean of student affairs of a college is arranging for a rock group to perform a concert on campus. The group charges a flat fee of \$2440 or, instead, a fee of \$1000 plus 40% of the gate. It is likely that 800 students will attend. At most, how much could the dean charge for a ticket so that the second arrangement is no more costly than the flat fee? If this maximum is charged, how much money will be left over to pay for publicity, guards, and other concert expenses?

OBJECTIVE

To solve equations and inequalities involving absolute values.

2.4 Absolute Value

Absolute-Value Equations

On the real-number line, the distance of a number x from 0 is called the **absolute value** of x and is denoted by $|x|$. For example, $|5| = 5$ and $|-5| = 5$ because both 5 and -5 are 5 units from 0. (See Fig. 2.17.) Similarly, $|0| = 0$. Notice that $|x|$ can never be negative; that is, $|x| \geq 0$.

Basically, the absolute value of a real number is its value when its sign is ignored.

If x is positive or zero, then $|x|$ is simply x itself, so we can omit the vertical bars and write $|x| = x$. On the other hand, consider the absolute value of a negative number, like $x = -5$.

$$|x| = |-5| = 5 = -(-5) = -x.$$

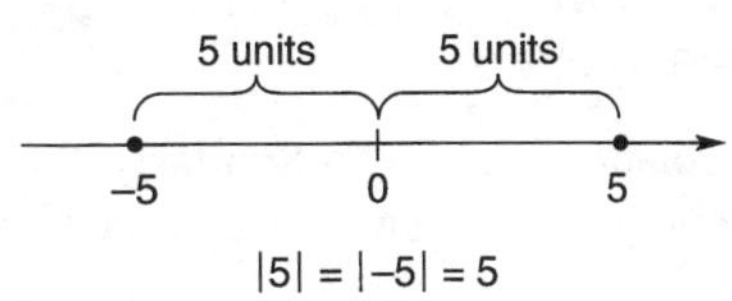

FIGURE 2.17 Absolute value.

Thus, if x is negative, then $|x|$ is the positive number $-x$. The minus sign indicates that we have changed the sign of x. Accordingly, aside from its geometrical interpretation, the absolute value can be defined as follows.

DEFINITION

*The **absolute value** of a real number x, written |x|, is defined as*

$$|x| = \begin{cases} x, & \text{if } x \geq 0, \\ -x, & \text{if } x < 0. \end{cases}$$

Applying the definition, we have $|3| = 3$, $|-8| = -(-8)$, and $|\frac{1}{2}| = \frac{1}{2}$. Also, $-|2| = -2$ and $-|-2| = -2$.

Pitfall ▼ $\sqrt{x^2}$ is not necessarily x, but

$$\sqrt{x^2} = |x|.$$

For example, $\sqrt{(-2)^2} = |-2| = 2$, not -2. This agrees with the fact that

$$\sqrt{(-2)^2} = \sqrt{4} = 2.$$

Also, $|-x| \neq x$ and

$$|-x - 1| \neq x + 1.$$

For example, if we let $x = -3$, then $|-(-3)| \neq -3$, and

$$|-(-3) - 1| \neq -3 + 1.$$

EXAMPLE 1 Solving Absolute-Value Equations

a. *Solve* $|x - 3| = 2$.

Solution: This equation states that $x - 3$ is a number 2 units from 0. Thus, either

$$x - 3 = 2 \quad \text{or} \quad x - 3 = -2.$$

Solving these equations gives $x = 5$ or $x = 1$.

b. *Solve* $|7 - 3x| = 5$.

Solution: The equation is true if $7 - 3x = 5$ or if $7 - 3x = -5$. Solving these equations gives $x = \frac{2}{3}$ or $x = 4$.

c. *Solve* $|x - 4| = -3$.

Solution: The absolute value of a number is never negative, so the solution set is $\varnothing$. ■

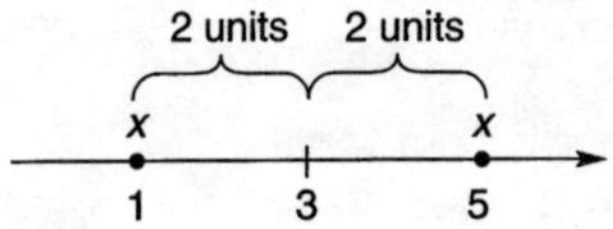

FIGURE 2.18 The solution of $|x - 3| = 2$ is 1 or 5.

We may interpret $|a - b|$ or $|b - a|$ as the distance between a and b. For example, the distance between 5 and 9 is either

$$|9 - 5| = |4| = 4,$$

$$\text{or} \quad |5 - 9| = |-4| = 4.$$

Similarly, the equation $|x - 3| = 2$ states that the distance between x and 3 is 2 units. Thus, x can be 1 or 5, as shown in Example 1(a) and Fig. 2.18.

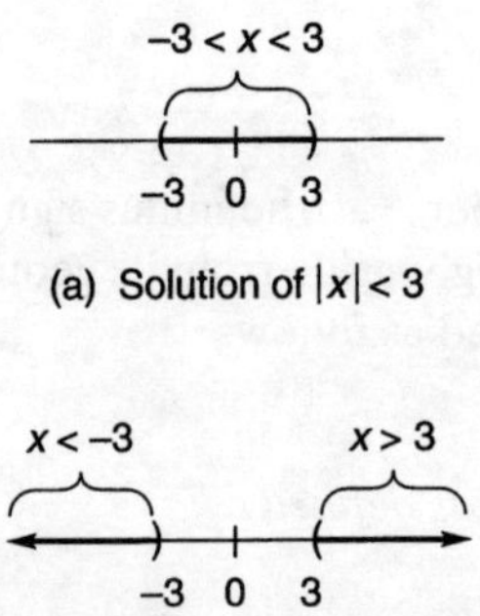

(a) Solution of $|x| < 3$

(b) Solution of $|x| > 3$

FIGURE 2.19 Solution of $|x| < 3$ and $|x| > 3$.

Absolute-Value Inequalities

Let us turn now to inequalities involving absolute values. If $|x| < 3$, then x is less than 3 units from 0. Hence, x must lie between -3 and 3, that is, on the in-

TABLE 2.1

Inequality ($d > 0$)	Solution
$|x| < d$	$-d < x < d$
$|x| \leq d$	$-d \leq x \leq d$
$|x| > d$	$x < -d$ or $x > d$
$|x| \geq d$	$x \leq -d$ or $x \geq d$

terval $-3 < x < 3$. [See Fig. 2.19(a).] On the other hand, if $|x| > 3$, then x must be greater than 3 units from 0. Hence, there are two intervals in the solution: Either $x < -3$ or $x > 3$. [See Fig. 2.19(b).] We can extend these ideas as follows: If $|x| \leq 3$, then $-3 \leq x \leq 3$; if $|x| \geq 3$, then $x \leq -3$ or $x \geq 3$. Table 2.1 gives a summary of the solutions to absolute-value inequalities.

EXAMPLE 2 Solving Absolute-Value Inequalities

a. *Solve* $|x - 2| < 4$.

Solution: The number $x - 2$ must be less than 4 units from 0. From the preceding discussion, this means that $-4 < x - 2 < 4$. We may set up the procedure for solving this inequality as follows:

$$-4 < x - 2 < 4,$$

$$-4 + 2 < x < 4 + 2 \quad \text{(adding 2 to each member)},$$

$$-2 < x < 6.$$

Thus, the solution is the open interval $(-2, 6)$. This means that all numbers between -2 and 6 satisfy the original inequality. (See Fig. 2.20.)

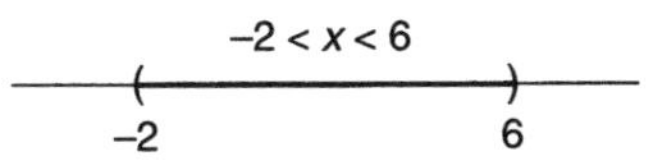

FIGURE 2.20 The solution of $|x - 2| < 4$ is the interval $(-2, 6)$.

b. *Solve* $|3 - 2x| \leq 5$.

Solution:

$$-5 \leq 3 - 2x \leq 5,$$

$$-5 - 3 \leq -2x \leq 5 - 3 \quad \text{(subtracting 3 from each member)},$$

$$-8 \leq -2x \leq 2,$$

$$4 \geq x \geq -1 \quad \text{(dividing each member by } -2\text{)},$$

$$-1 \leq x \leq 4 \quad \text{(rewriting)}.$$

Note that the sense of the original inequality was *reversed* when we divided by a negative number. The solution is the closed interval $[-1, 4]$. ■

EXAMPLE 3 Solving Absolute-Value Inequalities

a. *Solve* $|x + 5| \geq 7$.

Solution: Here $x + 5$ must be *at least* 7 units from 0. Thus, either $x + 5 \leq -7$ or $x + 5 \geq 7$. This means that either $x \leq -12$ or $x \geq 2$. Thus, the solution consists of two intervals: $(-\infty, -12]$ and $[2, \infty)$. We can abbreviate this collection of numbers by writing

$$(-\infty, -12] \cup [2, \infty).$$

where the connecting symbol $\cup$ is called the *union* symbol. (See Fig. 2.21.) More formally, the **union** of sets A and B is the set consisting of all elements that are in either A or B (or in both A and B).

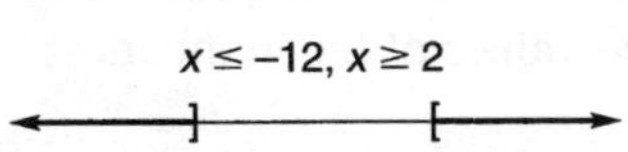

FIGURE 2.21 The union $(-\infty, -12] \cup [2, \infty)$.

b. *Solve* $|3x - 4| > 1$.

Solution: Either $3x - 4 < -1$ or $3x - 4 > 1$. Thus, either $3x < 3$ or $3x > 5$. Therefore, $x < 1$ or $x > \frac{5}{3}$, so the solution consists of all numbers in the set $(-\infty, 1) \cup (\frac{5}{3}, \infty)$. ■

The inequalities $x < 1$ and $x > \frac{5}{3}$ cannot be combined into a single inequality, although you might like to do so. It is incorrect to combine $\frac{5}{3} < x$ and $x < 1$ as $\frac{5}{3} < x < 1$, since this implies that $\frac{5}{3} < 1$.

Principles in Practice 1

Absolute-Value Notation

Express the following statement using absolute-value notation:

The actual weight w of a box of cereal must be within 0.3 oz of the weight stated on the box, which is 22 oz.

EXAMPLE 4 Absolute-Value Notation

Using absolute-value notation, express the following statements:

a. x is less than 3 units from 5.

Solution:

$$|x - 5| < 3.$$

b. x differs from 6 by at least 7.

Solution:

$$|x - 6| \geq 7.$$

c. $x < 3$ and $x > -3$ simultaneously.

Solution:

$$|x| < 3.$$

d. x is more than 1 unit from -2.

Solution:

$$|x - (-2)| > 1,$$

$$|x + 2| > 1.$$

e. x is less than σ (a Greek letter read "sigma") units from μ (a Greek letter read "mu").

Solution:

$$|x - \mu| < \sigma.$$

■

Properties of the Absolute Value

Four basic properties of the absolute value are as follows:

1. $|ab| = |a| \cdot |b|.$

2. $\left|\frac{a}{b}\right| = \frac{|a|}{|b|}.$

3. $|a - b| = |b - a|.$

4. $-|a| \leq a \leq |a|.$

For example, Property 1 states that the absolute value of the product of two numbers is equal to the product of the absolute values of the numbers.

EXAMPLE 5 Properties of Absolute Value

a. $|(-7) \cdot 3| = |-7| \cdot |3| = 21.$

b. $|4 - 2| = |2 - 4| = 2.$

c. $|7 - x| = |x - 7|.$

d. $\left|\frac{-7}{3}\right| = \frac{|-7|}{|3|} = \frac{7}{3}$; $\quad \left|\frac{-7}{-3}\right| = \frac{|-7|}{|-3|} = \frac{7}{3}.$

e. $\left|\frac{x - 3}{-5}\right| = \frac{|x - 3|}{|-5|} = \frac{|x - 3|}{5}.$

f. $-|2| \leq 2 \leq |2|.$

■

■ Exercise 2.4

In Problems 1–10, write an equivalent form without the absolute-value symbol.

1. $|-13|$. **2.** $|2^{-1}|$. **3.** $|8 - 2|$. **4.** $|(-4 - 6)/2|$.

5. $|3(-\frac{5}{3})|$. **6.** $|2 - 7| - |7 - 2|$. **7.** $|x| < 3$. **8.** $|x| < 10$.

9. $|2 - \sqrt{5}|$. **10.** $|\sqrt{5} - 2|$.

11. Using the absolute value symbol, express each fact.

a. x is less than 3 units from 7.

b. x differs from 2 by less than 3.

c. x is no more than 5 units from 7.

d. The distance between 7 and x is 4.

e. $x + 4$ is less than 2 units from 0.

f. x is between -3 and 3, but is not equal to 3 or -3.

g. $x < -6$ or $x > 6$.

h. $x - 6 > 4$ or $x - 6 < -4$.

i. The number x of hours that a machine will operate efficiently differs from 105 by less than 3.

j. The average monthly income x (in dollars) of a family differs from 850 by less than 100.

12. Use absolute-value notation to indicate that x and μ differ by no more than σ.

13. Use absolute-value notation to indicate that the prices p_1 and p_2 of two products may differ by no more than 2 (dollars).

14. Find all values of x such that $|x - \mu| \leq 2\sigma$.

In Problems 15–36, solve the given equation or inequality.

15. $|x| = 7$. **16.** $|-x| = 2$. **17.** $\left|\frac{x}{3}\right| = 2$. **18.** $\left|\frac{4}{x}\right| = 8$.

19. $|x - 5| = 8$. **20.** $|4 + 3x| = 2$. **21.** $|5x - 2| = 0$. **22.** $|7x + 3| = x$.

23. $|7 - 4x| = 5$. **24.** $|1 - 2x| = 1$. **25.** $|x| < 4$. **26.** $|-x| < 3$.

27. $\left|\frac{x}{4}\right| > 2$. **28.** $\left|\frac{x}{3}\right| > \frac{1}{2}$. **29.** $|x + 7| < 2$. **30.** $|5x - 1| < -6$.

31. $|x - \frac{1}{2}| > \frac{1}{2}$. **32.** $|1 - 3x| > 2$. **33.** $|5 - 2x| \leq 1$. **34.** $|4x - 1| \geq 0$.

35. $\left|\frac{3x - 8}{2}\right| \geq 4$. **36.** $\left|\frac{x - 8}{4}\right| \leq 2$.

In Problems 37–38, express the statement using absolute value notation.

37. In a science experiment, the measurement of a distance d is 10 m, and is accurate to ± 1 cm.

38. The difference in temperature between two chemicals which are to be mixed must be no less than 5 degrees and no more than 10 degrees.

39. Statistics In statistical analysis, the Chebyshev inequality asserts that if x is a random variable, μ is its mean, and σ is its standard deviation, then

$$(\text{probability that } |x - \mu| > h\sigma) \geq \frac{1}{h^2}.$$

Find those values of x such that $|x - \mu| > h\sigma$.

40. Manufacturing Tolerance In the manufacture of widgets, the average dimension of a part is 0.01 cm. Using the absolute-value symbol, express the fact that an individual measurement x of a part does not differ from the average by more than 0.005 cm.

2.5 REVIEW

IMPORTANT TERMS AND SYMBOLS

Section 2.1 fixed cost overhead variable cost total cost total revenue profit

Section 2.2 $a < b$ $a \leq b$ $a > b$ $a \geq b$ $a < x < b$ inequality sense of inequality equivalent inequality linear inequality $-\infty < x < \infty$ open interval closed interval endpoints interval notation

Section 2.4 absolute value, $|x|$ union, $\cup$

SUMMARY

With a word problem, an equation is not handed to you. Instead, you must set it up by translating verbal statements into an equation (or inequality). This is *mathematical modeling.* It is important that you first read the problem more than once so that you clearly understand what facts are given and what you are asked to find. Then choose a letter to represent the unknown quantity that you want to find. Use the relationships and facts given in the problem and translate them into an equation involving the letter. Finally, solve the equation, and see if your solution answers what was asked. Sometimes the solution to the *equation* will not be the answer to the *problem,* but it may be useful in obtaining that answer.

Some basic relationships that are used in solving business problems are:

$$\text{total cost} = \text{variable cost} + \text{fixed cost},$$

$$\text{total revenue} = (\text{price per unit})(\text{number of units sold}),$$

$$\text{profit} = \text{total revenue} - \text{total cost}.$$

The inequality symbols $<, \leq, >$, and $\geq$ are used to represent an inequality, which is a statement that one number is, for example, less than another number. Three basic operations that, when applied to an inequality, guarantee an equivalent inequality are:

1. Adding (or subtracting) the same number to (or from) both sides.
2. Multiplying (or dividing) both sides by the same positive number.
3. Multiplying (or dividing) both sides by the same negative number and reversing the sense of the inequality.

These operations are useful in solving a linear inequality (an inequality that can be put in the form $ax + b < 0$ or $ax + b \leq 0$, where $a \neq 0$).

$$|x| = x, \text{ if } x \geq 0 \quad \text{and} \quad |x| = -x, \text{ if } x < 0.$$

An algebraic definition of absolute value is

We interpret $|a - b|$ or $|b - a|$ as the distance between a and b. If $d > 0$, then the solution to the inequality $|x| < d$ is the interval $(-d, d)$. The solution to $|x| > d$ consists of two intervals and is given by $(-\infty, -d) \cup (d, \infty)$. Some basic properties of the absolute value are:

1. $|ab| = |a| \cdot |b|$,
2. $\left|\dfrac{a}{b}\right| = \dfrac{|a|}{|b|}$,
3. $|a - b| = |b - a|$.
4. $-|a| \leq a \leq |a|$.

REVIEW PROBLEMS

In Problems 1–15, solve the equation or inequality.

1. $3x - 8 \geq 4(x - 2)$.

2. $2x - (7 + x) \leq x$.

3. $-(5x + 2) < -(2x + 4)$.

4. $-2(x + 6) > x + 4$.

5. $3p(1 - p) > 3(2 + p) - 3p^2$.

6. $2(4 - \frac{3}{2}q) < 5$.

7. $\dfrac{x + 1}{3} - \dfrac{1}{2} \leq 2$.

8. $\dfrac{x}{2} + \dfrac{x}{3} > \dfrac{x}{4}$.

9. $\dfrac{1}{4}s - 3 \leq \dfrac{1}{8}(3 + 2s)$.

10. $\dfrac{1}{3}(t + 2) \geq \dfrac{1}{4}t + 4$.

11. $|3 - 2x| = 7$.

12. $\left|\dfrac{5x - 8}{13}\right| = 0$.

13. $|4t - 1| < 1$.

14. $4 < \left|\dfrac{2}{3}x + 5\right|$.

15. $|3 - 2x| \geq 4$.

16. Profit A profit of 40% on the selling price of a product is equivalent to what percent profit on the cost?

17. Stock Exchange On a certain day, there were 1132 different issues traded on the New York Stock Exchange. There were 48 more issues showing an increase than showing a decline, and no issues remained the same. How many issues suffered a decline?

18. Sales Tax The sales tax in a certain state is 6%. If a total of $3017.29 in purchases, including tax, is made in the course of a year, how much of it is tax?

19. Production Allocation A company will manufacture a total of 10,000 units of its product at plants A and B. Available data are as follows:

	Plant A	Plant B
Unit cost for labor and material	$5	$5.50
Fixed cost	$30,000	$35,000

Between the two plants the company has decided to allot no more than \$117,000 for total costs. What is the minimum number of units that must be produced at plant A?

20. **Storage Tanks** A company is replacing two cylindrical oil-storage tanks with one new tank. The old tanks are each 16 ft high. One has a radius of 15 ft and the other a radius of 20 ft. The new tank will also be 16 ft high. Find its radius if it is to have the same volume as the old tanks combined. [*Hint:* The volume V of a cylindrical tank is $V = \pi r^2 h$, where r is the radius of the circular base and h is the height of the tank.]

MATHEMATICAL *SNAPSHOT*

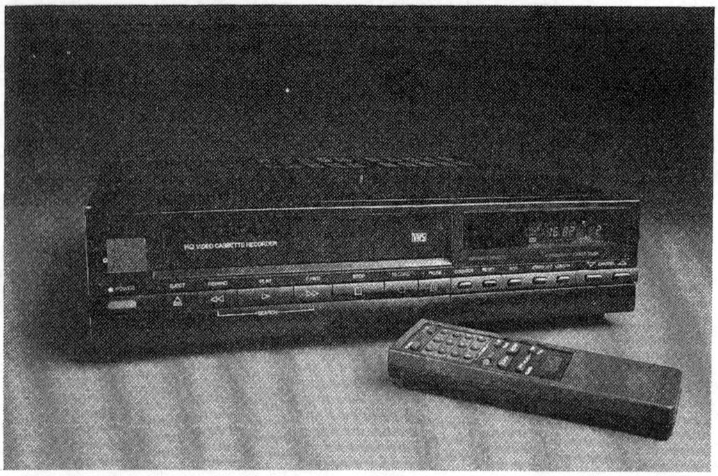

QUALITY VCR RECORDING[2]

The following is an entertaining discussion of VCR recording and illustrates mathematical concepts of this chapter.

If you are like the millions of others who have a VCR, you have seen how convenient it is to record television shows for future viewing. Here you will learn how to get the highest quality recording from this marvel of home entertainment hardware.

With a VHS format, you may have your choice of standard play speed (SP), long play speed (LP), or extended play speed (EP). SP is the fastest speed and provides the best picture quality. LP, a slower speed, gives lower quality recording, and EP, which is the slowest speed, gives the lowest picture quality.

With the common T-120 videotape, the maximum recording time at SP is 2 hours. At LP it is 4 hours, and at EP it is 6 hours. In the following discussion you may assume that these recording times are exact and that the amount of tape used changes uniformly with recording time.

If you wish to record a movie that is no more than 2 hours long, obviously SP should be used to get the best picture quality. However, for recording a 3-hour movie on a single T-120 tape, using only SP speed would cause the tape to be filled 1 hour before the movie is over. You can overcome this difficulty by using SP in conjunction with another speed, making sure that the time at SP is maximized.

For example, you can begin at LP and then complete the recording at SP. Obviously your problem is to determine when the changeover to SP should be made. Let t be the time, in hours, that LP is used. Then $3 - t$ hours of the movie remain to be recorded at SP. Since the tape speed in the LP mode is $\frac{1}{4}$ tape per hour and the speed in the SP mode is $\frac{1}{2}$ tape per hour, the portion of the tape used at LP is $t/4$ and the portion at SP is $(3 - t)/2$. The sum of these portions must be 1 because the whole tape would be used up. Therefore, you need to solve a linear equation.

$$\frac{t}{4} + \frac{3 - t}{2} = 1,$$

$$t + 2(3 - t) = 4,$$

$$6 - t = 4,$$

$$t = 2.$$

Thus you should record at LP for 2 hours and then switch to SP for the remaining $3 - t = 3 - 2 = 1$ hour. This means that one-third of the movie will be recorded with the best picture quality.

Instead of restricting yourself to a 3-hour movie, you can generalize the above problem to handle a movie of length l hours, where $2 < l \leq 4$. This situation gives

$$\frac{t}{4} + \frac{l - t}{2} = 1,$$

whose solution is

$$t = 2l - 4.$$

Alternatively, you might feel that there is not much difference between the picture qualities at LP

[2]Adapted from Gregory N. Fiore, "An Application of Linear Equations to the VCR," *Mathematics Teacher,* 81 (October 1988), 570–72. By permission of the National Council of Teachers of Mathematics.

and EP speeds. If you desire to begin at EP and end with SP, you could handle a movie of length l, where $2 < l \leq 6$. Let t be the time, in hours, that EP is used. Then

$$\frac{t}{6} + \frac{l - t}{2} = 1,$$

$$t + 3(l - t) = 6,$$

$$-2t + 3l = 6,$$

$$3l - 6 = 2t,$$

$$t = \frac{3}{2}l - 3.$$

For example, with a 3-hour movie you would record at EP for $t = \frac{3}{2}(3) - 3 = 1\frac{1}{2}$ hours and then at SP for $3 - 1\frac{1}{2} = 1\frac{1}{2}$ hours. This shows that using EP instead of LP gives you $\frac{1}{2}$ hour more of quality recording at SP. As a second example, consider recording a 4-hour-and-20-minute movie. Here $l = 4\frac{1}{3}$ hours, so you would use EP for

$$t = \frac{3}{2}\left(\frac{13}{3}\right) - 3 = 3\tfrac{1}{2} \text{ hours}$$

and use SP for the rest of the movie.

Finally, suppose that a television show has commercial interruptions, but you want to record it with the commercials eliminated. You can handle this by stopping the VCR at the beginning of a commercial and restarting when the commercial is over. Assume you estimate that every hour of an l-hour show contains c minutes of commercials. Then the total commercial time, in hours, is $lc/60$. Thus the noncommercial length of the show, in hours, is

$$l - \frac{lc}{60}.$$

If you begin recording at EP and then switch to SP, you will have

$$\frac{t}{6} + \frac{l - \frac{lc}{60} - t}{2} = 1,$$

where t is the time, in hours, at EP. Solving gives

$$t + 3\left(l - \frac{lc}{60} - t\right) = 6,$$

$$t + 3l - \frac{lc}{20} - 3t = 6,$$

$$-2t = -3l + \frac{lc}{20} + 6$$

$$t = \frac{3}{2}l - \frac{c}{40}l - 3.$$

For example, take the case of a 3-hour show that has 6 minutes of commercials each hour. Then $l = 3$ and $c = 6$, so

$$t = \frac{3}{2}(3) - \frac{6}{40}(3) - 3$$

$$= \frac{21}{20} \text{ hour.}$$

$$= 1 \text{ hour and 3 minutes.}$$

This means that you should record 1 hour and 3 minutes of the noncommercial part of the show at EP and the remainder of the noncommercial time at SP.

■ Exercises

1. If LP and SP modes are used to record a $2\frac{1}{2}$-hour movie, how long after the start of the movie should the switch from LP to SP be made?

2. If EP and SP modes are used to record a $2\frac{1}{2}$-hour show, how many minutes after the start of the show should the switch from EP to SP be made?

3. If EP and SP modes are used to record a movie of length 2 hours and 40 minutes, how long after the start of the movie should the switch from EP to SP be made?

4. EP and SP modes are used to record a 3-hour movie. How long after the start of the movie should the switch from EP to SP be made if the viewer eliminates 8 minutes of commercials each hour?

5. Repeat Problem 4 if LP and SP modes are to be used.

Chapter 4
Functions and Graphs

OBJECTIVE

To understand what a function is and to determine domains and function values.

3.1 FUNCTIONS

In the 17th century, Gottfried Wilhelm Leibniz, one of the inventors of calculus, introduced the term *function* into the mathematical vocabulary. The concept of a function is one of the most basic in all of mathematics, and it is essential to the study of calculus.

Briefly, a function is a special type of relation that expresses how one quantity (the *output*) depends on another quantity (the *input*). For example, when money is invested at some interest rate, the interest I (output) depends on the length of time t (input) that the money is invested. To express this dependence, we say that I is a "function of" t. Functional relations like this are usually specified by a formula that shows what must be done to the input to find the output.

To illustrate, suppose \$100 earns simple interest at an annual rate of 6%. Then it can be shown that interest and time are related by the formula

$$I = 100(0.06)t, \tag{1}$$

where I is in dollars and t is in years. For example,

$$\text{if } t = \tfrac{1}{2}, \qquad \text{then} \qquad I = 100(0.06)(\tfrac{1}{2}) = 3. \tag{2}$$

Thus, Formula (1) assigns the output 3 to the input $\frac{1}{2}$. We can think of Formula (1) as defining a *rule:* Multiply t by 100(0.06). The rule assigns to each input number t exactly one output number I, which we symbolize by the following arrow notation:

$$t \to I \qquad \text{or} \qquad t \to 100(0.06)t.$$

This rule is an example of a *function* in the following sense:

DEFINITION

A ***function*** *is a rule that assigns to each input number exactly one output number. The set of all input numbers to which the rule applies is called the* ***domain*** *of the function. The set of all output numbers is called the* ***range.***

For the interest function defined by Formula (1), the input number t cannot be negative, because negative time makes no sense. Thus, the domain consists of all nonnegative numbers—that is, all $t \geq 0$. From (2), we see that when the input is $\frac{1}{2}$, the output is 3. Thus, 3 is in the range.

We have been using the term *function* in a restricted sense because, in general, the inputs or outputs do not have to be numbers. For example, a list of states and their capitals assigns to each state its capital (exactly one output). Hence, a function is implied. However, for the time being, we shall consider only functions whose domains and ranges consist of real numbers.

A variable that represents input numbers for a function is called an **independent variable.** A variable that represents output numbers is called a **dependent variable** because its value *depends* on the value of the independent variable. We say that the dependent variable is a *function of* the independent variable. That is, output is a function of input. Thus, for the interest formula $I = 100(0.06)t$, the independent variable is t, the dependent variable is I, and I is a function of t.

As another example, the equation (or formula)

$$y = x + 2 \tag{3}$$

defines y as a function of x. The equation gives the rule, "Add 2 to x." This rule assigns to each input x exactly one output $x + 2$, which is y. If $x = 1$, then $y = 3$; if $x = -4$, then $y = -2$. The independent variable is x and the dependent variable is y.

In $y^2 = x$, x and y are related, but the relationship is not a function of x.

Not all equations in x and y define y as a function of x. For example, let $y^2 = x$. If x is 9, then $y^2 = 9$, so $y = \pm 3$. Hence, to the input 9, there are assigned not one, but *two,* output numbers: 3 and -3. This violates the definition of a function, so y is **not** a function of x.

On the other hand, some equations in two variables define either variable as a function of the other variable. For example, if $y = 2x$, then for each input x, there is exactly one output, $2x$. Thus, y is a function of x. However, solving the equation for x gives $x = y/2$. For each input y, there is exactly one output, $y/2$. Consequently, x is a function of y.

Usually, the letters f, g, h, F, G, and so on are used to represent function rules. For example, Eq. (3), $y = x + 2$, defines y as a function of x, where the rule is "Add 2 to the input." Suppose we let f represent this rule. Then we say that f is the function. To indicate that f assigns the output 3 to the input 1, we write $f(1) = 3$, which is read "f of 1 equals 3." Similarly, $f(-4) = -2$. More generally, if x is any input, we have the following notation:

$f(x)$ is an output number.

$f(x)$, which is read "f of x," means the output number in the range of f that corresponds to the input number x in the domain.

$$\underbrace{\overset{\text{input}}{\overset{\downarrow}{f(x)}}}_{\substack{\uparrow \\ \text{output}}}$$

Thus, the output $f(x)$ is the same as y. But since $y = x + 2$, we may write $y = f(x) = x + 2$, or simply,

$$f(x) = x + 2.$$

For example, to find $f(3)$, which is the output corresponding to the input 3, we replace each x in $f(x) = x + 2$ by 3:

$$f(3) = 3 + 2 = 5.$$

Likewise,

$$f(8) = 8 + 2 = 10,$$

$$f(-4) = -4 + 2 = -2.$$

Output numbers such as $f(-4)$ are called **function values** (or functional values). Keep in mind that they are in the range of f.

Pitfall ▼ $f(x)$ does **not** mean f times x. $f(x)$ is the output that corresponds to the input x.

Functional notation is used extensively in calculus.

Quite often, functions are defined by "functional notation." For example, the equation $g(x) = x^3 + x^2$ defines the function g that assigns the output number $x^3 + x^2$ to an input number x:

$$g: x \rightarrow x^3 + x^2.$$

In other words, g adds the cube and the square of an input number. Some function values are

$$g(2) = 2^3 + 2^2 = 12,$$
$$g(-1) = (-1)^3 + (-1)^2 = -1 + 1 = 0,$$
$$g(t) = t^3 + t^2,$$
$$g(x + 1) = (x + 1)^3 + (x + 1)^2.$$

The idea of *replacement* is very important in determining function values.

Note that $g(x + 1)$ was found by replacing each x in $x^3 + x^2$ by the input $x + 1$.

When we refer to the function g defined by $g(x) = x^3 + x^2$, we shall feel free to call the equation itself a function. Thus, we speak of "the function $g(x) = x^3 + x^2$," and, similarly, "the function $y = x + 2$."

Let's be specific about the domain of a function. Unless otherwise stated, the domain consists of all real numbers for which the rule of the function makes sense; that is, the rule gives function values that are real numbers.

For example, suppose

$$h(x) = \frac{1}{x - 6}.$$

Here any real number can be used for x except 6, because the denominator is 0 when x is 6. So the domain of h is understood to be all real numbers except 6.

Principles in Practice 1
Finding Domains

The area of a circle depends on the length of the radius of the circle.

a. Write a function $a(r)$ for the area of a circle when the length of the radius is r.

b. What is the domain of this function out of context?

c. What is the domain of this function in the given context?

EXAMPLE 1 Finding Domains

Find the domain of each function.

a. $f(x) = \dfrac{x}{x^2 - x - 2}$.

Solution: We cannot divide by zero, so we must find any values of x that make the denominator 0. These *cannot* be input numbers. Thus, we set the denominator equal to 0 and solve for x:

$$x^2 - x - 2 = 0 \quad \text{(quadratic equation)},$$
$$(x - 2)(x + 1) = 0 \quad \text{(factoring)},$$
$$x = 2, -1.$$

Therefore, the domain of f is all real numbers *except* 2 and -1.

b. $g(t) = \sqrt{2t - 1}$.

Solution: $\sqrt{2t - 1}$ is a real number if $2t - 1$ is greater than or equal to 0. If $2t - 1$ is negative, then $\sqrt{2t - 1}$ is not a real number. (It is an *imaginary number.*) Since function values must be real numbers, we must assume that

$$2t - 1 \geq 0,$$

$$2t \geq 1 \quad \text{(adding 1 to both sides),}$$

$$t \geq \frac{1}{2} \quad \text{(dividing both sides by 2).}$$

Thus, the domain is the interval $[\frac{1}{2}, \infty)$. ■

Principles in Practice 2
Finding Domain and Function Values

The time it takes to go a given distance depends on the speed at which one is traveling.

a. Write a function $t(r)$ for the time it takes if the distance is 300 miles and the speed is r.
b. What is the domain of this function out of context?
c. What is the domain of this function in the given context?
d. Find $t(x)$, $t\left(\frac{x}{2}\right)$, and $t\left(\frac{x}{4}\right)$.
e. What happens to the time if the speed is reduced (divided) by a constant c? Describe this situation using an equation.

EXAMPLE 2 Finding Domain and Function Values

Let $g(x) = 3x^2 - x + 5$. Any real number can be used for x, so the domain of g is all real numbers.

a. *Find* $g(z)$.

Solution: Replacing each x in $g(x) = 3x^2 - x + 5$ by z gives

$$g(z) = 3(z)^2 - z + 5 = 3z^2 - z + 5.$$

b. *Find* $g(r^2)$.

Solution: Replacing each x in $g(x) = 3x^2 - x + 5$ by r^2 gives

$$g(r^2) = 3(r^2)^2 - r^2 + 5 = 3r^4 - r^2 + 5.$$

c. *Find* $g(x + h)$.

Solution:

$$\begin{aligned} g(x + h) &= 3(x + h)^2 - (x + h) + 5 \\ &= 3(x^2 + 2hx + h^2) - x - h + 5 \\ &= 3x^2 + 6hx + 3h^2 - x - h + 5. \end{aligned}$$ ■

Pitfall ▼ Don't be confused by notation. In Example 2(c), we found $g(x + h)$ by replacing each x in $g(x) = 3x^2 - x + 5$ by the input $x + h$. **Don't** write the function and then add h. That is, $g(x + h) \neq g(x) + h$:

$$g(x + h) \neq 3x^2 - x + 5 + h.$$

Also, **don't** use the distributive law on $g(x + h)$. The parentheses do **not** stand for multiplication. That is,

$$g(x + h) \neq g(x) + g(h).$$

EXAMPLE 3 Finding a Difference Quotient

If $f(x) = x^2$, *find* $\dfrac{f(x + h) - f(x)}{h}$.

The difference quotient of a function is an important mathematical concept.

Solution: The expression $\dfrac{f(x + h) - f(x)}{h}$ is referred to as a **difference quotient.** Here the numerator is a difference of function values. We have

$$\begin{aligned} \frac{f(x + h) - f(x)}{h} &= \frac{(x + h)^2 - x^2}{h} \\ &= \frac{x^2 + 2hx + h^2 - x^2}{h} = \frac{2hx + h^2}{h} \\ &= \frac{h(2x + h)}{h} = 2x + h. \end{aligned}$$ ■

In some cases, the domain of a function is restricted for physical or economic reasons. For example, the previous interest function $I = 100(0.06)t$ has $t \geq 0$ because t represents time. Example 4 will give another illustration.

Principles in Practice 3
Demand Function

Suppose the weekly demand function for large pizzas at a local pizza parlor is $p = 26 - \frac{q}{40}$.

a. If the current price is \$18.50 per pizza, how many pizzas are sold each week?
b. If 200 pizzas are sold each week, what is the current price?
c. If the owner wants to double the number of large pizzas sold each week (to 400), what should the price be?

EXAMPLE 4 Demand Function

Suppose that the equation $p = 100/q$ describes the relationship between the price per unit p of a certain product and the number of units q of the product that consumers will buy (that is, demand) per week at the stated price. This equation is called a *demand equation* for the product. If q is an input number, then to each value of q, there is assigned exactly one output number p:

$$q \rightarrow \frac{100}{q} = p.$$

For example,

$$20 \rightarrow \frac{100}{20} = 5;$$

that is, when q is 20, p is 5. Thus, price p is a function of quantity demanded, q. This function is called a **demand function.** The independent variable is q, and p is the dependent variable. Since q cannot be 0 (division by 0 is not defined) and cannot be negative (q represents quantity), the domain is all values of q such that $q > 0$. ■

We have seen that a function is essentially a *correspondence* whereby to each input number in the domain there is assigned exactly one output number in the range. For the correspondence given by $f(x) = x^2$, some sample assignments are shown by the arrows in Fig. 3.1. The next example discusses a functional correspondence that is not given by an algebraic formula.

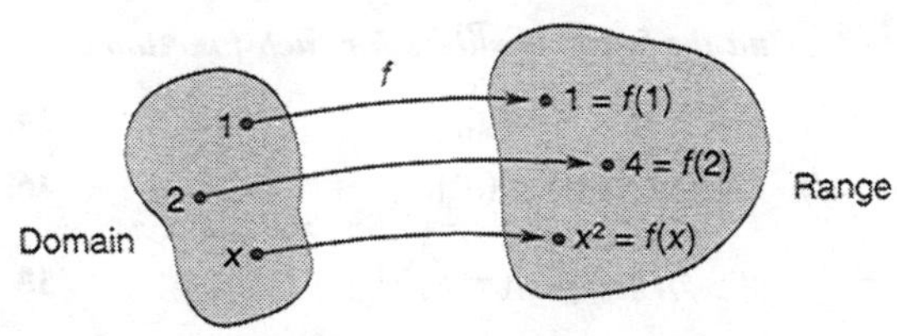

FIGURE 3.1 Functional correspondence for $f(x) = x^2$.

Supply Schedule

p Price per Unit in Dollars	q Quantity Supplied per Week
500	11
600	14
700	17
800	20

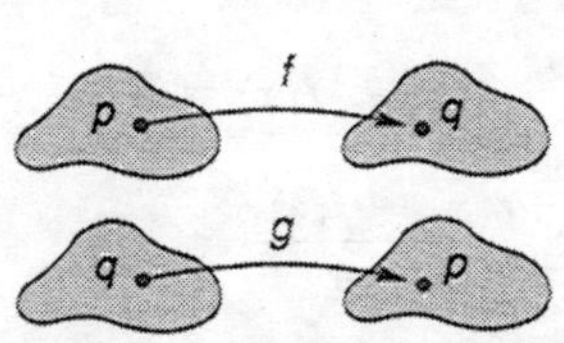

FIGURE 3.2 Supply schedule and supply functions.

EXAMPLE 5 Supply Schedule

The table in Fig. 3.2 is a *supply schedule*. Such a table gives a correspondence between the price p of a certain product and the quantity q the producers will supply per week at that price. For each price, there corresponds exactly one quantity, and vice versa.

If p is the independent variable, then q is a function of p, say, $q = f(p)$, and

$$f(500) = 11, \quad f(600) = 14, \quad f(700) = 17, \quad \text{and} \quad f(800) = 20.$$

Notice that as price per unit increases, the producers are willing to supply more units per week.

On the other hand, if q is the independent variable, then p is a function of q, say $p = g(q)$, and

$$g(11) = 500, \quad g(14) = 600, \quad g(17) = 700, \quad \text{and} \quad g(20) = 800.$$

We speak of f and g as **supply functions.** ∎

TECHNOLOGY

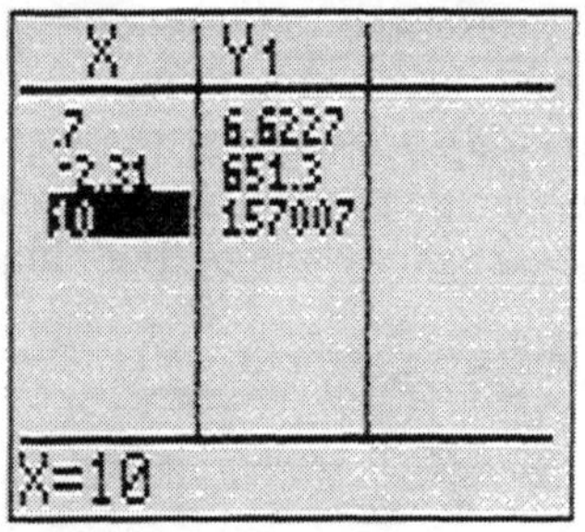

FIGURE 3.3 Table of function values for $f(x) = 17x^4 - 13x^3 + 7$.

Function values are easily computed with a graphics calculator. For example, suppose

$$f(x) = 17x^4 - 13x^3 + 7,$$

and we wish to find $f(0.7)$, $f(-2.31)$, and $f(10)$. With a TI-82, we first enter the function as Y_1:

$$Y_1 = 17X^\wedge 4 - 13X^\wedge 3 + 7.$$

After pressing the TABLE key, we successively enter the x-values .7, −2.31, and 10. The results are shown in Fig. 3.3. We remark that there are other methods of determining function values with the TI-82.

■ Exercise 3.1

In Problems 1–12, give the domain of each function.

1. $f(x) = \frac{8}{x}$.

2. $g(x) = \frac{x}{5}$.

3. $h(x) = \sqrt{x - 3}$.

4. $H(z) = \frac{1}{\sqrt{z}}$.

5. $F(t) = 4t^2 - 6$.

6. $H(x) = \frac{x}{x + 8}$.

7. $f(x) = \frac{3x - 1}{2x + 5}$.

8. $g(x) = \sqrt{4x + 3}$.

9. $G(y) = \frac{4}{y^2 - y}$.

10. $f(x) = \frac{x + 1}{x^2 + 6x + 5}$.

11. $h(s) = \frac{4 - s^2}{2s^2 - 7s - 4}$.

12. $G(r) = \frac{2}{r^2 + 1}$.

In Problems 13–24, find the function values for each function.

13. $f(x) = 2x + 1$; $f(0), f(3), f(-4)$.

14. $H(s) = s^2 - 3$; $H(4), H(\sqrt{2}), H(\frac{2}{3})$.

15. $G(x) = 2 - x^2$; $G(-8), G(u), G(u^2)$.

16. $f(x) = 7x$; $f(s), f(t + 1), f(x + 3)$.

17. $g(u) = u^2 + u$; $g(-2), g(2v), g(-x^2)$.

18. $h(v) = \frac{1}{\sqrt{v}}$; $h(16), h\left(\frac{1}{4}\right), h(1 - x)$.

19. $f(x) = x^2 + 2x + 1$; $f(1), f(-1), f(x + h)$.

20. $H(x) = (x + 4)^2$; $H(0), H(2), H(t - 4)$.

21. $g(x) = \frac{x - 5}{x^2 + 4}$; $g(5), g(3x), g(x + h)$.

22. $H(x) = \sqrt{4 + x}$; $H(-4), H(-3), H(x + 1) - H(x)$.

23. $f(x) = x^{4/3}$; $f(0), f(64), f(\frac{1}{8})$.

24. $g(x) = x^{2/5}$; $g(32), g(-64), g(t^{10})$.

In Problems 25–32, find (a) $f(x + h)$ and (b) $\frac{f(x + h) - f(x)}{h}$; simplify your answers.

25. $f(x) = 4x - 5$.

26. $f(x) = \frac{x}{2}$.

27. $f(x) = x^2 + 2x$.

28. $f(x) = 2x^2 - 3x - 5$.

29. $f(x) = 2 - 4x - 3x^2$.

30. $f(x) = x^3$.

31. $f(x) = \frac{1}{x}$.

32. $f(x) = \frac{x + 1}{x}$.

33. If $f(x) = 9x + 7$, find $\frac{f(2 + h) - f(2)}{h}$.

34. If $f(x) = x^2 - x$, find $\frac{f(x) - f(4)}{x - 4}$.

In Problems 35–38, is y a function of x? Is x a function of y?

35. $y - 3x - 4 = 0$.

36. $x^2 + y = 0$.

37. $y = 7x^2$.

38. $x^2 + y^2 = 1$.

39. The formula for the area of a circle of radius r is $A = \pi r^2$. Is the area a function of the radius?

40. Suppose $f(b) = ab^2 + a^2b$. (a) Find $f(a)$. (b) Find $f(ab)$.

41. **Value of Business** A business with an original capital of \$10,000 has income and expenses each week of \$2000 and \$1600, respectively. If all profits are retained in the business, express the value V of the business at the end of t weeks as a function of t.

42. **Depreciation** If a \$30,000 machine depreciates 2% of its original value each year, find a function f that expresses the machine's value V after t years have elapsed.

43. **Profit Function** If q units of a certain product are sold (q is nonnegative), the profit P is given by the equation $P = 1.25q$. Is P a function of q? What is the dependent variable? the independent variable?

44. **Demand Function** Suppose the yearly demand function for a particular actor to star in a film is $p = \frac{600{,}000}{q}$, where q is the number of films he stars in during the year. If the actor currently charges \$300,000 per film, how many films does he star in each year? If he wants to star in four films per year, what should his price be?

45. **Supply Function** Suppose the weekly supply function for a pound of your house-blend coffee at a local coffee shop is $p = \frac{q}{50}$, where q is the number of pounds of coffee supplied per week. How many pounds of coffee per week will be supplied if the price is \$8.00 a pound? How many pounds of coffee per week will be supplied if the price is \$20.00 a pound? How does the amount supplied change as the price increases?

46. **Hospital Discharges** An insurance company examined the records of a group of individuals hospitalized for a particular illness. It was found that the total proportion discharged at the end of t days of hospitalization is given by

$$f(t) = 1 - \left(\frac{300}{300 + t}\right)^3.$$

Evaluate (a) $f(0)$, (b) $f(100)$, and (c) $f(300)$. (d) At the end of how many days was 99.9% (0.999) of the group discharged?

47. **Psychology** A psychophysical experiment was conducted to analyze human response to electrical shocks.[1] The subjects received a shock of a certain intensity. They were told to assign a magnitude of 10 to this particular shock, called the standard stimulus. Then other shocks (stimuli) of various intensities were given. For each one, the response R was to be a number that indicated the perceived magnitude of the shock relative to that of the standard stimulus. It was found that R was a function of the intensity I of the shock (I in microamperes) and was estimated by

$$R = f(I) = \frac{I^{4/3}}{2500}, \qquad 500 \le I \le 3500.$$

Evaluate (a) $f(1000)$ and (b) $f(2000)$. (c) Suppose that I_0 and $2I_0$ are in the domain of f. Express $f(2I_0)$ in terms of $f(I_0)$. What effect does the doubling of intensity have on response?

48. **Psychology** In a paired-associate learning experiment,[2] the probability of a correct response as a function of the number n of trials has the form

$$P(n) = 1 - \frac{1}{2}(1 - c)^{n-1}, \qquad n \ge 1,$$

where the estimated value of c is 0.344. Find $P(1)$ and $P(2)$ by using this value of c.

49. **Demand Schedule** The following table is called a *demand schedule*. It gives a correspondence between the price p of a product and the quantity q that consumers will demand (that is, purchase) at that price. (a) If $p = f(q)$, list the numbers in the domain of f. Find $f(2900)$ and $f(3000)$, (b) If $q = g(p)$, list the numbers in the domain of g. Find $g(10)$ and $g(17)$.

Price per Unit, p	Quantity Demanded per Week, q
\$10	3000
12	2900
17	2300
20	2000

In Problems 50–53, use your calculator to find the indicated values for the given function. Round answers to two decimal places.

50. $f(x) = 2.03x^3 - 5.27x^2 - 13.71$; (a) $f(1.73)$, (b) $f(-5.78)$, (c) $f(\sqrt{2})$.

51. $f(x) = \frac{14.7x^2 - 3.95x - 15.76}{24.3 - x^3}$; (a) $f(4)$, (b) $f(-17/4)$, (c) $f(\pi)$.

52. $f(x) = (20 - 3x)(2.25x^2 - 17.1x - 13)^4$; (a) $f(0.1)$, (b) $f(-0.01)$, (c) $f(6.6)$.

53. $f(x) = \sqrt{\frac{\sqrt{2}x^2 + 47.62(x + 1)}{9.07}}$; (a) $f(15.93)$, (b) $f(-146)$, (c) $f(0)$.

[1] Adapted from H. Babkoff, "Magnitude Estimation of Short Electrocutaneous Pulses," *Psychological Research*, 39, no. 1 (1976), 39–49.

[2] D. Laming, *Mathematical Psychology* (New York: Academic Press, 1983).

OBJECTIVE

To introduce constant functions, polynomial functions, rational functions, compound functions, the absolute-value function, and factorial notation.

3.2 Special Functions

In this section, we shall look at functions having special forms and representations. We begin with perhaps the simplest type of function there is: a *constant function.*

EXAMPLE 1 Constant Function

Let $h(x) = 2$. The domain of h is all real numbers. All function values are 2. For example,

$$h(10) = 2, \qquad h(-387) = 2, \qquad h(x + 3) = 2.$$

We call h a *constant function* because all the function values are the same. More generally, we have this definition:

A function of the form $h(x) = c$, where c is a *constant,* is called a **constant function.**

■

Principles in Practice 1
Constant Function

Suppose the monthly health insurance premiums for an individual are \$125.00.
a. Write the monthly health insurance premiums as a function of the number of visits the individual makes to the doctor.
b. How do the health insurance premiums change as the number of visits to the doctor increases?
c. What kind of function is this?

A constant function belongs to a broader class of functions, called *polynomial functions.* In general, a function of the form

$$f(x) = c_n x^n + c_{n-1}x^{n-1} + \cdots + c_1 x + c_0,$$

where n is a nonnegative integer and $c_n, c_{n-1}, \ldots, c_0$ are constants with $c_n \neq 0$, is called a **polynomial function** (in x). The number n is called the **degree** of the polynomial, and c_n is the **leading coefficient.** Thus,

Each term in a polynomial function is either a constant or a constant times a positive integral power of x.

$$f(x) = 3x^2 - 8x + 9$$

is a polynomial function of degree 2 with leading coefficient 3. Likewise, $g(x) = 4 - 2x$ has degree 1 and leading coefficient -2. Polynomial functions of degree 1 or 2 are called **linear** or **quadratic functions,** respectively. For example, $g(x) = 4 - 2x$ is linear and $f(x) = 3x^2 - 8x + 9$ is quadratic. Note that a nonzero constant function, such as $f(x) = 5$ [which can be written as $f(x) = 5x^0$], is a polynomial function of degree 0. The constant function $f(x) = 0$ is also considered a polynomial function, but has no degree assigned to it. The domain of any polynomial function is all real numbers.

EXAMPLE 2 Polynomial Functions

a. $f(x) = x^3 - 6x^2 + 7$ is a polynomial (function) of degree 3 with leading coefficient 1.

b. $g(x) = \dfrac{2x}{3}$ is a linear function with leading coefficient $\dfrac{2}{3}$.

c. $f(x) = \dfrac{2}{x^3}$ is *not* a polynomial function. Because $f(x) = 2x^{-3}$ and the exponent for x is not a nonnegative integer, this function does not have the proper form for a polynomial. Similarly, $g(x) = \sqrt{x}$ is not a polynomial, because $g(x) = x^{1/2}$. ■

Principles in Practice 2
Polynomial Functions

The function $d(t) = 3t^2$ represents the distance in meters a car will go in t seconds when it is has a constant acceleration of 6 meters per second.
a. What kind of function is this?
b. What is its degree?
c. What is its leading coefficient?

Another type of function is a *rational function,* which involves polynomials.

A function that is a quotient of polynomial functions is called a **rational function.**

EXAMPLE 3 Rational Functions

a. $f(x) = \dfrac{x^2 - 6x}{x + 5}$ is a rational function, since the numerator and denominator are each polynomials. Note that this rational function is not defined for $x = -5$.

b. $g(x) = 2x + 3$ is a rational function, since $2x + 3 = \dfrac{2x + 3}{1}$. In fact, every polynomial function is also a rational function. ■

Every polynomial function is a rational function.

Sometimes more than one expression is needed to define a function, as Example 4 shows.

Principles in Practice 3
Compound Function

To reduce inventory, a department store charges three rates. If you buy 0–5 pairs of socks, the price is \$3.50 per pair. If you buy 6–10 pairs of socks, the price is \$3.00 per pair. If you buy more than 10 pairs, the price is \$2.75 per pair. Write a compound function to represent the cost of buying n pairs of socks

EXAMPLE 4 Compound Function

Let

$$F(s) = \begin{cases} 1, & \text{if } -1 \le s < 1, \\ 0, & \text{if } 1 \le s \le 2, \\ s - 3 & \text{if } 2 < s \le 8. \end{cases}$$

This is called a **compound function** because the rule for specifying it is given by more than one expression. Here s is the independent variable, and the domain of F is all s such that $-1 \le s \le 8$. The value of s determines which expression to use.

Find $F(0)$: Since $-1 \le 0 < 1$, we have $F(0) = 1$.

Find $F(2)$: Since $1 \le 2 \le 2$, we have $F(2) = 0$.

Find $F(7)$: Since $2 < 7 \le 8$, we substitute 7 for s in $s - 3$.

$$F(7) = 7 - 3 = 4.$$

■

TECHNOLOGY

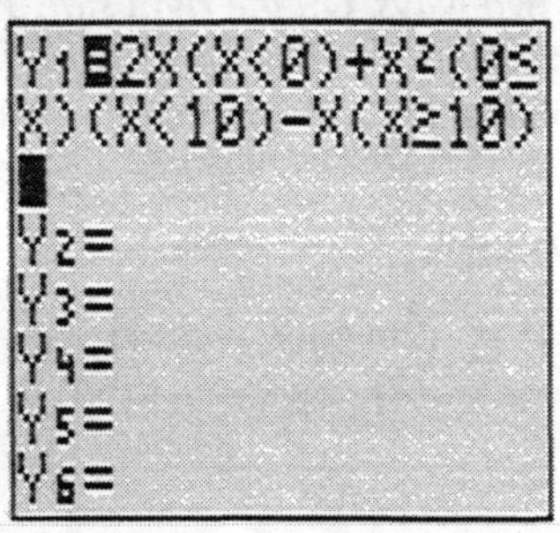

FIGURE 3.4 Entering a compound function.

To illustrate how to enter a compound function with a TI-82, Fig. 3.4 shows a key sequence for the function

$$f(x) = \begin{cases} 2x, & \text{if } x < 0, \\ x^2, & \text{if } 0 \le x < 10, \\ -x, & \text{if } x \ge 10. \end{cases}$$

EXAMPLE 5 Absolute-Value Function

The function $f(x) = |x|$ is called the *absolute-value function.* Recall that the **absolute value,** or **magnitude,** of a real number x is denoted $|x|$ and is defined by

The absolute-value function can be considered a compound function.

$$|x| = \begin{cases} x, & \text{if } x \geq 0, \\ -x, & \text{if } x < 0. \end{cases}$$

Thus, the domain of f is all real numbers. Some function values are

$$f(16) = |16| = 16,$$
$$f(-\tfrac{4}{3}) = |-\tfrac{4}{3}| = -(-\tfrac{4}{3}) = \tfrac{4}{3},$$
$$f(0) = |0| = 0.$$

In our next examples, we make use of *factorial notation*.

The symbol $r!$, with r a positive integer, is read "**r factorial.**" It represents the product of the first r positive integers:

$$r! = 1 \cdot 2 \cdot 3 \cdots r.$$

We define 0! to be 1.

Principles in Practice 4
Factorials

Seven different books are to be placed on a shelf. How many ways can they be arranged? Represent the question as a factorial problem and give the solution.

EXAMPLE 6 Factorials

a. $5! = 1 \cdot 2 \cdot 3 \cdot 4 \cdot 5 = 120.$

b. $3!(6 - 5)! = 3! \cdot 1! = (3 \cdot 2 \cdot 1)(1) = (6)(1) = 6.$

c. $\dfrac{4!}{0!} = \dfrac{1 \cdot 2 \cdot 3 \cdot 4}{1} = \dfrac{24}{1} = 24.$

EXAMPLE 7 Genetics

Suppose two black guinea pigs are bred and produce exactly five offspring. Under certain conditions, it can be shown that the probability P that exactly r of the offspring will be brown and the others black is a function of r, say, $P = P(r)$, where

Factorials occur frequently in probability theory.

$$P(r) = \frac{5!\left(\frac{1}{4}\right)^r\left(\frac{3}{4}\right)^{5-r}}{r!(5-r)!}, \qquad r = 0, 1, 2, \ldots, 5.$$

The letter P in $P = P(r)$ is used in two ways. On the right side, P represents the function rule. On the left side, P represents the dependent variable. The domain of P is all integers from 0 to 5, inclusive. Find the probability that exactly three guinea pigs will be brown.

Solution: We want to find $P(3)$. We have

$$P(3) = \frac{5!\left(\frac{1}{4}\right)^3\left(\frac{3}{4}\right)^2}{3!2!} = \frac{120\left(\frac{1}{64}\right)\left(\frac{9}{16}\right)}{6(2)} = \frac{45}{512}.$$

Exercise 3.2

In Problems 1–4, determine whether the given function is a polynomial function.

1. $f(x) = x^2 - x^4 + 4.$ **2.** $f(x) = \dfrac{x^2 + 7}{3}.$ **3.** $g(x) = \dfrac{3}{x^2 + 7}.$ **4.** $g(x) = 3^{-2}x^2.$

In Problems 5–8, determine whether the given function is a rational function.

5. $f(x) = \dfrac{x^2 + x}{x^3 + 4}.$ **6.** $f(x) = \dfrac{3}{2x + 1}.$ **7.** $g(x) = 8x.$ **8.** $g(x) = 4x^{-4}.$

In Problems 9–12, find the domain of each function.

9. $H(z) = 16.$

10. $f(t) = \pi.$

11. $f(x) = \begin{cases} 5x, & \text{if } x > 1, \\ 4, & \text{if } x \le 1. \end{cases}$

12. $f(x) = \begin{cases} 4, & \text{if } x = 3, \\ x^2, & \text{if } 1 \le x < 3. \end{cases}$

In Problems 13–16, state (a) the degree and (b) the leading coefficient of the given polynomial function.

13. $F(x) = 7x^3 - 2x^2 + 6.$

14. $f(x) = x.$

15. $f(x) = 2 - 3x^4 + 2x.$

16. $f(x) = 9.$

In Problems 17–22, find the function values for each function.

17. $f(x) = 8;\ f(2),\ f(t + 8),\ f(-\sqrt{17}).$

18. $g(x) = |x - 3|;\ g(10),\ g(3),\ g(-3).$

19. $F(t) = \begin{cases} 1, & \text{if } t > 0 \\ 0, & \text{if } t = 0; \\ -1, & \text{if } t < 0 \end{cases}$

$F(10),\ F(-\sqrt{3}),\ F(0),\ F(-\frac{18}{5}).$

20. $f(x) = \begin{cases} 4, & \text{if } x \ge 0 \\ 3, & \text{if } x < 0 \end{cases};$

$f(3),\ f(-4),\ f(0).$

21. $G(x) = \begin{cases} x, & \text{if } x \ge 3 \\ 2 - x, & \text{if } x < 3 \end{cases};$

$G(8),\ G(3),\ G(-1),\ G(1).$

22. $h(r) = \begin{cases} 3r - 1, & \text{if } r > 2 \\ r^2 - 4r + 7, & \text{if } r < 2 \end{cases};$

$h(3),\ h(-3),\ h(2).$

In Problems 23–28, determine the value of each expression.

23. $6!.$

24. $0!.$

25. $(4 - 2)!.$

26. $5! \cdot 3!.$

27. $\dfrac{5!}{4!}.$

28. $\dfrac{8!}{5!(8-5)!}.$

29. Train Trip A daily round-trip train ticket to the city costs \$3.50. Write the cost of a daily round-trip ticket as a function of a passenger's income. What kind of function is this?

30. Geometry A rectangular prism has length three more than its width and height one less than twice the width. Write the volume of the rectangular prism as a function of the width. What kind of function is this?

31. Cost Function In manufacturing a component for a machine, the initial cost of a die is \$850 and all other additional costs are \$3 per unit produced. (a) Express the total cost C (in dollars) as a linear function of the number q of units produced. (b) How many units are produced if the total cost is \$1600?

32. Investment If a principal of P dollars is invested at a simple annual interest rate of r for t years, express the total accumulated amount of the principal and interest as a function of t. Is your result a linear function of t?

33. Sales To encourage large group sales, a theater charges two rates. If your group is less than 10, each ticket costs \$7.50. If your group is 10 or more, each ticket costs \$7.00. Write a compound function to represent the cost of buying n tickets.

34. Factorials At an amusement park, a group of friends wants to ride the log ride in every possible order. How many rides will a group of three have to take? How about a group of four? Five?

35. Genetics Under certain conditions, if two brown-eyed parents have exactly three children, the probability that there will be exactly r blue-eyed children is given by the function $P = P(r)$, where

$$P(r) = \frac{3!(\frac{1}{4})^r(\frac{3}{4})^{3-r}}{r!(3-r)!}, \qquad r = 0, 1, 2, 3.$$

Find the probability that exactly two of the children will be blue-eyed.

36. Genetics In Example 7, find the probability that all five offspring will be brown.

37. Bacteria Growth Bacteria are growing in a culture. The time t (in hours) for the bacteria to double in number (the generation time) is a function of the temperature T (in °C) of the culture. If this function is given by[3]

$$t = f(T) = \begin{cases} \dfrac{1}{24}T + \dfrac{11}{4}, & \text{if } 30 \le T \le 36, \\ \dfrac{4}{3}T - \dfrac{175}{4}, & \text{if } 36 < T \le 39, \end{cases}$$

(a) determine the domain of f, and (b) find $f(30)$, $f(36)$, and $f(39)$.)

[3] Adapted from F. K. E. Imrie and A. J. Vlitos, "Production of Fungal Protein from Carob," in *Single-Cell Protein II*, ed. S. R. Tannenbaum and D. I. C. Wang (Cambridge, MA. MIT Press, 1975).

In Problems 38–41, use your calculator to find the indicated function values for the given function. Round answers to two decimal places.

38. $f(x) = \begin{cases} 0.08x^5 - 47.98, & \text{if } x \geq 7.98 \\ 0.67x^6 - 37.41, & \text{if } x < 7.98 \end{cases}$;
(a) $f(7.98)$, (b) $f(2.26)$, (c) $f(9)$.

39. $f(x) = \begin{cases} 47.1x^5 + 30.4, & \text{if } x > 0 \\ 9.4x^3 - x, & \text{if } x \leq 0 \end{cases}$
(a) $f(5.5)$, (b) $f(-3.6)$, (c) $f(6/7)$.

40. $f(x) = \begin{cases} 4.07x - 2.3 & \text{if } x < -8 \\ 19.12, & \text{if } -8 \leq x < -2; \\ x^2 - 4x^{-2}, & \text{if } x \geq -2 \end{cases}$
(a) $f(-6.3)$, (b) $f(-13.2)$, (c) $f(13.2)$.

41. $f(x) = \begin{cases} x/(x+3), & \text{if } x < -5 \\ x(x-4)^2, & \text{if } -5 \leq x < 0; \\ \sqrt{2.1x + 3}, & \text{if } x \geq 0 \end{cases}$
(a) $f(-\sqrt{30})$, (b) $f(46)$, (c) $f(-2/3)$.

OBJECTIVE

To combine functions by means of addition, subtraction, multiplication, division, and composition.

3.3 Combinations of Functions

There are different ways of combining two functions to create a new function. Suppose f and g are the functions given by

$$f(x) = x^2 \qquad \text{and} \qquad g(x) = 3x.$$

Adding $f(x)$ and $g(x)$ gives

$$f(x) + g(x) = x^2 + 3x.$$

This operation defines a new function called the *sum* of f and g, denoted $f + g$. Its function value at x is $f(x) + g(x)$. That is,

$$(f + g)(x) = f(x) + g(x) = x^2 + 3x.$$

For example,

$$(f + g)(2) = 2^2 + 3(2) = 10.$$

In general, for any functions f and g, we define the **sum** $f + g$, the **difference** $f - g$, the **product** fg, and the **quotient** $\frac{f}{g}$ as follows:[4]

$$(f + g)(x) = f(x) + g(x),$$
$$(f - g)(x) = f(x) - g(x),$$
$$(fg)(x) = f(x) \cdot g(x),$$
$$\frac{f}{g}(x) = \frac{f(x)}{g(x)}.$$

Thus, for $f(x) = x^2$ and $g(x) = 3x$, we have

$$(f + g)(x) = f(x) + g(x) = x^2 + 3x,$$
$$(f - g)(x) = f(x) - g(x) = x^2 - 3x,$$
$$(fg)(x) = f(x) \cdot g(x) = x^2(3x) = 3x^3,$$
$$\frac{f}{g}(x) = \frac{f(x)}{g(x)} = \frac{x^2}{3x} = \frac{x}{3}, \qquad \text{for } x \neq 0.$$

[4]In each of the four combinations, we assume that x is in the domains of both f and g. In the quotient, we also do not allow any value of x for which $g(x)$ is 0.

EXAMPLE 1 Combining Functions

If $f(x) = 3x - 1$ *and* $g(x) = x^2 + 3x$, *find*

a. $(f + g)(x)$, **b.** $(f - g)(x)$,

c. $(fg)(x)$, **d.** $\frac{f}{g}(x)$.

Solution:

a. $(f + g)(x) = f(x) + g(x) = (3x - 1) + (x^2 + 3x) = x^2 + 6x - 1$.

b. $(f - g)(x) = f(x) - g(x) = (3x - 1) - (x^2 + 3x) = -1 - x^2$.

c. $(fg)(x) = f(x)g(x) = (3x - 1)(x^2 + 3x) = 3x^3 + 8x^2 - 3x$.

d. $\frac{f}{g}(x) = \frac{f(x)}{g(x)} = \frac{3x - 1}{x^2 + 3x}$. ■

Composition

We may also combine two functions by first applying one function to a number and then applying the other function to the result. For example, suppose $g(x) = 3x$, $f(x) = x^2$, and $x = 2$. Then $g(2) = 3(2) = 6$. Thus, g sends the input 2 into the output 6:

$$2 \xrightarrow{g} 6.$$

Next, we let the output 6 become input to f:

$$f(6) = 6^2 = 36.$$

So f sends 6 into 36:

$$6 \xrightarrow{f} 36.$$

By first applying g and then f, we send 2 into 36:

$$2 \xrightarrow{g} 6 \xrightarrow{f} 36.$$

To be more general, let's replace the 2 by x, where x is in the domain of g. (See Fig. 3.5). Applying g to x, we get the number $g(x)$, which we shall assume is in the domain of f. By applying f to $g(x)$, we get $f(g(x))$, read "f of g of x," which is in the range of f. The operation of applying g and then applying f to the result defines a so-called composite function denoted $f \circ g$. This function assigns the output number $f(g(x))$ to the input number x. (See the bottom arrow in Fig. 3.5.) Thus, $(f \circ g)(x) = f(g(x))$. We can think of $f(g(x))$ as a function of a function.

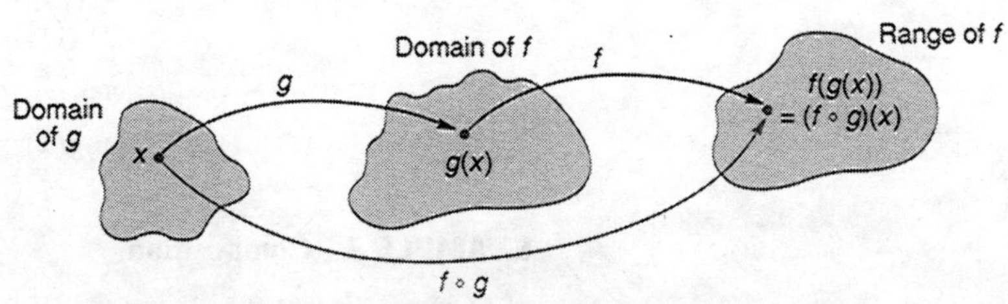

FIGURE 3.5 Composition of f with g.

DEFINITION

If f and g are functions, the ***composition of f with g*** *is the function $f \circ g$ defined by*

$$(f \circ g)(x) = f(g(x)),$$

where the domain of $f \circ g$ is the set of all x in the domain of g such that $g(x)$ is in the domain of f.

For $f(x) = x^2$ and $g(x) = 3x$, we can get a simple form for $f \circ g$:

$$(f \circ g)(x) = f(g(x)) = f(3x) = (3x)^2 = 9x^2.$$

For example, $(f \circ g)(2) = 9(2)^2 = 36$, as we saw before.

Principles in Practice 1
Composition

A CD costs x dollars wholesale. The price the store pays is given by the function $s(x) = x + 3$. The price the customer pays is $c(x) = 2x$, where x is the price the store pays. Write a composite function to find the customer's price as a function of the wholesale price.

EXAMPLE 2 Composition

Let $f(x) = \sqrt{x}$ and $g(x) = x + 1$. Find

a. $(f \circ g)(x)$, **b.** $(g \circ f)(x)$.

Solution:

a. $(f \circ g)(x)$ is $f(g(x))$. Now g adds 1 to x, and f takes the square root of the result. Thus,

$$(f \circ g)(x) = f(g(x)) = f(x + 1) = \sqrt{x + 1}.$$

The domain of g is all real numbers x, and the domain of f is all nonnegative reals. Hence, the domain of the composition is all x for which $g(x) = x + 1$ is nonnegative. That is, the domain is all $x \geq -1$, or equivalently, the interval $[-1, \infty)$.

b. $(g \circ f)(x)$ is $g(f(x))$. Now f takes the square root of x, and g adds 1 to the result. Thus, g adds 1 to $\sqrt{x}$, and we have

$$(g \circ f)(x) = g(f(x)) = g(\sqrt{x}) = \sqrt{x} + 1.$$

The domain of f is all $x \geq 0$, and the domain of g is all reals. Hence, the domain of the composition is all $x \geq 0$ for which $f(x) = \sqrt{x}$ is real, namely, all $x \geq 0$. ■

Pitfall ▼ Generally, $f \circ g \neq g \circ f$. In Example 2,

$$(f \circ g)(x) = \sqrt{x + 1},$$

but we have

$$(g \circ f)(x) = \sqrt{x} + 1.$$

Also, do not confuse $f(g(x))$ with $(fg)(x)$, which is the product $f(x)g(x)$. Here

$$f(g(x)) = \sqrt{x + 1},$$

but

$$f(x)g(x) = \sqrt{x}(x + 1).$$

EXAMPLE 3 Composition

If $F(p) = p^2 + 4p - 3$ and $G(p) = 2p + 1$, find

a. $F(G(p))$, **b.** $G(F(1))$.

Solution:

a. $F(G(p)) = F(2p + 1) = (2p + 1)^2 + 4(2p + 1) - 3$

$= 4p^2 + 12p + 2.$

b. $G(F(1)) = G(1^2 + 4 \cdot 1 - 3) = G(2) = 2 \cdot 2 + 1 = 5.$ ■

In calculus, it is necessary at times to think of a particular function as a composition of two simpler functions, as the next example shows.

Principles in Practice 2

Expressing a Function as a Composition

Suppose the area of a square garden is $g(x) = (x + 3)^2$. Express g as a composition of two functions, and explain what each function represents.

EXAMPLE 4 Expressing a Function as a Composition

Express $h(x) = (2x - 1)^3$ *as a composition.*

Solution:
We note that $h(x)$ is obtained by finding $2x - 1$ and cubing the result. Suppose we let $g(x) = 2x - 1$ and $f(x) = x^3$. Then

$$h(x) = (2x - 1)^3 = [g(x)]^3 = f(g(x)) = (f \circ g)(x),$$

which gives h as a composition of two functions. ■

TECHNOLOGY

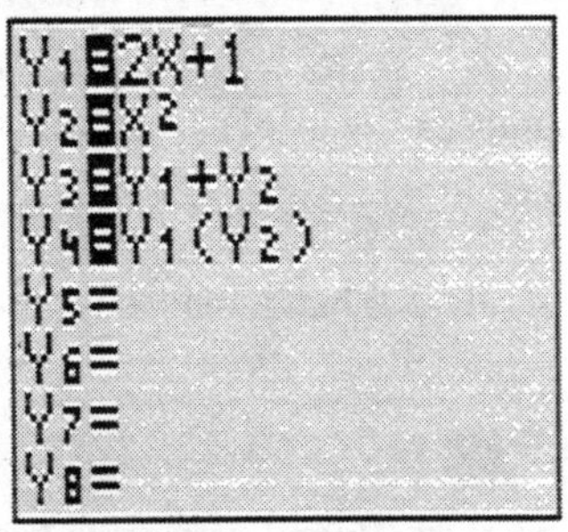

FIGURE 3.6 Y_3 and Y_4 are combinations of Y_1 and Y_2.

Two functions can be combined by using a graphics calculator. Consider the functions

$$f(x) = 2x + 1 \quad \text{and} \quad g(x) = x^2,$$

which we enter as Y_1 and Y_2, as shown in Fig. 3.6. The sum of f and g is given by $Y_3 = Y_1 + Y_2$ and the composition $f \circ g$ by $Y_4 = Y_1(Y_2)$. For example, $f(g(3))$ is obtained by evaluating Y_4 at 3.

■ Exercise 3.3

1. If $f(x) = x + 3$ and $g(x) = x + 5$, find the following.

a. $(f + g)(x)$. **b.** $(f + g)(0)$. **c.** $(f - g)(x)$.

d. $(fg)(x)$. **e.** $(fg)(-2)$. **f.** $\frac{f}{g}(x)$.

g. $(f \circ g)(x)$. **h.** $(f \circ g)(3)$. **i.** $(g \circ f)(x)$.

2. If $f(x) = 2x$ and $g(x) = 4 + x$, find the following.

a. $(f + g)(x)$. **b.** $(f - g)(x)$. **c.** $(f - g)(4)$.

d. $(fg)(x)$. **e.** $\frac{f}{g}(x)$. **f.** $\frac{f}{g}(2)$.

g. $(f \circ g)(x)$. **h.** $(g \circ f)(x)$. **i.** $(g \circ f)(2)$.

3. If $f(x) = x^2$ and $g(x) = x^2 + x$, find the following.

a. $(f + g)(x)$. **b.** $(f - g)(x)$. **c.** $(f - g)(-\frac{1}{2})$.

d. $(fg)(x)$. **e.** $\frac{f}{g}(x)$. **f.** $\frac{f}{g}(-\frac{1}{2})$.

g. $(f \circ g)(x)$. **h.** $(g \circ f)(x)$. **i.** $(g \circ f)(-3)$.

4. If $f(x) = x^2 - 1$ and $g(x) = 4$, find the following.

a. $(f + g)(x)$. **b.** $(f + g)(\frac{1}{2})$. **c.** $(f - g)(x)$.

d. $(fg)(x)$. **e.** $(fg)(4)$. **f.** $\frac{f}{g}(x)$.

g. $(f \circ g)(x)$. **h.** $(f \circ g)(100)$. **i.** $(g \circ f)(x)$.

5. If $f(x) = 3x^2 + 6$ and $g(x) = 4 - 2x$, find $f(g(2))$ and $g(f(2))$.

6. If $f(p) = \frac{4}{p}$ and $g(p) = \frac{p - 2}{3}$, find both $(f \circ g)(p)$ and $(g \circ f)(p)$.

7. If $F(t) = t^2 + 3t + 1$ and $G(t) = \frac{2}{t-1}$, find $(F \circ G)(t)$ and $(G \circ F)(t)$.

8. If $F(s) = \sqrt{s}$ and $G(t) = 3t^2 + 4t + 2$, find $(F \circ G)(t)$ and $(G \circ F)(t)$.

9. If $f(w) = \frac{1}{w^2+1}$ and $g(v) = \sqrt{v+2}$, find $(f \circ g)(v)$ and $(g \circ f)(w)$.

10. If $f(x) = x^2 + 3$, find $(f \circ f)(x)$.

In Problems 11–16, find functions f and g such that h(x) = f(g(x)).

11. $h(x) = (4x - 3)^5$.

12. $h(x) = \sqrt{x^2 - 2}$.

13. $h(x) = \frac{1}{x^2 - 2}$.

14. $h(x) = (3x^3 - 2x)^3 - (3x^3 - 2x)^2 + 7$.

15. $h(x) = \sqrt[5]{\frac{x+1}{3}}$.

16. $h(x) = \frac{x+1}{(x+1)^2 + 2}$.

17. **Profit** A coffeehouse sells a pound of coffee for \$9.75. Expenses are \$4500 each month, plus \$4.25 for each pound of coffee sold.
(a) Write a function $r(x)$ for the total monthly revenue as a function of the number of pounds of coffee sold.
(b) Write a function $e(x)$ for the total monthly expenses as a function of the number of pounds of coffee sold.
(c) Write a function $(r - e)(x)$ for the total monthly profit as a function of the number of pounds of coffee sold.

18. **Geometry** Suppose the volume of a cube is $v(x) = (4x - 2)^3$. Express v is a composition of two functions, and explain what each function represents.

19. **Business** A manufacturer determines that the total number of units of output per day, q, is a function of the number of employees, m, where

$$q = f(m) = \frac{(40m - m^2)}{4}.$$

The total revenue r that is received for selling q units is given by the function g, where $r = g(q) = 40q$. Find $(g \circ f)(m)$. What does this composition function describe?

20. **Sociology** Studies have been conducted concerning the statistical relations between a person's status, education, and income.[5] Let S denote a numerical value of status based on annual income I. For a certain population, suppose

$$S = f(I) = 0.45(I - 1000)^{0.53}.$$

Furthermore, suppose a person's income I is a function of the number of years of education E, where

$$I = g(E) = 7202 + 0.29E^{3.68}.$$

Find $(f \circ g)(E)$. What does this function describe?

In Problems 21–24, for the given functions f and g, find the indicated function values. Round answers to two decimal places.

21. $f(x) = (4x - 13)^2$, $g(x) = 0.2x^2 - 4x + 3$;
(a) $(f + g)(4.1)$, (b) $(f \circ g)(3)$.

22. $f(x) = \sqrt{\frac{x+2}{x}}$, $g(x) = 13.4x + 7.31$;
(a) $\frac{f}{g}(10)$, (b) $(g \circ f)(-6)$.

23. $f(x) = x^{2/3}$, $g(x) = x^3 - 7$; (a) $(fg)(5)$,
(b) $(g \circ f)(2.25)$.

24. $f(x) = \frac{5}{x+3}$, $g(x) = \frac{2}{x^2}$; (a) $(f - g)(7.3)$,
(b) $(f \circ g)(-4.17)$.

[5] R. K. Leik and B. F. Meeker, *Mathematical Sociology* (Englewood Cliffs, NJ; Prentice-Hall, 1975)

OBJECTIVE

To graph equations and functions in rectangular coordinates, to determine intercepts, to apply the vertical-line test, and to determine the domain and range of a function from a graph.

3.4 Graphs in Rectangular Coordinates

A **rectangular** (or **Cartesian**) **coordinate system** allows us to specify and locate points in a plane. It also provides a geometric way to represent equations in two variables as well as functions.

In a plane, two real-number lines, called *coordinate axes,* are constructed perpendicular to each other so that their origins coincide, as in Fig. 3.7. Their point of intersection is called the *origin* of the coordinate system. For now, we shall call the horizontal line the *x-axis* and the vertical line the *y-axis.* The unit distance on the x-axis need not be the same as on the y-axis.

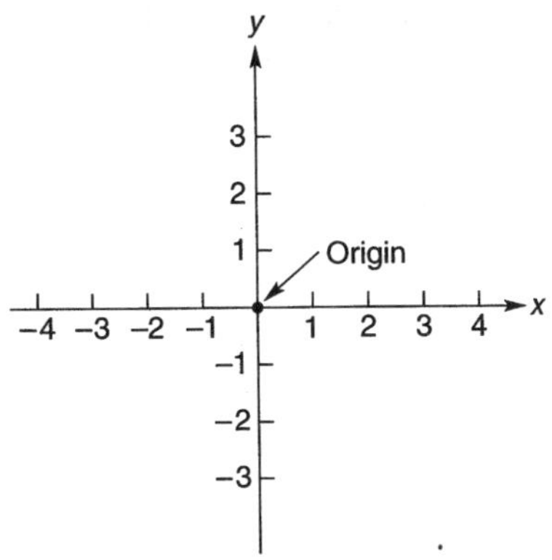

FIGURE 3.7 Coordinate axes.

The plane on which the coordinate axes are placed is called a *rectangular coordinate plane* or, more simply, an *x,y-plane.* Every point in the *x,y*-plane can be labeled to indicate its position. To label point P in Fig. 3.8(a), we draw perpendiculars from P to the x-axis and y-axis. They meet these axes at 4 and 2, respectively. Thus, P determines two numbers, 4 and 2. We say that the **rectangular coordinates** of P are given by the **ordered pair** (4, 2). The word "ordered" is important. In Fig. 3.8(b), the point corresponding to (4, 2) is not the same as that for (2, 4):

$$(4, 2) \neq (2, 4).$$

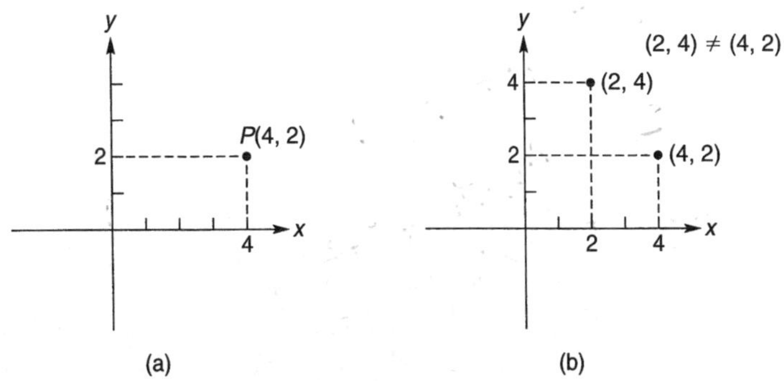

FIGURE 3.8 Rectangular coordinates.

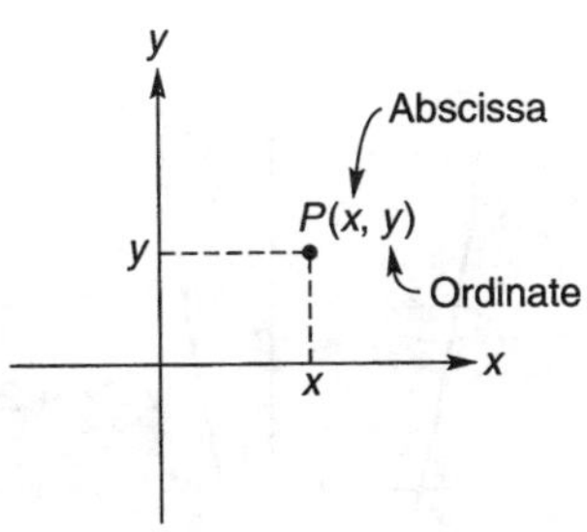

FIGURE 3.9 Coordinates of *P*.

In general, if P is any point, then its rectangular coordinates will be given by an ordered pair of the form (x, y). (See Fig. 3.9.) We call x the *abscissa* or *x-coordinate* of *P*, and y the *ordinate* or *y-coordinate* of *P*.

Accordingly, with each point in a given coordinate plane, we can associate exactly one ordered pair (x, y) of real numbers. Also, it should be clear that with each ordered pair (x, y) of real numbers, we can associate exactly one point in that plane. Since there is a *one-to-one correspondence* between the points in the plane and all ordered pairs of real numbers, we shall refer to a point P with abscissa x and ordinate y simply as the point (x, y), or as $P(x, y)$. Moreover, we shall use the words "point" and "ordered pair" interchangeably.

In Fig. 3.10, the coordinates of various points are indicated. For example, the point (1, −4) is located one unit to the right of the y-axis and four units below the x-axis. The origin is (0, 0). The x-coordinate of every point on the y-axis is 0, and the y-coordinate of every point on the x-axis is 0.

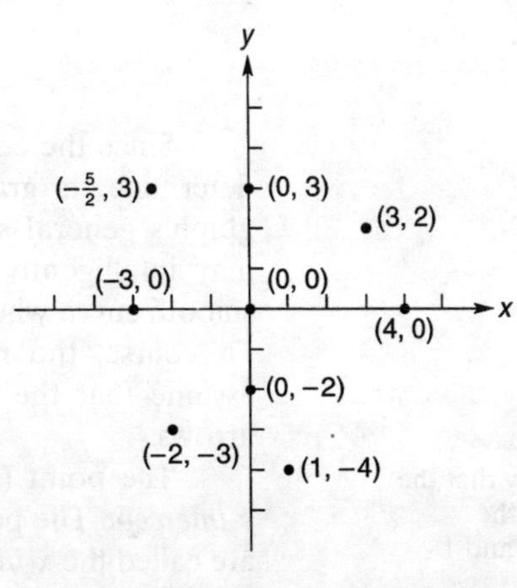

FIGURE 3.10 Coordinates of points.

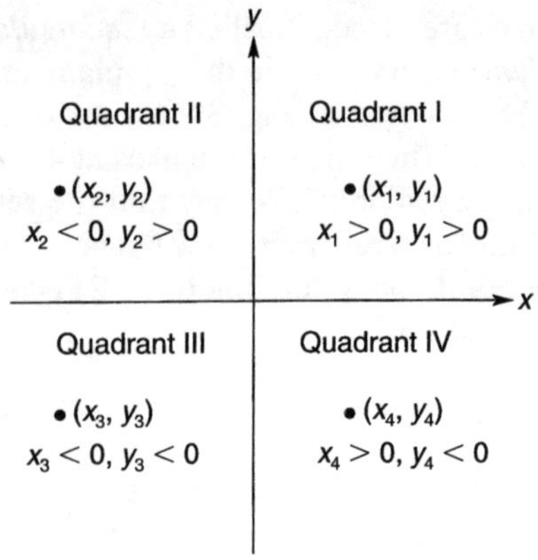

FIGURE 3.11 Quadrants.

The coordinate axes divide the plane into four regions called *quadrants* (Fig. 3.11). For example, quadrant I consists of all points (x_1, y_1) with $x_1 > 0$ and $y_1 > 0$. The points on the axes do not lie in any quadrant.

Using a rectangular coordinate system, we can geometrically represent equations in two variables. For example, let us consider

$$y = x^2 + 2x - 3. \tag{1}$$

A solution of this equation is a value of x and a value of y that make the equation true. For example, if $x = 1$, substituting into Eq. (1) gives

$$y = 1^2 + 2(1) - 3 = 0.$$

Thus, $x = 1$, $y = 0$ is a solution. Similarly,

$$\text{if } x = -2, \quad \text{then} \quad y = (-2)^2 + 2(-2) - 3 = -3,$$

and so $x = -2$, $y = -3$ is also a solution. By choosing other values for x, we can get more solutions. [See Fig. 3.12(a).] It should be clear that there are infinitely many solutions of Eq. (1).

Each solution gives rise to a point (x, y). For example, to $x = 1$ and $y = 0$ corresponds $(1, 0)$. The **graph** of $y = x^2 + 2x - 3$ is the geometric representation of all its solutions. In Fig. 3.12(b), we have plotted the points corresponding to the solutions in the table.

x	y
−4	5
−3	0
−2	−3
−1	−4
0	−3
1	0
2	5

(a)

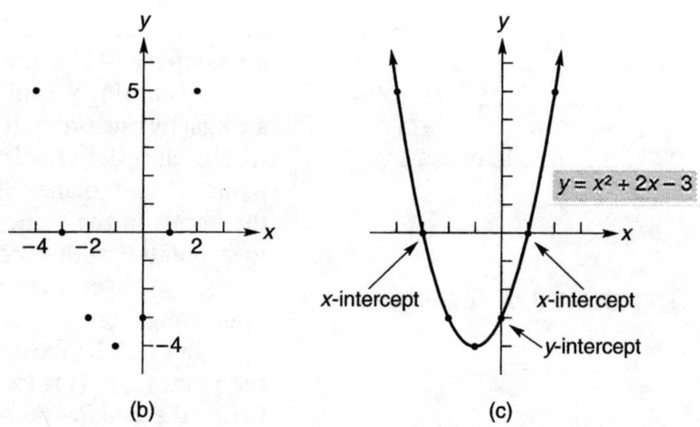

FIGURE 3.12 Graphing $y = x^2 + 2x - 3$.

Since the equation has infinitely many solutions, it seems impossible to determine its graph precisely. However, we are concerned only with the graph's general shape. For this reason, we plot enough points so that we may intelligently guess its proper shape. Then we join these points by a smooth curve wherever conditions permit. This gives the curve in Fig. 3.12(c). Of course, the more points we plot, the better our graph is. Here we assume that the graph extends indefinitely upward, which is indicated by arrows.

Often, we simply say that the y-intercept is 3 and the x-intercepts are -3 and 1.

The point $(0, -3)$ where the curve intersects the y-axis is called the *y-intercept.* The points $(-3, 0)$ and $(1, 0)$ where the curve intersects the x-axis are called the *x-intercepts.* In general, we have the following definition.

DEFINITION

An ***x*-intercept** *of the graph of an equation in x and y is a point where the graph intersects the x-axis. A* ***y*-intercept** *is a point where the graph intersects the y-axis.*

To find the x-intercepts of the graph of an equation in x and y, we first set $y = 0$ and solve the resulting equation for x. To find the y-intercepts, we first set $x = 0$ and solve for y. For example, let us find the x-intercepts for the graph of $y = x^2 + 2x - 3$. Setting $y = 0$ and solving for x gives

$$0 = x^2 + 2x - 3,$$

$$0 = (x + 3)(x - 1),$$

$$x = -3, 1.$$

Thus, the x-intercepts are $(-3, 0)$ and $(1, 0)$, as we saw before. If $x = 0$, then

$$y = 0^2 + 2(0) - 3 = -3.$$

So $(0, -3)$ is the y-intercept. Keep in mind that an x-intercept has its y-coordinate 0, and a y-intercept has its x-coordinate 0. Intercepts are useful because they indicate precisely where the graph intersects the axes.

Principles in Practice 1
Intercepts and Graph

Rachel has saved \$7250 for college expenses. She plans to spend \$600 a month from this account. Write an equation to represent the situation, and identify the intercepts.

EXAMPLE 1 Intercepts and Graph

Find the x- and y-intercepts of the graph of $y = 2x + 3$, and sketch the graph.

Solution: If $y = 0$, then

$$0 = 2x + 3 \quad \text{or} \quad x = -\frac{3}{2}.$$

Thus, the x-intercept is $(-\frac{3}{2}, 0)$. If $x = 0$, then

$$y = 2(0) + 3 = 3.$$

So the y-intercept is $(0, 3)$. Figure 3.13 shows a table of other points on the graph and a sketch of the graph. ■

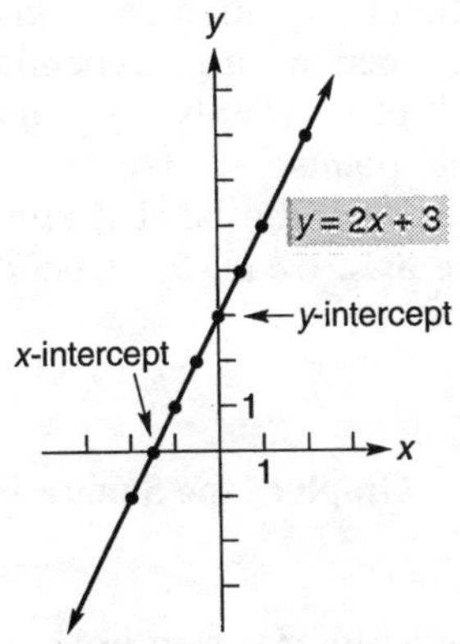

x	0	$-\frac{3}{2}$	$\frac{1}{2}$	$-\frac{1}{2}$	1	−1	2	−2
y	3	0	4	2	5	1	7	−1

FIGURE 3.13 Graph of $y = 2x + 3$.

EXAMPLE 2 Intercepts and Graph

Determine the intercepts, if any, of the graph of $s = \frac{100}{t}$, and sketch the graph.

Principles in Practice 2
Intercepts and Graph

The price of admission to an amusement park is \$24.95. This fee allows the customer to ride all the rides at the park as often as he or she likes. Write an equation that represents the relationship between the number of rides, x, that a customer takes and the cost per ride, y, to that customer. Describe the graph of this equation, and identify the intercepts. Assume $x > 0$.

Solution: For the graph, we shall label the horizontal axis t and the vertical axis s (Fig. 3.14). Because t cannot equal 0 (division by 0 is not defined), there is no s-intercept. Thus, the graph has no point corresponding to $t = 0$. Moreover, there is no t-intercept, because if $s = 0$, then the equation

$$0 = \frac{100}{t}$$

has no solution. Figure 3.14 shows the graph. In general, the graph of $s = k/t$, where k is a nonzero constant, is called a *hyperbola*. ■

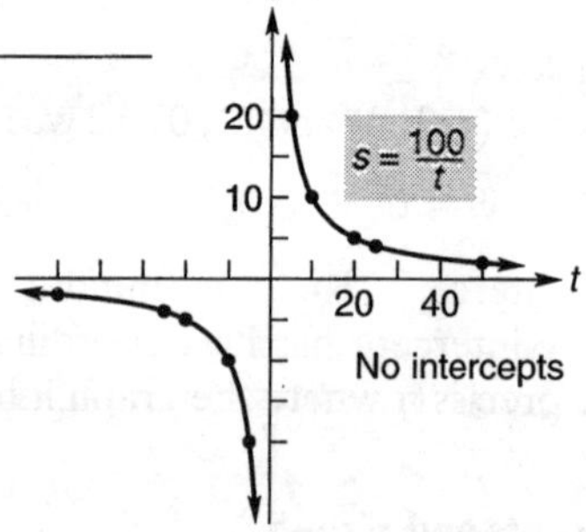

t	5	−5	10	−10	20	−20	25	−25	50	−50
s	20	−20	10	−10	5	−5	4	−4	2	−2

FIGURE 3.14 Graph of $s = \dfrac{100}{t}$.

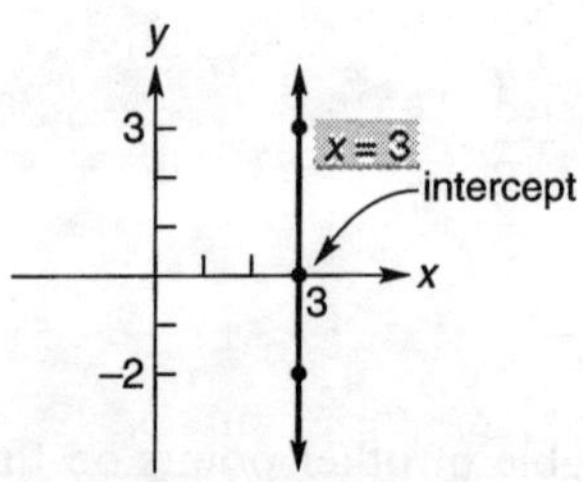

x	3	3	3
y	0	3	−2

FIGURE 3.15 Graph of $x = 3$.

EXAMPLE 3 Intercepts and Graph

Determine the intercepts of the graph of $x = 3$, and sketch the graph.

Solution: We can think of $x = 3$ as an equation in the variables x and y if we write it as $x = 3 + 0y$. Here y can be any value, but x must be 3. Because $x = 3$ when $y = 0$, the x-intercept is $(3, 0)$. There is no y-intercept, because x cannot be 0. (See Fig. 3.15.) The graph is a vertical line. ■

In addition to equations, we can also represent functions in a coordinate plane. If f is a function with independent variable x and dependent variable y, then the graph of f is simply the graph of the equation $y = f(x)$. This graph consists of all points (x, y) or $(x, f(x))$, where x is in the domain of f. The vertical axis can be labeled either y or $f(x)$ and is referred to as the **function-value axis.** *We always label the horizontal axis with the independent variable.*

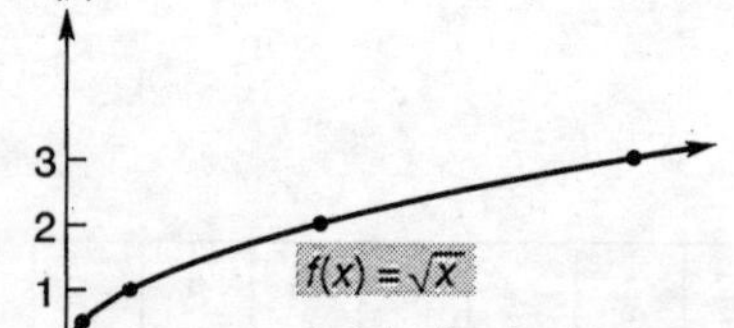

x	0	$\frac{1}{4}$	1	4	9
f(x)	0	$\frac{1}{2}$	1	2	3

FIGURE 3.16 Graph of $f(x) = \sqrt{x}$.

EXAMPLE 4 Graph of the Square-Root Function

Graph $f(x) = \sqrt{x}$.

Solution: The graph is shown in Fig. 3.16. We label the vertical axis as $f(x)$. Recall that $\sqrt{x}$ denotes the *principal* square root of x. Thus, $f(9) = \sqrt{9} = 3$, not ± 3. Also, we cannot choose negative values for x, because we don't want imaginary numbers for $\sqrt{x}$. That is, we must have $x \geq 0$. Let us now consider intercepts. If $f(x) = 0$, then $\sqrt{x} = 0$, or $x = 0$. Also, if $x = 0$, then $f(x) = 0$. Thus, the x-intercept and the vertical-axis intercept are the same, namely, $(0, 0)$. ■

Principles in Practice 3

Graph of the Absolute-Value Function

Brett rented a bike from a rental shop, rode at a constant rate of 12 mph for 2.5 hours along a bike path, and then returned along the same path. Graph the absolute-value function to represent Brett's distance from the rental shop as a function of time over the appropriate domain.

EXAMPLE 5 Graph of the Absolute-Value Function

Graph $p = G(q) = |q|$.

Solution: We use the independent variable q to label the horizontal axis. The function-value axis can be labeled either $G(q)$ or p. (See Fig. 3.17.) Notice that the q- and p-intercepts are the same point, $(0, 0)$. ■

q	0	1	−1	3	−3	5	−5
p	0	1	1	3	3	5	5

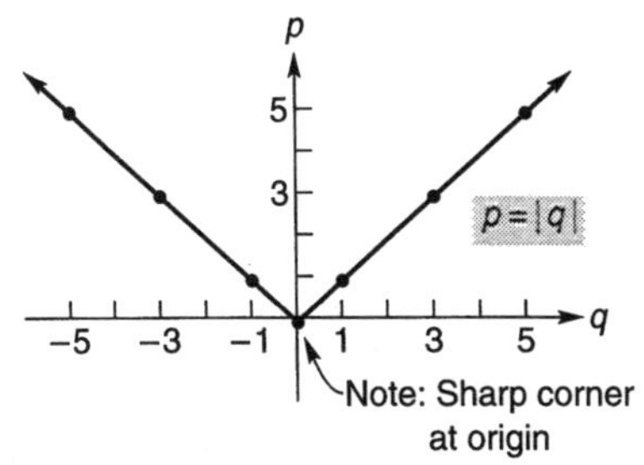

FIGURE 3.17 Graph of $p = |q|$.

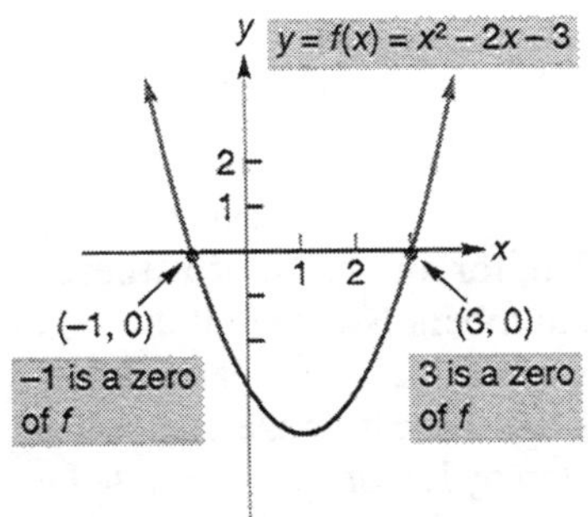

FIGURE 3.18 Zeros of function.

In general, the real solutions to an equation $f(x) = 0$ are the real zeros of f.

The notion of a *zero* is important in the study of functions.

DEFINITION

*A **zero** of a function f is any value of x for which* $f(x) = 0$.

For example, a zero of the function $f(x) = 2x - 6$ is 3 because $f(3) = 2(3) - 6 = 0$. Here we call 3 a *real* zero because it is a real number. We note that zeros of f can be found by setting $f(x) = 0$ and solving for x. Thus, the real zeros of a function are precisely the x-intercepts of its graph, because it is at these points that $f(x) = 0$.

To further illustrate, Fig. 3.18 shows the graph of the function $y = f(x) = x^2 - 2x - 3$. The x-intercepts of the graph are -1 and 3. Hence, -1 and 3 are zeros of f, or equivalently, -1 and 3 are solutions to the equation $x^2 - 2x - 3 = 0$.

TECHNOLOGY

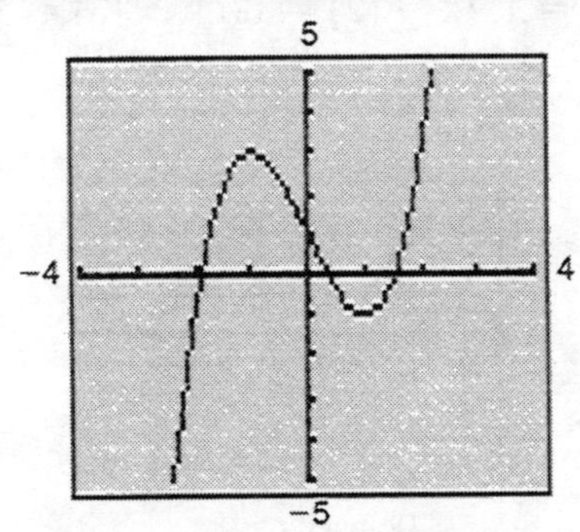

FIGURE 3.19 The roots of $x^3 - 3x + 1 = 0$ are approximately -1.88, 0.35, and 1.53.

To solve the equation $x^3 = 3x - 1$ with a graphics calculator, we first express the equation in the form $f(x) = 0$

$$f(x) = x^3 - 3x + 1 = 0.$$

Next we graph f and then estimate the x-intercepts, either by using zoom and trace or by using the root operation. (See Fig. 3.19.) Note that we defined our window for $-4 \le x \le 4$ and $-5 \le y \le 5$.

Figure 3.20 shows the graph of some function $y = f(x)$. The point $(x, f(x))$ implies that corresponding to the input number x on the horizontal axis is the output number $f(x)$ on the vertical axis, as indicated by the arrow. For example,

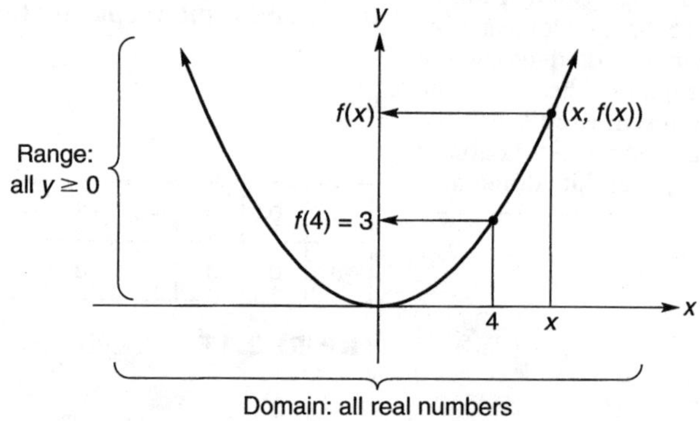

FIGURE 3.20 Domain, range, function values.

corresponding to the input 4 is the output 3, so $f(4) = 3$. From the shape of the graph, it seems reasonable to assume that, for any value of x, there is an output number, so the domain of f is all real numbers. Notice that the set of all y-coordinates of points on the graph is the set of all nonnegative numbers. Thus, the range of f is all $y \geq 0$. This shows that we can make an "educated" guess about the domain and range of a function by looking at its graph. *In general, the domain consists of all x-values that are included in the graph, and the range is all y-values that are included.* For example, Fig. 3.16 implies that both the domain and range of $f(x) = \sqrt{x}$ are all nonnegative numbers. From Fig. 3.17, it is clear that the domain of $p = G(q) = |q|$ is all real numbers and the range is all $p \geq 0$.

EXAMPLE 6 Domain, Range, and Function Values

Figure 3.21 shows the graph of a function F. To the right of 4, assume that the graph repeats itself indefinitely. Then the domain of F is all $t \geq 0$. The range is $-1 \leq s \leq 1$. Some function values are

$$F(0) = 0, \quad F(1) = 1, \quad F(2) = 0, \quad F(3) = -1.$$

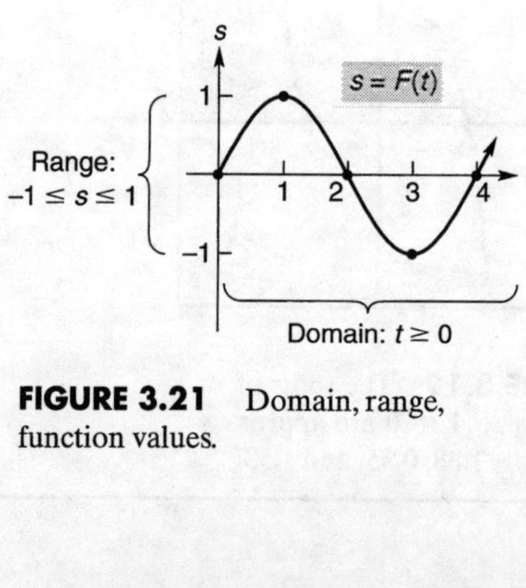

FIGURE 3.21 Domain, range, function values.

TECHNOLOGY

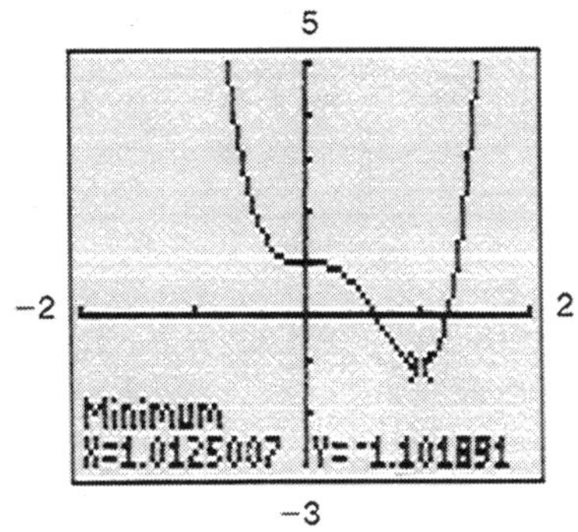

FIGURE 3.22 The range of $f(x) = 6x^4 - 8.1x^3 + 1$ is approximately $[-1.10, \infty)$.

Using a graphics calculator, we can estimate the range of a function. The graph of

$$f(x) = 6x^4 - 8.1x^3 + 1$$

is shown in Fig. 3.22. The lowest point on the graph corresponds to the minimum value of $f(x)$, and the range is all reals greater than or equal to this minimum. We can estimate this minimum y-value either by using trace and zoom or by selecting the "minimum" operation.

Principles in Practice 4
Graph of a Compound Function

To encourage conservation, a gas company charges two rates. You pay \$0.53 per therm for 0–70 therms and \$0.74 for each therm over 70. Graph the compound function which represents the monthly cost of t therms of gas.

EXAMPLE 7 Graph of a Compound Function

Graph the compound function

$$f(x) = \begin{cases} x, & \text{if } 0 \le x < 3, \\ x - 1, & \text{if } 3 \le x \le 5, \\ 4, & \text{if } 5 < x \le 7. \end{cases}$$

Solution: The domain of f is $0 \le x \le 7$. The graph is given in Fig. 3.23, where the *hollow dot* means that the point is *not* included in the graph. Notice that the range of f is all real numbers y such that $0 \le y \le 4$. ■

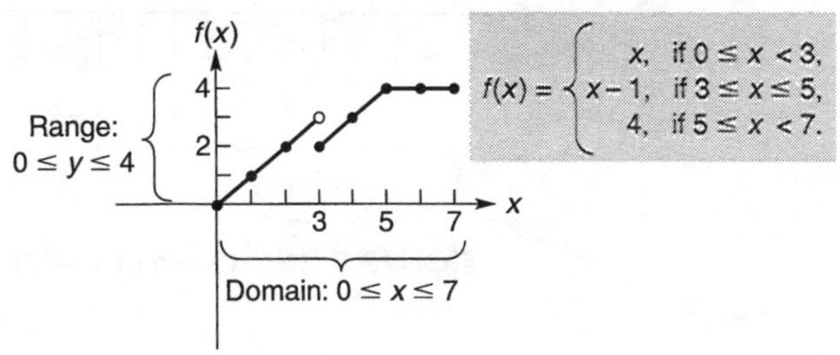

x	0	1	2	3	4	5	6	7
f(x)	0	1	2	2	3	4	4	4

FIGURE 3.23 Graph of compound function.

There is an easy way to tell whether a curve is the graph of a function. In Fig. 3.24(a), notice that with the given x there are associated *two* values of y: y_1 and y_2. Thus, the curve is *not* the graph of a function of x. Looking at it another way, we have the following general rule, called the **vertical-line test.** If a *vertical* line L can be drawn that intersects a curve in at least two points, then the curve is *not* the graph of a function of x. When no such vertical line can be drawn, the curve *is* the graph of a function of x. Consequently, the curves in Fig. 3.24 do not represent functions of x, but those in Fig. 3.25 do.

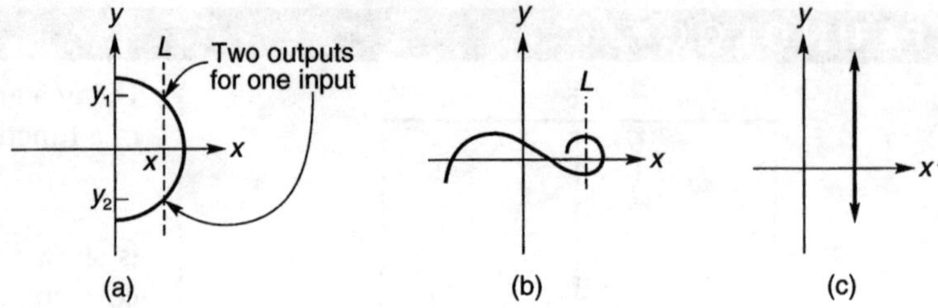

FIGURE 3.24 y is not a function of x.

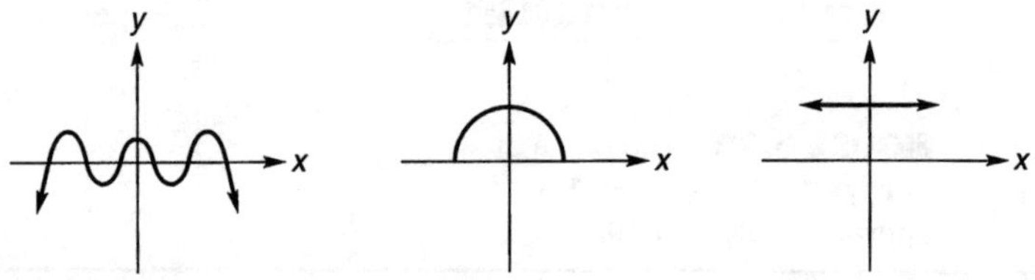

FIGURE 3.25 Functions of x.

EXAMPLE 8 A Graph That Does Not Represent a Function of x

Graph $x = 2y^2$.

Solution: Here it is easier to choose values of y and then find the corresponding values of x. Figure 3.26 shows the graph. By the vertical-line test, the equation $x = 2y^2$ does not define a function of x.

x	0	2	2	8	8	18	18
y	0	1	−1	2	−2	3	−3

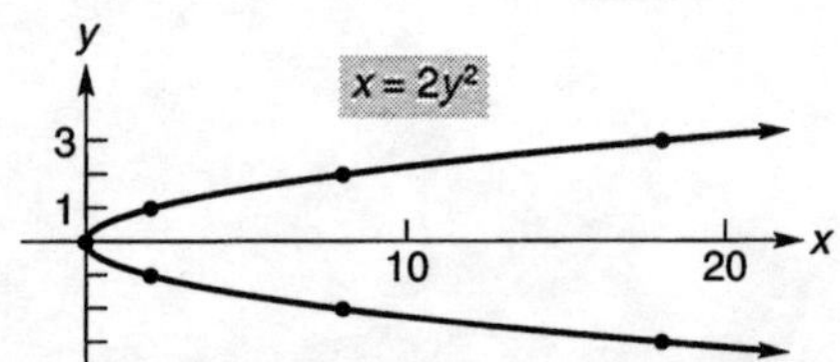

FIGURE 3.26 Graph of $x = 2y^2$.

■ Exercise 3.4

In Problems 1 and 2, locate and label each of the points, and give the quadrant, if possible, in which each point lies.

1. $(2, 7), (8, -3), (-\frac{1}{2}, -2), (0, 0)$.

2. $(-4, 5), (3, 0), (1, 1), (0, -6)$.

3. Figure 3.27(a) shows the graph of $y = f(x)$.

a. Estimate $f(0), f(2), f(4)$, and $f(2)$.

b. What is the domain of f?

c. What is the range of f?

d. What is a real zero of f?

4. Figure 3.27(b) shows the graph of $y = f(x)$.

a. Estimate $f(0)$ and $f(2)$.

b. What is the domain of f?

c. What is the range of f?

d. What is a real zero of f?

5. Figure 3.28(a) shows the graph of $y = f(x)$.

a. Estimate $f(0), f(1)$, and $f(-1)$.

b. What is the domain of f?

c. What is the range of f?

d. What is a real zero of f?

6. Figure 3.28(b) shows the graph of $y = f(x)$.

a. Estimate $f(0), f(2), f(3)$, and $f(4)$.

b. What is the domain of f?

c. What is the range of f?

d. What is a real zero of f?

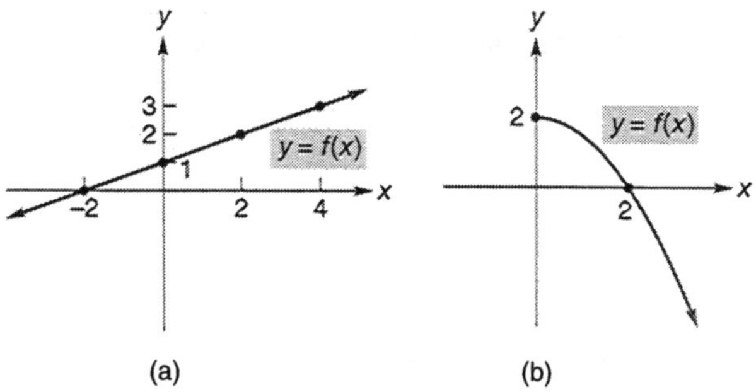

FIGURE 3.27 Diagram for Problems 3 and 4.

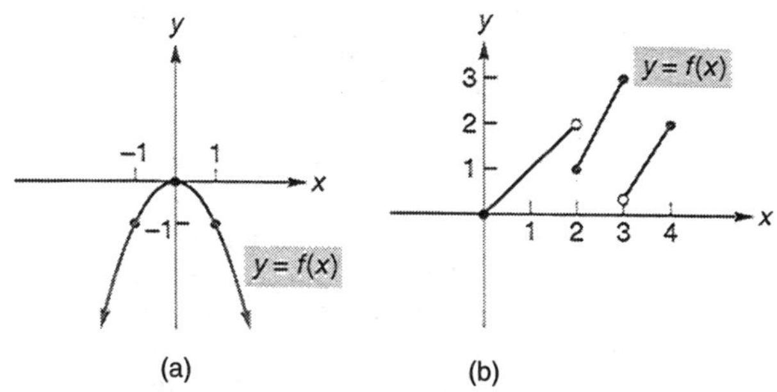

FIGURE 3.28 Diagram for Problems 5 and 6.

In Problems 7–20, determine the intercepts of the graph of each equation, and sketch the graph. Based on your graph, is y a function of x, and if so, what are the domain and range?

7. $y = x$.

8. $y = x + 1$.

9. $y = 3x - 5$.

10. $y = 3 - 2x$.

11. $y = x^2$.

12. $y = \dfrac{3}{x}$.

13. $x = 0$.

14. $y = x^2 - 9$.

15. $y = x^3$.

16. $x = -4$.

17. $x = -3y^2$.

18. $x^2 = y^2$.

19. $2x + y - 2 = 0$.

20. $x + y = 1$.

In Problems 21–34, graph each function and give the domain and range. Also, determine the intercepts.

21. $s = f(t) = 4 - t^2$.

22. $f(x) = 5 - 2x^2$.

23. $y = g(x) = 2$.

24. $G(s) = -8$.

25. $y = h(x) = x^2 - 4x + 1$.

26. $y = f(x) = x^2 + 2x - 8$.

27. $f(t) = -t^3$.

28. $p = h(q) = q(2 - q)$.

29. $s = F(r) = \sqrt{r - 5}$.

30. $F(r) = -\dfrac{1}{r}$.

31. $f(x) = |2x - 1|$.

32. $v = H(u) = |u - 3|$.

33. $F(t) = \dfrac{16}{t^2}$.

34. $y = f(x) = \dfrac{2}{x - 4}$.

In Problems 35–38, graph each compound function and give the domain and range.

35. $c = g(p) = \begin{cases} p, & \text{if } 0 \le p < 2, \\ 2, & \text{if } p \ge 2. \end{cases}$

36. $f(x) = \begin{cases} 2x + 1, & \text{if } -1 \le x < 2, \\ 9 - x^2, & \text{if } x \ge 2. \end{cases}$

37. $g(x) = \begin{cases} x + 6, & \text{if } x \ge 3, \\ x^2, & \text{if } x < 3. \end{cases}$

38. $f(x) = \begin{cases} x + 1, & \text{if } 0 < x \le 3, \\ 4, & \text{if } 3 < x \le 5, \\ x - 1, & \text{if } x > 5. \end{cases}$

39. Which of the graphs in Fig. 3.29 represent functions of x?

40. Debt Payments Jane has charged $800 on her credit cards. She plans to pay them off at the rate of $75 per month. Write an equation to represent the amount she owes, excluding any finance charges, and identify the intercepts.

41. Pricing To encourage an even flow of customers, a restaurant varies the price of an item throughout the day. From 6:00 P.M. to 8:00 P.M., customers pay full price. At lunch, from 10:30 A.M. until 2:30 P.M., customers pay half price. From 2:30 P.M. until 4:30 P.M., customers get a

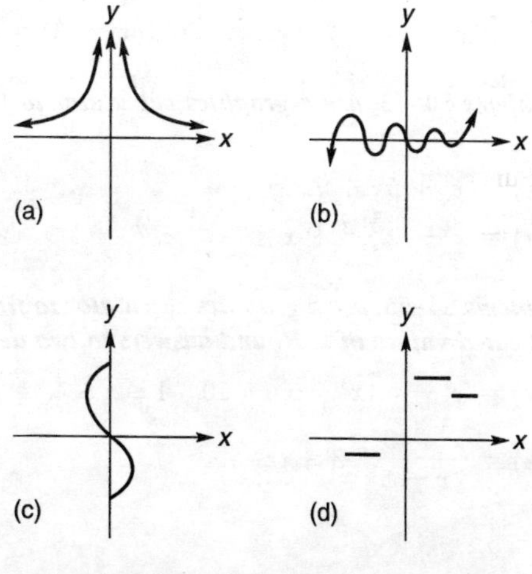

FIGURE 3.29 Diagram for Problem 39.

dollar off the lunch price. From 4:30 P.M. until 6:00 P.M., customers get \$5.00 off the dinner price. From 8:00 P.M. until closing time at 10:00 P.M., customers get \$5.00 off the dinner price. Graph the compound function to represent the cost of an item throughout the day for a dinner price of \$18.

42. **Supply Schedule** Given the following supply schedule (see Example 5 of Sec. 3.1), plot each quantity–price pair by choosing the horizontal axis for the possible quantities. Approximate the points in between the data by connecting the data points with a smooth curve. The result is a *supply curve.* From the graph, determine the relationship between price and supply. (That is, as price increases, what happens to the quantity supplied?) Is price per unit a function of quantity supplied?

Quantity Supplied per Week, q	Price per Unit, p
30	\$10
100	20
150	30
190	40
210	50

43. **Demand Schedule** The following table is called a *demand schedule.* It indicates the quantities of brand X that consumers will demand (that is, purchase) each week at certain prices per unit (in dollars). Plot each quantity–price pair by choosing the vertical axis for the possible prices. Connect the points with a smooth curve. In this way, we approximate points in between the given data. The result is called a *demand curve.* From the graph, determine the relationship between the price of brand X and the amount that will be demanded. (That is, as price decreases, what happens to the quantity demanded?) Is price per unit a function of quantity demanded?

Quantity Demanded, q	Price per Unit, p
5	\$20
10	10
20	5
25	4

44. **Inventory** Sketch the graph of

$$y = f(x) = \begin{cases} -100x + 600, & \text{if } 0 \le x < 5, \\ -100x + 1100, & \text{if } 5 \le x < 10, \\ -100x + 1600, & \text{if } 10 \le x < 15. \end{cases}$$

A function such as this might describe the inventory y of a company at time x.

45. **Psychology** In a psychological experiment on visual information, a subject briefly viewed an array of letters and was then asked to recall as many letters as possible from the array. The procedure was repeated several times. Suppose that y is the average number of letters recalled from arrays with x letters. The graph of the results approximately fits the graph of

$$y = f(x) = \begin{cases} x, & \text{if } 0 \le x \le 4, \\ \frac{1}{2}x + 2, & \text{if } 4 < x \le 5, \\ 4.5, & \text{if } 5 < x \le 12. \end{cases}$$

Plot this function.[6]

In Problems 46–49, use a graphics calculator to find all real roots of the given equation. Round answers to two decimal places.

46. $7x^3 + 2x = 3.$

47. $x(x^2 - 3) = x^4 + 1.$

48. $(9x + 3.1)^2 = 7.4 - 4x^2.$

49. $(x - 3)^3 = x^2 - 8.$

In Problems 50–53, use a graphics calculator to find all real zeros of the given function. Round answers to two decimal places.

50. $f(x) = x^3 + 5x + 7.$

51. $f(x) = x^4 - 2.5x^3 - 2.$

52. $g(x) = x^4 - 2.5x^3 + x.$

53. $g(x) = \sqrt{3}x^5 - 4x^2 + 1.$

In Problems 54–56, use a graphics calculator to find (a) the maximum value of f(x) and (b) the minimum value of f(x) for the indicated values of x. Round answers to two decimal places.

54. $f(x) = x^4 - 4.1x^3 + x^2 + 10, \quad 1 \le x \le 4.$

55. $f(x) = x(2.1x^2 - 3)^2 - x^3 + 1, \quad -1 \le x \le 1.$

56. $f(x) = \dfrac{x^2 - 3}{2x - 5}, \quad 3 \le x \le 5.$

[6]Adapted from G. R. Loftus and E. F. Loftus, *Human Memory: The Processing of Information* (New York: Lawrence Erlbaum Associates, Inc., distributed by the Halsted Press, Division of John Wiley & Sons, Inc., 1976).

57. From the graph of $f(x) = \sqrt{2}x^3 + 1.1x^2 + 4$, find (a) the range and (b) the intercepts. Round values to two decimal places.

58. From the graph of $f(x) = 2 - 3x^3 - x^4$, find (a) the maximum value of $f(x)$, (b) the range of f, and (c) the real zeros of f. Round values to two decimal places.

59. From the graph of $f(x) = \dfrac{x^2 + 5.2}{2.1 + \sqrt{x}}$, find (a) the minimum value of $f(x)$, (b) the range of f, and (c) the intercepts. (d) Does f have any real zeros? Round values to two decimal places.

60. Graph $f(x) = \dfrac{4.1x^3 + \sqrt{2}}{x^2 - 3}$ for $2 \leq x \leq 5$. Determine (a) the maximum value of $f(x)$, (b) the minimum value of $f(x)$, (c) the range of f, and (d) all intercepts. Round values to two decimal places.

OBJECTIVE

To study symmetry about the x-axis, the y-axis, and the origin, and to apply symmetry to curve sketching.

3.5 SYMMETRY

Examining the graphical behavior of equations is a basic part of mathematics. In this section, we shall examine equations to determine whether their graphs have *symmetry.* In a later chapter, you will see that calculus is a *great* aid in graphing because it helps determine the shape of a graph. It provides powerful techniques for determining whether or not a curve "wiggles" between points.

Consider the graph of $y = x^2$ in Fig. 3.30. The portion to the left of the y-axis is the reflection (or mirror image) through the y-axis of that portion to the right of the y-axis, and vice versa. More precisely, if (x_0, y_0) is any point on this graph, then the point $(-x_0, y_0)$ must also lie on the graph. We say that this graph is *symmetric about the y-axis.*

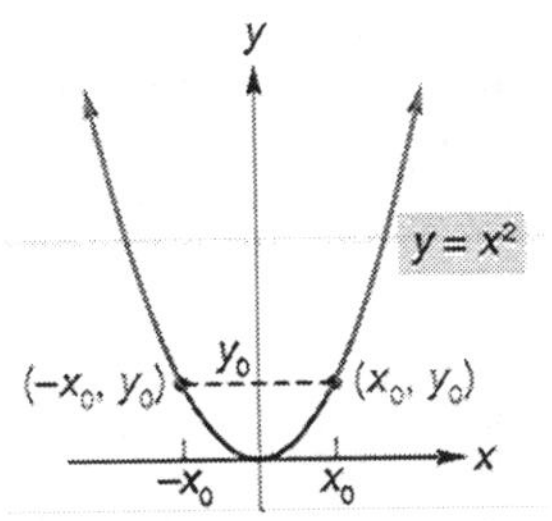

FIGURE 3.30 Symmetry about y-axis.

DEFINITION

A graph is ***symmetric about the y-axis*** *if and only if $(-x_0, y_0)$ lies on the graph when (x_0, y_0) does.*

EXAMPLE 1 y-**Axis Symmetry**

Use the preceding definition to show that the graph of $y = x^2$ is symmetric about the y-axis.

Solution: Suppose (x_0, y_0) is *any* point on the graph of $y = x^2$. Then

$$y_0 = x_0^2.$$

We must show that the coordinates of $(-x_0, y_0)$ satisfy $y = x^2$:

$$y_0 = (-x_0)^2?$$

$$y_0 = x_0^2?$$

But we already know that $y_0 = x_0^2$. Thus, the graph *is* symmetric about the y-axis. ■

When one is testing for symmetry in Example 1, (x_0, y_0) can be any point on the graph. In the future, for convenience, we shall omit the subscripts. This means that a graph is symmetric about the y-axis if replacing x by $-x$ in its equation results in an equivalent equation.

Another type of symmetry is shown by the graph of $x = y^2$ in Fig. 3.31. Here the portion below the x-axis is the reflection through the x-axis of that portion above the x-axis, and vice versa. If the point (x, y) lies on the graph, then $(x, -y)$ also lies on it. This graph is said to be *symmetric about the x-axis.*

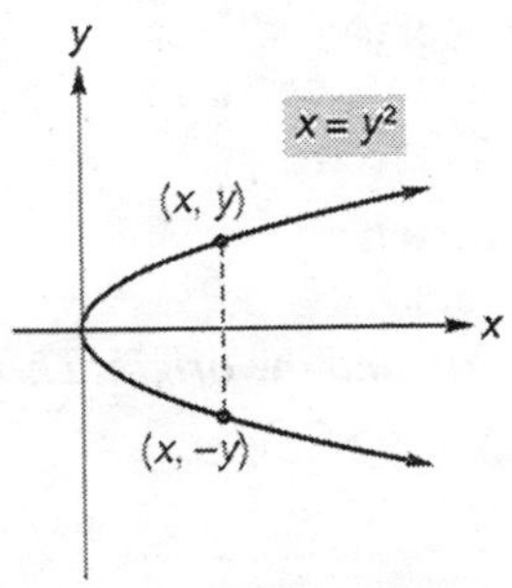

FIGURE 3.31 Symmetry about x-axis.

DEFINITION

A graph is ***symmetric about the x-axis*** *if and only if $(x, -y)$ lies on the graph when (x, y) does.*

Thus, the graph of an equation in x and y has x-axis symmetry if replacing y by $-y$ results in an equivalent equation. For example, applying this test to the graph of $x = y^2$ shown in Fig. 3.31 gives

$$x = (-y)^2,$$
$$x = y^2,$$

which is equivalent to the original equation. Hence, the graph is symmetric about the x-axis.

A third type of symmetry, *symmetry about the origin,* is illustrated by the graph of $y = x^3$ (Fig. 3.32). Whenever the point (x, y) lies on the graph, $(-x, -y)$ also lies on it. As a result, the line segment joining points (x, y) and $(-x, -y)$ is bisected by the origin.

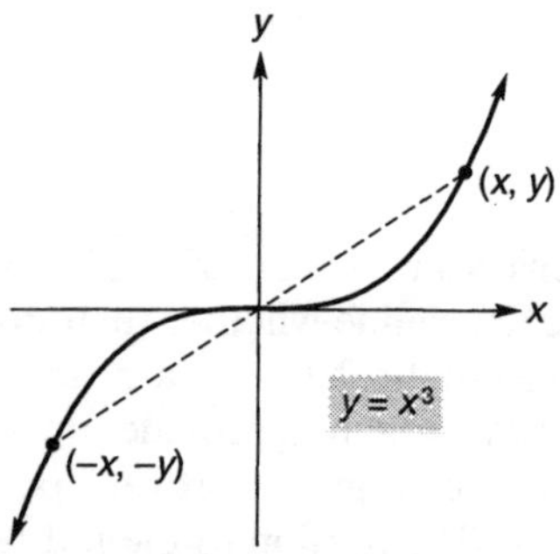

FIGURE 3.32 Symmetry about origin.

DEFINITION

A graph is ***symmetric about the origin*** *if and only if $(-x, -y)$ lies on the graph when (x, y) does.*

Thus, the graph of an equation in x and y has symmetry about the origin if replacing x by $-x$ and $-y$ by y results in an equivalent equation. For example, applying this test to the graph of $y = x^3$ shown in Fig. 3.32 gives

$$-y = (-x)^3,$$
$$-y = -x^3,$$
$$y = x^3,$$

which is equivalent to the original equation. Accordingly, the graph is symmetric about the origin.

Table 3.1 summarizes the tests for symmetry. When we know that a graph has symmetry, we can sketch it by plotting fewer points than would otherwise be needed.

TABLE 3.1 Tests for Symmetry

Symmetry about x-axis	Replace y by $-y$ in given equation. Symmetric if equivalent equation is obtained.
Symmetry about y-axis	Replace x by $-x$ in given equation. Symmetric if equivalent equation is obtained.
Symmetry about origin	Replace x by $-x$ and y by $-y$ in given equation. Symmetric if equivalent equation is obtained.

EXAMPLE 2 Graphing with Intercepts and Symmetry

Test $y = \frac{1}{x}$ for symmetry about the x-axis, the y-axis, and the origin. Then find the intercepts and sketch the graph.

Solution:

***Symmetry** x-axis:* Replacing y by $-y$ in $y = 1/x$ gives

$$-y = \frac{1}{x} \quad \text{or} \quad y = -\frac{1}{x},$$

which is not equivalent to the given equation. Thus, the graph is *not* symmetric about the x-axis.

y-axis: Replacing x by $-x$ in $y = 1/x$ gives

$$y = \frac{1}{-x} \quad \text{or} \quad y = -\frac{1}{x},$$

which is not equivalent to the given equation. Hence, the graph is *not* symmetric about the y-axis.

Origin: Replacing x by $-x$ and y by $-y$ in $y = 1/x$ gives

$$-y = \frac{1}{-x} \quad \text{or} \quad y = \frac{1}{x},$$

which is equivalent to the given equation. Consequently, the graph *is* symmetric about the origin.

Intercepts Since x cannot be 0, the graph has no y-intercept. If y is 0, then $0 = 1/x$, and this equation has no solution. Thus, no x-intercept exists.

Discussion Because no intercepts exist, the graph cannot intersect either axis. If $x > 0$, we obtain points only in quadrant I. Fig. 3.33 shows the portion of the graph in quadrant I. By symmetry, we reflect that portion through the origin to obtain the entire graph.

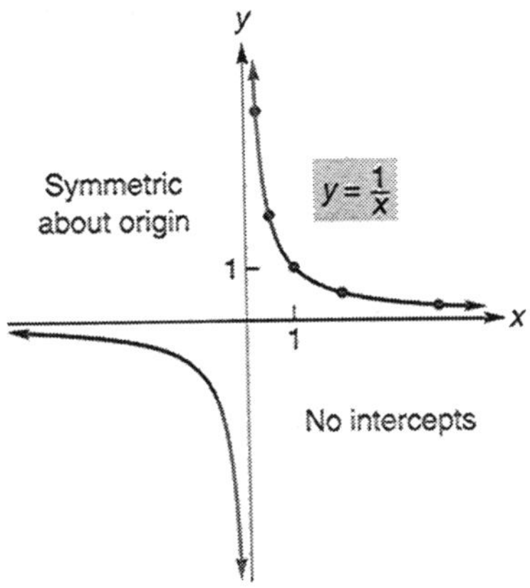

x	$\frac{1}{4}$	$\frac{1}{2}$	1	2	4
y	4	2	1	$\frac{1}{2}$	$\frac{1}{4}$

FIGURE 3.33 Graph of $y = \dfrac{1}{x}$.

EXAMPLE 3 Graphing with Intercepts and Symmetry

Test $y = f(x) = 1 - x^4$ for symmetry about the x-axis, the y-axis, and the origin. Then find the intercepts and sketch the graph.

Solution:
Symmetry *x-axis:* Replacing y by $-y$ in $y = 1 - x^4$ gives

$$-y = 1 - x^4 \quad \text{or} \quad y = -1 + x^4,$$

which is not equivalent to the given equation. Thus, the graph is *not* symmetric about the x-axis.

y-axis: Replacing x by $-x$ in $y = 1 - x^4$ gives

$$y = 1 - (-x)^4 \quad \text{or} \quad y = 1 - x^4,$$

which is equivalent to the given equation. Hence, the graph *is* symmetric about the y-axis.

Origin: Replacing x by $-x$ and y by $-y$ in $y = 1 - x^4$ gives

$$-y = 1 - (-x)^4, \quad -y = 1 - x^4, \quad y = -1 + x^4,$$

which is not equivalent to the given equation. Thus, the graph is *not* symmetric about the origin.

Intercepts Testing for x-intercepts, we set $y = 0$ in $y = 1 - x^4$. Then

$$1 - x^4 = 0,$$
$$(1 - x^2)(1 + x^2) = 0,$$
$$(1 - x)(1 + x)(1 + x^2) = 0,$$
$$x = 1 \quad \text{or} \quad x = -1.$$

The x-intercepts are therefore $(1, 0)$ and $(-1, 0)$. Testing for y-intercepts, we set $x = 0$. Then $y = 1$, so $(0, 1)$ is the only y-intercept.

Discussion If the intercepts and some points (x, y) to the right of the y-axis are plotted, we can sketch the *entire* graph by using symmetry about the y-axis (Fig. 3.34). ■

x	y
0	1
$\frac{1}{2}$	$\frac{15}{16}$
$\frac{3}{4}$	$\frac{175}{256}$
1	0
$\frac{3}{2}$	$-\frac{65}{16}$

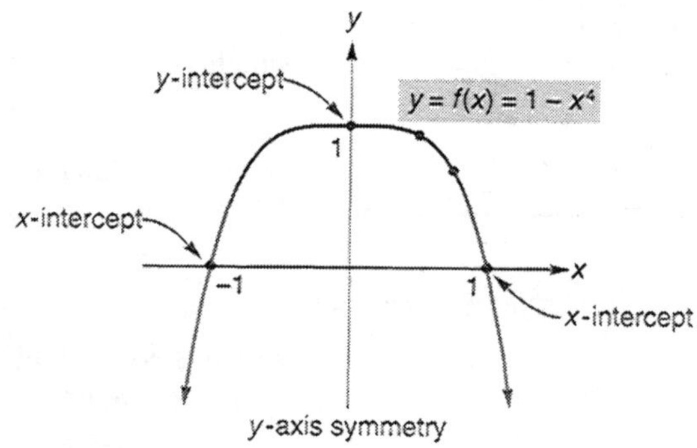

FIGURE 3.34 Graph of $y = 1 - x^4$.

In Example 3, we showed that the graph of $y = f(x) = 1 - x^4$ does not have x-axis symmetry. With the exception of the constant function $f(x) = 0$, *the graph of any **function** $y = f(x)$ cannot be symmetric about the x-axis,* because such symmetry implies two y-values with the same x-value, which violates the definition of a function.

EXAMPLE 4 Graphing with Intercepts and Symmetry

Test the graph of $4x^2 + 9y^2 = 36$ for intercepts and symmetry. Sketch the graph.

Solution:

Intercepts If $y = 0$, then $4x^2 = 36$, so $x = \pm 3$. Thus, the x-intercepts are $(3, 0)$ and $(-3, 0)$. If $x = 0$, then $9y^2 = 36$, so $y = \pm 2$. Hence, the y-intercepts are $(0, 2)$ and $(0, -2)$.

Symmetry *x-axis:* Replacing y by $-y$ in $4x^2 + 9y^2 = 36$ gives

$$4x^2 + 9(-y)^2 = 36, \qquad \text{or} \qquad 4x^2 + 9y^2 = 36.$$

Since we obtain the original equation, there is symmetry about the x-axis.

y-axis: Replacing x by $-x$ in $4x^2 + 9y^2 = 36$ gives

$$4(-x)^2 + 9y^2 = 36, \qquad \text{or} \qquad 4x^2 + 9y^2 = 36.$$

Again we have the original equation, so there is also symmetry about the y-axis.

Origin: Replacing x by $-x$ and y by $-y$ in $4x^2 + 9y^2 = 36$ gives

$$4(-x)^2 + 9(-y)^2 = 36, \qquad \text{or} \qquad 4x^2 + 9y^2 = 36.$$

Since this is the original equation, the graph is also symmetric about the origin.

Discussion In Fig. 3.35, the intercepts and some points in the first quadrant are plotted. The points in that quadrant are then connected by a smooth curve. By symmetry about the x-axis, the points in the fourth quadrant are obtained. Then, by symmetry about the y-axis, the complete graph is found. There are other ways of graphing the equation by using symmetry. For example, after plotting the intercepts and some points in the first quadrant, we can obtain the

x	y
± 3	0
0	± 2
1	$\frac{4\sqrt{2}}{3}$
2	$\frac{2\sqrt{5}}{3}$
$\frac{5}{2}$	$\frac{\sqrt{11}}{3}$

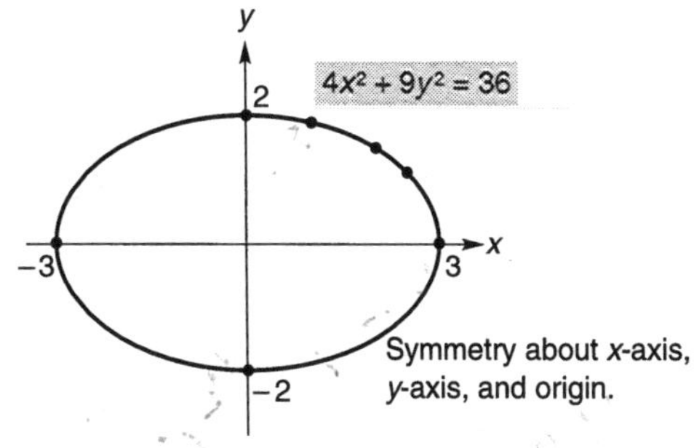

FIGURE 3.35 Graph of $4x^2 + 9y^2 = 36$.

points in the third quadrant by symmetry about the origin. By symmetry about the x-axis (or y-axis), we can then obtain the entire graph. ■

This fact can be a timesaving device in checking for symmetry.

In Example 4, the graph is symmetric about the x-axis, the y-axis, and the origin. It can be shown that **for any graph, if any two of the three types of symmetry exist, then the remaining type must also exist.**

■ Exercise 3.5

In Problems 1–16, find the x- and y-intercepts of the graphs of the equations. Also, test for symmetry about the x-axis, the y-axis, and the origin. Do not sketch the graphs.

1. $y = 5x$.

2. $y = f(x) = x^2 - 4$.

3. $2x^2 + y^2x^4 = 8 - y$.

4. $x = y^3$.

5. $4x^2 - 9y^2 = 36$.

6. $y = 7$.

7. $x = -2$.

8. $y = |2x| - 2$.

9. $x = -y^{-4}$.

10. $y = \sqrt{x^2 - 4}$.

11. $x - 4y - y^2 + 21 = 0$.

12. $x^3 - xy + y^2 = 0$.

13. $y = f(x) = \dfrac{x^3}{x^2 + 5}$.

14. $x^2 + xy + y^2 = 0$.

15. $y = \dfrac{1}{x^3 + 1}$.

16. $y = \dfrac{x^4}{x + y}$.

In Problems 17–24, find the x- and y-intercepts of the graphs of the equations. Also, test for symmetry about the x-axis, the y-axis, and the origin. Then sketch the graphs.

17. $2x + y^2 = 4$.

18. $x = y^4$.

19. $y = f(x) = x^3 - 4x$.

20. $y = x - x^3$.

21. $|x| - |y| = 0$.

22. $x^2 + y^2 = 16$.

23. $4x^2 + y^2 = 16$.

24. $x^2 - y^2 = 1$.

25. Prove that the graph of $y = f(x) = 2 - 0.03x^2 - x^4$ is symmetric about the y-axis, and then graph the function. (a) Make use of symmetry, where possible, to find all intercepts. Determine (b) the maximum value of $f(x)$ and (c) the range of f. Round all values to two decimal places.

26. Prove that the graph of $y = f(x) = 2x^4 - 7x^2 + 5$ is symmetric about the y-axis, and then graph the function. Find all real zeros of f. Round your answers to two decimal places.

OBJECTIVE

To be familiar with the shapes of the graphs of six basic functions and to consider translation, reflection, and vertical stretching or shrinking of the graph of a function.

3.6 Translations and Reflections

Up to now, our approach to graphing has been based on plotting points and making use of any symmetry that exists. But this technique is not necessarily the preferred way. Later in this book, we shall analyze graphs by using other techniques. However, some functions and their associated graphs occur so frequently—for example, for illustrative purposes—that we find it worthwhile to memorize them. Figure 3.36 shows six such functions.

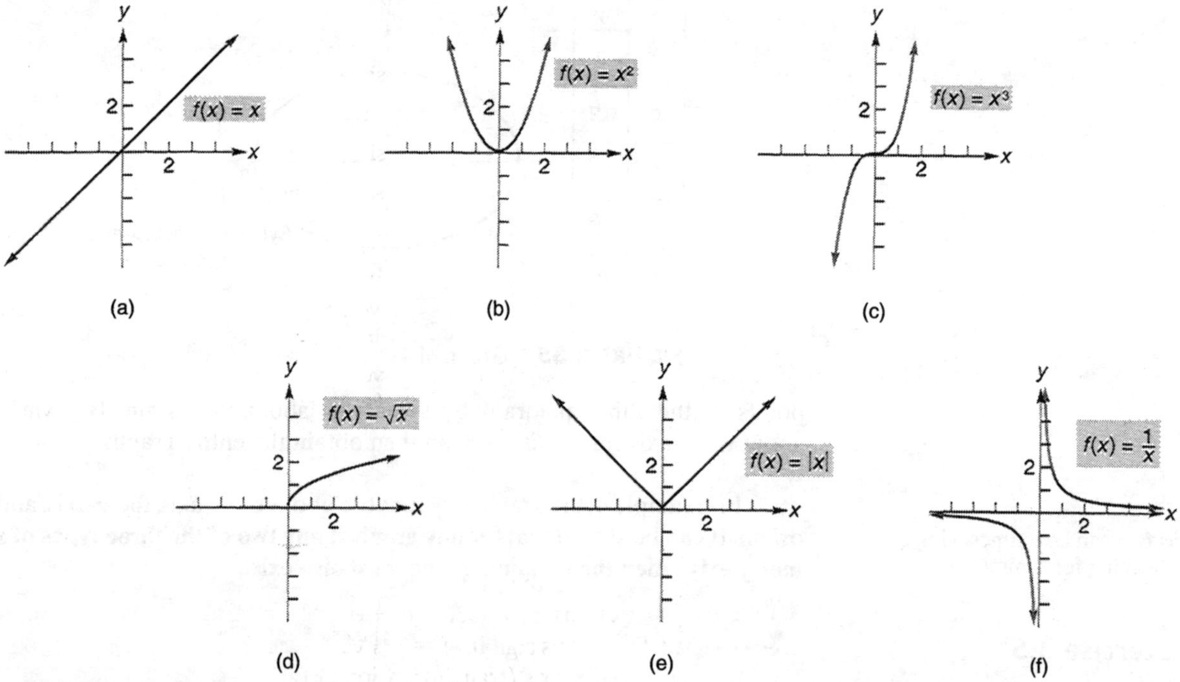

FIGURE 3.36 Functions frequently used.

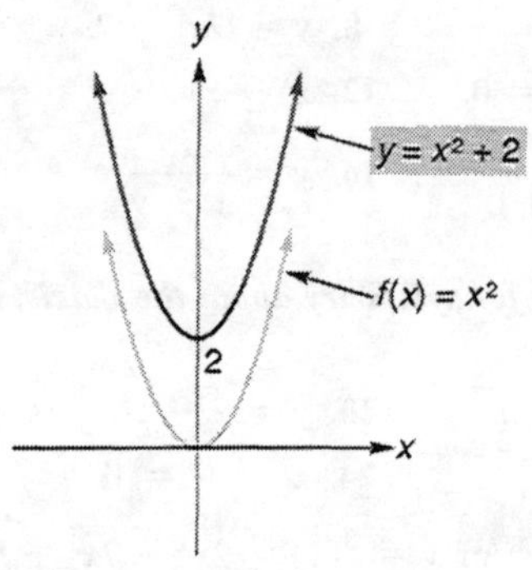

FIGURE 3.37 Graph of $y = x^2 + 2$.

At times, by altering a function through an *algebraic* manipulation, the graph of the new function can be obtained from the graph of the original function by performing a *geometric* manipulation. For example, we can use the graph of $f(x) = x^2$ to graph $y = x^2 + 2$. Note that $y = f(x) + 2$. Thus, for each x, the corresponding ordinate for the graph of $y = x^2 + 2$ is 2 more than the ordinate for the graph of $f(x) = x^2$. This means that the graph of $y = x^2 + 2$ is simply the graph of $f(x) = x^2$, shifted, or *translated,* 2 units upward. (See Fig. 3.37.) We say that the graph of $y = x^2 + 2$ is a *transformation* of the graph of $f(x) = x^2$. Table 3.2 gives a list of basic types of transformations.

EXAMPLE 1 Horizontal Translation

Sketch the graph of $y = (x - 1)^3$.

Solution: We observe that $(x - 1)^3$ is x^3 with x replaced by $x - 1$. Thus, if $f(x) = x^3$, then $y = (x - 1)^3 = f(x - 1)$, which has the form $f(x - c)$, where $c = 1$. From Table 3.2, the graph of $y = (x - 1)^3$ is the graph of $f(x) = x^3$, shifted 1 unit to the right. (See Fig. 3.38.) ■

EXAMPLE 2 Shrinking and Reflection

Sketch the graph of $y = -\frac{1}{2}\sqrt{x}$.

Solution: We can do this problem in two steps. First, observe that $\frac{1}{2}\sqrt{x}$ is $\sqrt{x}$ multiplied by $\frac{1}{2}$. Thus, if $f(x) = \sqrt{x}$, then $\frac{1}{2}\sqrt{x} = \frac{1}{2}f(x)$, which has the form

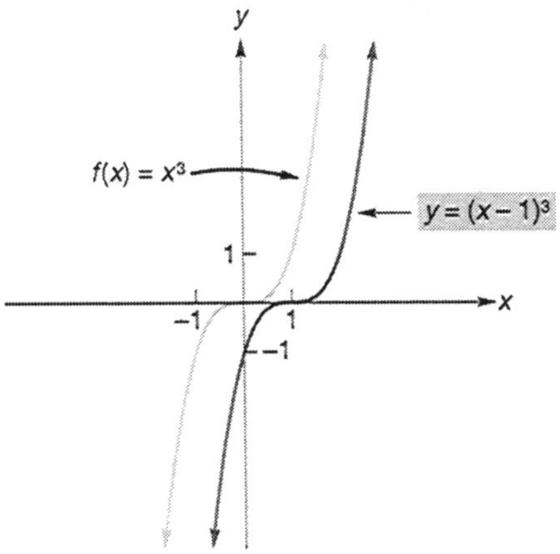

FIGURE 3.38 Graph of $y = (x - 1)^3$.

TABLE 3.2 Transformations, $c > 0$

Equation	How to transform graph of $y = f(x)$ to obtain graph of equation
1. $y = f(x) + c$	shift c units upward
2. $y = f(x) - c$	shift c units downward
3. $y = f(x - c)$	shift c units to right
4. $y = f(x + c)$	shift c units to left
5. $y = -f(x)$	reflect about x-axis
6. $y = f(-x)$	reflect about y-axis
7. $y = cf(x)$, $c > 1$	vertically stretch away from x-axis by a factor of c
8. $y = cf(x)$, $c < 1$	vertically shrink toward x-axis by a factor of c

$cf(x)$, where $c = \frac{1}{2}$. So the graph of $y = \frac{1}{2}\sqrt{x}$ is the graph of f shrunk vertically toward the x-axis by a factor of $\frac{1}{2}$ (transformation 8, Table 3.2; see Fig. 3.39). Second, the minus sign in $y = -\frac{1}{2}\sqrt{x}$ causes a reflection in the graph of $y = \frac{1}{2}\sqrt{x}$ about the x-axis (transformation 5, Table 3.2; see Fig. 3.39). ■

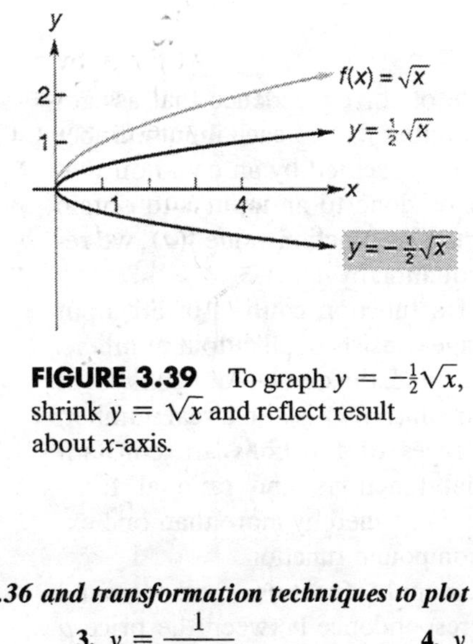

FIGURE 3.39 To graph $y = -\frac{1}{2}\sqrt{x}$, shrink $y = \sqrt{x}$ and reflect result about x-axis.

■ Exercise 3.6

In Problems 1–12, use the graphs of the functions in Fig. 3.36 and transformation techniques to plot the given functions.

1. $y = x^2 - 2$. **2.** $y = -x^2$. **3.** $y = \dfrac{1}{x-2}$. **4.** $y = \sqrt{x+2}$.

5. $y = \dfrac{2}{x}$. **6.** $y = |x| + 1$. **7.** $y = |x+1| - 2$. **8.** $y = -\frac{1}{2}x^3$.

9. $y = 1 - (x-1)^2$. **10.** $y = (x-1)^2 + 1$. **11.** $y = \sqrt{-x}$. **12.** $y = \dfrac{1}{2-x}$.

In Problems 13–16, describe what must be done to the graph of $y = f(x)$ to obtain the graph of the given equation.

13. $y = f(x-4) + 3$. **14.** $y = f(x+3) - 4$. **15.** $y = f(-x) + 3$. **16.** $y = -f(x+3)$.

17. Graph the function $y = \sqrt[3]{x} + k$ for $k = 0, 1, 2, 3, -1, -2$, and -3. Observe the vertical translations compared to the first graph.

18. Graph the function $y = \sqrt[3]{x + k}$ for $k = 0, 1, 2, 3, -1, -2$, and -3. Observe the horizontal translations compared to the first graph.

19. Graph the function $y = k\sqrt[3]{x}$ for $k = 1, 2, \frac{1}{2}$, and 3. Observe the vertical stretching and shrinking compared to the first graph. Graph the function for $k = -2$. Observe that the graph is the same as that obtained by stretching the reflection of $y = \sqrt[3]{x}$ about the x-axis by a factor of 2.

3.7 REVIEW

IMPORTANT TERMS AND SYMBOLS

Section 3.1	function; domain; range; independent variable; dependent variable; $f(x)$; function value; difference quotient, $\dfrac{f(x+h) - f(x)}{h}$; demand function; supply function
Section 3.2	constant function; polynomial function; linear function; quadratic function; compound function; absolute value, $\lvert x \rvert$; factorial, $r!$; rational function
Section 3.3	$f + g$; $f - g$; fg; f/g; $f \circ g$; composition function
Section 3.4	rectangular coordinate system; coordinate axes; origin; x, y-plane; ordered pair, (x, y); coordinates of a point; x-coordinate; y-coordinate; abscissa; ordinate; quadrant; graph of equation; y-intercept; x-intercept; graph of function; function-value axis; zeros of function; vertical-line test
Section 3.5	x-axis symmetry; y-axis symmetry; symmetry about origin

SUMMARY

A function f is a rule of correspondence that assigns exactly one output number $f(x)$ to each input number x. Usually, a function is specified by an equation that indicates what must be done to an input x to obtain $f(x)$. To obtain a particular function value $f(a)$, we replace each x in the equation by a.

The domain of a function consists of all input numbers, and the range consists of all output numbers. Unless otherwise specified, the domain of f consists of all real numbers x for which $f(x)$ is also a real number.

Some special types of functions are constant functions, polynomial functions, and rational functions. A function that is defined by more than one expression is called a compound function.

In economics, supply functions and demand functions give a correspondence between the price p of a product and the number of units q of the product that producers (or consumers) will supply (or buy) at that price.

Two functions f and g can be combined to form a sum, difference, product, quotient, or composition as follows:

$$(f + g)(x) = f(x) + g(x),$$
$$(f - g)(x) = f(x) - g(x),$$
$$(fg)(x) = f(x)g(x),$$
$$\frac{f}{g}(x) = \frac{f(x)}{g(x)},$$
$$(f \circ g)(x) = f(g(x)).$$

A rectangular coordinate system allows us to geometrically represent equations in two variables, as well as functions. The graph of an equation in x and y consists of all points (x, y) that correspond to the solutions of the equation. We plot a sufficient number of points and connect them (where appropriate) so that the basic shape of the graph is apparent. Points where the graph intersects the x- and y-axes are called x-intercepts and y-intercepts, respectively. An x-intercept is found by letting y be 0 and solving for x; a y-intercept is found by letting x be 0 and solving for y.

The graph of a function f is the graph of the equation $y = f(x)$ and consists of all points $(x, f(x))$ such that x is in the domain of f. The zeros of f are the values of x for which $f(x) = 0$. From the graph of a function, it is easy to determine the domain and range.

The fact that a graph represents a function can be determined by using the vertical-line test. A vertical line cannot cut the graph of a function at more than one point.

When the graph of an equation has symmetry, the mirror-image effect allows us to sketch the graph by plotting fewer points than would otherwise be needed. The tests for symmetry are as follows:

Symmetry about x-axis	Replace y by $-y$ in given equation. Symmetric if equivalent equation is obtained.
Symmetry about y-axis	Replace x by $-x$ in given equation. Symmetric if equivalent equation is obtained.
Symmetry about origin	Replace x by $-x$ and y by $-y$ in given equation. Symmetric if equivalent equation is obtained.

Sometimes the graph of a function can be obtained from that of a familiar function by means of a vertical shift upward or downward, a horizontal shift to the right or left, a reflection about the x-axis or y-axis, or a vertical stretching or shrinking away from or toward the x-axis. Such transformations are indicated in Table 3.2 in Sec. 3.6.

REVIEW PROBLEMS

In Problems 1–6, give the domain of each function.

1. $f(x) = \dfrac{x}{x^2 - 3x + 2}$.

2. $g(x) = x^2 + 3x$.

3. $F(t) = 7t + 4t^2$.

4. $G(x) = 18$.

5. $h(x) = \dfrac{\sqrt{x}}{x - 1}$.

6. $H(s) = \dfrac{\sqrt{s - 5}}{4}$.

In Problems 7–14, find the function values for the given function.

7. $f(x) = 3x^2 - 4x + 7$; $f(0), f(-3), f(5), f(t)$.

8. $g(x) = 4$; $g(4)$, $g(\frac{1}{100})$, $g(-156)$, $g(x + 4)$.

9. $G(x) = \sqrt{x - 1}$; $G(1), G(10), G(t + 1), G(x^2)$.

10. $F(x) = \dfrac{x - 3}{x + 4}$; $F(-1), F(0), F(5), F(x + 3)$.

11. $h(u) = \dfrac{\sqrt{u + 4}}{u}$; $h(5), h(-4), h(x), h(u - 4)$.

12. $H(s) = \dfrac{(s - 4)^2}{3}$; $H(-2), H(7), H(\frac{1}{2}), H(x^2)$.

13. $f(x) = \begin{cases} 4, & \text{if } x < 2 \\ 8 - x^2, & \text{if } x > 2 \end{cases}$;

$f(4), f(-2), f(0), f(10)$.

14. $h(q) = \begin{cases} q, & \text{if } -1 \le q < 0 \\ 3 - q, & \text{if } 0 \le q < 3; \\ 2q^2, & \text{if } 3 \le g \le 5 \end{cases}$

$h(0)$, $h(4)$, $h(-\frac{1}{2})$, $h(\frac{1}{2})$.

In Problems 15–18, find (a) $f(x + h)$ and (b) $\dfrac{f(x + h) - f(x)}{h}$, and simplify your answers.

15. $f(x) = 3 - 7x$.

16. $f(x) = x^2 + 4$.

17. $f(x) = 4x^2 + 2x - 5$.

18. $f(x) = \dfrac{7}{x + 1}$.

19. If $f(x) = 3x - 1$ and $g(x) = 2x + 3$, find the following.

a. $(f + g)(x)$. **b.** $(f + g)(4)$. **c.** $(f - g)(x)$.

d. $(fg)(x)$. **e.** $(fg)(1)$. **f.** $\dfrac{f}{g}(x)$.

g. $(f \circ g)(x)$. **h.** $(f \circ g)(5)$. **i.** $(g \circ f)(x)$.

20. If $f(x) = x^2$ and $g(x) = 2x + 1$, find the following.

a. $(f + g)(x)$. **b.** $(f - g)(x)$. **c.** $(f - g)(-3)$.

d. $(fg)(x)$. **e.** $\dfrac{f}{g}(x)$. **f.** $\dfrac{f}{g}(2)$.

g. $(f \circ g)(x)$. **h.** $(g \circ f)(x)$. **i.** $(g \circ f)(-4)$.

In Problems 21–24, find $(f \circ g)(x)$ and $(g \circ f)(x)$.

21. $f(x) = \frac{1}{x}$, $g(x) = x - 1$.

22. $f(x) = \frac{x+1}{4}$, $g(x) = \sqrt{x}$.

23. $f(x) = x + 2$, $g(x) = x^3$.

24. $f(x) = 2$, $g(x) = 3$.

In Problems 25 and 26, find the intercepts of the graph of each equation, and test for symmetry about the x-axis, the y-axis, and the origin. Do not sketch the graph.

25. $y = 2x - 3x^3$.

26. $\frac{xy^2}{x^2+1} = 4$.

In Problems 27 and 28, find the x- and y-intercepts of the graphs of the equations. Also, test for symmetry about the x-axis, the y-axis, and the origin. Then sketch the graphs.

27. $y = 9 - x^2$.

28. $y = 3x - 7$.

In Problems 29–32, graph each function and give its domain and range. Also, determine the intercepts.

29. $G(u) = \sqrt{u+4}$.

30. $f(x) = |x| + 1$.

31. $y = g(t) = \frac{2}{t-4}$.

32. $g(t) = \sqrt{4t}$.

33. Graph the following compound function, and give its domain and range:

$$y = f(x) = \begin{cases} 1 - x, & \text{if } x \le 0, \\ 1, & \text{if } x > 0. \end{cases}$$

34. Use the graph of $f(x) = \sqrt{x}$ to sketch the graph of $y = \sqrt{x-2} - 1$.

35. Use the graph of $f(x) = x^2$ to sketch the graph of $y = -\frac{1}{2}x^2 + 2$.

36. Trend Equation The projected annual sales (in dollars) of a new product is given by the equation $S = 150{,}000 + 3000t$, where t is the time in years from 1996. Such an equation is called a *trend equation.* Find the projected annual sales for 2001. Is S a function of t?

37. In Fig. 3.40, which graphs represent functions of x?

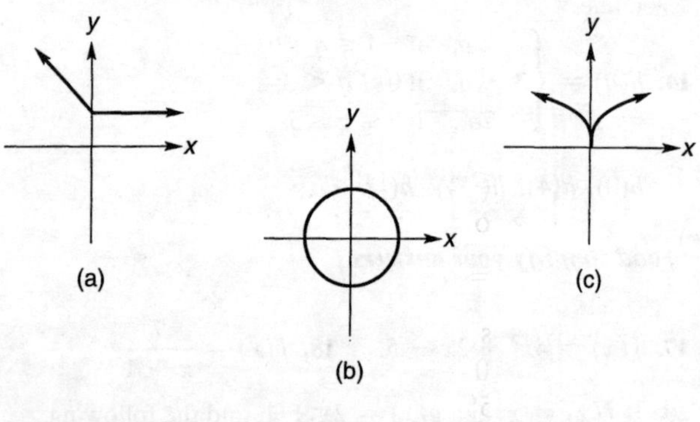

FIGURE 3.40 Diagram for Problem 37.

38. If $f(x) = (4x^{2.3} - 3x^3 + 7)^5$, find (a) $f(2)$ and (b) $f(2.3)$. Round your answers to two decimal places.

39. Find all real roots of the equation

$$5x^3 - 7x^2 = 4x - 2.$$

Round your answers to two decimal places.

40. Find all real roots of the equation

$$x^4 - 4x^3 = (2x-1)^2.$$

Round your answers to two decimal places.

41. Find all real zeros of

$$f(x) = x(2.1x^2 - 3)^2 - x^3 + 1.$$

Round your answers to two decimal places.

42. Determine the range of

$$f(x) = \begin{cases} -2.5x - 4, & \text{if } x < 0, \\ 6 + 4.1x - x^2, & \text{if } x \ge 0. \end{cases}$$

43. From the graph of $f(x) = -x^3 + 0.04x + 7$, find (a) the range and (b) the intercepts. Round values to two decimal places.

44. From the graph of $f(x) = \sqrt{x+1}(x^2 - 3)$, find (a) the minimum value of $f(x)$, (b) the range of f, and (c) all real zeros of f. Round values to two decimal places.

45. Graph $y = f(x) = x^3 + x^k$ for $k = 0, 1, 2, 3,$ and 4. For which values of k does the graph have (a) symmetry about the y-axis? (b) symmetry about the origin?

MATHEMATICAL *SNAPSHOT*

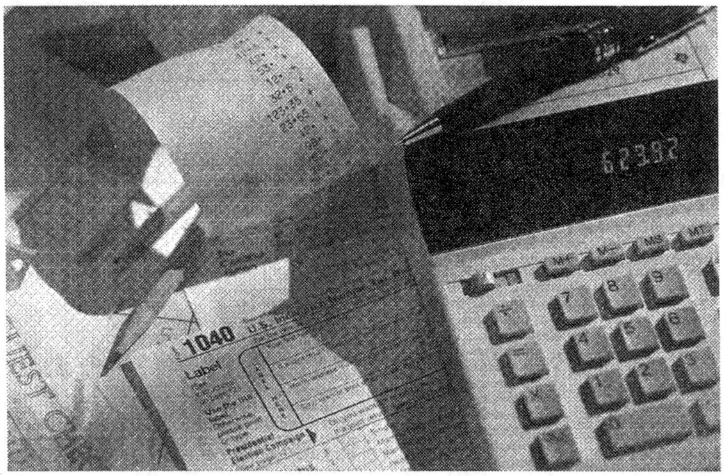

A TAXING EXPERIENCE!

You probably have heard the old saying; "There are only two things that are certain in life: death and taxes." Here you will see how functions can be applied to one of these "certainties," namely, taxes.

You will be using the 1997 federal tax rates for a married couple filing a joint return. Suppose that you want to determine a formula for the function f such that $f(x)$ is the tax in dollars on a taxable income of x dollars. The tax is based on various ranges of taxable income. According to Schedule Y-1 of the Internal Revenue Service (see Fig. 3.41):

- If x is \$0 or less, the tax is \$0.
- If x is over \$0, but not over \$41,200, the tax is 15% of x.
- If x is over \$41,200, but not over \$99,600, the tax is \$6,180, plus 28% of the amount over \$41,200.
- If x is over \$99,600, but not over \$151,750, the tax is \$22,532, plus 31% of the amount over \$99,600.
- If x is over \$151,750, but not over \$271,050, the tax is \$38,698.50, plus 36% of the amount over \$151,750.
- If x is over \$271,050, the tax is \$81,646.50, plus 39.6% of the amount over \$271,050.

Clearly, if $x \le 0$, then

$$f(x) = 0.$$

If $0 < x \le 41{,}200$, then

$$f(x) = 0.15x.$$

Notice that the tax on a taxable income of \$41,200 is

$$f(41{,}200) = 0.15(41{,}200) = \$6{,}180.$$

Schedule Y-1—Use if your filing status is **Married filing jointly or Qualifying widow(er)**

If the amount on Form 1040 line 38, is *Over–*	*But not over–*	Enter on Form 1040, line 39		*of the amount over–*
\$0	\$41,200		15%	\$0
41,200	99,600	**\$6,180.00 +**	28%	41,200
99,600	151,750	**22,532.00 +**	31%	99,600
151,750	271,050	**38,698.50 +**	36%	151,750
271,050	----------	**81,646.50 +**	39.6%	271,050

FIGURE 3.41 Internal Revenue Service 1997 Schedule Y-1

If $41{,}200 < x \le 99{,}600$, then the amount over 41,200 is $x - 41{,}200$, so

$$f(x) = 6180 + 0.28(x - 41{,}200).$$

Because 6180 is 15% of 41,200, for a taxable income between \$41,200 and \$99,600 you are essentially taxed at the rate of 15% for the first \$41,200 of income and at the rate of 28% for the remaining income. Notice that the tax on \$99,600 is

$$\begin{aligned} f(99{,}600) &= 6180 + 0.28(99{,}600 - 41{,}200 \\ &= 6180 + 0.28(58{,}400) \\ &= 6180 + 16{,}352 = \$22{,}532. \end{aligned}$$

If $99{,}600 < x \le 151{,}750$, then the amount over 99,600 is $x - 99{,}600$, so

$$f(x) = 22{,}532 + 0.31(x - 99{,}600).$$

Because $22,532 is the tax on $99,600, for a taxable income between $99,600 and $151,750 you are taxed at the rate of 15% for the first $41,200 of income, at the rate of 28% for the next $58,400 of income ($99{,}600 - 41{,}200 = 58{,}400$), and at the rate of 31% for the remaining income. Notice that the tax on $151,750 is

$$\begin{aligned} f(151{,}750) &= 22{,}532 + 0.31(151{,}750 - 99{,}600) \\ &= 22{,}532 + 0.31(52{,}150) \\ &= 22{,}532 + 16{,}166.50 \\ &= \$38{,}698.50. \end{aligned}$$

If $151{,}750 < x \le 271{,}050$, then the amount over 151,750 is $x - 151{,}750$, so

$$f(x) = 38{,}698.50 + 0.36(x - 151{,}750).$$

Because $38,698.50 is the tax on $151,750, for a taxable income between $151,750 and $271,050 you are taxed at the rate of 15% for the first $41,200 of income, at the rate of 28% for the next $58,400 of income, at the rate of 31% for the next $52,150 of income ($151{,}750 - 99{,}600 = 52{,}150$), and at the rate of 36% for the remaining income. Notice that the tax on $271,050 is

$$\begin{aligned} f(271{,}050) &= 38{,}698.50 + 0.36(271{,}050 - 151{,}750) \\ &= 38{,}698.50 + 0.36(119{,}300) \\ &= 38{,}698.50 + 42{,}948 = \$81{,}646.50. \end{aligned}$$

If $x > 271{,}050$, then the amount over 271,050 is $x - 271{,}050$, so

$$f(x) = 81{,}646.50 + 0.396(x - 271{,}050).$$

Because $81,646.50 is the tax on $271,050, for a taxable income over $271,050 you are taxed at the rate of 15% for the first $41,200 of income, at the rate of 28% for the next $58,400 of income, at the rate of 31% for the next $52,150 of income, at the rate of 36% for the next $119,300 in income, and at the rate of 39.6% for the remaining income.

Summarizing all of these results gives the compound function

$$f(x) = \begin{cases} 0, & \text{if } x \le 0, \\ 0.15x, & \text{if } 0 < x \le 41{,}200, \\ 6{,}180 + 0.28(x - 41{,}200), & \text{if } 41{,}200 < x \le 99{,}600, \\ 22{,}532 + 0.31(x - 99{,}600), & \text{if } 99{,}600 < x \le 151{,}750, \\ 38{,}698.50 + 0.36(x - 151{,}750), & \text{if } 151{,}750 < x \le 271{,}050, \\ 81{,}646.50 + 0.396(x - 271{,}050), & \text{if } x > 271{,}050. \end{cases}$$

For example, the tax on a taxable income of $60,000 is (using the third expression)

$$\begin{aligned} f(60{,}000) &= 6{,}180 + 0.28(60{,}000 - 41{,}200) \\ &= 6{,}180 + 0.28(18{,}800) \\ &= 6{,}180 + 5{,}264 = \$11{,}444. \end{aligned}$$

With these formulas, you can geometrically depict the income tax function, as in Fig. 3.42.

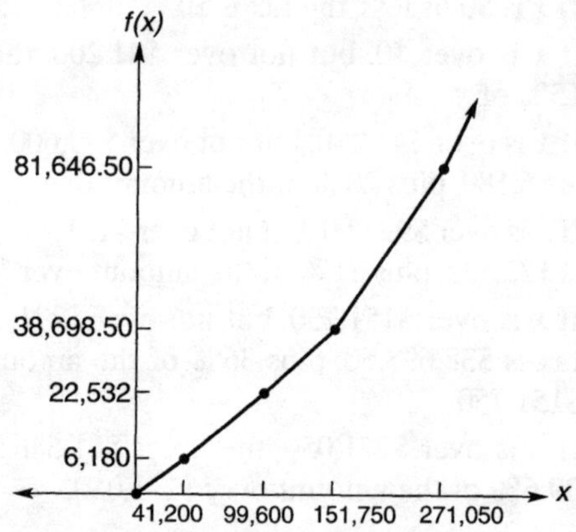

FIGURE 3.42 Income tax function.

■ Exercises

Use the preceding income tax function f to determine the tax on the given taxable income.

1. $100,000.

2. $25,350.

3. $280,000.

4. $162,700.

Chapter 5
Lines, Parabolas, and Systems

OBJECTIVE

To develop the notion of slope and different forms of equations of lines.

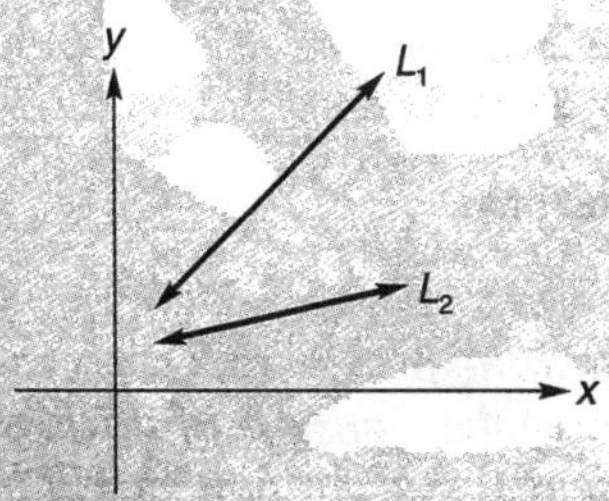

FIGURE 4.1 Line L_1 is "steeper" than L_2.

4.1 Lines

Slope of a Line

Many relationships between quantities can be represented conveniently by straight lines. One feature of a straight line is its "steepness." For example, in Fig. 4.1, line L_1 rises faster as it goes from left to right than does line L_2. In this sense, L_1 is steeper.

To measure the steepness of a line, we use the notion of *slope*. In Fig. 4.2, as we move along line L from $(1, 3)$ to $(3, 7)$, the x-coordinate increases

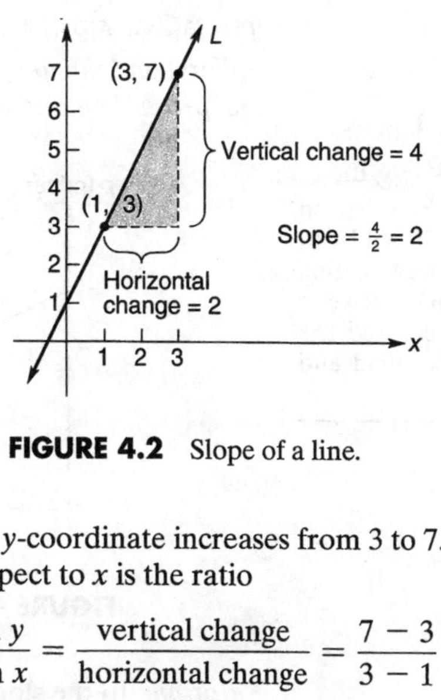

FIGURE 4.2 Slope of a line.

from 1 to 3, and the y-coordinate increases from 3 to 7. The average rate of change of y with respect to x is the ratio

$$\frac{\text{change in } y}{\text{change in } x} = \frac{\text{vertical change}}{\text{horizontal change}} = \frac{7-3}{3-1} = \frac{4}{2} = 2.$$

The ratio of 2 means that for each 1-unit increase in x, there is a 2-unit *increase* in y. Due to the increase, the line *rises* from left to right. It can be shown that, regardless of which two points on L are chosen to compute the ratio of the change in y to the change in x, the result is always 2, which we call the *slope* of the line.

DEFINITION

Let (x_1, y_1) and (x_2, y_2) be two different points on a nonvertical line. The slope of the line is

$$m = \frac{y_2 - y_1}{x_2 - x_1} \quad \left(= \frac{\text{vertical change}}{\text{horizontal change}}\right). \tag{1}$$

Having no slope does not mean having a slope of zero.

A vertical line does not have a slope, because any two points on it must have $x_1 = x_2$ [see Fig. 4.3(a)], which gives a denominator of zero in Eq. (1). For a horizontal line, any two points must have $y_1 = y_2$. [See Fig. 4.3(b).] This gives a numerator of zero in Eq. (1), and hence, the slope of the line is zero.

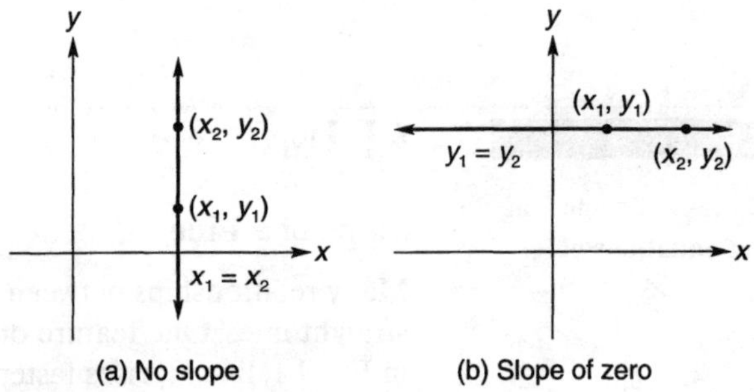

FIGURE 4.3 Vertical and horizontal lines.

This example shows how the slope can be interpreted.

EXAMPLE 1 Price–Quantity Relationship

The line in Fig. 4.4 shows the relationship between the price p of a widget (in dollars) and the quantity q of widgets (in thousands) that consumers will buy at that price. Find and interpret the slope.

Principles in Practice 1
Price–Time Relationship

A doctor purchased a new car in 1991 for $32,000. In 1994, he sold it to a friend for $26,000. Draw a line showing the relationship between the selling price of the car and the year in which it was sold. Find and interpret the slope.

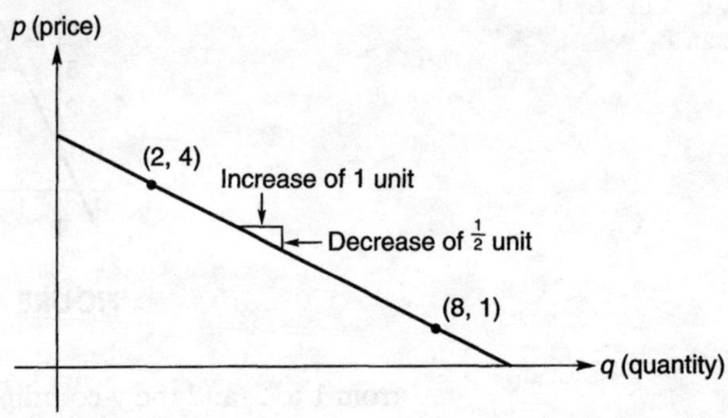

FIGURE 4.4 Price–quantity line.

Solution: In the slope formula (1), we replace the x's by q's and the y's by p's. Either point in Fig. 4.4 may be chosen as (q_1, p_1). Letting $(2, 4) = (q_1, p_1)$ and $(8, 1) = (q_2, p_2)$, we have

$$m = \frac{p_2 - p_1}{q_2 - q_1} = \frac{1 - 4}{8 - 2} = \frac{-3}{6} = -\frac{1}{2}.$$

The slope is negative, $-\frac{1}{2}$. This means that, for each 1-unit increase in quantity (one thousand widgets), there corresponds a **decrease** in price of $\frac{1}{2}$ (dollar per widget). Due to this decrease, the line **falls** from left to right. ■

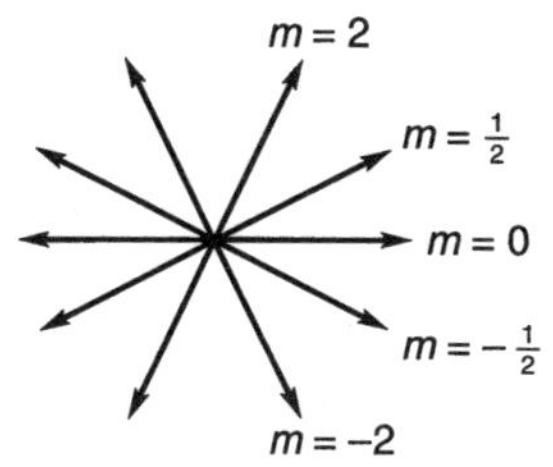

FIGURE 4.5 Slopes of lines.

In summary, we can characterize the orientation of a line by its slope:

Zero slope:	horizontal line.
Undefined slope:	vertical line.
Positive slope:	line rises from left to right.
Negative slope:	line falls from left to right.

Lines with different slopes are shown in Fig. 4.5. Notice that *the closer the slope is to 0, the more nearly horizontal is the line. The greater the absolute value of the slope, the more nearly vertical is the line.* We remark that two lines are parallel if and only if they have the same slope or are vertical.

Equations of Lines

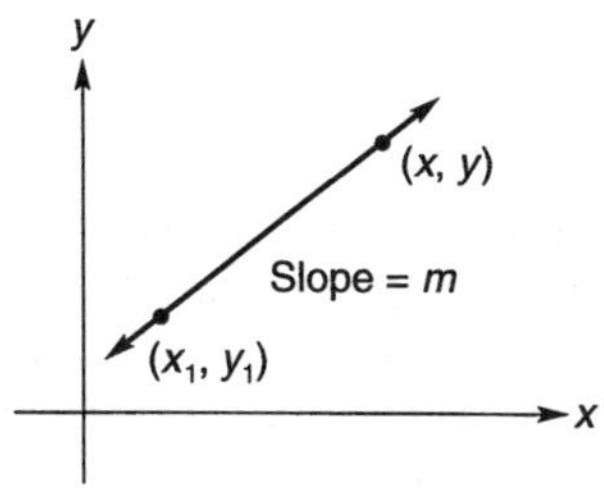

FIGURE 4.6 Line through (x_1, y_1) with slope m.

If we know a point on a line and the slope of the line, we can find an equation whose graph is that line. Suppose that line L has slope m and passes through the point (x_1, y_1). If (x, y) is *any* other point on L (see Fig. 4.6), we can find an algebraic relationship between x and y. Using the slope formula on the points (x_1, y_1) and (x, y) gives

$$\frac{y - y_1}{x - x_1} = m,$$

$$y - y_1 = m(x - x_1). \tag{2}$$

Every point on L satisfies Eq. (2). It is also true that every point satisfying Eq. (2) must lie on L. Thus, Eq. (2) is an equation for L and is given a special name:

$$y - y_1 = m(x - x_1)$$

is a **point–slope form** of an equation of the line through (x_1, y_1) with slope m.

Principles in Practice 2
Point–Slope Form

A new applied mathematics program at a university has grown in enrollment by 14 students per year for the last five years. If the program enrolled 50 students in its third year, what is an equation for the number of students S in the program as a function of the number of years T since its inception?

EXAMPLE 2 Point–Slope Form

Find an equation of the line that has slope 2 and passes through $(1, -3)$.

Solution: Using a point–slope form with $m = 2$ and $(x_1, y_1) = (1, -3)$ gives

$$y - y_1 = m(x - x_1),$$

$$y - (-3) = 2(x - 1),$$

$$y + 3 = 2x - 2,$$

which can be rewritten as

$$2x - y - 5 = 0.$$

■

An equation of the line passing through two given points can be found easily, as Example 3 shows.

EXAMPLE 3 Determining a Line from Two Points

Find an equation of the line passing through $(-3, 8)$ and $(4, -2)$.

Principles in Practice 3
Determining a Line from Two Points

Find an equation of the line passing through the given points. A temperature of 41°F is equivalent to 5°C, and a temperature of 77°F is equivalent to 25°C.

Solution:

Strategy: First we find the slope of the line from the given points. Then we substitute the slope and one of the points into a point–slope form.

The line has slope

$$m = \frac{-2 - 8}{4 - (-3)} = -\frac{10}{7}.$$

Using a point–slope form with $(-3, 8)$ as (x_1, y_1) gives

$$y - 8 = -\tfrac{10}{7}[x - (-3)],$$
$$y - 8 = -\tfrac{10}{7}(x + 3),$$
$$7y - 56 = -10x - 30,$$

or

$$10x + 7y - 26 = 0.$$ ■

Choosing $(4, -2)$ as (x_1, y_1) would give an equivalent result.

Recall that a point $(0, b)$ where a graph intersects the y-axis is called a y-intercept (Fig. 4.7). If the slope m and y-intercept b of a line are known, an equation for the line is [by using a point–slope form with $(x_1, y_1) = (0, b)$]

$$y - b = m(x - 0).$$

Solving for y gives $y = mx + b$, called the *slope–intercept form* of an equation of the line:

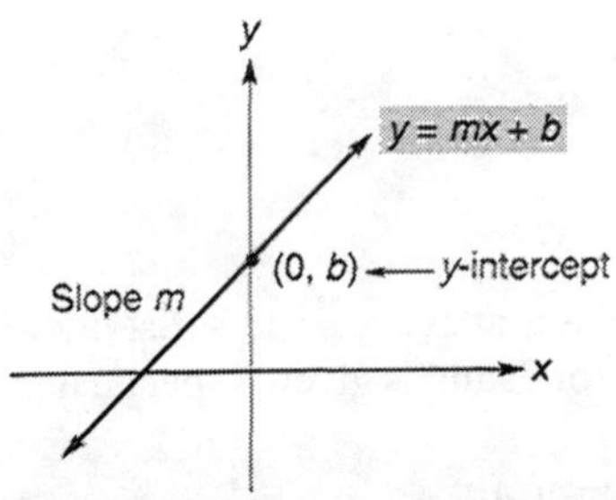

FIGURE 4.7 Line with slope m and y-intercept b.

$$y = mx + b$$

is the **slope–intercept form** of an equation of the line with slope m and y-intercept b.

EXAMPLE 4 Slope–Intercept Form

Find an equation of the line with slope 3 and y-intercept −4.

Solution: Using the slope–intercept form $y = mx + b$ with $m = 3$ and $b = -4$ gives

$$y = 3x + (-4),$$
$$y = 3x - 4.$$ ■

Principles in Practice 4
Finding the Slope and y-Intercept of a Line

One formula for the recommended dosage (in milligrams) of medication for a child t years old is $y = \frac{1}{24}(t + 1)a$, where a is the adult dosage. For an over-the-counter pain reliever, $a = 1000$. Find the slope and y-intercept of this equation.

EXAMPLE 5 Find the Slope and y-Intercept of a Line

Find the slope and y-intercept of the line with equation $y = 5(3 - 2x)$.

Solution:

Strategy: We shall rewrite the equation so it has the slope–intercept form $y = mx + b$. Then the slope is the coefficient of x and the y-intercept is the constant term.

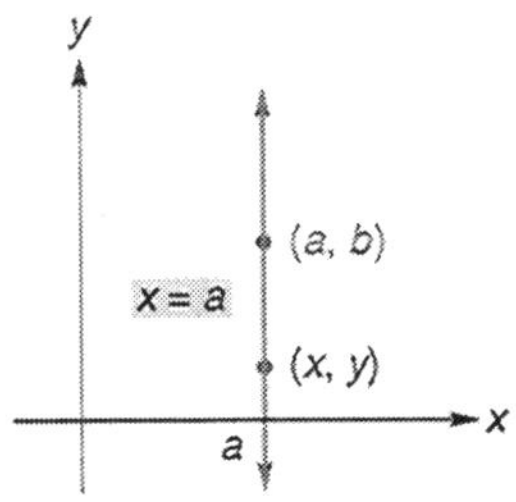

FIGURE 4.8 Vertical line through (a, b).

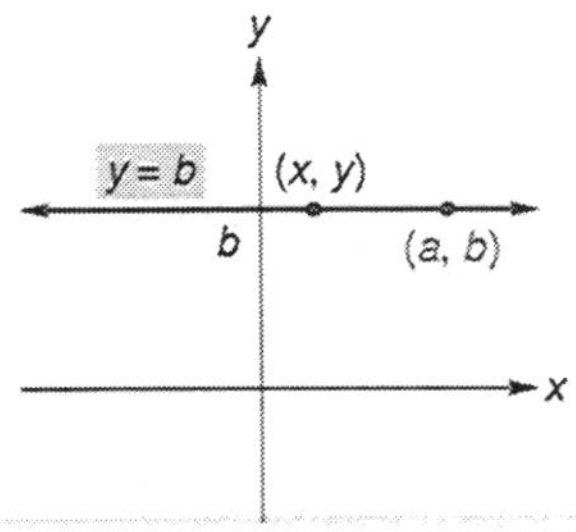

FIGURE 4.9 Horizontal line through (a, b).

We have

$$y = 5(3 - 2x),$$
$$y = 15 - 10x,$$
$$y = -10x + 15.$$

Thus, $m = -10$ and $b = 15$, so the slope is -10 and the y-intercept is 15. ■

If a *vertical* line passes through (a, b) (see Fig. 4.8), then any other point (x, y) lies on the line if and only if $x = a$. The y-coordinate can have any value. Hence, an equation of the line is $x = a$. Similarly, an equation of the *horizontal* line passing through (a, b) is $y = b$. (See Fig. 4.9.) Here the x-coordinate can have any value.

EXAMPLE 6 Equations of Horizontal and Vertical Lines

a. An equation of the vertical line through $(-2, 3)$ is $x = -2$. An equation of the horizontal line through $(-2, 3)$ is $y = 3$.

b. The x- and y-axes are horizontal and vertical lines, respectively. Because $(0, 0)$ lies on both axes, an equation of the x-axis is $y = 0$, and an equation of the y-axis is $x = 0$. ■

From our discussions, we can show that every straight line is the graph of an equation of the form $Ax + By + C = 0$, where A, B, and C are constants and A and B are not both zero. We call this a **general linear equation** (or an *equation of the first degree*) **in the variables *x* and *y*,** and x and y are said to be **linearly related.** For example, a general linear equation for $y = 7x - 2$ is $(-7)x + (1)y + (2) = 0$. Conversely, the graph of a general linear equation is a straight line. Table 4.1 gives the various forms of equations of straight lines.

Table 4.1 provides a good summary for you.

Do not confuse the forms of equations of horizontal and vertical lines. Remember which one has the form $x =$ constant and which one has the form $y =$ constant.

TABLE 4.1 Forms of Equations of Straight Lines

Point-slope form	$y - y_1 = m(x - x_1)$
Slope-intercept form	$y = mx + b$
General linear form	$Ax + By + C = 0$
Vertical line	$x = a$
Horizontal line	$y = b$

Principles in Practice 5
Converting Forms of Equations of Lines

Find a general linear form of the Fahrenheit–Celsius conversion equation whose slope–intercept form is $F = \frac{9}{5}C + 32$.

EXAMPLE 7 Converting Forms of Equations of Lines

a. *Find a general linear form of the line whose slope–intercept form is*

$$y = -\tfrac{2}{3}x + 4.$$

Solution: Getting one side to be 0, we obtain

$$\frac{2}{3}x + y - 4 = 0,$$

which is a general linear form with $A = \frac{2}{3}$, $B = 1$, and $C = -4$. An alternative general form can be obtained by clearing fractions:

$$2x + 3y - 12 = 0.$$

This illustrates that a general linear form of a line is not unique.

b. *Find the slope–intercept form of the line having a general linear form* $3x + 4y - 2 = 0$.

Solution: We want the form $y = mx + b$, so we solve the given equation for y. We have

$$3x + 4y - 2 = 0,$$
$$4y = -3x + 2,$$
$$y = -\frac{3}{4}x + \frac{1}{2},$$

which is the slope–intercept form. Note that the line has slope $-\frac{3}{4}$ and y-intercept $\frac{1}{2}$. ■

Principles in Practice 6

Graphing a General Linear Equation

Sketch the graph of the Fahrenheit–Celsius conversion equation that you found in Principles in Practice 5. How could you use this graph to convert a Celsius temperature to Fahrenheit?

EXAMPLE 8 Graphing a General Linear Equation

Sketch the graph of $2x - 3y + 6 = 0$.

Solution:

Strategy: Since this is a general linear equation, its graph is a straight line. Thus, we need only determine two different points on the graph in order to sketch it. We shall find the intercepts.

If $x = 0$, then $-3y + 6 = 0$, so the y-intercept is 2. If $y = 0$, then $2x + 6 = 0$, so the x-intercept is -3. We now draw the line passing through $(0, 2)$ and $(-3, 0)$. (See Fig. 4.10.) ■

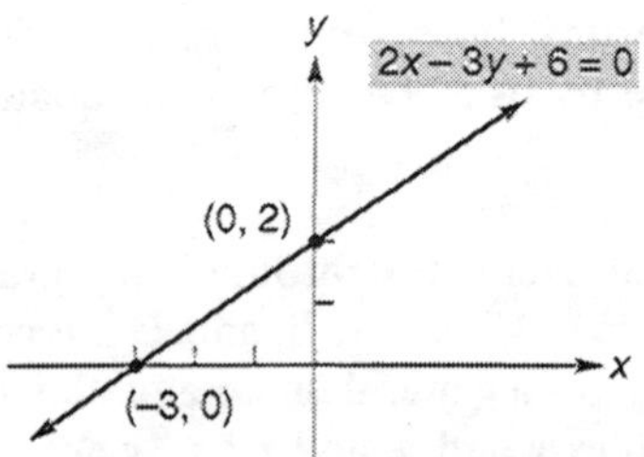

FIGURE 4.10 Graph of $2x - 3y + 6 = 0$.

TECHNOLOGY

To graph the equation of Example 8 with a graphics calculator, we first express y in terms of x:

$$2x - 3y + 6 = 0,$$
$$3y = 2x + 6,$$
$$y = \tfrac{1}{3}(2x + 6).$$

Essentially, y is expressed as a function of x; the graph is shown in Fig. 4.11.

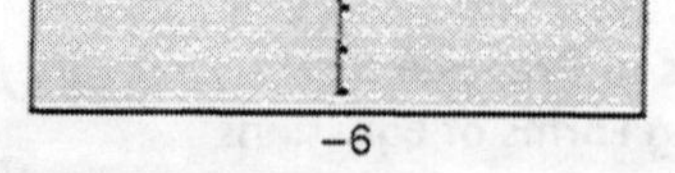

FIGURE 4.11 Calculator graph of $2x - 3y + 6 = 0$.

Parallel and Perpendicular Lines

As stated previously, there is a rule for parallel lines:

Parallel Lines

Two lines are parallel if and only if they have the same slope or are vertical.

There is also a rule for perpendicular lines. Look back to Fig. 4.5 and observe that the line with slope $-\frac{1}{2}$ is perpendicular to the line with slope 2. The fact that the slope of either of these lines is the negative reciprocal of the slope of the other line is not a coincidence, as the following rule states.

Perpendicular Lines

Two lines with slopes m_1 and m_2 are perpendicular to each other if, and only if,

$$m_1 = -\frac{1}{m_2}.$$

Moreover, a horizontal line and a vertical line are perpendicular to each other.

Principles in Practice 7

Parallel and Perpendicular Lines

Show that a triangle with vertices at $A(0, 0)$, $B(6, 0)$, and $C(7, 7)$ is not a right triangle.

EXAMPLE 9 Parallel and Perpendicular Lines

Figure 4.12 shows two lines passing through $(3, -2)$. *One is parallel to the line* $y = 3x + 1$, *and the other is perpendicular to it. Find equations of these lines.*

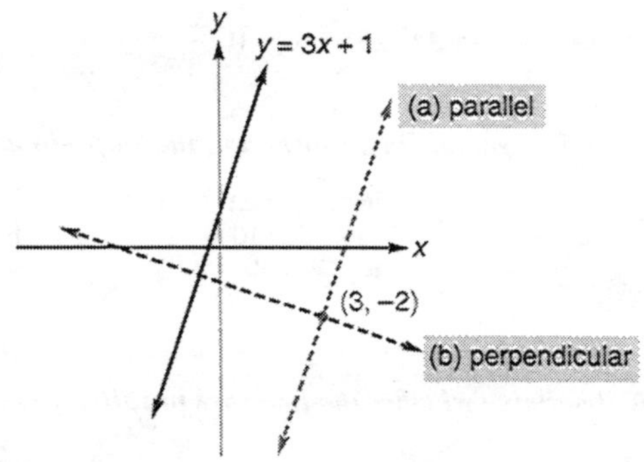

FIGURE 4.12 Lines parallel and perpendicular to $y = 3x + 1$ (Example 9).

Solution: The slope of $y = 3x + 1$ is 3. Thus, the line through $(3,-2)$ that is *parallel* to $y = 3x + 1$ also has slope 3. Using a point–slope form, we get

$$y - (-2) = 3(x - 3),$$
$$y + 2 = 3x - 9,$$
$$y = 3x - 11.$$

The slope of a line *perpendicular* to $y = 3x + 1$ must be $-\frac{1}{3}$ (the negative reciprocal of 3). Using a point–slope form, we get

$$y - (-2) = -\tfrac{1}{3}(x - 3),$$
$$y + 2 = -\tfrac{1}{3}x + 1,$$
$$y = -\tfrac{1}{3}x - 1.$$

■

■ Exercise 4.1

In Problems 1–8, find the slope of the straight line that passes through the given points.

1. $(4, 1), (7, 10)$.
2. $(-3, 11), (2, 1)$.
3. $(4, -2), (-6, 3)$.
4. $(2, -4), (3, -4)$.
5. $(5, 3), (5, -8)$.
6. $(0, -6), (3, 0)$.
7. $(5, -2), (4, -2)$.
8. $(1, -6), (1, 0)$.

In Problems 9–24, find a general linear equation ($Ax + By + C = 0$) of the straight line that has the indicated properties, and sketch each line.

9. Passes through $(2, 8)$ and has slope 6.
10. Passes through origin and has slope -5.
11. Passes through $(-2, 5)$ and has slope $-\frac{1}{4}$.
12. Passes through $(-\frac{5}{2}, 5)$ and has slope $\frac{1}{3}$.
13. Passes through $(-6, 1)$ and $(1, 4)$.
14. Passes through $(7, 1)$ and $(7, -5)$.
15. Passes through $(3, -1)$ and $(-2, -9)$.
16. Passes through $(0, 0)$ and $(2, 3)$.
17. Has slope 2 and y-intercept 4.
18. Has slope 7 and y-intercept -5.
19. Has slope $-\frac{1}{2}$ and y-intercept -3.
20. Has slope 0 and y-intercept $-\frac{1}{2}$.
21. Is horizontal and passes through $(-3, -2)$.
22. Is vertical and passes through $(-1, 4)$.
23. Passes through $(2, -3)$ and is vertical.
24. Passes through the origin and is horizontal.

In Problems 25–34, find, if possible, the slope and y-intercept of the straight line determined by the equation, and sketch the graph.

25. $y = 2x - 1$.
26. $x - 1 = 5$.
27. $x + 2y - 3 = 0$.
28. $y + 4 = 7$.
29. $x = -5$.
30. $x - 1 = 5y + 3$.
31. $y = 3x$.
32. $y - 7 = 3(x - 4)$.
33. $y = 1$.
34. $2y - 3 = 0$.

In Problems 35–40, find a general linear form and the slope–intercept form of the given equation.

35. $2x = 5 - 3y$.
36. $3x + 2y = 6$.
37. $4x + 9y - 5 = 0$.
38. $2(x - 3) - 4(y + 2) = 8$.
39. $\frac{x}{2} - \frac{y}{3} = -4$.
40. $y = \frac{1}{300}x + 8$.

In Problems 41–50, determine whether the lines are parallel, perpendicular, or neither.

41. $y = 7x + 2, \quad y = 7x - 3$.
42. $y = 4x + 3, \quad y = 5 + 4x$.
43. $y = 5x + 2, \quad -5x + y - 3 = 0$.
44. $y = x, \quad y = -x$.
45. $x + 2y + 1 = 0, \quad y = -2x$.
46. $x + 2y = 0, \quad x + y - 4 = 0$.
47. $y = 3, \quad x = -\frac{1}{3}$.
48. $x = 3, \quad x = -4$.
49. $3x + y = 4, \quad x - 3y + 1 = 0$.
50. $x - 1 = 0, \quad y = 0$.

In Problems 51–60, find an equation of the line satisfying the given conditions. Give the answer in slope–intercept form if possible.

51. Passing through $(-1, 3)$ and parallel to $y = 4x - 5$.
52. Passing through $(2, -8)$ and parallel to $x = -4$.
53. Passing through $(2, 1)$ and parallel to $y = 2$.
54. Passing through $(3, -4)$ and parallel to $y = 3 + 2x$.
55. Perpendicular to $y = 3x - 5$ and passing through $(3, 4)$.
56. Perpendicular to $y = -4$ and passing through $(1, 1)$.
57. Passing through $(7, 4)$ and perpendicular to $y = -4$.
58. Passing through $(-5, 4)$ and perpendicular to the line $2y = -x + 1$.
59. Passing through $(-7, -5)$ and parallel to the line $2x + 3y + 6 = 0$.
60. Passing through $(-2, 1)$ and parallel to the y-axis.

---

61. A straight line passes through $(1, 2)$ and $(-3, 8)$. Find the point on it that has an x-coordinate of 5.

62. A straight line has slope 2 and y-intercept $(0, 1)$. Does the point $(-1, -1)$ lie on the line?

63. **Stock** In 1986, the stock in a biotechnology company traded for \$30 per share. In 1996 the company started having trouble, and the stock price dropped to \$10 per share. Draw a line showing the relationship between the price per share and the year in which it traded, with years on the x-axis and price on the y-axis. Find and interpret the slope.

64. **Speed of Sound** A graph of the speed of sound, S (in meters per second), at sea level versus the temperature T of the air (in Celsius) has a slope of 0.61. The equation that describes the relationship between speed of sound and air temperature is $S = 0.61T + b$. When the temperature is 15°C, a researcher measures the speed of sound as 340.55 meters per second. Determine b to complete the equation.

In Problems 65–66, find an equation of the line describing the following information.

65. **Home Runs** In one season, a major league baseball player has hit 14 home runs by the end of the third month and 20 home runs by the end of the fifth month.

66. **Business** A delicatessen owner starts her business with debts of \$100,000. After operating for five years, she has accumulated a profit of \$40,000.

67. **Due Date** The length, L, of a human fetus more than 12 weeks old can be estimated by the formula $L = 1.53t - 6.7$, where L is in centimeters and t is in weeks from conception. An obstetrician uses the length of a fetus, measured by ultrasound, to determine the approximate age of the fetus and establish a due date for the mother. The formula must be rewritten to result in an age t, given a fetal length L. Find the slope and L-intercept of the equation.

68. **Discus Throw** A mathematical model can approximate the winning distance for the Olympic discus throw by the formula $d = 175 + 1.75t$, where d is in feet and $t = 0$ corresponds to the year 1948. Find a general linear form of this equation.

69. **Campus Map** A coordinate map of a college campus gives the coordinates (x, y) of three major buildings as follows: computer center, $(3.5, -1)$; engineering lab, $(0.5, 0)$; and library $(-1, -4.5)$. Find the equations (in slope–intercept form) of the straight-line paths connecting (a) the engineering lab with the computer center and (b) the engineering lab with the library. Show that these two paths are perpendicular to each other.

70. **Geometry** Show that the points $A(0, 0)$, $B(0, 4)$, $C(2, 3)$, and $D(2, 7)$ are the vertices of a parallelogram. (Opposite sides of a parallelogram are parallel.)

71. **Approach Angle** A small plane is landing at an airport with an approach angle of 45 degrees, or slope of -1. The plane begins its descent when it has an elevation of 3300 feet. Find the equation that describes the relationship between the craft's altitude and distance traveled, assuming that at distance 0 it starts the approach angle. Graph your equation on a graphics calculator. What does the graph tell you about the approach if the airport is 4000 feet from where the plane starts its landing.

72. **Cost Equation** The average daily cost C for a room at a city hospital has risen by \$42.50 per year for the years 1980 through 1990. If the average cost in 1987 was \$542.50, what is an equation which describes the average cost during this decade, as a function of the number of years T since 1980.

73. **Revenue Equation** A small business predicts its revenue growth by a straight-line method with a slope of \$50,000 per year. In its fifth year it had revenues of \$330,000. Find an equation that describes the relationship between revenues R and the number of years T since it opened for business.

74. Graph $y = 1.3x + 7$ and verify that the y-intercept is 7.

75. Graph the lines whose equations are

$$y = 1.5x + 1,$$

$$y = 1.5x - 1,$$

and

$$y = 1.5x + 2.5.$$

What do you observe about the orientation of these lines? Why would you expect this result from the equations of the lines themselves?

76. Graph the line $y = 3.4x - 2.3$. Find the coordinates of any two points on the line, and use them to estimate the slope. What is the actual slope of the line?

77. Using the standard window, graph the lines with equations

$$0.1875x - 0.3y + 0.94 = 0$$

and

$$0.32x + 0.2y + 1.01 = 0$$

on the same viewing rectangle. Now, change the window to a square window (for example, on a TI-82, use ZOOM, ZSquare). Notice that the lines appear to be perpendicular to each other. Prove that this is indeed the case.

OBJECTIVE

To develop the notion of demand and supply curves and to introduce linear functions.

4.2 Applications and Linear Functions

Many situations in economics can be described by using straight lines, as Example 1 shows.

EXAMPLE 1 Production Levels

Suppose that a manufacturer uses 100 lb of material to produce products A and B, which require 4 lb and 2 lb of material per unit, respectively. If x and y denote the number of units produced of A and B, respectively, then all levels of production are given by the combinations of x and y that satisfy the equation

$$4x + 2y = 100, \qquad \text{where } x, y \geq 0.$$

Thus, the levels of production of A and B are linearly related. Solving for y gives

$$y = -2x + 50 \qquad \text{(slope–intercept form)},$$

so the slope is -2. The slope reflects the rate of change of the level of production of B with respect to the level of production of A. For example, if 1 more unit of A is to be produced, it will require 4 more pounds of material, resulting in $\frac{4}{2} = 2$ *fewer* units of B. Accordingly, as x increases by 1 unit, the corresponding value of y decreases by 2 units. To sketch the graph of $y = -2x + 50$, we can use the y-intercept (0, 50) and the fact that when $x = 10$, $y = 30$. (See Fig. 4.13.) ■

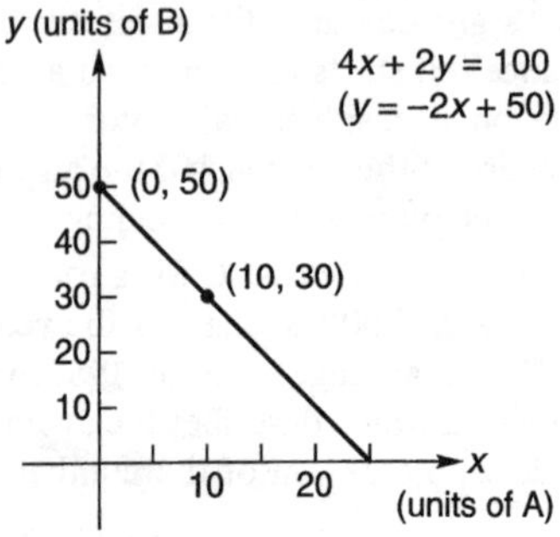

FIGURE 4.13 Linearly related production levels.

Principles in Practice 1
Production Levels

A sporting-goods manufacturer allocates 1000 units of time per day to make skis and ski boots. If it takes 8 units of time to make a ski, and 14 units of time to make a boot, find an equation to describe all possible production levels of the two products.

Demand and Supply Curves

For each price level of a product, there is a corresponding quantity of that product which consumers will demand (that is, purchase) during some time period. Usually, the higher the price, the smaller is the quantity demanded; as the price falls, the quantity demanded increases. If the price per unit of the product is given by p and the corresponding quantity (in units) is given by q, then an equation relating p and q is called a **demand equation.** Its graph is called a **demand curve.** Figure 4.14(a) shows a demand curve. In keeping with the practice of most economists, the horizontal axis is the q-axis and the vertical axis is the p-axis. We shall assume that the price per unit is given in dollars and the period is one week. Thus, the point (a, b) in Fig. 4.14(a) indicates that, at a price of b dollars per unit, consumers will demand a units per week. Since negative prices or quantities are not meaningful, both a and b must be nonnegative. For most products, an increase in the quantity demanded corresponds to a decrease in price. Thus, a demand curve typically falls from left to right, as in Fig. 4.14(a).

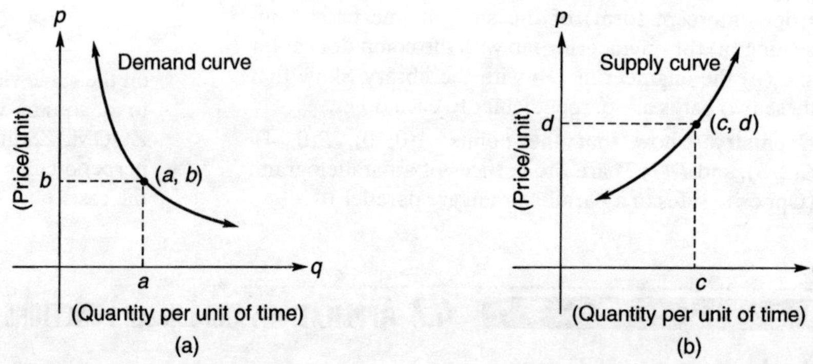

FIGURE 4.14 Demand and supply curves.

In response to various prices, there is a corresponding quantity of product that *producers* are willing to supply to the market during some time period. Usually, the higher the price per unit, the larger is the quantity that producers are willing to supply; as the price falls, so will the quantity supplied. If p denotes the price per unit and q denotes the corresponding quantity, then an equation relating p and q is called a **supply equation,** and its graph is called a **supply curve.** Figure 4.14(b) shows a supply curve. If p is in dollars and the period is one week, then the point (c, d) indicates that, at a price of d dollars each, producers will supply c units per week. As before, c and d are nonnegative. A supply curve usually rises from left to right, as in Fig. 4.14(b). This indicates that a producer will supply more of a product at higher prices.

Typically, a demand curve falls from left to right and a supply curve rises from left to right. However, there are exceptions. For example, the demand for insulin could be represented by a vertical line, since this demand can remain constant regardless of price.

We shall now focus on demand and supply curves that are straight lines (Fig. 4.15). They are called *linear* demand and *linear* supply curves. Such curves have equations in which p and q are linearly related. Because a demand curve typically falls from left to right, a linear demand curve has a negative slope. [See Fig. 4.15(a).] However, the slope of a linear supply curve is positive, because the curve rises from left to right. [See Fig. 4.15(b).]

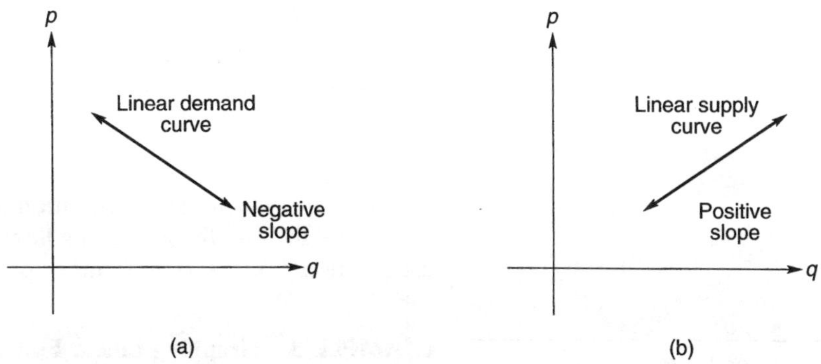

FIGURE 4.15 Linear demand and supply curves.

Principles in Practice 2
Finding a Demand Equation

The demand per week for 26-inch television sets is 1200 units when the price is \$575 each and 800 units when the price is \$725 each. Find the demand equation for the sets, assuming that it is linear.

EXAMPLE 2 Finding a Demand Equation

Suppose the demand per week for a product is 100 *units when the price is* \$58 *per unit and* 200 *units at* \$51 *each. Determine the demand equation, assuming that it is linear.*

Solution:

Strategy: Since the demand equation is linear, the demand curve must be a straight line. We are given that quantity q and price p are linearly related such that $p = 58$ when $q = 100$ and $p = 51$ when $q = 200$. Thus, the given data can be represented in a q, p-coordinate plane [see Fig. 4.15(a)] by points (100, 58) and (200, 51). With these points, we can find an equation of the line—that is, the demand equation.

The slope of the line passing through (100, 58) and (200, 51) is

$$m = \frac{51 - 58}{200 - 100} = -\frac{7}{100}.$$

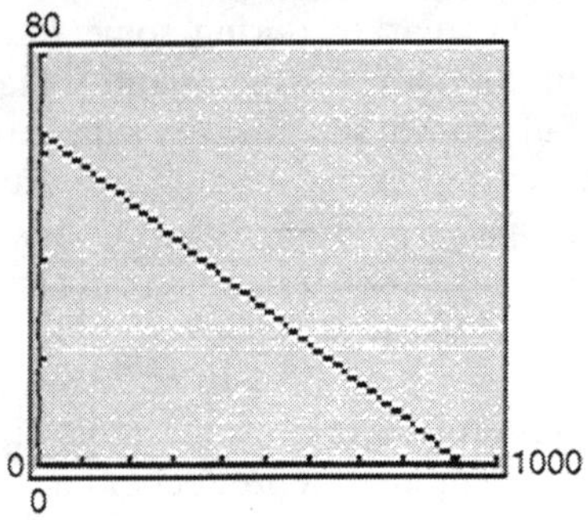

FIGURE 4.16 Graph of demand function $p = -\frac{7}{100}q + 65$.

An equation of the line (point–slope form) is

$$p - p_1 = m(q - q_1),$$

$$p - 58 = -\frac{7}{100}(q - 100).$$

Simplifying gives the demand equation

$$p = -\frac{7}{100}q + 65. \qquad (1)$$

Customarily, a demand equation (as well as a supply equation) expresses p in terms of q and actually defines a function of q. For example, Eq. (1) defines p as a function of q and is called the *demand function* for the product. (See Fig. 4.16). ■

Linear Functions

A *linear function* was described in Sec. 3.2. Here is a formal definition.

DEFINITION

*A function f is a **linear function** if and only if $f(x)$ can be written in the form $f(x) = ax + b$, where a and b are constants and $a \neq 0$.*

Suppose that $f(x) = ax + b$ is a linear function, and let $y = f(x)$. Then $y = ax + b$, which is an equation of a straight line with slope a and y-intercept b. Thus, **the graph of a linear function is a straight line.** We say that the function $f(x) = ax + b$ has slope a.

Principles in Practice 3
Graphing Linear Functions

A computer repair company charges a fixed amount plus an hourly rate for a service call. If x is the number of hours needed for a service call, the total cost of a call is described by the function $f(x) = 40x + 60$. Graph the function by finding and plotting two points.

EXAMPLE 3 Graphing Linear Functions

a. *Graph* $f(x) = 2x - 1$.

Solution: Here f is a linear function (with slope 2), so its graph is a straight line. Since two points determine a straight line, we need only plot two points and then draw a line through them. [See Fig. 4.17(a).] Note that one of the points plotted is the vertical-axis intercept, -1, which occurs when $x = 0$.

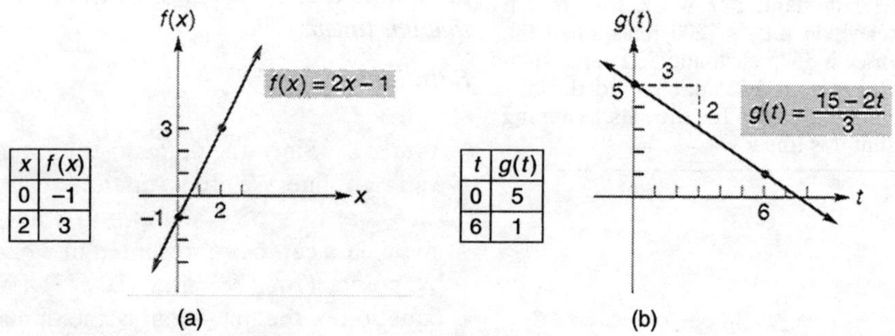

FIGURE 4.17 Graphs of linear functions.

b. *Graph* $g(t) = \dfrac{15 - 2t}{3}$.

Solution: Notice that g is a linear function, because we can express it in the form $g(t) = at + b$.

$$g(t) = \frac{15 - 2t}{3} = \frac{15}{3} - \frac{2t}{3} = -\frac{2}{3}t + 5.$$

The graph of g is shown in Fig. 4.17(b). Since the slope is $-\frac{2}{3}$, observe that as t increases by 3 units, $g(t)$ *decreases* by 2. ■

Principles in Practice 4
Determining a Linear Function

The height of children between the ages of 6 years and 10 years can be modeled by a linear function of age t in years. The height of one child changes by 2.3 inches per year, and she is 50.6 inches tall at age 8. Find a function that describes the height of this child at age t.

EXAMPLE 4 Determining a Linear Function

Suppose f is a linear function with slope 2 and $f(4) = 8$. Find $f(x)$.

Solution: Since f is linear, it has the form $f(x) = ax + b$. The slope is 2, so $a = 2$, and we have

$$f(x) = 2x + b. \qquad (2)$$

Now we determine b. Since $f(4) = 8$, we replace x by 4 in Eq. (2) and solve for b:

$$f(4) = 2(4) + b,$$
$$8 = 8 + b,$$
$$0 = b.$$

Hence, $f(x) = 2x$. ■

Principles in Practice 5
Determining a Linear Function

An antique necklace is expected to be worth \$360 after 3 years and \$640 after 7 years. Find a function that describes the value of the necklace after x years.

EXAMPLE 5 Determining a Linear Function

If $y = f(x)$ is a linear function such that $f(-2) = 6$ and $f(1) = -3$, find $f(x)$.

Solution:

Strategy: The function values correspond to points on the graph of f. With these points we can determine an equation of the line and hence the linear function.

The condition that $f(-2) = 6$ means that when $x = -2$, then $y = 6$. Thus, $(-2, 6)$ lies on the graph of f, which is a straight line. Similarly, $f(1) = -3$ implies that $(1, -3)$ also lies on the line. If we set $(x_1, y_1) = (-2, 6)$ and $(x_2, y_2) = (1, -3)$, the slope of the line is given by

$$m = \frac{y_2 - y_1}{x_2 - x_1} = \frac{-3 - 6}{1 - (-2)} = \frac{-9}{3} = -3.$$

We can find an equation of the line by using a point-slope form:

$$y - y_1 = m(x - x_1),$$
$$y - 6 = -3[x - (-2)],$$
$$y - 6 = -3x - 6,$$
$$y = -3x.$$

Because $y = f(x)$, $f(x) = -3x$. Of course, the same result is obtained if we set $(x_1, y_1) = (1, -3)$. ■

In many studies, data are collected and plotted on a coordinate system. An analysis of the results may indicate a functional relationship between the variables involved. For example, the data points may be approximated by points on a straight line. This would indicate a linear functional relationship, such as the one in the next example.

EXAMPLE 6 Diet for Hens

In testing an experimental diet for hens, it was determined that the average live weight w (in grams) of a hen was statistically a linear function of the number of days d after the diet began, where $0 \le d \le 50$. Suppose the average weight of a hen beginning the diet was 40 grams and 25 days later it was 675 grams.

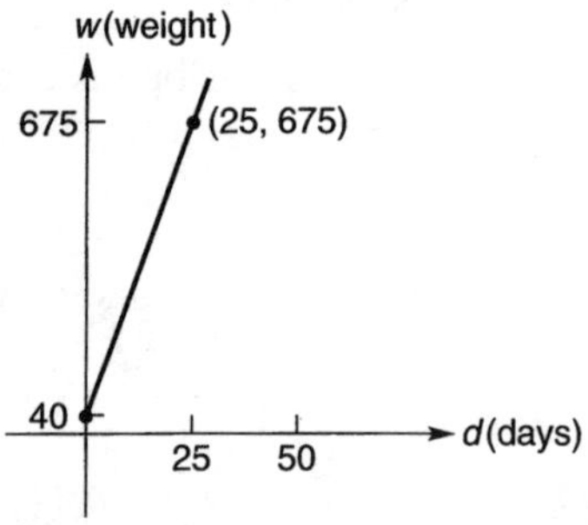

FIGURE 4.18 Linear function describing diet for hens.

a. *Determine w as a linear function of d.*

Solution: Since w is a linear function of d, its graph is a straight line. When $d = 0$ (the beginning of the diet), $w = 40$. Thus, (0, 40) lies on the graph. (See Fig. 4.18.) Similarly, (25, 675) lies on the graph. If we set $(d_1, w_1) = (0, 40)$ and $(d_2, w_2) = (25, 675)$, the slope of the line is

$$m = \frac{w_2 - w_1}{d_2 - d_1} = \frac{675 - 40}{25 - 0} = \frac{635}{25} = \frac{127}{5}.$$

Using a point–slope form, we have

$$w - w_1 = m(d - d_1),$$

$$w - 40 = \frac{127}{5}(d - 0),$$

$$w - 40 = \frac{127}{5}d,$$

$$w = \frac{127}{5}d + 40,$$

which expresses w as a linear function of d.

b. *Find the average weight of a hen when $d = 10$.*

Solution: When $d = 10$, $w = \frac{127}{5}(10) + 40 = 254 + 40 = 294$. Thus, the average weight of a hen 10 days after the beginning of the diet is 294 grams.

▪ Exercise 4.2

In Problems 1–6, find the slope and vertical-axis intercept of the linear function, and sketch the graph.

1. $y = f(x) = -4x$.

2. $y = f(x) = x + 1$.

3. $g(t) = 2t - 4$.

4. $g(t) = 2(4 - t)$.

5. $h(q) = \frac{7 - q}{2}$.

6. $h(q) = 0.5q + 0.25$.

In Problems 7–14, find f(x) if f is a linear function that has the given properties.

7. slope $= 4$, $f(2) = 8$.

8. $f(0) = 3$, $f(4) = -5$.

9. $f(1) = 2$, $f(-2) = 8$.

10. slope $= -6$, $f(\frac{1}{2}) = -2$.

11. slope $= -\frac{1}{2}, f(-\frac{1}{2}) = 4$.

12. $f(1) = 1$, $f(2) = 2$.

13. $f(-2) = -1$, $f(-4) = -3$.

14. slope $= 0.01$, $f(0.1) = 0.01$.

15. Demand Equation Suppose consumers will demand 40 units of a product when the price is \$12 per unit and 25 units when the price is \$18 each. Find the demand equation, assuming that it is linear. Find the price per unit when 30 units are demanded.

16. **Demand Equation** The demand per week for a best-selling book is 26,000 books when the price is \$16 each, and 10,000 books when the price is \$24 each. Find the demand equation for the book, assuming that it is linear.

17. **Supply Equation** A refrigerator manufacturer will produce 300 units when the price is \$840, and 220 units when the price is \$640. Assume that price p and quantity q produced are linearly related and find the supply equation.

18. **Supply Equation** Suppose a manufacturer of shoes will place on the market 50 (thousand pairs) when the price is 35 (dollars per pair) and 35 when the price is 30. Find the supply equation, assuming that price p and quantity q are linearly related.

19. **Cost Equation** Suppose the cost to produce 10 units of a product is \$40 and the cost of 20 units is \$70. If cost c is linearly related to output q, find a linear equation relating c and q. Find the cost to produce 35 units.

20. **Cost Equation** An advertiser goes to a printer and is charged \$73 for 100 copies of one flyer and \$82 for 400 copies of another flyer. This printer charges a fixed setup cost plus a charge for every copy of single-page flyers. Find a function that describes the cost of a printing job, if x is the number of copies made.

21. **Electric Rates** An electric utility company charges residential customers 11.5 cents per kilowatt hour plus a base charge each month. One customer's monthly bill comes to \$51.65 for 380 kilowatt hours. Find a linear function that describes the total monthly charges for electricity if x is the number of kilowatt hours used in a month.

22. **Radiation Therapy** A cancer patient is to receive drug and radiation therapies. Each cubic centimeter of the drug to be used contain 200 curative units, and each minute of radiation exposure gives 300 curative units. The patient requires 2400 curative units. If d cubic centimeters of the drug and r minutes of radiation are administered, determine an equation relating d and r. Graph the equation for $d \geq 0$, and $r \geq 0$; label the horizontal axis as d.

23. **Depreciation** Suppose the value of a piece of machinery decreases each year by 10% of is original value. If the original value is \$8000, find an equation that expresses the value v of the machinery t years after purchase, where $0 \leq t \leq 10$. Sketch the equation, choosing t as the horizontal axis and v as the vertical axis. What is the slope of the resulting line? This method of considering the value of equipment is called *straight-line depreciation.*

24. **Depreciation** A new television depreciates \$120 per year, and it is worth \$340 after four years. Find a function that describes the value of this television, if x is the age of the television in years.

25. **Appreciation** A new apartment building was sold for \$760,000 five years after it was purchased. The original owners calculated that the building appreciated \$40,000 per year while they owned it. Find a linear function that describes the appreciation of the building, if x is the number of years since the original purchase.

26. **Appreciation** A house purchased for \$198,000 is expected to double in value in 18 years. Find a linear equation that describes the house's value after x years.

27. **Repair Charges** A business-copier repair company charges a fixed amount plus an hourly rate for a service call. If a customer is billed \$150 for a one-hour service call and \$280 for a three-hour service call, find a linear function that describes the price of a service call, where x is the number of hours of service.

28. **Sheep Wool Length** For sheep maintained at high environmental temperatures, respiratory rate r (per minute) increases as wool length l (in centimeters) decreases.[1] Suppose sheep with a wool length of 2 cm have an (average) respiratory rate of 160, and those with a wool length of 4 cm have a respiratory rate of 125. Assume that r and l are linearly related. (a) Find an equation that gives r in terms of l. (b) Find the respiratory rate of sheep with a wool length of 1 cm.

29. **Isocost Line** In production analysis, an *isocost line* is a line whose points represent all combinations of two factors of production that can be purchased for the same amount. Suppose a farmer has allocated \$20,000 for the purchase of x tons of fertilizer (costing \$200 per ton) and y acres of land (costing \$2000 per acre). Find an equation of the isocost line which describes the various combinations that can be purchased for \$20,000. Observe that neither x nor y can be negative.

30. **Isoprofit Line** A manufacturer produces products X and Y for which the profits per unit are \$4 and \$6, respectively. If x units of X and y units of Y are sold, then the total profit P is given by $P = 4x + 6y$, where $x, y \geq 0$. (a) Sketch the graph of this equation for $P = 240$. The result is called an *isoprofit line,* and its points represent all combinations of sales that produce a profit of \$240. (b) Determine the slope for $P = 240$. (c) If $P = 600$, determine the slope. (d) Are isoprofit lines for products X and Y parallel?

31. **Grade Scaling** For reasons of comparison, a professor wants to rescale the scores on a set of test papers so that the maximum score is still 100, but the mean (average) is 80 instead of 56. (a) Find a linear equation that will do

[1] Adapted from G. E. Folk, Jr., *Textbook of Environmental Physiology,* 2d ed. (Philadelphia: Lea & Febiger, 1974).

this. [*Hint:* You want 56 to become an 80 and 100 to remain 100. Consider the points (56, 80) and (100, 100) and, more generally, (x, y), where x is the old score and y is the new score. Find the slope and use a point–slope form. Express y in terms of x.] (b) If 60 on the new scale is the lowest passing score, what was the lowest passing score on the original scale?

32. Psychology The result of Sternberg's psychological experiment[2] on information retrieval is that a person's reaction time R, in milliseconds, is statistically a linear function of memory set size N as follows:

$$R = 38N + 397.$$

Sketch the graph for $1 \leq N \leq 5$. What is the slope?

33. Psychology In a certain learning experiment involving repetition and memory,[3] the proportion p of items recalled was estimated to be linearly related to the effective study time t (in seconds), where t is between 5 and 9. For an effective study time of 5 seconds, the proportion of items recalled was 0.32. For each 1-second increase in study time, the proportion recalled increased by 0.059. (a) Find an equation that gives p in terms of t. (b) What proportion of items was recalled with 9 seconds of effective study time?

34. Diet for Pigs In testing an experimental diet for pigs, it was determined that the (average) live weight w (in kilograms) of a pig was statistically a linear function of the number of days d after the diet was initiated, where $0 \leq d \leq 100$. If the weight of a pig beginning the diet was 20 kg, and thereafter the pig gained 6.6 kg every 10 days, determine w as a function of d, and find the weight of a pig 50 days after the beginning of the diet.

[2] G. R. Loftus and E. F. Loftus, *Human Memory: The Processing of Information* (New York: Lawrence Erlbaum Associates, Inc., distributed by the Halsted Press, Division of John Wiley & Sons, Inc., 1976).

[3] D. L. Hintzman, "Repetition and Learning," in *The Psychology of Learning*, Vol. 10, ed. G. H. Bower (New York: Academic Press, Inc., 1976), p. 77.

35. Cricket Chirps Biologists have found that the number of chirps made per minute by crickets of a certain species is related to the temperature. The relationship is very close to being linear. At 68°F, those crickets chirp about 124 times a minute. At 80°F, they chirp about 172 times a minute. (a) Find an equation that gives Fahrenheit temperature t in terms of the number of chirps, c, per minute. (b) If you count chirps for only 15 seconds, how can you quickly estimate the temperature?

36. Electrical Circuit In a circuit, the voltage V (in volts) and current i (in amperes) are linearly related. When $i = 4, V = 2$; when $i = 12, V = 6$.

a. Determine V as a function of i.

b. Find the voltage when the current is 10.

37. Physics The pressure P of a fixed volume of gas, in centimeters of mercury, is linearly related to the temperature T, in degrees Celsius. In an experiment with dry air, it was found that $P = 90$ when $T = 40$ and that $P = 100$ when $T = 80$. Express P as a function of T.

38. Electrical Theory When a graph of the terminal potential difference V, in volts, of a Daniell cell is plotted as a function of the current i, in amperes, delivered to an external resistor, a straight line is obtained. The slope of this line is the negative of the internal resistance of the cell. For a particular cell with an internal resistance of 0.06 ohms it was found that $V = 0.6$ V when $i = 0.12$ A. Express V as a function of i.

39. Hydraulics A formula used in hydraulics is

$$Q = 3.340b^3 + 1.8704b^2x,$$

where b is a constant.

a. Is the graph of this equation a straight line?

b. If so, what is the slope of the line when $b = 1$?

OBJECTIVE

To sketch parabolas arising from quadratic functions.

4.3 QUADRATIC FUNCTIONS

In Sec. 3.2, a *quadratic function* was described as a polynomial function of degree 2. Here is a formal definition.

DEFINITION

*A function f is a **quadratic function** if and only if $f(x)$ can be written in the form $f(x) = ax^2 + bx + c$, where a, b, and c are constants and $a \neq 0$.*

For example, the functions $f(x) = x^2 - 3x + 2$ and $F(t) = -3t^2$ are quadratic. However, $g(x) = \frac{1}{x^2}$ is *not* quadratic, because it cannot be written in the form $g(x) = ax^2 + bx + c$.

The graph of the quadratic function $y = f(x) = ax^2 + bx + c$ is called a **parabola** and has a shape like the curves in Fig. 4.19. If $a > 0$, the graph extends upward indefinitely, and we say that the parabola *opens upward* [Fig. 4.19(a)]. If $a < 0$, the parabola *opens downward* [Fig. 4.19(b)].

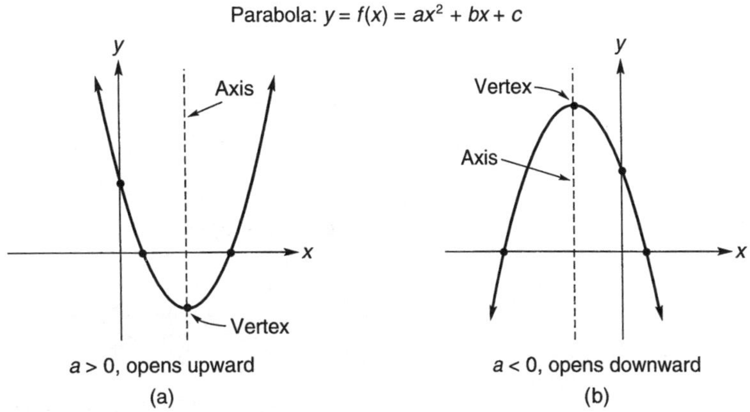

FIGURE 4.19 Parabolas.

Each parabola in Fig. 4.19 is *symmetric* about a vertical line, called the **axis of symmetry** of the parabola. That is, if the page were folded on one of these lines, then the two halves of the corresponding parabola would coincide. The axis (of symmetry) is *not* part of the parabola, but is a useful aid in sketching the parabola.

Figure 4.19 also shows points labeled **vertex,** where the axis cuts the parabola. If $a > 0$, the vertex is the "lowest" point on the parabola. This means that $f(x)$ has a minimum value at this point. By performing algebraic manipulations on $ax^2 + bx + c$ (referred to as *completing the square*), we can determine not only this minimum value, but also where it occurs. We have

$$f(x) = ax^2 + bx + c = (ax^2 + bx) + c.$$

Adding and subtracting $\frac{b^2}{4a}$ gives

$$\begin{aligned} f(x) &= \left(ax^2 + bx + \frac{b^2}{4a}\right) + c - \frac{b^2}{4a} \\ &= a\left(x^2 + \frac{b}{a}x + \frac{b^2}{4a^2}\right) + c - \frac{b^2}{4a}, \end{aligned}$$

so that

$$f(x) = a\left(x + \frac{b}{2a}\right)^2 + c - \frac{b^2}{4a}.$$

Since $\left(x + \frac{b}{2a}\right)^2 \geq 0$ and $a > 0$, it follows that $f(x)$ has a minimum value when $x + \frac{b}{2a} = 0$, that is, when $x = -\frac{b}{2a}$. The y-coordinate corresponding to this value of x is $f\left(-\frac{b}{2a}\right)$. Thus, the vertex is given by

$$\text{vertex} = \left(-\frac{b}{2a}, f\left(-\frac{b}{2a}\right)\right).$$

This is also the vertex of a parabola that opens downward ($a < 0$), but in this case $f\left(-\frac{b}{2a}\right)$ is the maximum value of $f(x)$. [See Fig. 4.19(b).]

The point where the parabola $y = ax^2 + bx + c$ intersects the y-axis (that is, the y-intercept) occurs when $x = 0$. The y-coordinate of this point is c, so the y-intercept is $(0, c)$, or, more simply, c. In summary, we have the following.

Graph of Quadratic Function

The graph of the quadratic function $y = f(x) = ax^2 + bx + c$ is a parabola.

1. If $a > 0$, the parabola opens upward. If $a < 0$, it opens downward.
2. The vertex is $\left(-\frac{b}{2a}, f\left(-\frac{b}{2a}\right)\right)$.
3. The y-intercept is c.

We can quickly sketch the graph of a quadratic function by first locating the vertex, the y-intercept, and a few other points, such as those where the parabola intersects the x-axis. These *x-intercepts* are found by setting $y = 0$ and solving for x. Once the intercepts and vertex are found, it is then relatively easy to pass the appropriate parabola through these points. In the event that the x-intercepts are very close to the vertex or that no x-intercepts exist, we find a point on each side of the vertex, so that we can give a reasonable sketch of the parabola. Keep in mind that passing a (dashed) vertical line through the vertex gives the axis of symmetry. By plotting points to one side of the axis, we can use symmetry and obtain corresponding points on the other side.

Principles in Practice 1
Graphing a Quadratic Function

The daily profit for a car dealership from the sale of a minivan is given by $P(x) = -x^2 + 2x + 399$, where x is the number of minivans sold. Find the function's vertex and intercepts, and graph the function.

EXAMPLE 1 Graphing a Quadratic Function

Graph the quadratic function $y = f(x) = -x^2 - 4x + 12$.

Solution: Here $a = -1, b = -4$, and $c = 12$. Since $a < 0$, the parabola opens downward and thus has a highest point. The x-coordinate of the vertex is

$$-\frac{b}{2a} = -\frac{-4}{2(-1)} = -2.$$

The y-coordinate is $f(-2) = -(-2)^2 - 4(-2) + 12 = 16$. Thus, the vertex is $(-2, 16)$, so the maximum value of $f(x)$ is 16. Since $c = 12$, the y-intercept is 12. To find the x-intercepts, we let y be 0 in $y = -x^2 - 4x + 12$ and solve for x:

$$0 = -x^2 - 4x + 12,$$

$$0 = -(x^2 + 4x - 12),$$

$$0 = -(x + 6)(x - 2).$$

Hence, $x = -6$ or $x = 2$, so the x-intercepts are -6 and 2. Now we plot the vertex, axis of symmetry, and intercepts. [See Fig. 4.20(a).] Since $(0, 12)$ is *two* units to the *right* of the axis, there is a corresponding point *two* units to the *left* of the axis with the same y-coordinate. Thus, we get the point $(-4, 12)$. Through all points, we draw a parabola opening downward. [See Fig. 4.20(b).]

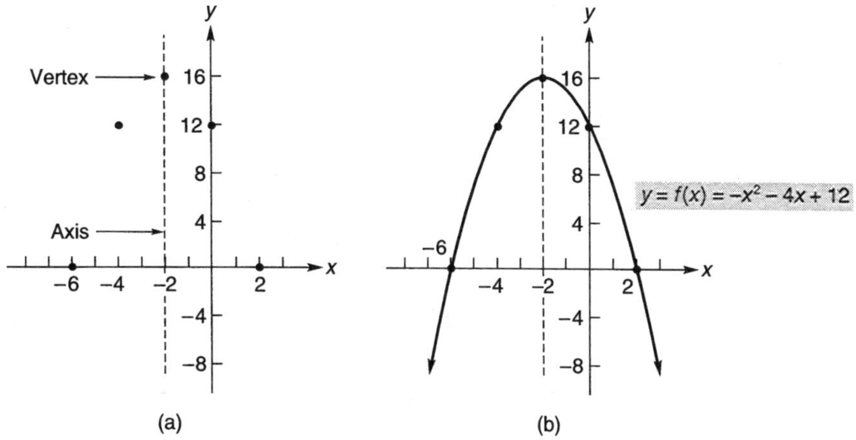

FIGURE 4.20 Graph of parabola $y = f(x) = -x^2 - 4x + 12$.

EXAMPLE 2 Graphing a Quadratic Function

Graph $p = 2q^2$.

Solution: Here p is a quadratic function of q, where $a = 2, b = 0$, and $c = 0$. Since $a > 0$, the parabola opens upward and thus has a lowest point. The q-coordinate of the vertex is

$$-\frac{b}{2a} = -\frac{0}{2(2)} = 0,$$

and the p-coordinate is $2(0)^2 = 0$. Consequently, the *minimum* value of p is 0 and the vertex is $(0, 0)$. In this case, the p-axis is the axis of symmetry. A parabola opening upward with vertex at $(0, 0)$ cannot have any other intercepts. Hence, to draw a reasonable graph, we plot a point on each side of the vertex. If $q = 2$, then $p = 8$. This gives the point $(2, 8)$ and, by symmetry, the point $(-2, 8)$. (See Fig. 4.21.)

q	p
2	8
−2	8

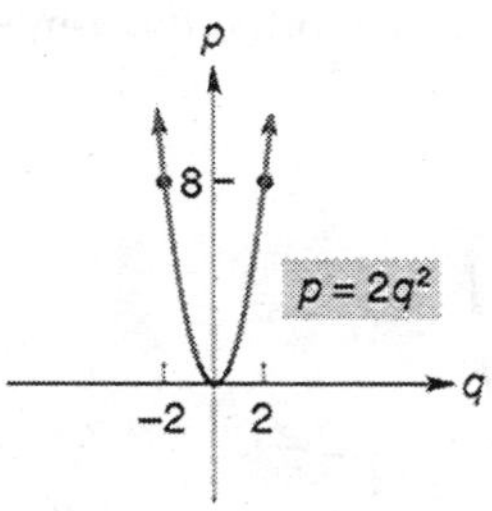

FIGURE 4.21 Graph of parabola $p = 2q^2$.

Example 3 illustrates that finding intercepts may require use of the quadratic formula.

EXAMPLE 3 Graphing a Quadratic Function

Graph $g(x) = x^2 - 6x + 7$.

Solution: Here g is a quadratic function, where $a = 1, b = -6$, and $c = 7$. The parabola opens upward, because $a > 0$. The x-coordinate of the vertex (lowest point) is

$$-\frac{b}{2a} = -\frac{-6}{2(1)} = 3.$$

and $g(3) = 3^2 - 6(3) + 7 = -2$, which is the minimum value of $g(x)$. Thus, the vertex is $(3, -2)$. Since $c = 7$, the vertical-axis intercept is 7. To find x-intercepts, we set $g(x) = 0$.

$$0 = x^2 - 6x + 7.$$

Principles in Practice 2
Graphing a Quadratic Function

A man standing on a pitcher's mound throws a ball straight up with an initial velocity of 32 feet per second. The height h of the ball in feet t seconds after it was thrown is described by the function $h(t) = -16t^2 + 32t + 8$, for $t \geq 0$. Find the function's vertex and intercepts, and graph the function.

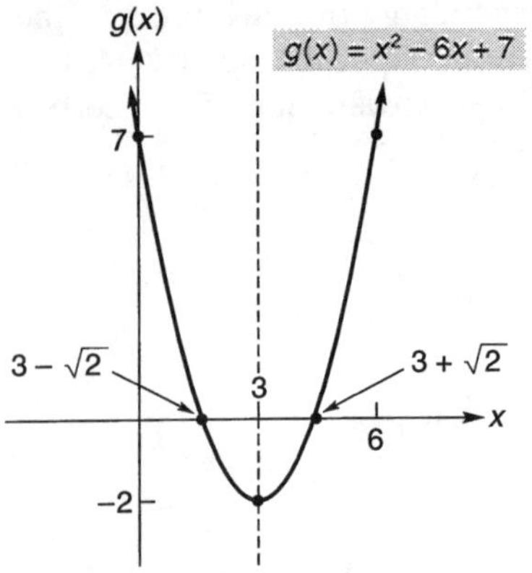

FIGURE 4.22 Graph of parabola $g(x) = x^2 - 6x + 7$.

The right side does not factor easily, so we shall use the quadratic formula to solve for x:

$$x = \frac{-b \pm \sqrt{b^2 - 4ac}}{2a} = \frac{-(-6) \pm \sqrt{(-6)^2 - 4(1)(7)}}{2(1)}$$

$$= \frac{6 \pm \sqrt{8}}{2} = \frac{6 \pm \sqrt{4 \cdot 2}}{2} = \frac{6 \pm 2\sqrt{2}}{2}$$

$$= \frac{6}{2} \pm \frac{2\sqrt{2}}{2} = 3 \pm \sqrt{2}.$$

Therefore, the x-intercepts are $3 + \sqrt{2}$ and $3 - \sqrt{2}$. After plotting the vertex, intercepts, and (by symmetry) the point (6, 7), we draw a parabola opening upward in Fig. 4.22. ■

EXAMPLE 4 Graphing a Quadratic Function

Graph $y = f(x) = 2x^2 + 2x + 3$ and find the range of f.

Solution: This function is quadratic with $a = 2, b = 2$, and $c = 3$. Since $a > 0$, the graph is a parabola opening upward. The x-coordinate of the vertex is

$$-\frac{b}{2a} = -\frac{2}{2(2)} = -\frac{1}{2}.$$

and the y-coordinate is $2(-\frac{1}{2})^2 + 2(-\frac{1}{2}) + 3 = \frac{5}{2}$. Thus, the vertex is $(-\frac{1}{2}, \frac{5}{2})$. Since $c = 3$, the y-intercept is 3. A parabola opening upward with its vertex above the x-axis has no x-intercepts. In Fig. 4.23 we plotted the y-intercept, the

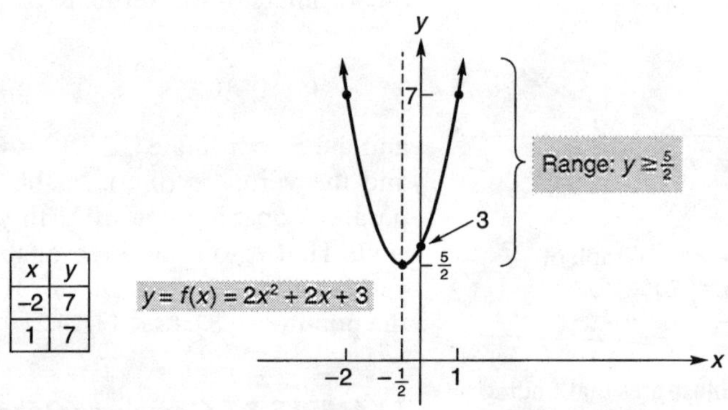

FIGURE 4.23 Graph of $y = f(x) = 2x^2 + 2x + 3$.

vertex, and an additional point $(-2, 7)$ to the left of the vertex. By symmetry, we also get the point $(1, 7)$. Passing a parabola through these points gives the desired graph. From the figure, we see that the range of f is all $y \geq \frac{5}{2}$, that is, the interval $[\frac{5}{2}, \infty)$. ■

EXAMPLE 5 Maximum Revenue

The demand function for a manufacturer's product is $p = 1000 - 2q$, where p is the price (in dollars) per unit when q units are demanded (per week) by consumers. Find the level of production that will maximize the manufacturer's total revenue, and determine this revenue.

Solution:

Strategy: To maximize revenue, we must determine the revenue function, $r = f(q)$. Using the relation

$$\textbf{total revenue} = \textbf{(price)(quantity)},$$

we have

$$r = pq.$$

Using the demand equation, we can express p in terms of q, so r will be strictly a function of q.

The formula for total revenue should be added to your repertoire of relationships in business and economics.

We have

$$\begin{aligned} r &= pq \\ &= (1000 - 2q)q. \\ r &= 1000q - 2q^2. \end{aligned}$$

Note that r is a quadratic function of q, with $a = -2$, $b = 1000$, and $c = 0$. Since $a < 0$ (the parabola opens downward), r is maximum at the vertex (q, r), where

$$q = -\frac{b}{2a} = -\frac{1000}{2(-2)} = 250.$$

The maximum value of r is given by

$$\begin{aligned} r &= 1000(250) - 2(250)^2 \\ &= 250{,}000 - 125{,}000 = 125{,}000. \end{aligned}$$

Thus, the maximum revenue that the manufacturer can receive is \$125,000, which occurs at a production level of 250 units. Figure 4.24(a) shows the graph of the revenue function. Only that portion for which $q \geq 0$ and $r \geq 0$ is drawn, since quantity and revenue cannot be negative. ■

Principles in Practice 3
Maximum Revenue

The demand function for a publisher's line of cookbooks is $p = 6 - 0.003q$, where p is the price (in dollars) per unit when q units are demanded (per day) by consumers. Find the level of production that will maximize the manufacturer's total revenue, and determine this revenue.

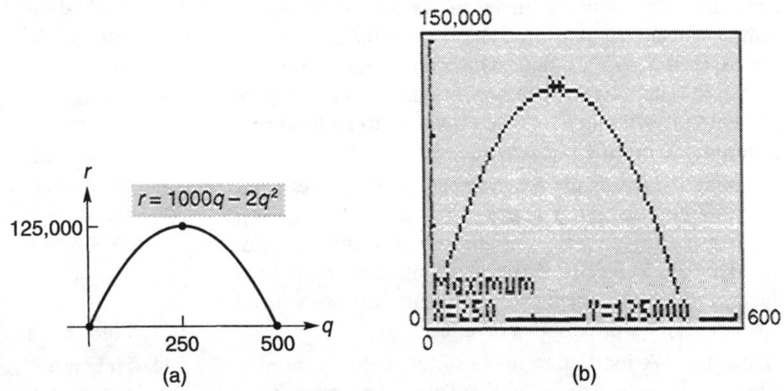

FIGURE 4.24 Graph of revenue function.

TECHNOLOGY

The maximum (or minimum) value of a function can be conveniently found with a graphics calculator either by using trace and zoom or by using the "maximum" (or "minimum") feature. Figure 4.24(b) shows the display for the revenue function of Example 5, namely, the graph of $y = 1000x - 2x^2$. Note that we replaced r by y and q by x.

■ Exercise 4.3

In Problems 1–8, state whether the function is quadratic.

1. $f(x) = 5x^2$.

2. $g(x) = \dfrac{1}{2x^2 - 4}$.

3. $g(x) = 7 - 6x$.

4. $h(s) = 2s^2(s^2 + 1)$.

5. $h(q) = (q + 4)^2$.

6. $f(t) = 2t(3 - t) + 4t$.

7. $f(s) = \dfrac{s^2 - 4}{2}$.

8. $g(t) = (t^2 - 1)^2$.

In Problems 9–12, do not include a graph.

9. (a) For the parabola $y = f(x) = -4x^2 + 8x + 7$, find the vertex. (b) Does the vertex correspond to the highest point or the lowest point on the graph?

10. Repeat Problem 9 if $y = f(x) = 8x^2 + 4x - 1$.

11. For the parabola $y = f(x) = x^2 + 2x - 8$, find (a) the y-intercept, (b) the x-intercepts, and (c) the vertex.

12. Repeat Problem 11 if $y = f(x) = 3 + x - 2x^2$.

In Problems 13–22, graph each function. Give the vertex and intercepts, and state the range.

13. $y = f(x) = x^2 - 6x + 5$.

14. $y = f(x) = -3x^2$.

15. $y = g(x) = -2x^2 - 6x$.

16. $y = f(x) = x^2 - 1$.

17. $s = h(t) = t^2 + 2t + 1$.

18. $s = h(t) = 2t^2 + 3t - 2$.

19. $y = f(x) = -9 + 8x - 2x^2$.

20. $y = H(x) = 1 - x - x^2$.

21. $t = f(s) = s^2 - 8s + 13$.

22. $t = f(s) = s^2 + 6s + 11$.

In Problems 23–26, state whether f(x) has a maximum value or a minimum value, and find that value.

23. $f(x) = 100x^2 - 20x + 25$.

24. $f(x) = -2x^2 - 16x + 3$.

25. $f(x) = 4x - 50 - 0.1x^2$.

26. $f(x) = x(x + 3) - 12$.

27. Revenue The demand function for a manufacturer's product is $p = f(q) = 1200 - 3q$, where p is the price (in dollars) per unit when q units are demanded (per week). Find the level of production that maximizes the manufacturer's total revenue and determine this revenue.

28. Revenue The demand function for an office supply company's line of plastic rulers is $p = 0.8 - 0.0004q$, where p is the price (in dollars) per unit when q units are demanded (per day) by consumers. Find the level of production that will maximize the manufacturer's total revenue, and determine this revenue.

29. Revenue The demand function for an electronics company's laptop computer line is $p = 2400 - 6q$, where p is the price (in dollars) per unit when q units are demanded (per week) by consumers. Find the level of production that will maximize the manufacturer's total revenue, and determine this revenue.

30. Marketing A marketing firm estimates that n months after the introduction of a client's new product, $f(n)$ thousand households will use it, where

$$f(n) = \tfrac{10}{9}n(12 - n), \qquad 0 \le n \le 12.$$

Estimate the maximum number of households that will use the product.

31. Profit The daily profit for the garden department of a store from the sale of trees is given by $P(x) = -x^2 + 18x + 144$, where x is the number of trees sold. Find the function's vertex and intercepts, and graph the function.

32. Psychology A prediction made by early psychology relating the magnitude of a stimulus x to the magnitude of a response y is expressed by the equation $y = kx^2$, where k is a constant of the experiment. In an experiment on pattern recognition, $k = 2$. Find the function's vertex and graph the equation. (Assume no restriction on x.)

33. Biology Biologists studied the nutritional effects on rats that were fed a diet containing 10% protein.[4] The protein consisted of yeast and corn flour. By varying the percentage P of yeast in the protein mix, the group estimated that the average weight gain (in grams) of a rat over a period of time was

$$f(P) = -\tfrac{1}{50}P^2 + 2P + 20, \qquad 0 \le P \le 100.$$

Find the maximum weight gain.

34. Height of Ball Suppose that the height s of a ball thrown vertically upward from the ground is given by

$$s = -4.9t^2 + 58.8t,$$

where s is in meters and t is elapsed time in seconds. (See Fig. 4.25.) After how many seconds will the ball reach its maximum height? What is the maximum height?

[4]Adapted from R. Bressani, "The Use of Yeast in Human Foods," in *Single-Cell Protein*, ed. R. I. Mateles and S. R. Tannenbaum (Cambridge, MA: MIT Press, 1968).

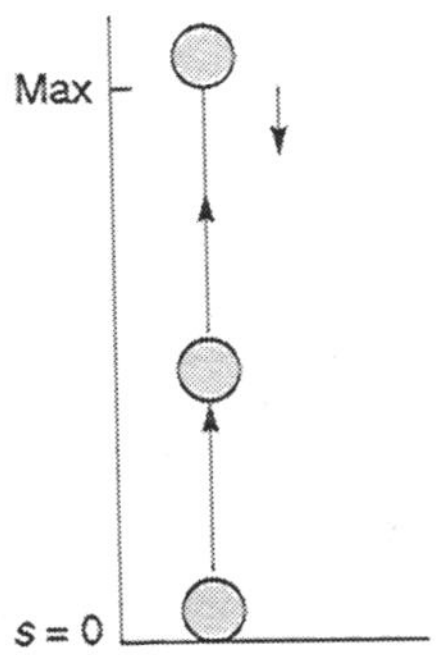

FIGURE 4.25
Ball thrown upward (Problem 34).

35. **Archery** A boy standing on a hill shoots an arrow straight up with an initial velocity of 64 feet per second. The height h of the arrow in feet t seconds after it was released is described by the function $h(t) = -16t^2 + 64t + 32$. What is the maximum height reached by the arrow? How many seconds after release does it take to reach this height?

36. **Toy Toss** A 6 year old girl standing on a toy chest throws a doll straight up with an initial velocity of 16 feet per second. The height h of the doll in feet t seconds after it was released is described by the function $h(t) = -16t^2 + 16t + 4$. How long does it take the doll to reach its maximum height? What is the maximum height?

37. **Rocket Launch** A toy rocket is launched straight up from the roof of a garage with an initial velocity of 80 feet per second. The height h of the rocket in feet t seconds after it was released is described by the function $h(t) = -16t^2 + 80t + 16$. Find the function's vertex and intercepts, and graph the function.

38. **Suspension Cable** The shape of the main cable on a suspension bridge can be described by the function

$$y = f(x) = \frac{1}{500}x^2 + \frac{1}{250}x + 10, \quad -100 \le x \le 100,$$

where $f(x)$ is the height of the cable (in feet) above the roadbed and x is the horizontal distance (in feet) from the center of the bridge. Graph the function and find its range.

39. **Physics** The displacement of an object from a reference point at time t is given by

$$s = 3.2t^2 - 16t + 28.7,$$

where s is in meters and t is in seconds.

a. For what value of t does the minimum displacement occur?

b. What is the minimum displacement of the object from the reference point?

40. **Force** During a collision, the force F (in newtons) that acted on an object varied with time t according to the equation $F = 87t - 21t^2$, where t is in seconds.

a. For what value of t was the force a maximum?

b. What was the maximum value of the force?

41. **Loaded Beam** When a horizontal beam of length l is uniformly loaded, the moment equation is

$$M = \frac{wlx}{2} - \frac{wx^2}{2},$$

where w is related to the load and x is measured from the left end of the beam.

a. For what value of x is M a maximum? (Assume that $w > 0$.)

b. What is the maximum value of M?

c. For what values of x does $M = 0$?

42. **Area** Express the area of the rectangle shown in Fig. 4.26 as a quadratic function of x. For what value of x will the area be a maximum?

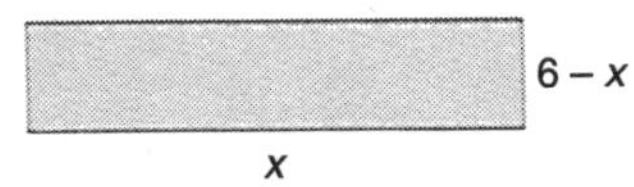

FIGURE 4.26 Diagram for Problem 42.

43. **Enclosing Plot** A building contractor wants to fence in a rectangular plot adjacent to a straight stream by using the stream for one side of the enclosed area. (See Fig. 4.27.) If the contractor has 200 feet of fence, find the dimensions of the maximum enclosed area.

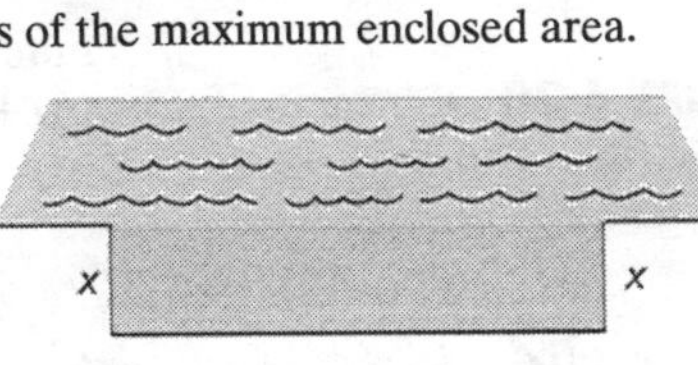

FIGURE 4.27 Diagram for Problem 43.

44. Find two numbers whose sum is 40 and whose product is a maximum.

45. From the graph of $y = 1.4x^2 - 3.1x + 4.6$, determine the coordinates of the vertex. Round values to two decimal places. Verify your answer by using the vertex formula.

46. Find the zeros of $f(x) = -\sqrt{2}x^2 + 3x + 8.5$ by examining the graph of f. Round values to two decimal places.

47. Determine the number of real zeros for each of the following quadratic functions:

a. $f(x) = 4.2x^2 - 8.1x + 10.4$.

b. $f(x) = 5x^2 - 2\sqrt{35}x + 7$.

c. $f(x) = \dfrac{5.1 - 7.2x - x^2}{4.8}$.

48. Find the maximum value (rounded to two decimal places) of the function $f(x) = 5.4 + 12x - 4.1x^2$ from its graph.

49. Find the minimum value (rounded to two decimal places) of the function $f(x) = 20x^2 - 3x + 7$ from its graph.

OBJECTIVE

To solve systems of linear equations in both two and three variables by using the technique of elimination by addition or by substitution. (In Chapter 6, other methods are shown.)

4.4 Systems of Linear Equations

Two-Variable Systems

When a situation must be described mathematically, it is not unusual for a *set* of equations to arise. For example, suppose that the manager of a factory is setting up a production schedule for two models of a new product. Model A requires 4 widgets and 9 klunkers. Model B requires 5 widgets and 14 klunkers. From its suppliers, the factory gets 335 widgets and 850 klunkers each day. How many of each model should the manager plan to make each day so that all the widgets and klunkers are used?

It's a good idea to construct a table that summarizes the important information. Table 4.2 shows the number of widgets and klunkers required for each model, as well as the total number available.

TABLE 4.2

	Model A	Model B	Total Available
Widgets	4	5	335
Klunkers	9	14	850

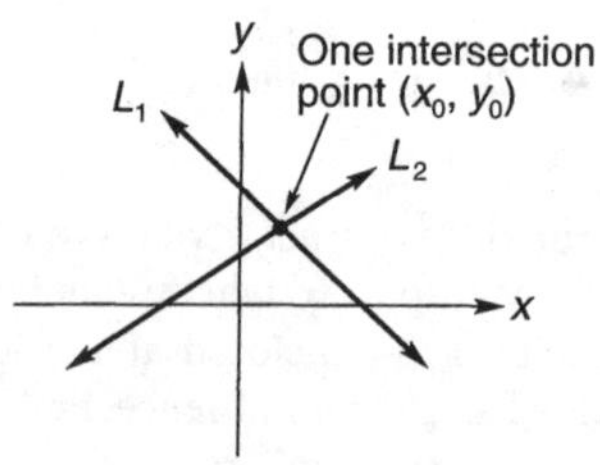

FIGURE 4.28 Linear system (one solution).

Suppose we let x be the number of model A made each day and y be the number of model B. Then these require a total of $4x + 5y$ widgets and $9x + 14y$ klunkers. Since 335 widgets and 850 klunkers are available, we have

$$\begin{cases} 4x + 5y = 335, & (1) \\ 9x + 14y = 850. & (2) \end{cases}$$

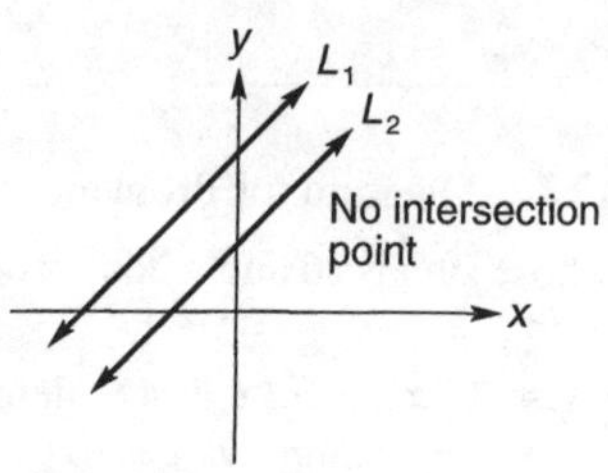

FIGURE 4.29 Linear system (no solution).

We call this set of equations a **system** of two linear equations in the variables (or unknowns) x and y. The problem is to find values of x and y for which *both* equations are true *simultaneously*. These are called *solutions* of the system.

Since Eqs. (1) and (2) are linear, their graphs are straight lines; call these lines L_1 and L_2. Now, the coordinates of any point on a line satisfy the equation of that line; that is, they make the equation true. Thus, the coordinates of any point of intersection of L_1 and L_2 will satisfy both equations. This means that a point of intersection gives a solution of the system.

If L_1 and L_2 are drawn on the same plane, there are three situations that could occur:

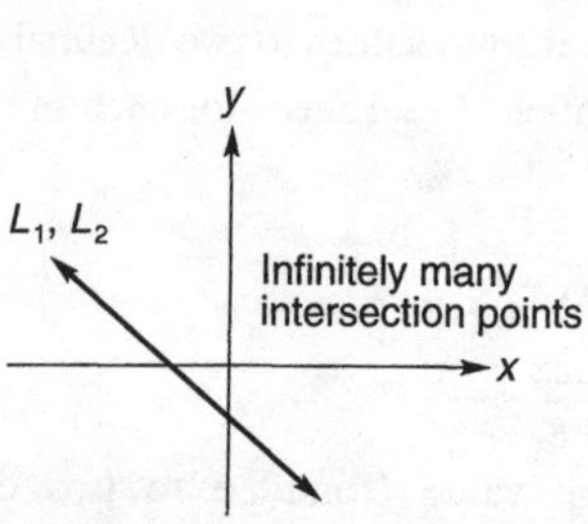

FIGURE 4.30 Linear system (infinitely many solutions).

1. L_1 and L_2 may intersect at exactly one point, say, (x_0, y_0). (See Fig., 4.28.) Thus, the system has the solution $x = x_0$ and $y = y_0$.
2. L_1 and L_2 may be parallel and have no points in common. (See Fig. 4.29.) In this case, there is no solution.
3. L_1 and L_2 may be the same line. (See Fig. 4.30.) Here the coordinates of any point on the line are a solution of the system. Consequently, there are infinitely many solutions.

Our main concern in this section is algebraic methods of solving a system of linear equations. Essentially, we successively replace the system by other systems that have the same solution (that is, we replace the original system by *equivalent systems*), but whose equations have a progressively more desirable form for determining the solution. More precisely, we seek an equivalent system containing an equation in which one of the variables does not appear. (That is, one of the variables is *eliminated.*) We shall illustrate this procedure for the system in the problem originally posed:

$$\begin{cases} 4x + 5y = 335, & (3) \\ 9x + 14y = 850. & (4) \end{cases}$$

To begin, we shall obtain an equivalent system in which x does not appear in one equation. First we find an equivalent system in which the coefficients of the x-terms in each equation are the same except for their sign. Multiplying Eq. (3) by 9 [that is, multiplying both sides of Eq. (3) by 9] and multiplying Eq. (4) by -4 gives

$$\begin{cases} 36x + 45y = 3015, & (5) \\ -36x - 56y = -3400. & (6) \end{cases}$$

The left and right sides of Eq. (6) are equal, so each side can be *added* to the corresponding side of Eq. (5). This results in

$$-11y = -385,$$

which has only one variable, as planned. Solving gives

$$y = 35,$$

so we obtain the equivalent system

$$\begin{cases} y = 35, & (7) \\ -36x - 56y = -3400. & (8) \end{cases}$$

Replacing y in Eq. (8) by 35, we get

$$\begin{aligned} -36x - 56(35) &= -3400, \\ -36x - 1960 &= -3400, \\ -36x &= -1440, \\ x &= 40. \end{aligned}$$

Thus, the original system is equivalent to

$$\begin{cases} y = 35, \\ x = 40. \end{cases}$$

We can check our answer by substituting $x = 40$ and $y = 35$ into *both* of the original equations. In Eq. (3), we get $4(40) + 5(35) = 335$, or $335 = 335$. In Eq. (4), we get $9(40) + 14(35) = 850$, or $850 = 850$. Hence, the solution is

$$x = 40 \quad \text{and} \quad y = 35.$$

Each day the manager should plan to make 40 of model A and 35 of model B. Our procedure is referred to as **elimination by addition.** Although we chose to eliminate x first, we could have done the same for y by a similar procedure.

EXAMPLE 1 Elimination-by-Addition Method

Use elimination by addition to solve the system.

$$\begin{cases} 3x - 4y = 13, \\ 3y + 2x = 3. \end{cases}$$

Principles in Practice 1
Elimination-by-Addition Method

A computer consultant has \$200,000 invested for retirement, part at 9% and part at 8%. If the total yearly income from the investments is \$17,200, how much is invested at each rate?

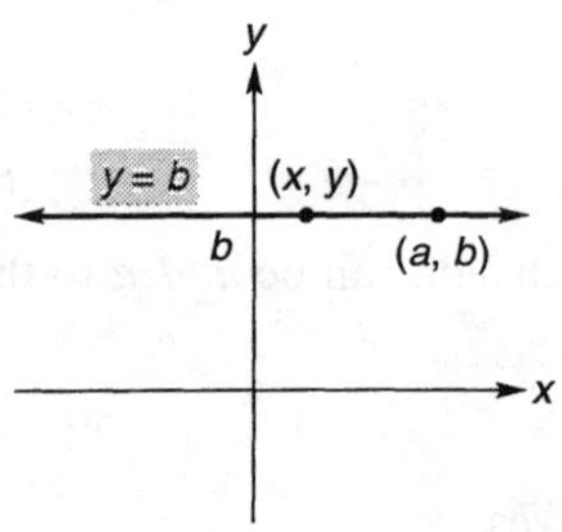

FIGURE 4.31 Linear system of Example 1: one solution.

Solution: Aligning the *x*- and *y*-terms for convenience gives

$$\begin{cases} 3x - 4y = 13, & (9) \\ 2x + 3y = 3. & (10) \end{cases}$$

To eliminate *y*, we multiply Eq. (9) by 3 and Eq. (10) by 4:

$$\begin{cases} 9x - 12y = 39, & (11) \\ 8x + 12y = 12. & (12) \end{cases}$$

Adding Eq. (11) to Eq. (12) gives $17x = 51$, from which $x = 3$. We have the equivalent system

$$\begin{cases} 9x - 12y = 39, & (13) \\ x = 3. & (14) \end{cases}$$

Replacing *x* by 3 in Eq. (13) results in

$$9(3) - 12y = 39,$$
$$-12y = 12,$$
$$y = -1,$$

so the original system is equivalent to

$$\begin{cases} y = -1, \\ x = 3. \end{cases}$$

The solution is $x = 3$ and $y = -1$. Figure 4.31 shows a graph of the system. ■

The system in Example 1,

$$\begin{cases} 3x - 4y = 13, & (15) \\ 2x + 3y = 3, & (16) \end{cases}$$

can be solved another way. We first choose one of the equations—for example, Eq. (15)—and solve it for one unknown in terms of the other, say *x* in terms of *y*. Hence Eq. (15) is equivalent to $3x = 4y + 13$, or

$$x = \frac{4}{3}y + \frac{13}{3},$$

and we obtain

$$\begin{cases} x = \frac{4}{3}y + \frac{13}{3}, & (17) \\ 2x + 3y = 3. & (18) \end{cases}$$

Substituting the right side of Eq. (17) for *x* in Eq. (18) gives

$$2\left(\frac{4}{3}y + \frac{13}{3}\right) + 3y = 3. \qquad (19)$$

Thus, *x* has been eliminated. Solving Eq. (19), we have

$$\frac{8}{3}y + \frac{26}{3} + 3y = 3,$$
$$8y + 26 + 9y = 9 \quad \text{(clearing fractions)},$$
$$17y = -17,$$
$$y = -1.$$

Replacing y in Eq. (17) by -1 gives $x = 3$, and the original system is equivalent to

$$\begin{cases} x = 3, \\ y = -1, \end{cases}$$

as before, This method is called **elimination by substitution.**

Principles in Practice 2
Method of Elimination by Substitution

Two species of deer, A and B, living in a wildlife refuge are given extra food in the winter. Each week, they receive 2 tons of food pellets and 4.75 tons of hay. Each deer of species A requires 4 pounds of the pellets and 5 pounds of hay. Each deer of species B requires 2 pounds of the pellets and 7 pounds of hay. How many of each species of deer will the food support so that all of the food is consumed each week?

EXAMPLE 2 Method of Elimination by Substitution

Use elimination by substitution to solve the system

$$\begin{cases} x + 2y - 8 = 0, \\ 2x + 4y + 4 = 0. \end{cases}$$

Solution: It is easy to solve the first equation for x. Doing so gives the equivalent system

$$\begin{cases} x = -2y + 8, & (20) \\ 2x + 4y + 4 = 0. & (21) \end{cases}$$

Substituting $-2y + 8$ for x in Eq. (21) yields

$$2(-2y + 8) + 4y + 4 = 0,$$
$$-4y + 16 + 4y + 4 = 0.$$

The latter equation simplifies to $20 = 0$. Thus, we have the system

$$\begin{cases} x = -2y + 8, & (22) \\ 20 = 0. & (23) \end{cases}$$

Since Eq. (23) is *never* true, there is **no solution** of the original system. The reason is clear if we observe that the original equations can be written in slope–intercept form as

$$y = -\frac{1}{2}x + 4$$

and

$$y = -\frac{1}{2}x - 1.$$

These equations represent straight lines having slopes of $-\frac{1}{2}$, but different y-intercepts, namely, 4 and -1. That is, they determine different parallel lines. (See Fig. 4.32.) ■

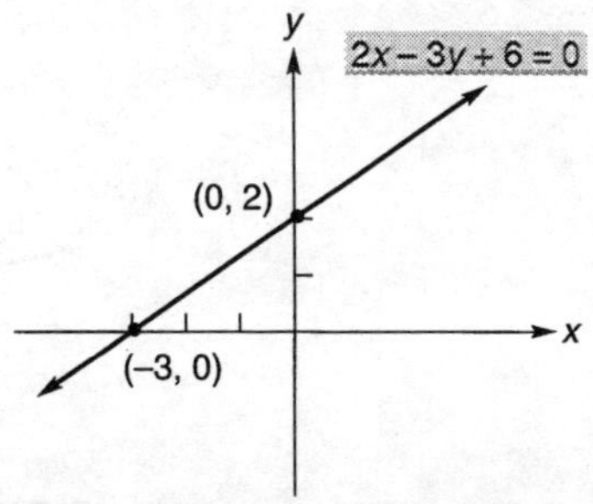

FIGURE 4.32 Linear system of Example 2: no solution.

Principles in Practice 3
A Linear System with Infinitely Many Solutions

Two species of fish, A and B, are raised in one pond at a fish farm where they are fed two vitamin supplements. Each day, they receive 100 grams of the first supplement and 200 grams of the second supplement. Each fish of species A requires 15 mg of the first supplement and 30 mg of the second supplement. Each fish of species B requires 20 mg of the first supplement and 40 mg of the second supplement. How many of each species of fish will the pond support so that all of the supplements are consumed each day?

EXAMPLE 3 A Linear System with Infinitely Many Solutions

Solve

$$\begin{cases} x + 5y = 2, & (24) \\ \frac{1}{2}x + \frac{5}{2}y = 1. & (25) \end{cases}$$

Solution: We begin by eliminating x from the second equation. Multiplying Eq. (25) by -2, we have

$$\begin{cases} x + 5y = 2, & (26) \\ -x - 5y = -2. & (27) \end{cases}$$

Adding Eq. (26) to Eq. (27) gives

$$\begin{cases} x + 5y = 2, & (28) \\ 0 = 0. & (29) \end{cases}$$

Because Eq. (29) is *always* true, any solution of Eq. (28) is a solution of the system. Now let us see how we can express our answer. From Eq. (28), we have $x = 2 - 5y$, where y can be any real number, say, r. Thus, we can write $x = 2 - 5r$. The complete solution is

$$x = 2 - 5r,$$
$$y = r,$$

where r is any real number. In this situation r is called a **parameter,** and we say that we have a one-parameter family of solutions. Each value of r determines a particular solution. For example, if $r = 0$, then $x = 2$ and $y = 0$ is a solution; if $r = 5$, then $x = -23$ and $y = 5$ is another solution. Clearly, the given system has infinitely many solutions.

It is worthwhile to note that by writing Eqs. (24) and (25) in their slope–intercept forms, we get the equivalent system

$$\begin{cases} y = -\frac{1}{5}x + \frac{2}{5}, \\ y = -\frac{1}{5}x + \frac{2}{5}, \end{cases}$$

in which both equations represent the same line. Hence, the lines coincide (Fig. 4.33), and Eqs. (24) and (25) are equivalent. The solution of the system consists of the coordinate pairs of all points on the line $x + 5y = 2$, and these points are given by our parametric solution. ■

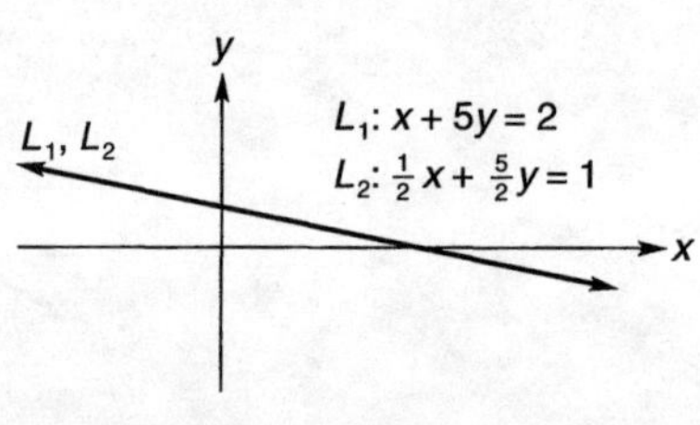

FIGURE 4.33 Linear system of Example 3: infinitely many solutions.

TECHNOLOGY

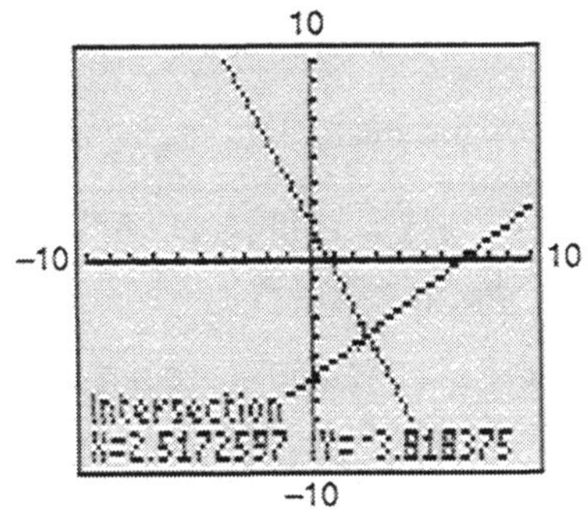

FIGURE 4.34 Graphical solution of system.

Graphically solve the system

$$\begin{cases} 9x + 4.1y = 7, \\ 2.6x - \quad 3y = 18. \end{cases}$$

Solution: First we solve each equation for y, so that each equation has the form $y = f(x)$:

$$y = \frac{1}{4.1}(7 - 9x),$$

$$y = -\frac{1}{3}(18 - 2.6x).$$

Next we enter these functions as Y_1 and Y_2 and display them on the same viewing rectangle. (See Fig. 4.34.) Finally, either using trace and zoom or using the intersection feature, we estimate the solution to be $x = 2.52$, $y = -3.82$.

EXAMPLE 4 Mixture

A chemical manufacturer wishes to fill an order for 500 liters of a 25% acid solution. (Twenty-five percent by volume is acid.) If solutions of 30% and 18% are available in stock, how many liters of each must be mixed to fill the order?

Solution: Let x and y respectively be the number of liters of the 30% and 18% solutions that should be mixed. Then

$$x + y = 500.$$

To help visualize the situation, we draw the diagram in Fig. 4.35. In 500 liters of a 25% solution, there will be $0.25(500) = 125$ liters of acid. This acid comes

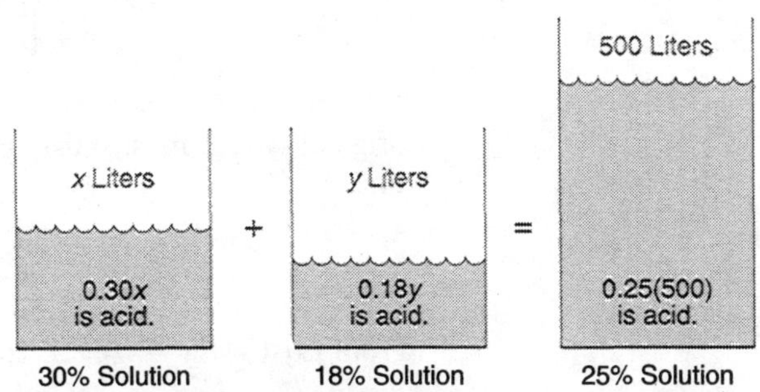

FIGURE 4.35 Mixture problem.

from two sources: $0.30x$ liters of it come from the 30% solution, and $0.18y$ liters of it come from the 18% solution. Hence,

$$0.30x + 0.18y = 125.$$

These two equations form a system of two linear equations in two unknowns. Solving the first for x gives $x = 500 - y$. Substituting in the second gives

$$0.30(500 - y) + 0.18y = 125.$$

Solving this equation for y, we find that $y = 208\frac{1}{3}$ liters. Thus, $x = 500 - 208\frac{1}{3} = 291\frac{2}{3}$ liters. (See Fig. 4.36.)

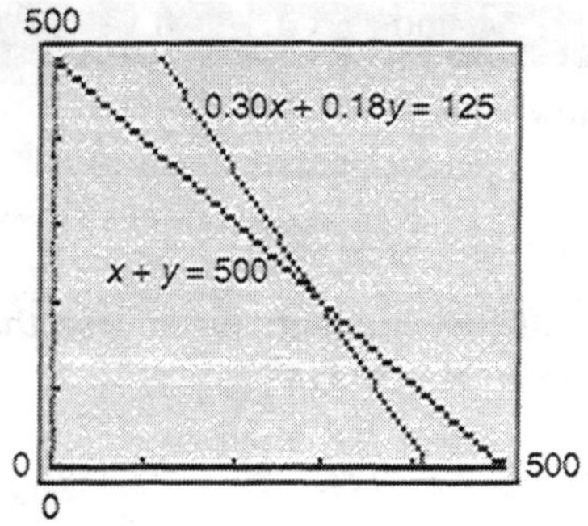

FIGURE 4.36 Graph for Example 4.

Three-Variable Systems

The methods used in solving a two-variable system of linear equations can be used to solve a three-variable system of linear equations. A **general linear equation in the three variables x, y, and z** is an equation having the form

$$Ax + By + Cz = D,$$

where A, B, C, and D are constants and A, B, and C are not all zero. For example, $2x - 4y + z = 2$ is such an equation. Geometrically, a general linear equation in three variables represents a *plane* in space, and a solution to a system of such equations is the intersection of planes. Example 5 shows how to solve a system of three linear equations in three variables.

Principles in Practice 4
Solving a Three-Variable Linear System

A coffee shop specializes in blending gourmet coffees. From type A, type B, and type C coffees, the owner wants to prepare a blend that will sell for \$8.50 for a 1-pound bag. The cost per pound of these coffees is \$12, \$9, and \$7, respectively. The amount of type B is to be twice the amount of type A. How much of each type of coffee will be in the final blend?

EXAMPLE 5 Solving a Three-Variable Linear System

Solve

$$\begin{cases} 2x + y + z = 3, & (30) \\ -x + 2y + 2z = 1, & (31) \\ x - y - 3z = -6. & (32) \end{cases}$$

Solution: This system consists of three linear equations in three variables. From Eq. (32), $x = y + 3z - 6$. By substituting for x in Eqs. (30) and (31), we obtain

$$\begin{cases} 2(y + 3z - 6) + y + z = 3, \\ -(y + 3z - 6) + 2y + 2z = 1, \\ x = y + 3z - 6. \end{cases}$$

Simplifying gives

$$\begin{cases} 3y + 7z = 15, & (33) \\ y - z = -5, & (34) \\ x = y + 3z - 6. & (35) \end{cases}$$

Note that x does not appear in Eqs. (33) and (34). Since any solution of the original system must satisfy Eqs. (33) and (34), we shall consider their solution first:

$$\begin{cases} 3y + 7z = 15, & (33) \\ y - z = -5. & (34) \end{cases}$$

From Eq. (34), $y = z - 5$. This means that we can replace Eq. (33) by

$$3(z - 5) + 7z = 15, \quad \text{or} \quad z = 3.$$

Since z is 3, we can replace Eq. (34) with $y = -2$. Hence, the previous system is equivalent to

$$\begin{cases} z = 3, \\ y = -2. \end{cases}$$

The original system becomes

$$\begin{cases} z = 3, \\ y = -2, \\ x = y + 3z - 6, \end{cases}$$

from which $x = 1$. The solution is $x = 1$, $y = -2$, and $z = 3$, which you may verify. ■

Just as a two-variable system may have a one-parameter family of solutions, a three-variable system may have a one- or two-parameter family of solutions.[5] The next two examples illustrate.

EXAMPLE 6 One-Parameter Family of Solutions

Solve

$$\begin{cases} x - 2y = 4, & (35) \\ 2x - 3y + 2z = -2, & (36) \\ 4x - 7y + 2z = 6. & (37) \end{cases}$$

Solution: Note that since Eq. (35) can be written $x - 2y + 0z = 4$, we can view Eqs. (35) to (37) as a system of three linear equations in the variables x, y, and z. From Eq. (35), we have $x = 2y + 4$. Using this equation and substitution, we can eliminate x from Eqs. (36) and (37):

$$\begin{cases} x = 2y + 4, \\ 2(2y + 4) - 3y + 2z = -2, \\ 4(2y + 4) - 7y + 2z = 6. \end{cases}$$

Or, more simply,

$$\begin{cases} x = 2y + 4, & (38) \\ y + 2z = -10, & (39) \\ y + 2z = -10. & (40) \end{cases}$$

Multiplying Eq. (40) by -1 gives

$$\begin{cases} x = 2y + 4, \\ y + 2z = -10, \\ -y - 2z = 10. \end{cases}$$

Adding the second equation to the third yields

$$\begin{cases} x = 2y + 4, \\ y + 2z = -10, \\ 0 = 0. \end{cases}$$

Since the equation $0 = 0$ is always true, we can essentially deal with the system

$$\begin{cases} x = 2y + 4, & (41) \\ y + 2z = -10. & (42) \end{cases}$$

Solving Eq. (42) for y, we have

$$y = -10 - 2z,$$

which expresses y in terms of z. We can also express x in terms of z. From Eq. (41),

$$\begin{aligned} x &= 2y + 4 \\ &= 2(-10 - 2z) + 4 \\ &= -16 - 4z. \end{aligned}$$

[5]Note to instructor: Examples 6 and 7 may be omitted without loss of continuity.

Thus, we have

$$\begin{cases} x = -16 - 4z, \\ y = -10 - 2z. \end{cases}$$

Since no restriction is placed on z, this suggests a parametric family of solutions. Setting $z = r$, we have the following family of solutions of the given system:

$$x = -16 - 4r,$$
$$y = -10 - 2r,$$
$$z = r,$$

Other parametric representations of the solution are possible.

where r can be any real number. We see, then, that the given system has infinitely many solutions. For example, setting $r = 1$ gives the particular solution $x = -20$, $y = -12$, and $z = 1$. ■

EXAMPLE 7 Two-Parameter Family of Solutions

Solve the system

$$\begin{cases} x + 2y + z = 4, \\ 2x + 4y + 2z = 8. \end{cases}$$

Solution: This is a system of two linear equations in three variables. We shall eliminate x from the second equation by first multiplying that equation by $-\frac{1}{2}$:

$$\begin{cases} x + 2y + z = 4, \\ -x - 2y - z = -4. \end{cases}$$

Adding the first equation to the second gives

$$\begin{cases} x + 2y + z = 4, \\ 0 = 0. \end{cases}$$

From the first equation, we obtain

$$x = 4 - 2y - z.$$

Since no restriction is placed on either y or z, they can be arbitrary real numbers, giving us a two-parameter family of solutions. Setting $y = r$ and $z = s$, we find that the solution of the given system is

$$x = 4 - 2r - s,$$
$$y = r,$$
$$z = s,$$

where r and s can be any real numbers. Each assignment of values to r and s results in a solution of the given system, so there are infinitely many solutions. For example, letting $r = 1$ and $s = 2$ gives the particular solution $x = 0$, $y = 1$, and $z = 2$.

■

■ Exercise 4.4

In Problems 1–24, solve the systems algebraically.

1. $\begin{cases} x + 4y = 3, \\ 3x - 2y = -5. \end{cases}$

2. $\begin{cases} 4x + 2y = 9, \\ 5y - 4x = 5. \end{cases}$

3. $\begin{cases} 3x - 4y = 13, \\ 2x + 3y = 3. \end{cases}$

4. $\begin{cases} 2x - y = 1, \\ -x + 2y = 7. \end{cases}$

5. $\begin{cases} 5v + 2w = 36, \\ 8v - 3w = -54. \end{cases}$

6. $\begin{cases} p + q = 3, \\ 3p + 2q = 19. \end{cases}$

7. $\begin{cases} x - 2y = 8, \\ 5x + 3y = 1. \end{cases}$

8. $\begin{cases} 3x + 5y = 7, \\ 5x + 9y = 7. \end{cases}$

9. $\begin{cases} 4x - 3y - 2 = 3x - 7y, \\ x + 5y - 2 = y + 4. \end{cases}$

10. $\begin{cases} 5x + 7y + 2 = 9y - 4x + 6, \\ \frac{21}{2}x - \frac{4}{3}y - \frac{11}{4} = \frac{3}{2}x + \frac{2}{3}y + \frac{5}{4}. \end{cases}$

11. $\begin{cases} \frac{2}{3}x + \frac{1}{2}y = 2, \\ \frac{3}{8}x + \frac{5}{6}y = -\frac{11}{2}. \end{cases}$

12. $\begin{cases} \frac{1}{2}z - \frac{1}{4}w = \frac{1}{6}, \\ z + \frac{1}{2}w = \frac{2}{3}. \end{cases}$

13. $\begin{cases} 4p + 12q = 6, \\ 2p + 6q = 3. \end{cases}$

14. $\begin{cases} 5x - 3y = 2, \\ -10x + 6y = 4. \end{cases}$

15. $\begin{cases} 2x + y + 6z = 3, \\ x - y + 4z = 1, \\ 3x + 2y - 2z = 2. \end{cases}$

16. $\begin{cases} x + y + z = -1, \\ 3x + y + z = 1, \\ 4x - 2y + 2z = 0. \end{cases}$

17. $\begin{cases} 5x - 7y + 4z = 2, \\ 3x + 2y - 2z = 3, \\ 2x - y + 3z = 4. \end{cases}$

18. $\begin{cases} 3x - 2y + z = 0, \\ -2x + y - 3z = 15, \\ \frac{3}{2}x + \frac{4}{5}y + 4z = 10. \end{cases}$

[6]19. $\begin{cases} x - 2z = 1, \\ y + z = 3. \end{cases}$

[6]20. $\begin{cases} 2y + 3z = 1, \\ 3x - 4z = 0. \end{cases}$

[6]21. $\begin{cases} x - y + 2z = 0, \\ 2x + y - z = 0, \\ x + 2y - 3z = 0. \end{cases}$

[6]22. $\begin{cases} x - 2y - z = 0, \\ 2x - 4y - 2z = 0, \\ -x + 2y + z = 0. \end{cases}$

[6]23. $\begin{cases} 2x + 2y - z = 3, \\ 4x + 4y - 2z = 6. \end{cases}$

[6]24. $\begin{cases} x + 2y - 3z = -4, \\ 2x + y - 3z = 4. \end{cases}$

25. **Mixture** A chemical manufacturer wishes to fill an order for 700 gallons of a 24% acid solution. Solutions of 20% and 30% are in stock. How many gallons of each solution must be mixed to fill the order?

26. **Mixture** A gardener has two fertilizers that contain different concentrations of nitrogen. One is 3% nitrogen and the other is 11% nitrogen. How many pounds of each should she mix to obtain 20 pounds of a 9% concentration?

27. **Fabric** A textile mill produces fabric made from different fibers. From cotton, polyester, and nylon, the owners want to produce a fabric blend that will cost \$3.25 per pound to make. The cost per pound of these fibers is \$4.00, \$3.00, and \$2.00, respectively. The amount of nylon is to be the same as the amount of polyester. How much of each fiber will be in the final fabric?

28. **Taxes** A company has taxable income of \$312,000. The federal tax is 25% of that portion left after the state tax has been paid. The state tax is 10% of that portion left after the federal tax has been paid. Find the federal and state taxes.

29. **Airplane Speed** An airplane travels 900 mi in 3 h with the aid of a tail wind. It takes 3 h, 36 min, for the return trip, flying against the same wind. Find the speed of the airplane in still air and the speed of the wind.

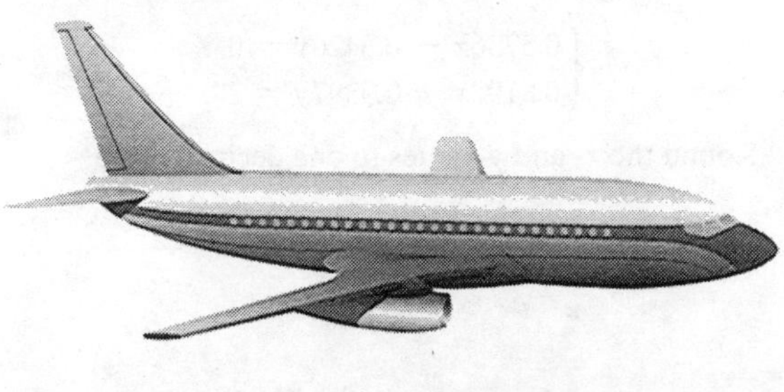

30. **Speed of Raft** On a trip on a raft, it took $\frac{3}{4}$ h to travel 12 mi downstream. The return trip took $1\frac{1}{2}$ h. Find the speed of the raft in still water and the speed of the current.

31. **Furniture Sales** A manufacturer of dining-room sets produces two styles: early American and contemporary. From past experience, management has determined that 20% more of the early American styles can be sold than the contemporary styles. A profit of \$250 is made on each early American set sold, whereas a profit of \$350 is made on each contemporary set. If, in the forthcoming year, management desires a total profit of \$130,000, how many units of each style must be sold?

32. **Survey** National Surveys was awarded a contract to perform a product-rating survey for Crispy Crackers. A total of 250 people were interviewed. National Surveys reported that 62.5% more people like Crispy Crackers than disliked them. However, the report did not indicate that 16% of those interviewed had no comment. How many of those surveyed liked Crispy Crackers? How many disliked them? How many had no comment?

33. **Equalizing Cost** United Products Co. manufactures calculators and has plants in the cities of Exton and

[6]Refers to concepts in Examples 6 and 7.

Whyton. At the Exton plant, fixed costs are $7000 per month, and the cost of producing each calculator is $7.50. At the Whyton plant, fixed costs are $8800 per month, and each calculator costs $6.00 to produce. Next month, United Products must produce 1500 calculators. How many must be made at each plant if the total cost at each plant is to be the same?

34. **Coffee Blending** A coffee wholesaler blends together three types of coffee that sell for $2.20, $2.30, and $2.60 per pound, so as to obtain 100 lb of coffee worth $2.40 per pound. If the wholesaler uses the same amount of the two higher priced coffees, how much of each type must be used in the blend?

35. **Commissions** A company pays its salespeople on a basis of a certain percentage of the first $100,000 in sales, plus a certain percentage of any amount over $100,000 in sales. If one salesperson earned $8500 on sales of $175,000 and another salesperson earned $14,800 on sales of $280,000, find the two percentages.

36. **Yearly Profits** In news reports, profits of a company this year (T) are often compared with those of last year (L), but actual values of T and L are not always given. This year, a company had profits of $20 million more than last year. The profits were up 25%. Determine T and L from these data.

37. **Production** Universal Control Co. makes industrial control units. Their new models are the Argon I and the Argon II. To make each Argon I unit, they use 6 doodles and 3 skeeters. To make each Argon II unit, they use 10 doodles and 8 skeeters. The company receives a total of 760 doodles and 500 skeeters each day from its supplier. How many units of each model of the Argon can the company make each day? Assume that all the parts are used.

38. **Investments** A person made two investments, and the percentage return per year on each was the same. Of the total amount invested, $\frac{3}{10}$ of it plus $600 was invested in one venture, and at the end of 1 year the person received a return of $384 from that venture. If the total return after 1 year was $1120, find the total amount invested.

39. **Production Run** A company makes three types of patio furniture: chairs, rockers, and chaise lounges. Each requires wood, plastic, and aluminum, in the amounts shown in the following table. The company has in stock 400 units of wood, 600 units of plastic, and 1500 units of aluminum. For its end-of-the-season production run, the company wants to use up all the stock. To do this, how many chairs, rockers, and chaise lounges should it make?

	Wood	Plastic	Aluminum
Chair	1 unit	1 unit	2 units
Rocker	1 unit	1 unit	3 units
Chaise lounge	1 unit	2 units	5 units

40. **Investments** A total of $35,000 was invested at three interest rates: 7, 8, and 9%. The interest for the first year was $2830, which was not reinvested. The second year the amount originally invested at 9% earned 10% instead, and the other rates remained the same. The total interest the second year was $2960. How much was invested at each rate?

41. **Hiring Workers** A company pays skilled workers in its assembly department $15 per hour. Semiskilled workers in that department are paid $9 per hour. Shipping clerks are paid $10 per hour. Because of an increase in orders, the company needs to hire a total of 70 workers in the assembly and shipping departments. It will pay a total of $760 per hour to these employees. Because of a union contract, twice as many semiskilled workers as skilled workers must be employed. How many semiskilled workers, skilled workers, and shipping clerks should the company hire?

42. **Solvent Storage** A 10,000-gallon railroad tank car is to be filled with solvent from two storage tanks, A and B. Solvent from A is pumped at the rate of 20 gal/min. Solvent from B is pumped at 30 gal/min. Usually, both pumps operate at the same time. However, because of a blown fuse, the pump on A is delayed 10 minutes. How many gallons from each storage tank will be used to fill the car?

43. Verify your answer to Problem 1 by using your graphics calculator.

44. Verify your answer to Problem 11 by using your graphics calculator.

45. Graphically solve the system.

$$\begin{cases} 0.24x - 0.34y = 0.28, \\ 0.11x + 0.21y = 0.86. \end{cases}$$

46. Graphically solve the system

$$\begin{cases} x + y = 2, \\ \frac{1}{4}x + \frac{2}{5}y = \frac{3}{5}. \end{cases}$$

Round the x- and y-values to two decimal places.

47. Graphically solve the system

$$\begin{cases} 0.5736x - 0.3420y = 0, \\ 0.8192x + 0.9397y = 20. \end{cases}$$

Round the x- and y-values to one decimal place

OBJECTIVE

To use substitution to solve nonlinear systems.

4.5 NONLINEAR SYSTEMS

A system of equations in which at least one equation is not linear is called a **nonlinear system.** We can often solve a nonlinear system by substitution, as was done with linear systems. The following examples illustrate.

EXAMPLE 1 Solving a Nonlinear System

Solve

$$\begin{cases} x^2 - 2x + y - 7 = 0, & (1) \\ 3x - y + 1 = 0. & (2) \end{cases}$$

Solution:

Strategy: If a nonlinear system contains a linear equation, we usually solve the linear equation for one variable and substitute for that variable in the other equation.

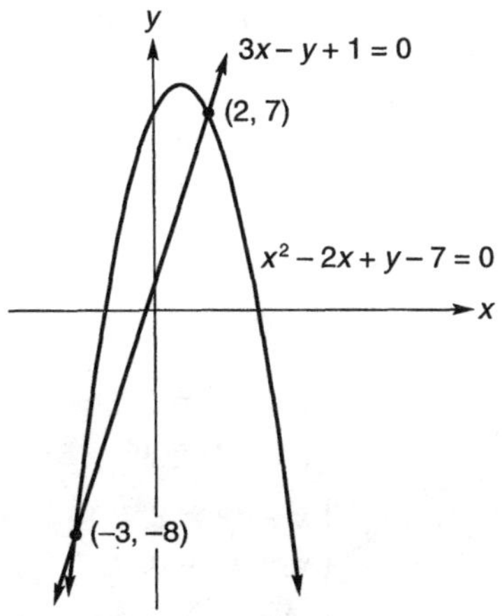

FIGURE 4.37 Nonlinear system of equations.

Solving Eq. (2) for y gives

$$y = 3x + 1. \tag{3}$$

Substituting into Eq. (1) and simplifying, we have

$$x^2 - 2x + (3x + 1) - 7 = 0,$$
$$x^2 + x - 6 = 0,$$
$$(x + 3)(x - 2) = 0,$$
$$x = -3 \quad \text{or} \quad x = 2.$$

If $x = -3$, then Eq. (3) implies that $y = -8$; if $x = 2$, then $y = 7$. You should verify that each pair of values satisfies the given system. Hence, the solutions are $x = -3$, $y = -8$ and $x = 2$, $y = 7$. These solutions can be seen geometrically in the graph of the system in Fig. 4.37. Notice that the graph of Eq. (1) is a parabola and the graph of Eq. (2) is a line. The solutions correspond to the intersection points $(-3, -8)$ and $(2, 7)$. ■

This example illustrates the need for checking all "solutions."

EXAMPLE 2 Solving a Nonlinear System

Solve

$$\begin{cases} y = \sqrt{x + 2}, \\ x + y = 4. \end{cases}$$

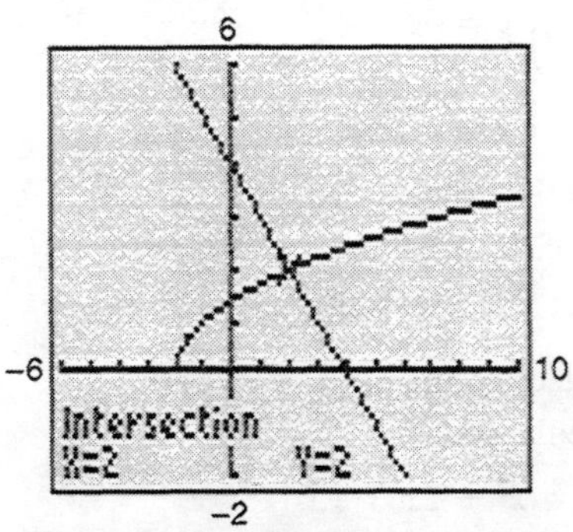

FIGURE 4.38 Nonlinear system of Example 2.

Solution: Solving the second equation, which is linear, for y gives

$$y = 4 - x. \tag{4}$$

Substituting into the first equation yields

$$4 - x = \sqrt{x + 2},$$
$$16 - 8x + x^2 = x + 2 \quad \text{(squaring both sides)},$$
$$x^2 - 9x + 14 = 0,$$
$$(x - 2)(x - 7) = 0.$$

Thus, $x = 2$ or $x = 7$. From Eq. (4), if $x = 2$, then $y = 2$; if $x = 7$, then $y = -3$. Since we performed the operation of squaring both sides, we must check our results. Although the pair $x = 2$, $y = 2$ satisfies both of the original equations, this is not the case for $x = 7$ and $y = -3$. Thus, the solution is $x = 2$, $y = 2$. (See Fig. 4.38.) ■

TECHNOLOGY

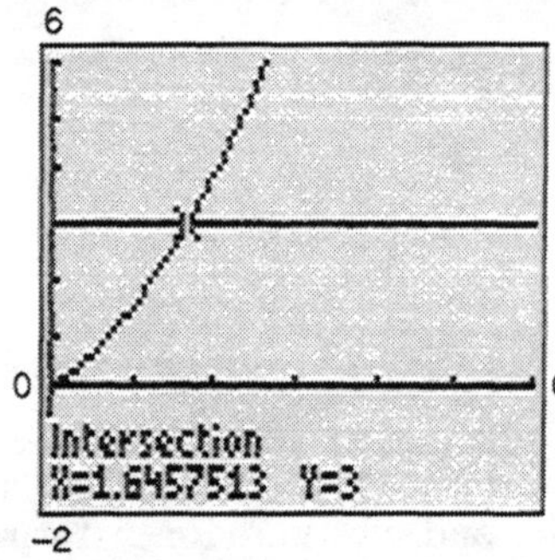

FIGURE 4.39 Solution of $0.5x^2 + x = 3$.

Graphically solve the equation $0.5x^2 + x = 3$, *where* $x \geq 0$.

Solution: To solve the equation, we could find zeros of the function $f(x) = 0.5x^2 + x - 3$. Alternatively, we can think of this problem as solving the nonlinear system

$$y = 0.5x^2 + x,$$

$$y = 3.$$

In Fig. 4.39, the intersection point is estimated to be $x = 1.65$, $y = 3$. Note that the graph of $y = 3$ is a horizontal line. The solution of the given equation is $x = 1.65$.

■ EXERCISE 4.5

In Problems 1–14, solve the given nonlinear system.

1. $\begin{cases} y = 4 - x^2, \\ 3x + y = 0. \end{cases}$

2. $\begin{cases} y = x^3, \\ x - y = 0. \end{cases}$

3. $\begin{cases} p^2 = 4 - q, \\ p = q + 2. \end{cases}$

4. $\begin{cases} y^2 - x^2 = 28, \\ x - y = 14. \end{cases}$

5. $\begin{cases} x = y^2, \\ y = x^2. \end{cases}$

6. $\begin{cases} p^2 - q = 0, \\ 3q - 2p - 1 = 0. \end{cases}$

7. $\begin{cases} y = 4x - x^2 + 8, \\ y = x^2 - 2x. \end{cases}$

8. $\begin{cases} x^2 - y = 8, \\ y - x^2 = 0. \end{cases}$

9. $\begin{cases} p = \sqrt{q}, \\ p = q^2. \end{cases}$

10. $\begin{cases} z = 4/w, \\ 3z = 2w + 2. \end{cases}$

11. $\begin{cases} x^2 = y^2 + 14, \\ y = x^2 - 16. \end{cases}$

12. $\begin{cases} x^2 + y^2 - 2xy = 1, \\ 3x - y = 5. \end{cases}$

13. $\begin{cases} x = y + 6, \\ y = 3\sqrt{x + 4}. \end{cases}$

14. $\begin{cases} y = \dfrac{x^2}{x - 1} + 1, \\ y = \dfrac{1}{x - 1}. \end{cases}$

15. Decorations The shape of a paper streamer suspended above a dance floor can be described by the function $y = 0.01x^2 + 0.01x + 7$, where y is the height of the streamer (in feet) above the floor and x is the horizontal distance (in feet) from the center of the room. A rope holding up other decorations touches the streamer and is described by the function $y = 0.01x + 8.0$. Where does the rope touch the paper streamer?

16. Awning The shape of a decorative awning over a storefront can be described by the function $y = 0.06x^2 + 0.012x + 8$, where y is the height of the edge of the awning (in feet) above the sidewalk and x is the distance (in feet) from the center of the store's doorway. A vandal pokes a stick through the awning, piercing it in two places. The position of the stick can be described by the function $y = 0.912x + 5$. Where are the holes in the awning caused by the vandal?

17. Graphically determine how many solutions there are to the system

$$\begin{cases} y = \dfrac{1}{x}, \\ y = x^2 - 4. \end{cases}$$

18. Graphically solve the system

$$\begin{cases} y = x^3, \\ y = 6 - x^2 \end{cases}$$

to one-decimal-place accuracy.

19. Graphically solve the system

$$\begin{cases} y = x^2 - 2x + 1, \\ y = x^3 + x^2 - 2x + 3 \end{cases}$$

to one-decimal-place accuracy.

20. Graphically solve the system

$$\begin{cases} y = x^3 + x, \\ y = 4x \end{cases}$$

to one-decimal-place accuracy.

In Problems 21–23, graphically solve the equation by treating it as a system. Round answers to two decimal places.

21. $0.8x^2 + 2x = 6$, where $x \geq 0$.

22. $\sqrt{x + 2} = 5 - x$.

23. $x^3 - 3x^2 = x - 8$.

OBJECTIVE

To solve systems describing equilibrium and break-even points.

4.6 Applications of Systems of Equations

Equilibrium

Recall from Sec. 4.2 that an equation that relates price per unit and quantity demanded (supplied) is called a *demand equation (supply equation).* Suppose that, for product Z, the demand equation is

$$p = -\frac{1}{180}q + 12 \tag{1}$$

and the supply equation is

$$p = \frac{1}{300}q + 8, \tag{2}$$

where $q, p \geq 0$. The corresponding demand and supply curves are the lines in Figs. 4.40 and 4.41, respectively. In analyzing Fig. 4.40, we see that consumers will purchase 540 units per week when the price is \$9 per unit, 1080 units when the price is \$6, and so on. Figure 4.41 shows that when the price is \$9 per unit

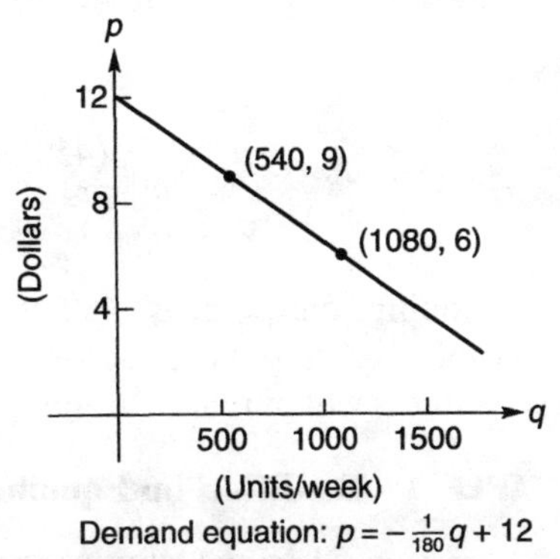

FIGURE 4.40 Demand curve.

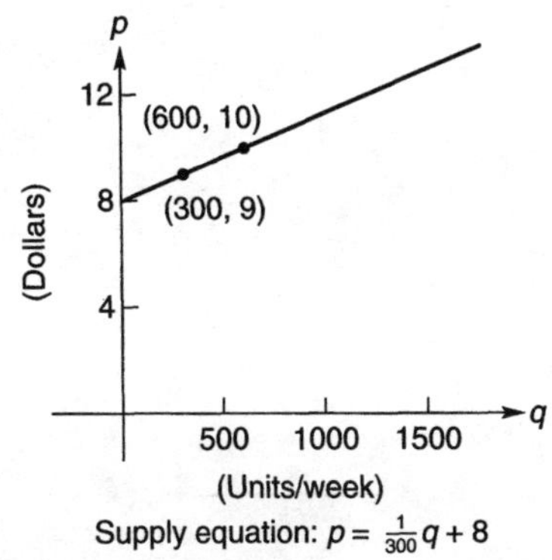

FIGURE 4.41 Supply curve.

producers will place 300 units per week on the market, at \$10 they will supply 600 units, and so on.

When the demand and supply curves of a product are represented on the same coordinate plane, the point (m, n) where the curves intersect is called the **point of equilibrium.** (See Fig. 4.42.) The price n, called the **equilibrium price,** is the price at which consumers will purchase the same quantity of a product that producers wish to sell at that price. In short, n is the price at which

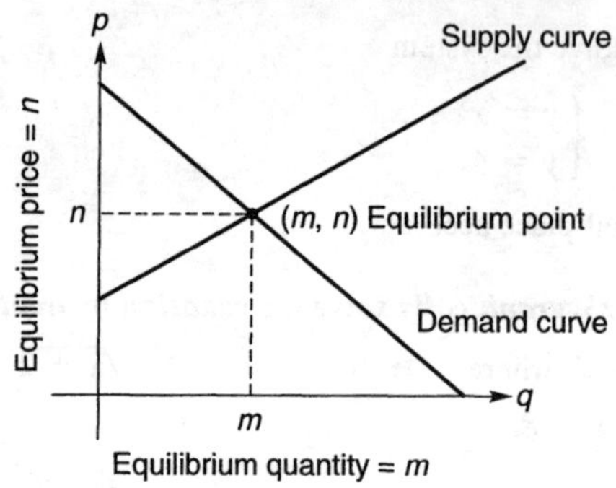

FIGURE 4.42 Equilibrium.

stability in the producer–consumer relationship occurs. The quantity m is called the **equilibrium quantity.**

To determine precisely the equilibrium point, we solve the system formed by the supply and demand equations. Let us do this for our previous data, namely, the system

$$\begin{cases} p = -\dfrac{1}{180}q + 12 & \text{(demand equation)},\\[2ex] p = \dfrac{1}{300}q + 8 & \text{(supply equation)}.\end{cases}$$

By substituting $\frac{1}{300}q + 8$ for p in the demand equation, we get

$$\frac{1}{300}q + 8 = -\frac{1}{180}q + 12,$$

$$\left(\frac{1}{300} + \frac{1}{180}\right)q = 4,$$

$$q = 450 \qquad \text{(equilibrium quantity)}.$$

Thus,

$$p = \frac{1}{300}(450) + 8$$

$$= 9.50 \qquad \text{(equilibrium price)},$$

and the equilibrium point is (450, 9.50). Therefore, at the price of \$9.50 per unit, manufacturers will produce exactly the quantity (450) of units per week that consumers will purchase at that price. (See Fig. 4.43.)

EXAMPLE 1 Tax Effect on Equilibrium

Let $p = \frac{8}{100}q + 50$ *be the supply equation for a manufacturer's product, and suppose the demand equation is* $p = -\frac{7}{100}q + 65$.

a. *If a tax of* \$1.50 *per unit is to be imposed on the manufacturer, how will the original equilibrium price be affected if the demand remains the same?*

Solution: Before the tax, the equilibrium price is obtained by solving the system

$$\begin{cases} p = \dfrac{8}{100}q + 50,\\[2ex] p = -\dfrac{7}{100}q + 65.\end{cases}$$

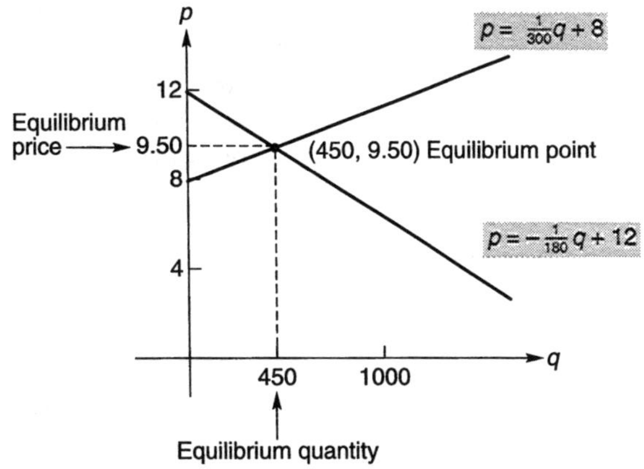

FIGURE 4.43 Equilibrium.

By substitution,

$$-\frac{7}{100}q + 65 = \frac{8}{100}q + 50,$$

$$15 = \frac{15}{100}q,$$

$$100 = q,$$

and

$$p = \frac{8}{100}(100) + 50 = 58.$$

Thus, \$58 is the original equilibrium price. Before the tax, the manufacturer supplies q units at a price of $p = \frac{8}{100}q + 50$ per unit. After the tax, he will sell the same q units for an additional \$1.50 per unit. The price per unit will be $(\frac{8}{100}q + 50) + 1.50$, so the new supply equation is

$$p = \frac{8}{100}q + 51.50.$$

Solving the system

$$\begin{cases} p = \frac{8}{100}q + 51.50, \\ p = -\frac{7}{100}q + 65 \end{cases}$$

will give the new equilibrium price:

$$\frac{8}{100}q + 51.50 = -\frac{7}{100}q + 65,$$

$$\frac{15}{100}q = 13.50,$$

$$q = 90,$$

$$p = \frac{8}{100}(90) + 51.50 = 58.70.$$

The tax of \$1.50 per unit increases the equilibrium price by \$0.70. (See Fig. 4.44.) Note that there is also a decrease in the equilibrium quantity from $q = 100$ to $q = 90$, because of the change in the equilibrium price. (In the exercises, you are asked to find the effect of a subsidy given to the manufacturer, which will reduce the price of the product.)

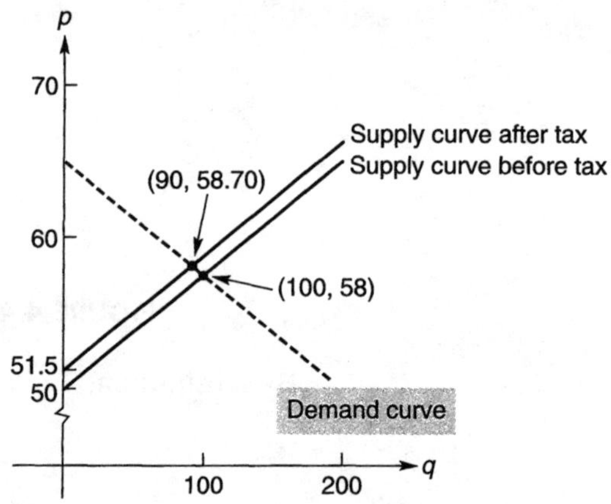

FIGURE 4.44 Equilibrium before and after tax.

b. *Determine the total revenue obtained by the manufacturer at the equilibrium point both before and after the tax.*

Solution: If q units of a product are sold at a price of p dollars each, then the total revenue is given by

$$y_{TR} = pq.$$

Before the tax, the revenue at (100, 58) is (in dollars)

$$y_{TR} = (58)(100) = 5800.$$

After the tax, it is

$$y_{TR} = (58.70)(90) = 5283,$$

which is a decrease. ■

EXAMPLE 2 Equilibrium with Nonlinear Demand

Find the equilibrium point if the supply and demand equations of a product are $p = \frac{q}{40} + 10$ *and* $p = \frac{8000}{q}$*, respectively.*

Solution: Here the demand equation is not linear. Solving the system

$$\begin{cases} p = \frac{q}{40} + 10, \\ p = \frac{8000}{q} \end{cases}$$

by substitution gives

$$\frac{8000}{q} = \frac{q}{40} + 10,$$

$$320{,}000 = q^2 + 400q \qquad \text{(multiplying both sides by } 40q\text{)},$$

$$q^2 + 400q - 320{,}000 = 0,$$
$$(q + 800)(q - 400) = 0,$$
$$q = -800 \qquad \text{or} \qquad q = 400.$$

We disregard $q = -800$, since q represents quantity. Choosing $q = 400$, we have $p = (8000/400) = 20$, so the equilibrium point is (400, 20). (See Fig. 4.45.) ■

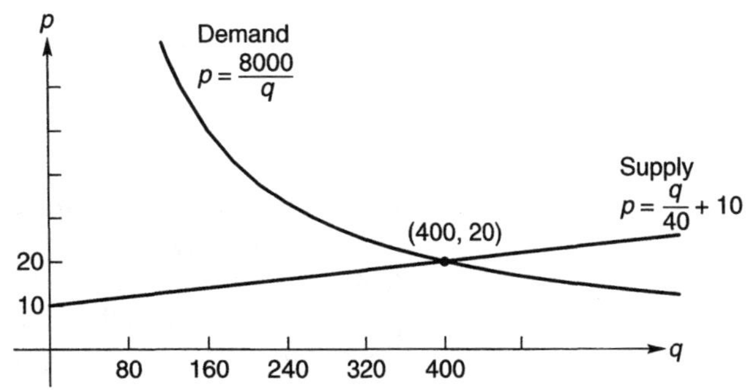

FIGURE 4.45 Equilibrium with nonlinear demand.

Break-even Points

Suppose a manufacturer produces product A and sells it at \$8 per unit. Then the total revenue y_{TR} received (in dollars) from selling q units is

$$y_{TR} = 8q \qquad \text{(total revenue)}.$$

The difference between the total revenue received for q units and the total cost of q units is the manufacturer's profit (or loss if the difference is negative):

$$\textbf{profit (or loss)} = \textbf{total revenue} - \textbf{total cost.}$$

Total cost, y_{TC}, is the sum of total variable costs y_{VC} and total fixed costs y_{FC}:

$$y_{TC} = y_{VC} + y_{FC}.$$

Fixed costs are those costs that, under normal conditions, do not depend on the level of production; that is, over some period of time they remain constant at all levels of output. (Examples are rent, officers' salaries, and normal maintenance.) **Variable costs** are those costs that vary with the level of production (such as the cost of materials, labor, maintenance due to wear and tear, etc.). For q units of product A, suppose that

$$y_{FC} = 5000 \qquad \text{(fixed cost)}$$

$$\text{and} \quad y_{VC} = \frac{22}{9}q \qquad \text{(variable cost)}.$$

Then

$$y_{TC} = \frac{22}{9}q + 5000 \qquad \text{(total cost)}.$$

The graphs of total cost and total revenue appear in Fig. 4.46. The horizontal axis represents the level of production, q, and the vertical axis represents the total dollar value, be it revenue or cost. The **break-even point** is the

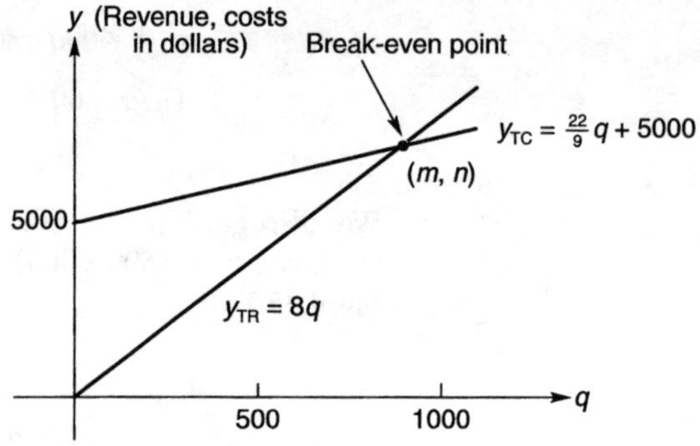

FIGURE 4.46 Break-even chart.

point at which total revenue equals total cost (TR = TC). It occurs when the levels of production and sales result in neither a profit nor a loss to the manufacturer. In the diagram, called a *break-even chart,* the break-even point is the point (m, n) at which the graphs of $y_{TR} = 8q$ and $y_{TC} = \frac{22}{9}q + 5000$ intersect. We call m the **break-even quantity** and n the **break-even revenue.** When total cost and revenue are linearly related to output, as in this case, for any production level greater than m, total revenue is greater than total cost, resulting in a profit. However, at any level less than m units, total revenue is less than total cost, resulting in a loss. At an output of m units, the profit is zero. In the following example, we shall examine our data in more detail.

EXAMPLE 3 Break-even Point, Profit, and Loss

A manufacturer sells a product at \$8 per unit, selling all that is produced. Fixed cost is \$5000 and variable cost per unit is $\frac{22}{9}$ (dollars).

a. *Find the total output and revenue at the break-even point.*

Solution: At an output level of q units, the variable cost is $y_{VC} = \frac{22}{9}q$ and the total revenue is $y_{TR} = 8q$. Hence,

$$y_{TR} = 8q,$$

$$y_{TC} = y_{VC} + y_{FC} = \frac{22}{9}q + 5000.$$

At the break-even point, total revenue equals total cost. Thus, we solve the system formed by the foregoing equations. Since

$$y_{TR} = y_{TC},$$

we have

$$8q = \frac{22}{9}q + 5000,$$

$$\frac{50}{9}q = 5000,$$

$$q = 900.$$

Hence, the desired output is 900 units, resulting in a total revenue (in dollars) of

$$y_{TR} = 8(900) = 7200.$$

(See Fig. 4.47.)

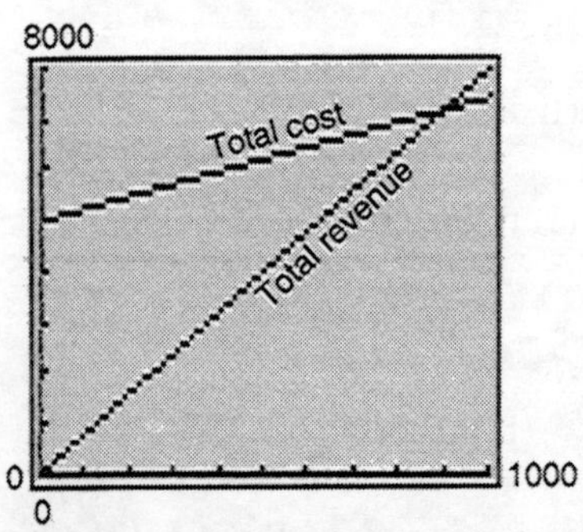

FIGURE 4.47 Equilibrium point (900, 7200).

b. *Find the profit when* 1800 *units are produced.*

Solution: Since profit = total revenue − total cost, when $q = 1800$ we have

$$y_{\text{TR}} - y_{\text{TC}} = 8(1800) - \left[\frac{22}{9}(1800) + 5000\right]$$
$$= 5000.$$

The profit when 1800 units are produced and sold is \$5000.

c. *Find the loss when* 450 *units are produced.*

Solution: When $q = 450$,

$$y_{\text{TR}} - y_{\text{TC}} = 8(450) - \left[\frac{22}{9}(450) + 5000\right] = -2500.$$

A loss of \$2500 occurs when the level of production is 450 units.

d. *Find the output required to obtain a profit of* \$10,000.

Solution: In order to obtain a profit of \$10,000, we have

$$\text{profit} = \text{total revenue} - \text{total cost},$$
$$10{,}000 = 8q - \left(\frac{22}{9}q + 5000\right),$$
$$15{,}000 = \frac{50}{9}q,$$
$$q = 2700.$$

Thus, 2700 units must be produced. ■

EXAMPLE 4 Break-even Quantity

Determine the break-even quantity of XYZ Manufacturing Co., given the following data: total fixed cost, \$1200; *variable cost per unit,* \$2; *total revenue for selling q units,* $y_{\text{TR}} = 100\sqrt{q}$.

Solution: For q units of output,

$$y_{\text{TR}} = 100\sqrt{q},$$
$$y_{\text{TC}} = 2q + 1200.$$

Equating total revenue to total cost gives

$$100\sqrt{q} = 2q + 1200,$$
$$50\sqrt{q} = q + 600 \quad \text{(dividing both sides by 2)}.$$

Squaring both sides, we have

$$2500q = q^2 + 1200q + (600)^2,$$
$$0 = q^2 - 1300q + 360{,}000.$$

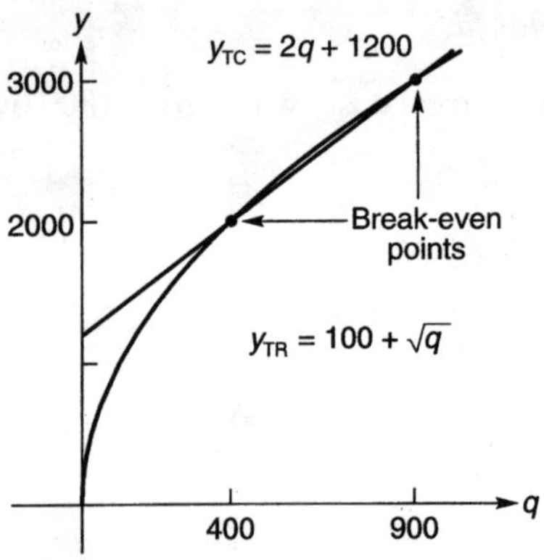

FIGURE 4.48 Two break-even points.

By the quadratic formula,

$$q = \frac{1300 \pm \sqrt{250{,}000}}{2},$$

$$q = \frac{1300 \pm 500}{2},$$

$$q = 400 \quad \text{or} \quad q = 900.$$

Although both $q = 400$ and $q = 900$ are break-even quantities, observe in Fig. 4.48 that when $q > 900$, total cost is greater than total revenue, so there will always be a loss. This occurs because here total revenue is not linearly related to output. Thus, producing more than the break-even quantity does not necessarily guarantee a profit. ■

■ Exercise 4.6

In Problems 1–8, you are given a supply equation and a demand equation for a product. If p represents price per unit in dollars and q represents the number of units per unit of time, find the equilibrium point. In Problems 1 and 2, sketch the system.

1. Supply: $p = \frac{3}{100}q + 2$,
Demand: $p = -\frac{7}{100}q + 12$.

2. Supply: $p = \frac{1}{2000}q + 3$,
Demand: $p = -\frac{1}{2500}q + \frac{42}{5}$.

3. Supply: $35q - 2p + 250 = 0$,
Demand: $65q + p - 537.5 = 0$.

4. Supply: $246p - 3.25q - 2460 = 0$,
Demand: $410p + 3q - 14{,}452.5 = 0$.

5. Supply: $p = 2q + 20$,
Demand: $p = 200 - 2q^2$.

6. Supply: $p = (q + 10)^2$,
Demand: $p = 388 - 16q - q^2$.

7. Supply: $p = \sqrt{q + 10}$,
Demand: $p = 20 - q$.

8. Supply: $p = \frac{1}{5}q + 5$,
Demand: $p = \dfrac{3000}{q + 200}$.

In Problems 9–14, y_{TR} represents total revenue in dollars and y_{TC} represents total cost in dollars for a manufacturer. If q represents both the number of units produced and the number of units sold, find the break-even quantity. Sketch a break-even chart in Problems 9 and 10.

9. $y_{TR} = 3q$,
$y_{TC} = 2q + 4500$,

10. $y_{TR} = 14q$,
$y_{TC} = \frac{40}{3}q + 1200$.

11. $y_{TR} = 0.05q$,
$y_{TC} = 0.85q + 600$.

12. $y_{TR} = 0.25q$,
$y_{TC} = 0.16q + 360$.

13. $y_{TR} = 100 - \dfrac{1000}{q + 10}$,
$y_{TC} = q + 40$.

14. $y_{TR} = 0.1q^2 + 7q$,
$y_{TC} = 2q + 500$.

15. Business Supply and demand equations for a certain product are

$$3q - 200p + 1800 = 0$$

and

$$3q + 100p - 1800 = 0,$$

respectively, where p represents the price per unit in dollars and q represents the number of units sold per time period.

a. Find the equilibrium price algebraically, and derive it graphically.

b. Find the equilibrium price when a tax of 27 cents per unit is imposed on the supplier.

16. Business A manufacturer of a product sells all that is produced. The total revenue is given by $y_{TR} = 7q$, and the total cost is given by $y_{TC} = 6q + 800$, where q represents the number of units produced and sold.

a. Find the level of production at the break-even point, and draw the break-even chart.

b. Find the level of production at the break-even point if the total cost increases by 5%.

17. Business A manufacturer sells a product at \$8.35 per unit, selling all produced. The fixed cost is \$2116 and the variable cost is \$7.20 per unit. At what level of production will there be a profit of \$4600? At what level of production will there be a loss of \$1150? At what level of production will the break-even point occur?

18. Business The market equilibrium point for a product occurs when 13,500 units are produced at a price of \$4.50 per unit. The producer will supply no units at \$1, and the consumers will demand no units at \$20. Find the supply and demand equations if they are both linear.

19. Business A manufacturer of widgets will break even at a sales volume of \$200,000. Fixed costs are \$40,000, and each unit of output sells for \$5. Determine the variable cost per unit.

20. Business The Footsie Sandal Co. manufactures sandals for which the material cost is \$0.80 per pair and the labor cost is \$0.90 per pair. Additional variable costs amount to \$0.30 per pair. Fixed costs are \$70,000. If each pair sells for \$2.50, how many pairs must be sold for the company to break even?

21. Business Find the break-even point for company Z, which sells all it produces, if the variable cost per unit is \$2, fixed costs are \$1050, and $y_{TR} = 50\sqrt{q}$, where q is the number of units of output produced.

22. Business A company has determined that the demand equation for its product is $p = 1000/q$, where p is the price per unit for q units produced and sold in some period. Determine the quantity demanded when the price per unit is (a) \$4, (b) \$2, and (c) \$0.50. For each of these prices, determine the total revenue that the company will receive. What will be the revenue regardless of the price? [*Hint:* Find the revenue when the price is p dollars.]

23. Business Using the data in Example 1, determine how the original equilibrium price will be affected if the company is given a government subsidy of \$1.50 per unit.

24. Business The Monroe Forging Company sells a corrugated steel product to the Standard Manufacturing Company and is in competition on such sales with other suppliers of the Standard Manufacturing Co. The vice president of sales of Monroe Forging Co. believes that by reducing the price of the product, a 40% increase in the volume of units sold to the Standard Manufacturing Co. could be secured. As the manager of the cost and analysis department, you have been asked to analyze the proposal of the vice president and submit your recommendations as to whether it is financially beneficial to the Monroe Forging Co. You are specifically requested to determine the following.

a. Net profit or loss based on the pricing proposal.

b. Unit sales volume under the proposed price that is required to make the same \$40,000 profit which is now earned at the current price and unit sales volume.

Use the following data in your analysis:

	Current Operations	Proposal of Vice President of Sales
Unit price	\$2.50	\$2.00
Unit sales volume	200,000 units	280,000 units
Variable cost		
Total	\$350,000	\$490,000
Per unit	\$1.75	\$1.75
Fixed cost	\$110,000	\$110,000
Profit	\$40,000	?

25. Business Suppose products A and B have demand and supply equations that are related to each other. If q_A and q_B are the quantities produced and sold of A and B, respectively, and p_A and p_B are their respective prices, the demand equations are

$$q_A = 8 - p_A + p_B$$

and

$$q_B = 26 + p_A - p_B,$$

and the supply equations are

$$q_A = -2 + 5p_A - p_B$$

and

$$q_B = -4 - p_A + 3p_B.$$

Eliminate q_A and q_B to get the equilibrium prices.

26. Business The supply equation for a product is

$$p = 0.3q^2 + 14.6,$$

and the demand equation is

$$p = \frac{34.2}{1 + 0.3q}.$$

Here p represents price per unit in dollars and q represents number of units (in thousands) per unit time. Graph both equations, and from your graph, determine the equilibrium price and equilibrium quantity to one decimal place.

27. Business For a manufacturer, the total-revenue equation is

$$y_{TR} = 20.5\sqrt{q + 4} - 41$$

and the total-cost equation is

$$y_{TC} = 0.02q^3 + 10.4,$$

where q represents (in thousands) both the number of units produced and the number of units sold. Graph a break-even chart and find the break-even quantity.

4.7 REVIEW

IMPORTANT TERMS AND SYMBOLS

Section 4.1	slope of a line; point–slope form; slope–intercept form; general linear equation in x and y; linearly related
Section 4.2	demand equation; demand curve; supply equation; supply curve; linear function
Section 4.3	quadratic function; parabola; axis of symmetry; vertex
Section 4.4	system of equations; equivalent systems; elimination by addition; elimination by substitution; parameter; general linear equation in x, y, and z
Section 4.5	nonlinear system
Section 4.6	point of equilibrium; equilibrium price; equilibrium quantity; profit; total cost; fixed cost; variable cost; break-even point; break-even quantity; break-even revenue

SUMMARY

The orientation of a nonvertical line is characterized by the slope of the line given by

$$m = \frac{y_2 - y_1}{x_2 - x_1},$$

where (x_1, y_1) and (x_2, y_2) are two different points on the line. The slope of a vertical line is not defined, and the slope of a horizontal line is zero. Rising lines have positive slopes; falling lines have negative slopes. Two lines are parallel if and only if they have the same slope or are vertical. Two lines with slopes m_1 and m_2 are perpendicular to each other if and only if $m_1 = -\frac{1}{m_2}$. A horizontal line and a vertical line are perpendicular to each other.

Basic forms of equations of lines are:

$y - y_1 = m(x - x_1)$	(point–slope form)
$y = mx + b$	(slope-intercept form)
$x = a$	(vertical line)
$y = b$	(horizontal line)
$Ax + By + C = 0$	(general)

The linear function $f(x) = ax + b\ (a \neq 0)$ has a straight line for its graph.

In economics, supply functions and demand functions have the form $p = f(q)$ and play an important role. Each gives a correspondence between the price p of a product and the number of units q of the product that manufacturers (or consumers) will supply (or purchase) at that price during some time period.

A quadratic function has the form

$$f(x) = ax^2 + bx + c \qquad (a \neq 0).$$

The graph of f is a parabola that opens upward if $a > 0$ and downward if $a < 0$. The vertex is

$$\left(-\frac{b}{2a}, f\left(-\frac{b}{2a}\right)\right),$$

and the y-intercept is c. The axis of symmetry, as well as the x- and y-intercepts, are useful in sketching the graph.

A system of linear equations may be solved with the method of elimination by addition or elimination by substitution. A solution may involve one or more parameters. Substitution is also useful in solving nonlinear systems.

Solving a system formed by the supply and demand equations for a product gives the equilibrium point, which indicates the price at which consumers will purchase the same quantity of a product that producers wish to sell at that price.

Profit is total revenue minus total cost, where total cost is the sum of fixed costs and variable costs. The break-even point is the point where total revenue equals total cost.

REVIEW PROBLEMS

1. The slope of the line through $(2, 5)$ and $(3, k)$ is 4. Find k.

2. The slope of the line through $(2, 3)$ and $(k, 3)$ is 0. Find k.

In Problems 3–9, determine the slope–intercept form and a general linear form of an equation of the straight line that has the indicated properties.

3. Passes through $(3, -2)$ and has y-intercept 1.

4. Passes through $(-1, -1)$ and is parallel to the line $y = 3x - 4$.

5. Passes through $(10, 4)$ and has slope $\frac{1}{2}$.

6. Passes through $(3, 5)$ and is vertical.

7. Passes through $(-2, 4)$ and is horizontal.

8. Passes through $(1, 2)$ and is perpendicular to the line $-3y + 5x = 7$.

9. Has y-intercept 2 and is perpendicular to $y + 3x = 2$.

10. Determine whether the point $(0, -7)$ lies on the line through $(1, -3)$ and $(4, 9)$.

In Problems 11–16, determine whether the lines are parallel, perpendicular, or neither.

11. $x + 4y + 2 = 0, \quad 8x - 2y - 2 = 0.$

12. $y - 2 = 2(x - 1), \quad 2x + 4y - 3 = 0.$

13. $x - 3 = 2(y + 4), \quad y = 4x + 2.$

14. $3x + 5y + 4 = 0, \quad 6x + 10y = 0.$

15. $y = \frac{1}{2}x + 5, \quad 2x = 4y - 3.$

16. $y = 7x, \quad y = 7.$

In Problems 17–20, write each line in slope–intercept form, and sketch. What is the slope of the line?

17. $3x - 2y = 4.$

18. $x = -3y + 4.$

19. $4 - 3y = 0.$

20. $y = 2x.$

In Problems 21–30, graph each function. For those that are linear, give the slope and the vertical-axis intercept. For those that are quadratic, give all intercepts and the vertex.

21. $y = f(x) = 4 - 2x.$

22. $s = g(t) = 8 - 2t - t^2.$

23. $y = f(x) = 9 - x^2.$

24. $y = f(x) = 3x - 7.$

25. $y = h(t) = t^2 - 4t - 5.$

26. $y = h(t) = 1 + 3t.$

27. $p = g(t) = 3t.$

28. $y = F(x) = (2x - 1)^2.$

29. $y = F(x) = -(x^2 + 2x + 3).$

30. $y = f(x) = \dfrac{x}{3} - 2.$

In Problems 31–44, solve the given system.

31. $\begin{cases} 2x - y = 6, \\ 3x + 2y = 5. \end{cases}$

32. $\begin{cases} 8x - 4y = 7, \\ y = 2x - 4. \end{cases}$

33. $\begin{cases} 4x + 5y = 3, \\ 3x + 4y = 2. \end{cases}$

34. $\begin{cases} 3x + 6y = 9, \\ 4x + 8y = 12. \end{cases}$

35. $\begin{cases} \frac{1}{4}x - \frac{3}{2}y = -4, \\ \frac{3}{4}x + \frac{1}{2}y = 8. \end{cases}$

36. $\begin{cases} \frac{1}{3}x - \frac{1}{4}y = \frac{1}{12}, \\ \frac{4}{3}x + 3y = \frac{5}{3}. \end{cases}$

37. $\begin{cases} 3x - 2y + z = -2, \\ 2x + y + z = 1, \\ x + 3y - z = 3. \end{cases}$

38. $\begin{cases} x + \dfrac{2y + x}{6} = 14, \\ y + \dfrac{3x + y}{4} = 20. \end{cases}$

39. $\begin{cases} x^2 - y + 2x = 7, \\ x^2 + y = 5. \end{cases}$

40. $\begin{cases} y = \dfrac{18}{x + 4}, \\ x - y + 7 = 0. \end{cases}$

[7]**41.** $\begin{cases} x + 2z = -2, \\ x + y + z = 5. \end{cases}$

[7]**42.** $\begin{cases} x + y + z = 0, \\ x - y + z = 0, \\ x + z = 0. \end{cases}$

[7]**43.** $\begin{cases} x - y - z = 0, \\ 2x - 2y + 3z = 0. \end{cases}$

[7]**44.** $\begin{cases} 2x - 5y + 6z = 1, \\ 4x - 10y + 12z = 2. \end{cases}$

45. Suppose a and b are linearly related so that $a = 1$ when $b = 2$ and $a = 2$ when $b = 1$. Find a general linear form of an equation that relates a and b. Also, find a when $b = 3$.

46. Temperature and Heart Rate When the temperature T (in degrees Celsius) of a cat is reduced, the cat's heart rate r (in beats per minute) decreases. Under laboratory conditions, a cat at a temperature of 37°C had a heart rate of 220, and at a temperature of 32°C its heart rate was 150. If r is linearly related to T, where T is between 26 and 38, (a) determine an equation for r in terms of T,

[7]Refer to concepts in Examples 6 and 7 in Sec. 4.4.

and (b) determine the cat's heart rate at a temperature of 28°C.

47. Suppose f is a linear function such that $f(1) = 5$ and $f(x)$ decreases by four units for every three-unit increase in x. Find $f(x)$.

48. If f is a linear function such that $f(-1) = 8$ and $f(2) = 5$, find $f(x)$.

49. Maximum Revenue The demand function for a manufacturer's product is $p = f(q) = 200 - 2q$, where p is the price (in dollars) per unit when q units are demanded. Find the level of production that maximizes the manufacturer's total revenue, and determine this revenue.

50. Sales Tax The difference in price of two items before a 5% sales tax is imposed is $4. The difference in price after the sales tax is imposed is $4.20. Find the price of each item before the sales tax.

51. Equilibrium Price If the supply and demand equations of a certain product are $125p - q - 250 = 0$ and $100p + q - 1100 = 0$, respectively, find the equilibrium price.

52. Break-even Point A manufacturer of a certain product sells all that is produced. Determine the break-even point if the product is sold at $16 per unit, fixed cost is $10,000, and variable cost is given by $y_{VC} = 8q$, where q is the number of units produced (y_{VC} expressed in dollars).

53. Psychology In psychology, the term *semantic memory* refers to our knowledge of the meaning and relationships of words, as well as the means by which we store and retrieve such information.[8] In a network model of semantic memory, there is a hierarchy of levels at which information is stored. In an experiment by Collins and Quillian based on a network model, data were obtained on the reaction time to respond to simple questions about nouns. The graph of the results shows that, on the average, the reaction time R (in milliseconds) is a linear function of the level L at which a characterizing property of the noun is stored. At level 0, the reaction time is 1310; at level 2, the reaction time is 1460. (a) Find the linear function. (b) Find the reaction time at level 1. (c) Find the slope and determine its significance.

54. Temperature Conversion Celsius temperature C is a linear function of Fahrenheit temperature F. Use the facts that 32°F is the same as 0°C and 212°F is the same as 100°C to find this function. Also, find C when F = 50.

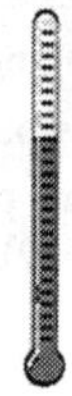

55. Enclosing Land A company has set aside $3000 to fence a rectangular portion of land adjacent to a straight stream by using the stream for one side of the enclosed area. The cost of the fencing parallel to the stream is $5 per foot installed, and the fencing for the remaining two sides is $3 per foot installed. Find the dimensions of the maximum enclosed area. [*Hint:* If the side opposite the stream has length x, first show that each of the other two sides has length $(3000 - 5x)/6$.]

56. Graphically solve the linear system

$$\begin{cases} 3x + 4y = 20, \\ 7x + 5y = 64. \end{cases}$$

57. Graphically solve the linear system

$$\begin{cases} 0.2x + 0.3y = 7, \\ 0.3x + 0.5y = 4. \end{cases}$$

Round x and y to two decimal places.

58. Graphically solve the nonlinear system

$$\begin{cases} y = \dfrac{2}{x}, \text{ where } x > 0, \\ y = x^2 - 6. \end{cases}$$

Round x and y to two decimal places.

59. Graphically solve the nonlinear system

$$\begin{cases} y = x^3 + 1, \\ y = 2 - x^2. \end{cases}$$

Round x and y to two decimal places.

60. Graphically solve the equation

$$x^2 + 4 = x^3 - 3x$$

by treating it as a system. Round x to two decimal places.

[8] G. R. Loftus and E. F. Loftus, *Human Memory: The Processing of Information* (New York: Lawrence Erlbaum Associates, Inc., distributed by the Halsted Press, Division of John Wiley & Sons, Inc., 1976).

MATHEMATICAL *SNAPSHOT*

ANYONE FOR TENNIS?[9]

Perhaps at some time you have made an appointment to see someone or to get something done, only to be met with a long wait. For example, people with appointments complain frequently about long waits to see physicians or to have their cars serviced. Even in sports, there are complaints about waiting. Tennis players may have to wait until 10:00 to begin their 9:00 match. Similarly for golfers, and so on. It appears that scheduling is at the heart of all these problems. Here you will learn one way to devise a time schedule for matches at a tennis tournament.

Suppose that 11 courts are available for a tennis tournament that begins at 8:00 A.M. Using an average match time of 1 hour and 30 minutes, a tournament director would typically schedule 11 matches at 8:00 A.M., 11 matches at 9:30 A.M., 11 matches at 11:00 A.M., and so on. However, the actual time length of a match varies. Some matches are completed after 30 minutes; others take more than 2 hours. Because of this, matches scheduled near the end of the tournament may run several hours late. Perhaps a better scheduling method is to have 11 matches begin at 8:00 A.M., then schedule some matches at 8:30 A.M., some at 9:00 A.M., some at 9:30 A.M., and continuing in this way, scheduling some matches every 30 minutes later. For convenience, refer to these 30-minute time intervals as periods 1, 2, 3, and so on, with period 1 beginning at 8:00 A.M. The problem is to determine the number of matches to be scheduled for each period.

Based on records of this tournament over the past several years, suppose that you estimate the average match time to be 1 hour and 37 minutes. These records also give rise to Table 4.3, which gives the average number of matches, as well as the cumulative total, played on the courts during each 30-minute interval. From Table 4.3 it appears that a schedule with 11 matches at 8:00 A.M., 1 match at 8:30 A.M., 4 matches at 9:00 A.M., and so on, is more reasonable than the typical schedule of 11 matches every $1\frac{1}{2}$ hours. In this situation, no one would have to wait more than 30 minutes. Denoting the period by x and the corresponding cumulative total number of matches by y, Fig. 4.49 gives a geometrical representation of the points (x, y). For example, the point $(2, 12)$ indicates that during period 2 the total number of matches played on the courts since the beginning of the tournament was 12.

It is obvious from Fig. 4.49 that the points almost lie on a straight line. Knowing the equation of such a line would allow you to predict the total number of matches to be scheduled by a given period. Because 11 matches must be scheduled for period 1, and the last match, the eighty-fifth, must be scheduled for period 23, it is not unreasonable to choose as your "predictor" line the one through the points $(1, 11)$ and $(23, 85)$, as shown in Fig. 4.49. The slope, m, is given by

$$m = \frac{85 - 11}{23 - 1} = \frac{74}{22} = \frac{37}{11}.$$

[9]Adapted from Brian Garman, "Applying a Linear Function to Schedule Tennis Matches," *The Mathematics Teacher,* 77, no. 7 (October 1984), 544–47. By permission of the National Council of Teachers of Mathematics.

TABLE 4.3

Period	Time	Matches per Period	Total Matches	Period	Time	Matches per Period	Total Matches
1	8:00	11	11	13	2:00	3	51
2	8:30	1	12	14	2:30	4	55
3	9:00	4	16	15	3:00	3	58
4	9:30	5	21	16	3:30	2	60
5	10:00	4	25	17	4:00	3	63
6	10:30	3	28	18	4:30	4	67
7	11:00	3	31	19	5:00	4	71
8	11:30	3	34	20	5:30	3	74
9	12:00	3	37	21	6:00	4	78
10	12:30	3	40	22	6:30	4	82
11	1:00	4	44	23	7:00	3	85
12	1:30	4	48				

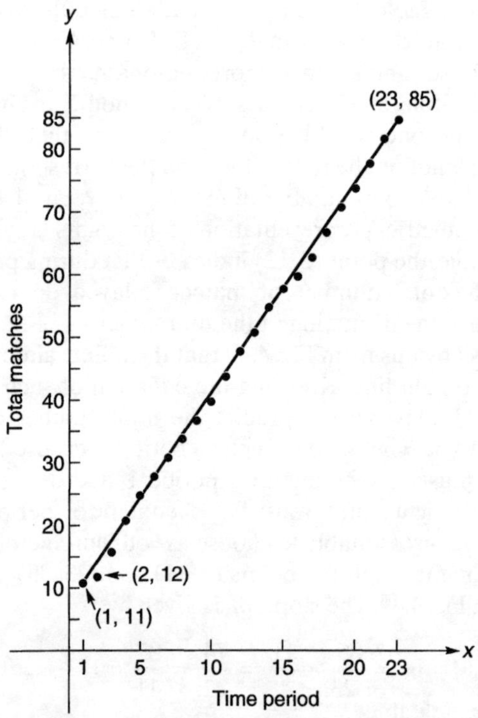

FIGURE 4.49 Total tennis matches as a function of time

A point–slope form of an equation of the line is

$$y - 11 = \frac{37}{11}(x - 1),$$

which can be rewritten as

$$y = \frac{37}{11}(x - 1) + 11. \tag{1}$$

For example, if $x = 2$, then Eq. (1) gives $y = \frac{158}{11} \approx 14.36$. Since fractional matches make no sense, the function in Eq. (1) must be refined. One method is to round the y-values in Eq. (1) to the nearest integer by making use of a function called the *greatest-integer function* and denoted by $[x]$. The notation $[x]$ means the greatest integer less than or equal to x. For example,

$$[4] = 4, \qquad [4.1] = 4, \qquad [5.9] = 5.$$

If you add 0.5 to any given number and then find the greatest integer in the sum, the result will be the given number rounded to the nearest integer. For instance, $[4.1 + 0.5] = [4.6] = 4$, which is 4.1 rounded to the nearest integer; $[5.9 + 0.5] = [6.4] = 6$, which is 5.9 rounded to the nearest integer. Thus the refined function, say f, that predicts the total number of matches up to and including period x is given by

$$f(x) = \left[\frac{37}{11}(x - 1) + 11.5\right].$$

In particular, if $x = 2$, then

$$\frac{37}{11}(x - 1) + 11.5 = \frac{37}{11}(2 - 1) + 11.5 \approx 14.86$$

and $[14.86] = 14$. Thus $f(2) = 14$. Table 4.4 gives the predicted totals, $f(x)$, for $x = 1, 2, 3, \ldots, 23$, as well as

TABLE 4.4

Period, x	Time	Actual Total	Predicted Total, $f(x)$	Adjusted Total, $F(x)$
1	8:00	11	11	11
2	8:30	12	14	12
3	9:00	16	18	16
4	9:30	21	21	21
5	10:00	25	24	24
6	10:30	28	28	28
7	11:00	31	31	31
8	11:30	34	35	35
9	12:00	37	38	38
10	12:30	40	41	41
11	1:00	44	45	45
12	1:30	48	48	48
13	2:00	51	51	51
14	2:30	55	55	55
15	3:00	58	58	58
16	3:30	60	61	61
17	4:00	63	65	65
18	4:30	67	68	68
19	5:00	71	72	72
20	5:30	74	75	75
21	6:00	78	78	78
22	6:30	82	82	82
23	7:00	85	85	85

the actual totals. Notice that f is a fairly good predictor of the actual totals. However, f predicts more matches for periods 2 and 3 than experience warrants. There are minor discrepancies for other periods. Since proper scheduling at the beginning of the tournament is of utmost importance, you can adjust the predicted totals to reflect 12 matches in period 2 and 16 matches in period 3. The adjusted totals are also shown in Table 4.4. It appears that a schedule based on adjusted totals is reasonable. The function, say, F, that describes this schedule is thus given by

$$F(x) = \begin{cases} 12, \text{ if } x = 2, \\ 16, \text{ if } x = 3, \\ [\frac{37}{11}(x - 1) + 11.5] & \text{otherwise,} \end{cases}$$

where $F(x)$ is the total number of matches scheduled through period x. Thus the number of matches assigned a starting time of period x is $F(x) - F(x - 1)$, where $x > 1$.

Of course not all tennis tournaments involve 85 matches on 11 courts. Also, the average match time varies and depends, for example, on the class of players or the type of court. Thus to handle other situations, suppose that you generalize the above scheduling function to a tournament involving m matches on c courts based on an average match time of h hours and t minutes.

First, you need to determine the number of 30-minute periods involved. Suppose that the tournament is to start at time $T = 0$ and E is the average time (in minutes) that a court is used throughout the day. Then

$$E = \frac{m(60h + t)}{c}.$$

The last match would end approximately at time $T = E$, so it would begin approximately at time $T = E - (60h + t)$. The number of 30-minute intervals from $T = 0$ to the starting time of the last match is

$$\frac{E - (60h + t)}{30} = \frac{\frac{m(60h + t)}{c} - (60h + t)}{30} = \frac{\frac{m(60h + t)}{c} - \frac{(60h + t)c}{c}}{30} = \frac{(m - c)(60h + t)}{30c}.$$

This number does not include the final period. The number of periods to be scheduled is the rounded-off value of

$$\frac{(m - c)(60h + t)}{30c} + 1.$$

Denoting the number of periods by z gives

$$z = \left[\frac{(m - c)(60h + t)}{30c} + 1.5\right]. \qquad (2)$$

Assume that the total number of matches, y, scheduled through period x is a linear function of x. Because c matches are scheduled for period 1, and by period z a total of m matches are scheduled, the points

$(1, c)$ and (z, m) lie on the graph of this function. The slope is

$$\frac{m-c}{z-1},$$

so a point–slope form of an equation of the line is

$$y - c = \frac{m-c}{z-1}(x-1).$$

Simplifying gives

$$y = \frac{m-c}{z-1}(x-1) + c.$$

This function must be refined so that the y values are rounded to the nearest integer. Furthermore, experience indicates that for periods 2 and 3, the total number of matches assigned should reflect the same proportions as in the original data based on 11 courts. That is, for period 2 there should be a total of $[\frac{12}{11}c + 0.5]$ matches assigned, and for period 3 there should be $[\frac{16}{11}c + 0.5]$. The function, say F, that describes this schedule is

$$F(x) = \begin{cases} [\frac{12}{11}c + 0.5], & \text{if } x = 2, \\ [\frac{16}{11}c + 0.5], & \text{if } x = 3, \\ \left[\frac{m-c}{z-1}(x-1) + c + 0.5\right] & \text{otherwise,} \end{cases}$$

where $F(x)$ is the total number of matches scheduled through period x, m is the number of matches in the tournament, c is the number of courts, and z is the number of periods, where z is given by Eq. (2).

This function describes the tennis scheduling system known as the "Garman system," which is now used in many championship tennis tournaments.

■ Exercises

In the following problems, use the Garman system of scheduling for a tennis tournament that involves 16 matches with an average match time of 1 hour and 45 minutes and four courts.

1. Determine (a) the value of z, the number of 30-minute periods, (b) $F(x)$, and (c) the values of $F(x)$ for $x = 1, 2, 3, \ldots, z$.

2. If the tournament is to begin at 10:00 A.M., for each period indicate the starting time and the number of matches to be played during the period.

Answers to Odd-Numbered Problems

EXERCISE 0.2

1. True. **3.** False; the natural numbers are 1, 2, 3, and so on. **5.** True. **7.** False; $\frac{4}{2} = 2$, a positive integer.
9. True. **11.** True.

EXERCISE 0.3

1. False. **3.** False. **5.** True. **7.** True. **9.** False.
11. Distributive. **13.** Associaive. **15.** Commutative.
17. Definition of subtraction. **19.** Distributive.

EXERCISE 0.4

1. -6. **3.** 2. **5.** 11. **7.** -2. **9.** -63.
11. -6. **13.** $6 - x$.
15. $-12x + 12y$ (or $12y - 12x$). **17.** $-\frac{1}{3}$. **19.** -2.
21. 18. **23.** 25. **25.** $3x - 12$. **27.** $-x + 2$.
29. $\frac{8}{11}$. **31.** $-\frac{5x}{7y}$. **33.** $\frac{2}{3x}$. **35.** 3. **37.** $\frac{7}{xy}$.
39. $\frac{5}{6}$. **41.** $-\frac{1}{6}$. **43.** $\frac{x - y}{9}$. **45.** $\frac{1}{24}$.
47. $\frac{x}{6y}$. **49.** Not defined. **51.** Not defined.

EXERCISE 0.5

1. $2^5 (= 32)$. **3.** w^{12}. **5.** $\frac{x^8}{x^{17}}$. **7.** $\frac{x^{10}}{y^{50}}$.
9. $8x^6y^9$. **11.** x^6. **13.** x^{14}. **15.** 5. **17.** -2.
19. $\frac{1}{2}$. **21.** 10. **23.** 8. **25.** $\frac{1}{4}$. **27.** $\frac{1}{32}$.
29. $4\sqrt{2}$. **31.** $x\sqrt[3]{2}$. **33.** $4x^2$. **35.** $-2\sqrt{3} + 4\sqrt[3]{2}$.
37. $3z^2$. **39.** $\frac{9t^2}{4}$. **41.** $\frac{x^3}{y^2z^2}$. **43.** $\frac{2}{x^4}$. **45.** $\frac{1}{9t^2}$.
47. $7^{1/3}s^{2/3}$. **49.** $x^{1/2} - y^{1/2}$. **51.** $\frac{x^{9/4}z^{3/4}}{y^{1/2}}$.
53. $\sqrt[5]{(8x - y)^4}$. **55.** $\frac{1}{\sqrt[5]{x^4}}$. **57.** $\frac{2}{\sqrt[5]{x^4}}$.
59. $\frac{3\sqrt{7}}{7}$. **61.** $\frac{2\sqrt{2x}}{x}$. **63.** $\frac{\sqrt[3]{9x^2}}{3x}$. **65.** 4.
67. $\frac{\sqrt[12]{8x^8y^4}}{xy}$. **69.** $\frac{2x^6}{y^3}$. **71.** $t^{2/3}$. **73.** $\frac{64y^6x^{1/2}}{x^2}$.
75. xyz. **77.** $\frac{1}{9}$. **79.** $\frac{4y^4}{x^2}$. **81.** $x^2y^{5/2}$. **83.** $\frac{y^{10}}{z^2}$.
85. x^8. **87.** $-\frac{4}{s^5}$. **89.** $\frac{4x^4z^4}{9y^4}$.

EXERCISE 0.6

1. $11x - 2y - 3$. **3.** $6t^2 - 2s^2 + 6$.
5. $2\sqrt{x} + \sqrt{2y} + \sqrt{3z}$.
7. $6x^2 - 9xy - 2z + \sqrt{2} - 4$.
9. $\sqrt{2y} - \sqrt{3z}$. **11.** $-7x + 14y - 19$.
13. $x^2 + 9y^2 + xy$. **15.** $6x^2 + 96$.
17. $-6x^2 - 18x - 18$. **19.** $x^2 + 9x + 20$.
21. $x^2 + x - 6$. **23.** $10x^2 + 19x + 6$.
25. $x^2 + 6x + 9$. **27.** $x^2 - 10x + 25$.
29. $2y + 6\sqrt{2y} + 9$. **31.** $4s - 1$.
33. $x^3 + 4x^2 - 3x - 12$.
35. $2x^4 + 2x^3 - 5x^2 - 2x + 3$. **37.** $5x^3 + 5x^2 + 6x$.
39. $3x^2 + 2y^2 + 5xy + 2x - 8$.
41. $x^3 + 15x^2 + 75x + 125$.
43. $8x^3 - 36x^2 + 54x - 27$. **45.** $z - 4$
47. $3x^3 + 2x - \frac{1}{2x^2}$. **49.** $x + \frac{-1}{x + 3}$.
51. $3x^2 - 8x + 17 + \frac{-37}{x + 2}$.
53. $t + 8 + \frac{64}{t - 8}$. **55.** $x - 2 + \frac{7}{3x + 2}$.

EXERCISE 0.7

1. $2(3x + 2)$. **3.** $5x(2y + z)$.
5. $4bc(2a^3 - 3ab^2d + b^3cd^2)$. **7.** $(x + 5)(x - 5)$.
9. $(p + 3)(p + 1)$. **11.** $(4x + 3)(4x - 3)$.
13. $(z + 4)(z + 2)$. **15.** $(x + 3)^2$.
17. $2(x + 4)(x + 2)$. **19.** $3(x - 1)(x + 1)$.
21. $(6y + 1)(y + 2)$. **23.** $2s(3s + 4)(2s - 1)$.
25. $x^{2/3}y(1 + 2xy)(1 - 2xy)$. **27.** $2x(x + 3)(x - 2)$.
29. $4(2x + 1)^2$. **31.** $x(xy - 5)^2$.
33. $(x - 2)^2(x + 2)$. **35.** $(y + 4)^2(y + 1)(y - 1)$.
37. $(x + 2)(x^2 - 2x + 4)$.
39. $(x + 1)(x^2 - x + 1)(x - 1)(x^2 + x + 1)$.
41. $2(x + 3)^2(x + 1)(x - 1)$. **43.** $P(1 + r)^2$
45. $(x^2 + 4)(x + 2)(x - 2)$.
47. $(y^4 + 1)(y^2 + 1)(y + 1)(y - 1)$.
49. $(x^2 + 2)(x + 1)(x - 1)$. **51.** $x(x + 1)^2(x - 1)^2$.

EXERCISE 0.8

1. $\frac{x+2}{x}$. **3.** $\frac{x-5}{x+5}$. **5.** $\frac{3x+2}{x+2}$.
7. $-\frac{y^2}{(y-3)(y+2)}$. **9.** $\frac{3-2x}{3+2x}$.
11. $\frac{2(x+4)}{(x-4)(x+2)}$. **13.** $\frac{x}{2}$. **15.** $\frac{1}{2n}$. **17.** $\frac{2}{3}$.
19. $-27x^2$. **21.** 1. **23.** $\frac{2x^2}{x-1}$. **25.** 1.
27. $-\frac{(2x+3)(1+x)}{x+4}$. **29.** $x+2$. **31.** $\frac{5}{3t}$.
33. $\frac{1}{1-p^2}$. **35.** $\frac{2x^2+3x+12}{(2x-1)(x+3)}$.
37. $\frac{2x-3}{(x-2)(x+1)(x-1)}$. **39.** $\frac{35-8x}{(x-1)(x+5)}$.
41. $\frac{x^2+2x+1}{x^2}$. **43.** $\frac{x}{1-xy}$. **45.** $\frac{x+1}{3x}$.
47. $\frac{(x+2)(6x-1)}{2x^2(x+3)}$. **49.** $\frac{2\sqrt{x}-2\sqrt{x+h}}{\sqrt{x}\sqrt{x+h}}$.
51. $2-\sqrt{3}$.
53. $-\frac{\sqrt{6}+2\sqrt{3}}{3}$. **55.** $-4-2\sqrt{6}$. **57.** $\frac{x-\sqrt{5}}{x^2-5}$.
59. $\frac{5\sqrt{3}-4\sqrt{2}-13}{2}$.

PRINCIPLES IN PRACTICE 1.1

1. $P = 2(w+2)+2w = 2w+4+2w = 4w+4$. **2.** 200 speciality coffees.
3. 46 weeks; $1715. **4.** $r = \frac{d}{t}$. **5.** $\sqrt{\frac{S}{\pi}}$.

EXERCISE 1.1

1. 0. **3.** $\frac{10}{3}$. **5.** 0. **7.** Adding 5 to both sides; equivalence guaranteed. **9.** Squaring both sides; equivalence *not* guaranteed. **11.** Dividing both sides by x; equivalence *not* guaranteed. **13.** Multiplying both sides by $x-1$; equivalence *not* guaranteed. **15.** Multiplying both sides by $(x-5)/x$; equivalence *not* guaranteed.
17. $\frac{5}{2}$. **19.** 0. **21.** 1. **23.** $\frac{12}{5}$. **25.** -1.
27. 2. **29.** $\frac{10}{3}$. **31.** 90. **33.** 8. **35.** $-\frac{26}{9}$.
37. $-\frac{37}{18}$. **39.** $\frac{60}{17}$. **41.** $\frac{14}{3}$. **43.** 3. **45.** $\frac{7}{8}$.
47. $P = \frac{I}{rt}$. **49.** $q = \frac{p+1}{8}$. **51.** $r = \frac{S-P}{Pt}$.
53. $a_1 = \frac{2S-na_n}{n}$. **55.** 120 m.

57. $c = x + 0.0825x = 1.0825x$. **59.** 3.
61. 29 hours. **63.** 0.00001. **65.** $\frac{1}{8}, -\frac{1}{14}$. **67.** $\frac{14}{61}$.

PRINCIPLES IN PRACTICE 1.2

1. $\frac{10}{r+2} = \frac{6}{r-2}$; 8 mph. **2.** $t = \frac{d}{r+w}$; $w = \frac{d}{t} - r$.
3. $\sqrt{x^2+16} - x = 2$; $x = 3$; ramp is 5 feet.

EXERCISE 1.2

1. $\frac{1}{5}$. **3.** $\varnothing$. **5.** $\frac{8}{3}$. **7.** $\frac{3}{2}$. **9.** 0. **11.** $\frac{5}{3}$.
13. $\frac{1}{8}$. **15.** 3. **17.** $\frac{5}{13}$. **19.** $\varnothing$. **21.** 3.
23. $\frac{262}{5}$. **25.** $-\frac{10}{9}$. **27.** 2. **29.** 7. **31.** $\frac{49}{36}$.
33. $-\frac{9}{4}$. **35.** $d = \frac{r}{1+rt}$. **37.** $n = \frac{2mI}{rB} - 1$.
39. 10. **41.** $t = \frac{d}{r-c}$; $r = \frac{d}{t} + c$.
43. Jill is 36 feet above the surface. Jack is 121 feet above the surface.

PRINCIPLES IN PRACTICE 1.3

1. The number is -5 or 6. **2.** 50 feet by 60 feet.
3. $1 \times 1 \times 5$. **4.** 15 items at $15 per item.
5. 2.5 seconds and 7.5 seconds. **6.** $100
7. Never, they only go up to 400 feet.

EXERCISE 1.3

1. 2. **3.** 4, 3. **5.** 3, -1. **7.** 6. **9.** ± 2.
11. 0, 8. **13.** $\frac{1}{2}$. **15.** 1, $-\frac{5}{2}$. **17.** 5, -2.
19. 0, $\frac{3}{2}$. **21.** 0, 1, -2. **23.** 0, ± 8. **25.** 0, $\frac{1}{2}$, $-\frac{4}{3}$.
27. -3, -1, 2. **29.** 3, 4. **31.** 4, -6. **33.** $\frac{3}{2}$.
35. $\frac{5 \pm \sqrt{13}}{2}$. **37.** No real roots. **39.** $\frac{1}{2}, -\frac{5}{3}$.
41. 4, $-\frac{5}{2}$. **43.** $\frac{-2 \pm \sqrt{14}}{2}$. **45.** $\pm\sqrt{3}, \pm\sqrt{2}$.
47. 2, $-\frac{1}{2}$. **49.** $\pm\frac{\sqrt{7}}{7}, \pm\frac{\sqrt{2}}{2}$. **51.** -4, 1.
53. $\frac{15}{7}, \frac{11}{5}$. **55.** $\frac{3}{2}, -1$. **57.** 6, -2. **61.** 5, -2.
63. $\frac{3}{2}$. **65.** -2. **67.** 7, 2. **69.** 4, 8. **71.** 2.
73. 0, 4. **75.** 4. **77.** 233.16, 1.84.
79. 6 inches by 8 inches. **83.** 1 year and 10 years.
85. 86.8 cm or 33.2 cm. **87.** **a.** 9 s, **b.** 3 s or 6 s.
89. 1.5, 0.75. **91.** No real root. **93.** 1.999, 0.963.

REVIEW PROBLEMS CHAPTER 1

1. $\frac{1}{4}$. **3.** $-\frac{2}{15}$. **5.** $-\frac{1}{2}$. **7.** $\varnothing$. **9.** $\frac{5}{2}$.

11. $\frac{1}{3}$. **13.** $-\frac{9}{7}$. **15.** $-\frac{5}{3}, 1$. **17.** $0, \frac{7}{5}$. **19.** 5.

21. $\pm\frac{\sqrt{15}}{3}$. **23.** $\frac{5}{8}, -3$. **25.** $\frac{5 \pm \sqrt{13}}{6}$. **27.** $4, \pm 3$.

29. $\frac{1}{2}$. **31.** $\frac{4 \pm \sqrt{13}}{3}$. **33.** 10. **35.** 5.

37. No solution. **39.** 10. **41.** 4, 8. **43.** −8, 1.

45. $Q = \frac{EA}{4\pi k}$. **47.** $C' = \lambda^2(n - 1 - C)$.

49. $T = \pm 2\pi\sqrt{\frac{L}{g}}$. **51.** $\omega = \pm\sqrt{\frac{2mgh - mv^2}{I}}$.

55. $6, \frac{5}{4}$. **57.** −0.709, 0.336.

MATHEMATICAL SNAPSHOT CHAPTER 1

1. a. \$111.40; **b.** \$1.14; **c.** 100 lb; **d.** 98.15 lb; **e.** −1.85%
2. 4.76%.

EXERCISE 2.1

1. 120. **3.** 48 of A, 80 of B. **5.** $5\frac{1}{3}$. **7.** 1 m.

9. 181,250. **11.** \$4000 at 6%, \$16,000 at $7\frac{1}{2}$%.

13. \$4.25. **15.** 4%. **17.** 80. **19.** \$8000.
21. 1600. **23.** \$116.25. **25.** 40. **27.** 46,000.
29. Either \$440 or \$460. **31.** \$100. **33.** 77.
35. 80 ft by 140 ft. **37.** 9 cm long, 4 cm wide.
39. \$112,000. **41.** 60. **43.** Either 125 units of A and 100 units of B, or 150 units of A and 125 units of B.

PRINCIPLES IN PRACTICE 2.2

1. 5375.
2. $150 - x_4 \geq 0$; $3x_4 - 210 \geq 0$; $x_4 + 60 \geq 0$; $x_4 \geq 0$.

EXERCISE 2.2

1. $(4, \infty)$.

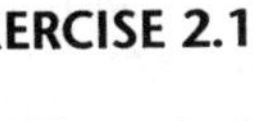

3. $(-\infty, 5]$.

5. $\left(-\infty, -\frac{1}{2}\right]$.

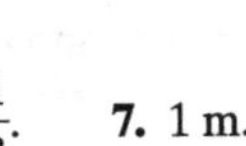

7. $\left(-\infty, -\frac{2}{5}\right)$.

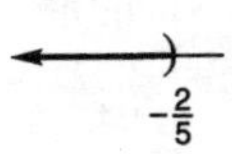

9. $(0, \infty)$.

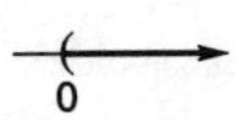

11. $\left[-\frac{7}{5}, \infty\right)$.

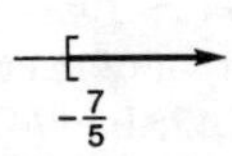

13. $\left(-\frac{2}{7}, \infty\right)$.

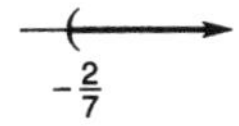

15. $\varnothing$.

17. $\left(-\infty, \frac{\sqrt{3} - 2}{2}\right)$.

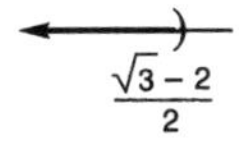

19. $(-\infty, 6)$.

21. $(-\infty, -5]$.

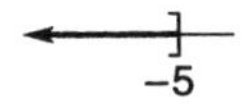

23. $(-\infty, \infty)$.

25. $\left(\frac{17}{9}, \infty\right)$.

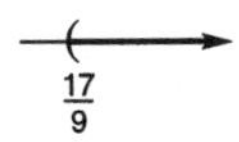

27. $\left[-\frac{14}{3}, \infty\right)$.

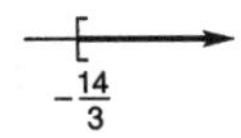

29. $(0, \infty)$.

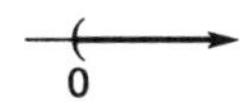

31. $(-\infty, 0)$.

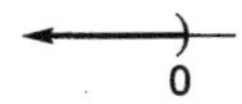

33. $(-\infty, -2]$.

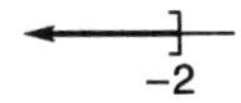

35. $444{,}000 < S < 636{,}000$. **37.** $x < 70$ degrees.

EXERCISE 2.3

1. 120,001. **3.** 22,500. **5.** 60,000.
7. \$25,714.29. **9.** 1000. **11.** $t > 36.5$.
13. \$4.50, \$1160.

PRINCIPLES IN PRACTICE 2.4

1. $|w - 22| \leq 0.3$ oz.

EXERCISE 2.4

1. 13. **3.** 6. **5.** 5. **7.** $-3 < x < 3$.
9. $\sqrt{5} - 2$.
11. a. $|x - 7| < 3$, **b.** $|x - 2| < 3$, **c.** $|x - 7| \leq 5$, **d.** $|x - 7| = 4$, **e.** $|x + 4| < 2$, **f.** $|x| < 3$, **g.** $|x| > 6$, **h.** $|x - 6| > 4$, **i.** $|x - 105| < 3$, **j.** $|x - 850| < 100$.
13. $|p_1 - p_2| \leq 2$. **15.** ± 7. **17.** ± 6. **19.** 13, −3.
21. $\frac{2}{5}$. **23.** $\frac{1}{2}, 3$. **25.** $(-4, 4)$.
27. $(-\infty, -8) \cup (8, \infty)$. **29.** $(-9, -5)$.
31. $(-\infty, 0) \cup (1, \infty)$. **33.** $[2, 3]$.
35. $(-\infty, 0] \cup \left[\frac{16}{3}, \infty\right)$. **37.** $|d - 10| \leq 0.01$ m
39. $(-\infty, \mu - h\sigma) \cup (\mu + h\sigma, \infty)$.

REVIEW PROBLEMS CHAPTER 2

1. $(-\infty, 0]$. **3.** $\left(\frac{2}{3}, \infty\right)$. **5.** $\varnothing$.
7. $\left(-\infty, \frac{13}{2}\right]$. **9.** $(-\infty, \infty)$. **11.** −2, 5.
13. $\left(0, \frac{1}{2}\right)$. **15.** $\left(-\infty, -\frac{1}{2}\right] \cup \left[\frac{7}{2}, \infty\right)$.
17. 542. **19.** 6000.

MATHEMATICAL SNAPSHOT CHAPTER 2

1. 1 hour. **3.** 1 hour. **5.** 1 hour and 12 minutes.

PRINCIPLES IN PRACTICE 3.1

1. a. $a(r) = \pi r^2$; **b.** all real numbers; **c.** $r \geq 0$.
2. a. $t(r) = \frac{300}{r}$; **b.** all real numbers except 0; **c.** $r > 0$;
d. $t(x) = \frac{300}{x}; t\left(\frac{x}{2}\right) = \frac{600}{x}; t\left(\frac{x}{4}\right) = \frac{1200}{x}$;
e. The time is scaled by a factor c; $t\left(\frac{x}{c}\right) = \frac{300c}{x}$.
3. a. 300 pizzas; **b.** \$21.00 per pizza; **c.** \$16.00 per pizza.

EXERCISE 3.1

1. All real numbers except 0. **3.** All real numbers ≥ 3.
5. All real numbers. **7.** All real numbers except $-\frac{5}{2}$.
9. All real numbers except 0 and 1.
11. All real numbers except 4 and $-\frac{1}{2}$.
13. $1, 7, -7$. **15.** $-62, 2 - u^2, 2 - u^4$.
17. $2, (2v)^2 + 2v = 4v^2 + 2v, (-x^2)^2 + (-x^2) = x^4 - x^2$.
19. $4, 0, (x + h)^2 + 2(x + h) + 1$
$= x^2 + 2xh + h^2 + 2x + 2h + 1$.
21. $0, \frac{3x - 5}{(3x)^2 + 4} = \frac{3x - 5}{9x^2 + 4}, \frac{(x + h) - 5}{(x + h)^2 + 4}$
$= \frac{x + h - 5}{x^2 + 2xh + h^2 + 4}$.
23. $0, 256, \frac{1}{16}$. **25. a.** $4x + 4h - 5$; **b.** 4.
27. a. $x^2 + 2hx + h^2 + 2x + 2h$; **b.** $2x + h + 2$.
29. a. $2 - 4x - 4h - 3x^2 - 6hx - 3h^2$;
b. $-4 - 6x - 3h$.
31. a. $\frac{1}{x + h}$; **b.** $-\frac{1}{x(x + h)}$. **33.** 9.
35. y is a function of x; x is a function of y.
37. y is a function of x; x is not a function of y.
39. Yes. **41.** $V = f(t) = 10{,}000 + 400t$.
43. Yes; P; q.
45. 400 pounds per week; 1000 pounds per week; amount supplied increases as the price increases.
47. a. 4; **b.** $8\sqrt[3]{2}$; **c.** $f(2I_0) = 2\sqrt[3]{2}f(I_0)$;doubling the intensity increases the response by a factor of $2\sqrt[3]{2}$.
49. a. 3000, 2900, 2300, 2000; 12, 10;
b. 10, 12, 17, 20; 3000, 2300.
51. a. −5.13; **b.** 2.64; **c.** −17.43.
53. a. 11.33; **b.** 50.62; **c.** 2.29.

PRINCIPLES IN PRACTICE 3.2

1. a. $p(n) = \$125$; **b.** The premiums do not change;
c. constant function.
2. a. quadratic function; **b.** 2; **c.** 3.
3. $c(n) = \begin{cases} 3.50n & \text{if } n \leq 5 \\ 3.00n & \text{if } 5 < n \leq 10 \\ 2.75n & \text{if } n > 10 \end{cases}$
4. $7! = 5040$.

EXERCISE 3.2

1. Yes. **3.** No. **5.** Yes. **7.** Yes.
9. All real numbers. **11.** All real numbers.
13. a. 3; **b.** 7. **15. a.** 4; **b.** −3. **17.** 8, 8, 8.
19. 1, −1, 0, −1. **21.** 8, 3, 3, 1. **23.** 720. **25.** 2.
27. 5.
29. $c(i) = \$3.50$; constant function.
31. a. $C = 850 + 3q$; **b.** 250.
33. $c(n) = \begin{cases} 7.50n & \text{if } n < 10 \\ 7.00n & \text{if } n \geq 10 \end{cases}$ **35.** $\frac{9}{64}$.
37. a. All T such that $30 \leq T \leq 39$; **b.** $4, \frac{17}{4}, \frac{33}{4}$.
39. a. 237,077.34; **b.** −434.97; **c.** 52.19.
41. a. 2.21; **b.** 9.98; **c.** −14.52.

PRINCIPLES IN PRACTICE 3.3

1. $c(s(x)) = c(x + 3) = 2(x + 3) = 2x + 6$.
2. Let the length of a side be represented by the function $l(x) = x + 3$ and the area of a square with sides of length x be represented by $a(x) = x^2$. Then $g(x) = (x + 3)^2 = [l(x)]^2 = a(l(x))$.

EXERCISE 3.3

1. a. $2x + 8$; **b.** 8; **c.** −2; **d.** $x^2 + 8x + 15$; **e.** 3;
f. $\frac{x + 3}{x + 5}$; **g.** $x + 8$; **h.** 11; **i.** $x + 8$.
3. a. $2x^2 + x$; **b.** $-x$; **c.** $\frac{1}{2}$; **d.** $x^4 + x^3$;
e. $\frac{x^2}{x^2 + x} = \frac{x}{x + 1}$ (for $x \neq 0$); **f.** −1;
g. $(x^2 + x)^2 = x^4 + 2x^3 + x^2$; **h.** $x^4 + x^2$; **i.** 90.
5. 6; −32. **7.** $\frac{4}{(t - 1)^2} + \frac{6}{t - 1} + 1; \frac{2}{t^2 + 3t}$.
9. $\frac{1}{v + 3}; \sqrt{\frac{2w^2 + 3}{w^2 + 1}}$. **11.** $f(x) = x^5, g(x) = 4x - 3$.
13. $f(x) = \frac{1}{x}, g(x) = x^2 - 2$.
15. $f(x) = \sqrt[5]{x}, g(x) = \frac{x + 1}{3}$.
17. a. $r(x) = 9.75x$ **b.** $e(x) = 4.25x + 4500$
c. $(r - e)(x) = 5.5x - 4500$.
19. $400m - 10m^2$; the total revenue received when the total output of m employees is sold.
21. a. 1.52; **b.** 1747.24. **23. a.** 345.03; **b.** −1.94.

PRINCIPLES IN PRACTICE 3.4

1. $y = -600x + 7250$; x-intercept $\left(12\frac{1}{12}, 0\right)$; y-intercept $(0, 7250)$.
2. $y = 24.95$; horizontal line; no x-intercept; y-intercept $(0, 24.95)$.
3.

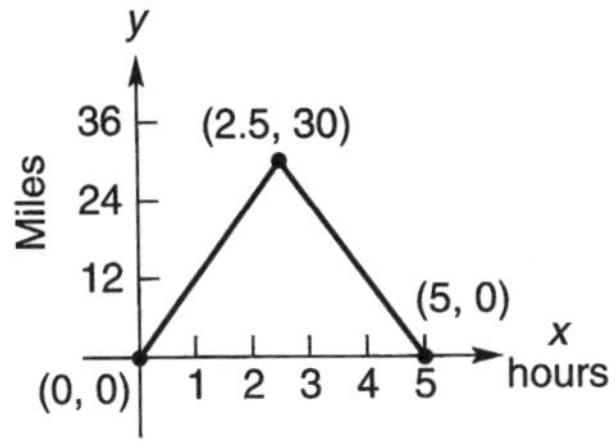

4.

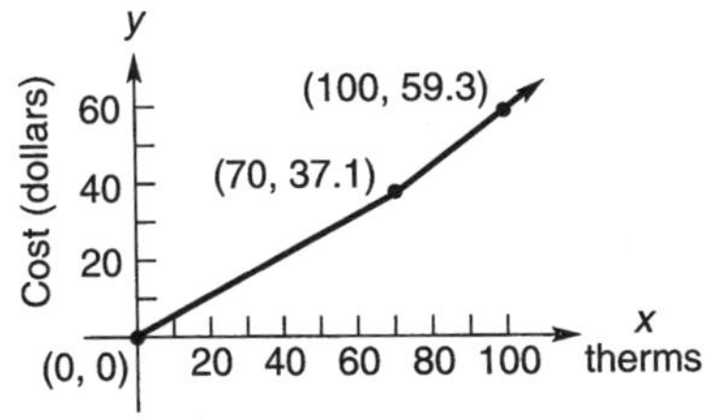

EXERCISE 3.4

1.

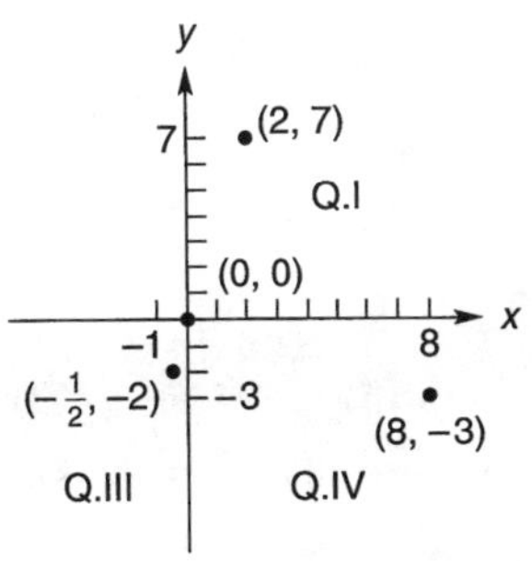

3. **a.** $1, 2, 3, 0$; **b.** all real numbers; **c.** all real numbers; **d.** -2.
5. **a.** $0, -1, -1$; **b.** all real numbers; **c.** all nonpositive real numbers; **d.** 0.
7. $(0, 0)$; function; all real numbers; all real numbers.

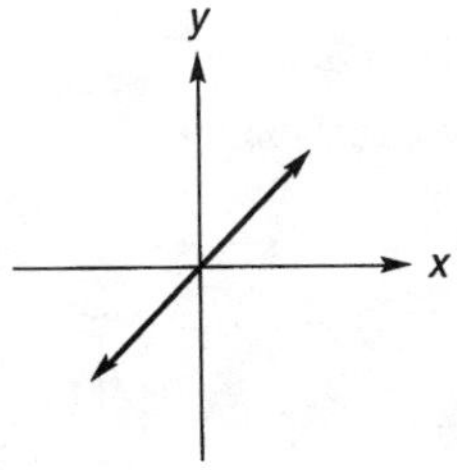

9. $(0, -5)$, $\left(\frac{5}{3}, 0\right)$; function; all real numbers; all real numbers.

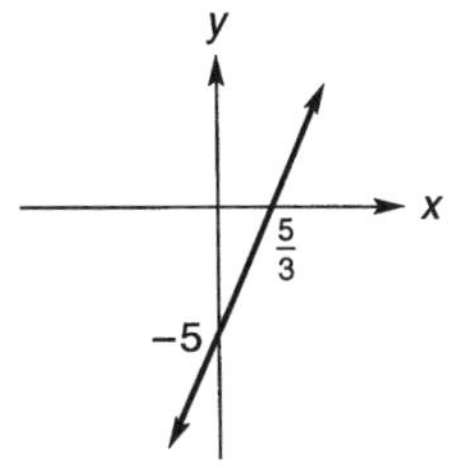

11. $(0, 0)$; function; all real numbers; all nonnegative real numbers.

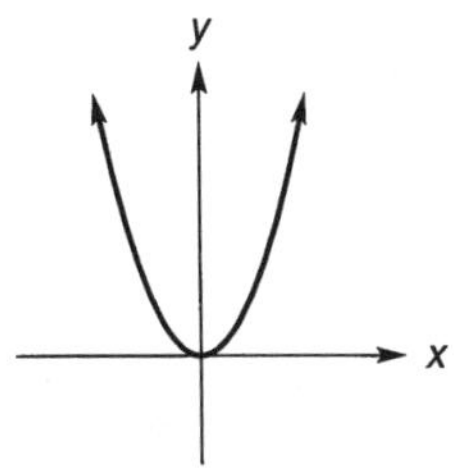

13. Every point on y-axis; not a function of x.

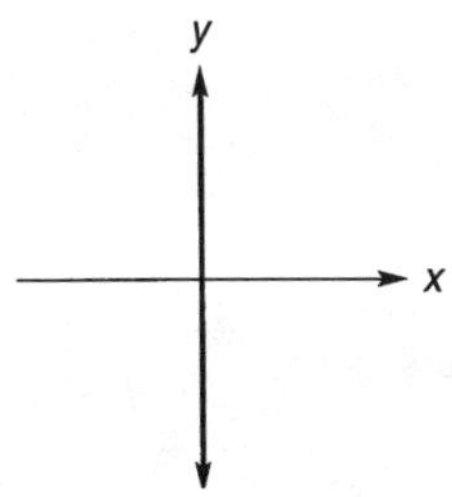

15. $(0, 0)$; function; all real numbers; all real numbers.

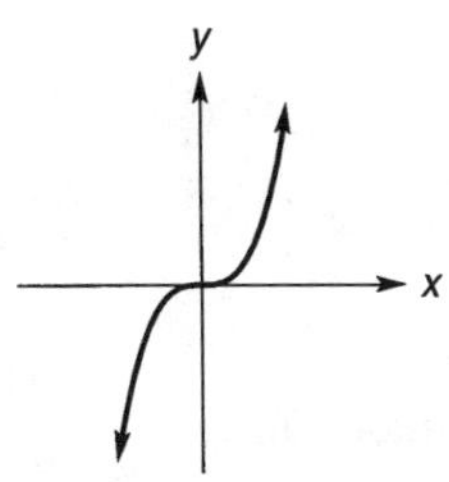

17. $(0, 0)$; not a function of x.

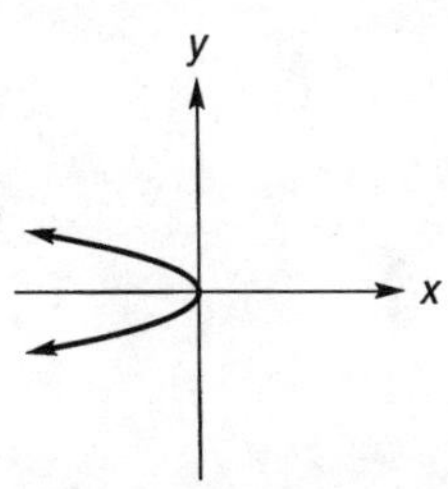

19. (0, 2), (1, 0); function; all real numbers; all real numbers.

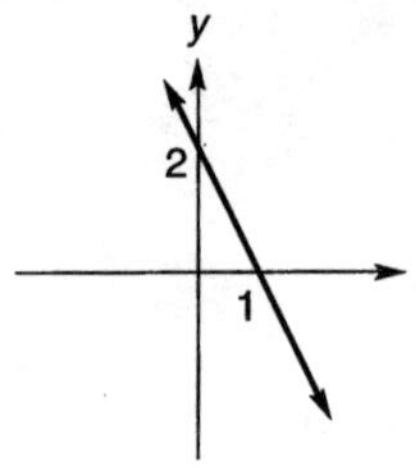

21. All real numbers; all real numbers ≤ 4; (0, 4), (2, 0), (−2, 0).

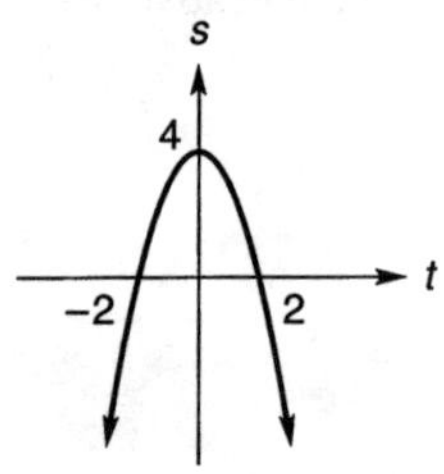

23. All real numbers; 2; (0, 2).

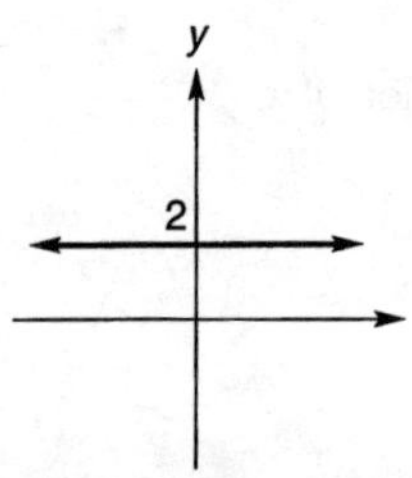

25. All real numbers; all real numbers ≥ -3; (0, 1), $(2 \pm \sqrt{3}, 0)$.

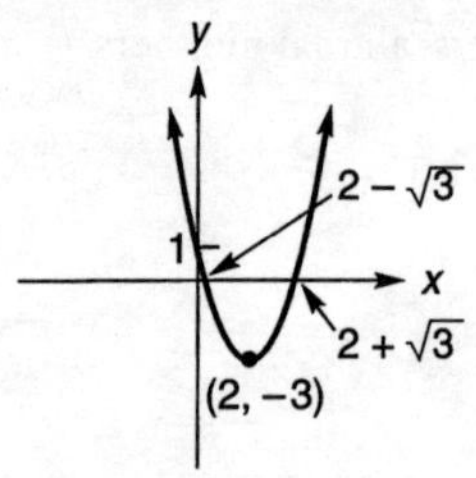

27. All real numbers; all real numbers; (0, 0).

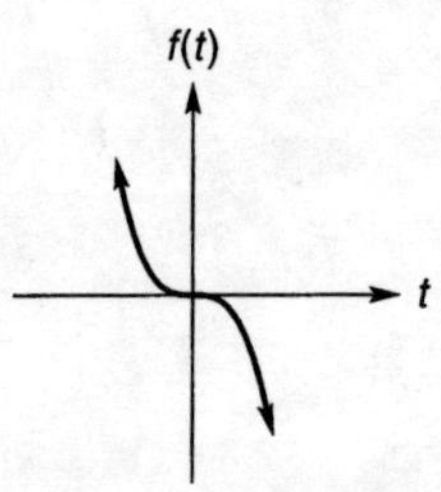

29. All real numbers ≥ 5; all nonnegative real numbers; (5, 0).

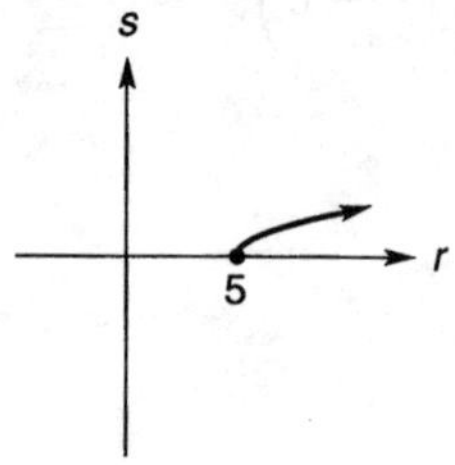

31. All real numbers; all nonnegative real numbers; (0, 1), $\left(\frac{1}{2}, 0\right)$.

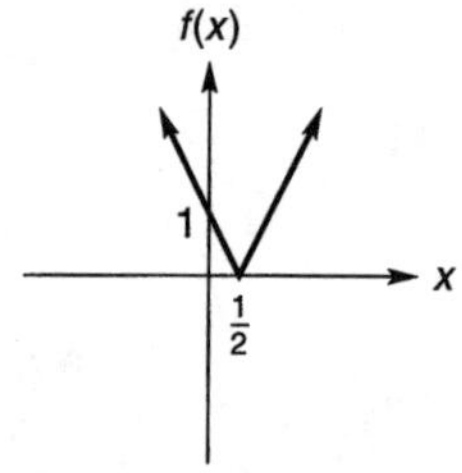

33. All nonzero real numbers; all positive real numbers; no intercepts.

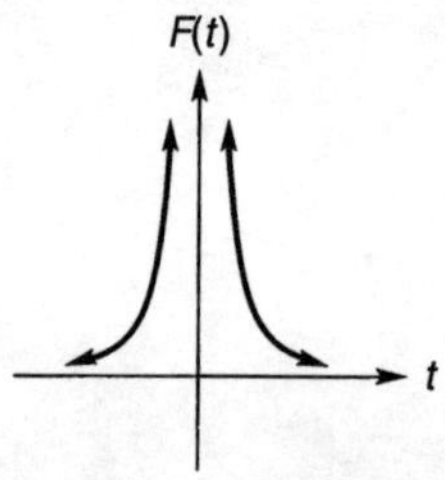

35. All nonnegative real numbers; all real numbers c where $0 \leq c \leq 2$.

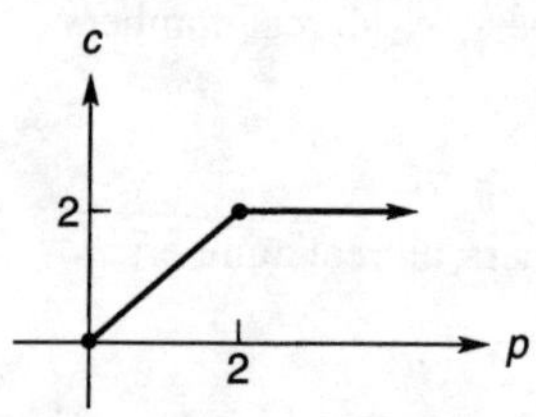

37. All real numbers; all nonnegative real numbers.

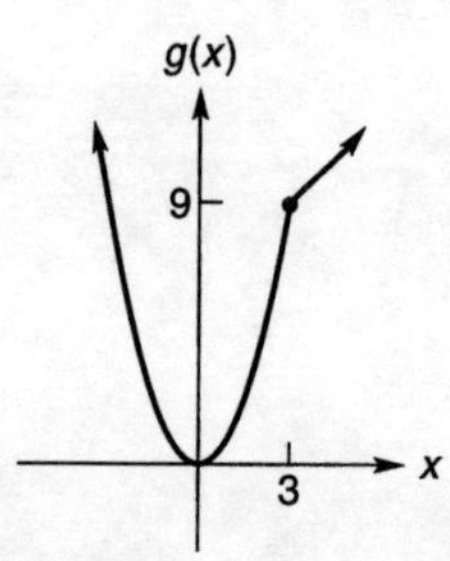

39. (a), (b), (d).

41.

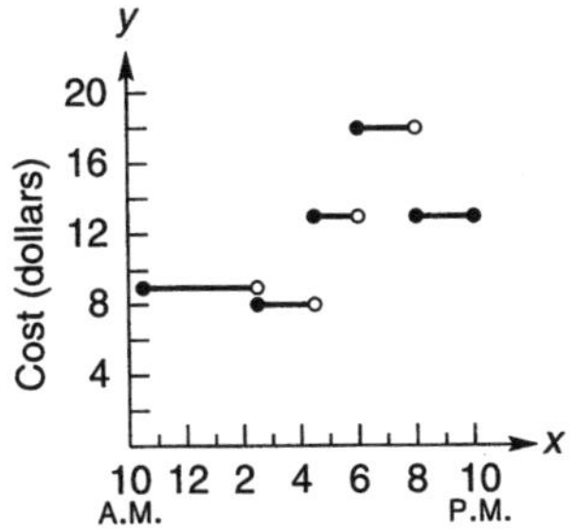

43. As price decreases, quantity increaes; p is a function of q.

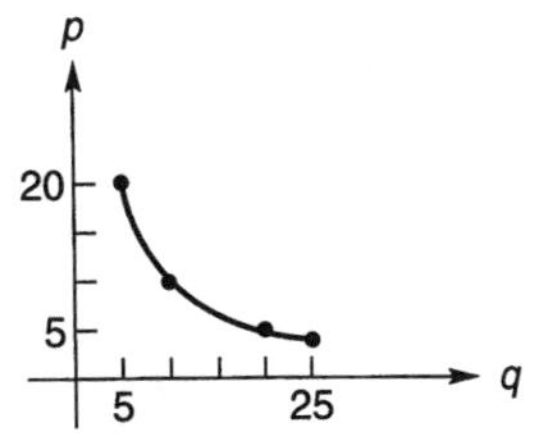

45.

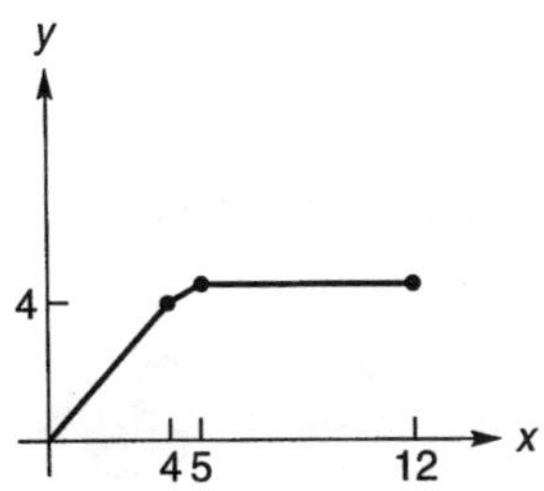

47. $-1, -0.35$. 49. $1.11, 2.83, 6.06$.
51. $-0.84, 2.61$. 53. $-0.49, 0.52, 1.25$.
55. a. 3.94; b. -1.94.
57. a. $(-\infty, \infty)$; b. $(-1.73, 0)$, $(0, 4.00)$.
59. a. 1.93; b. $[1.93, \infty)$; c. $(0, 2.48)$; d. no.

EXERCISE 3.5

1. $(0, 0)$; sym. about origin.
3. $(\pm 2, 0)$, $(0, 8)$; sym about y-axis.
5. $(\pm 3, 0)$; sym. about x-axis, y-axis, origin.
7. $(-2, 0)$; sym. about x-axis. 9. Sym. about x-axis.
11. $(-21, 0)$, $(0, -7)$, $(0, 3)$.
13. $(0, 0)$; sym. about origin. 15. $(0, 1)$.
17. $(2, 0)$, $(0, \pm 2)$; sym. about x-axis.

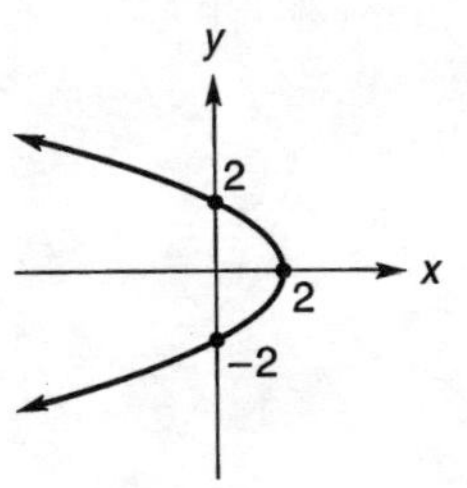

19. $(\pm 2, 0)$, $(0, 0)$; sym. about origin.

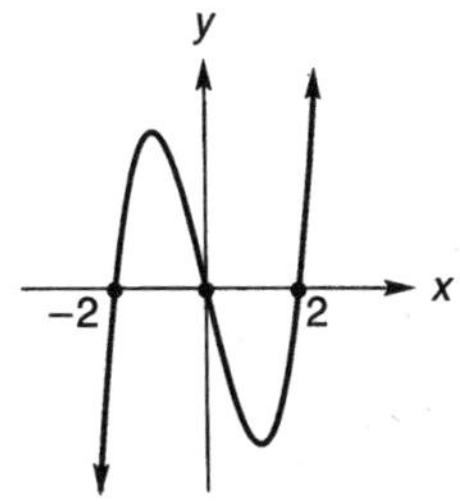

21. $(0, 0)$; sym. about x-axis, y-axis, origin.

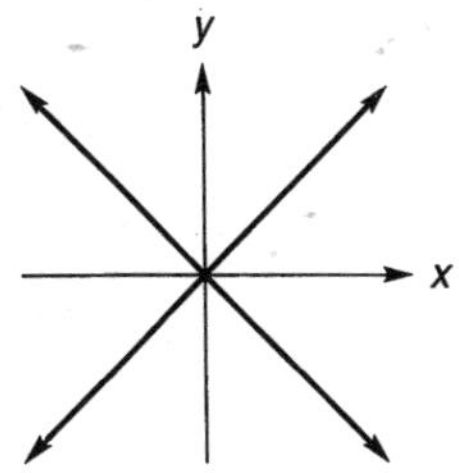

23. $(\pm 2, 0)$, $(0, \pm 4)$; sym. about x-axis, y-axis, origin.

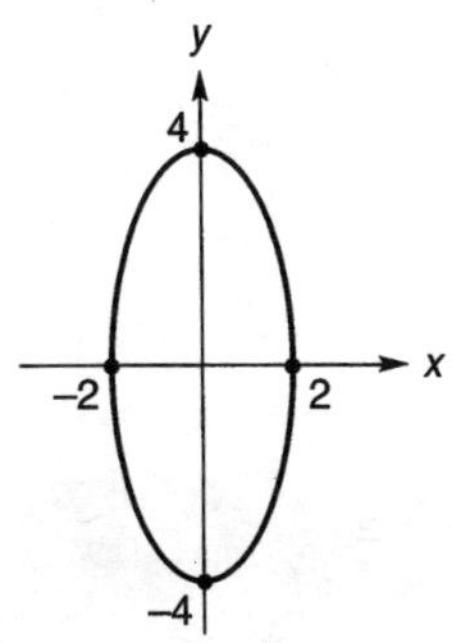

25. a. $(\pm 1.18, 0)$, $(0, 2)$; b. 2; c. $(-\infty, 2]$.

EXERCISE 3.6

1.

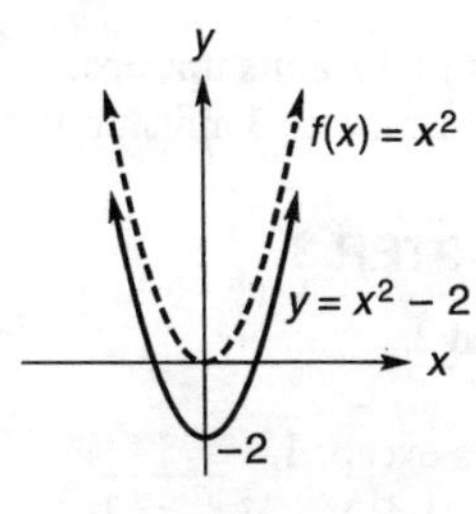

3.

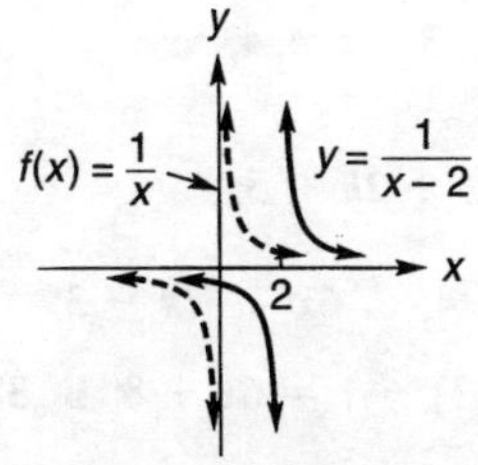

5.

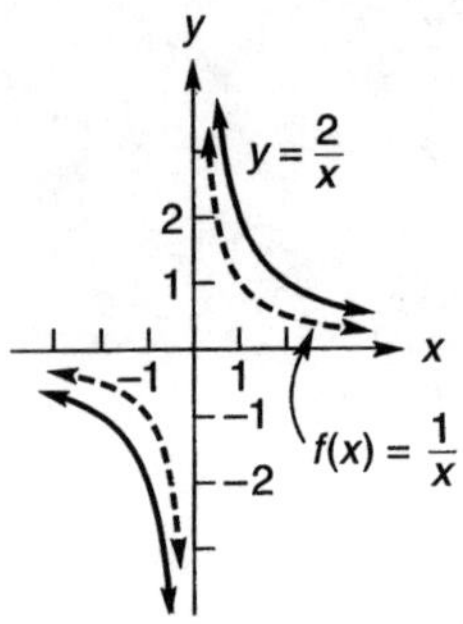

7.

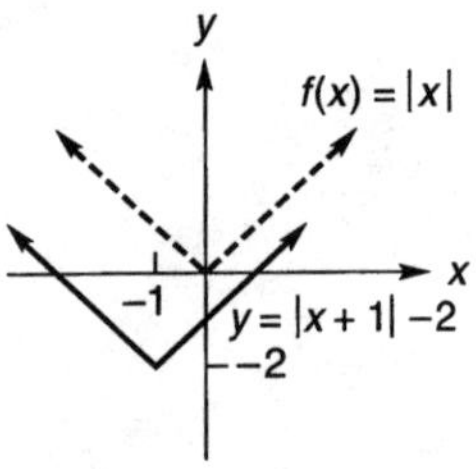

9.

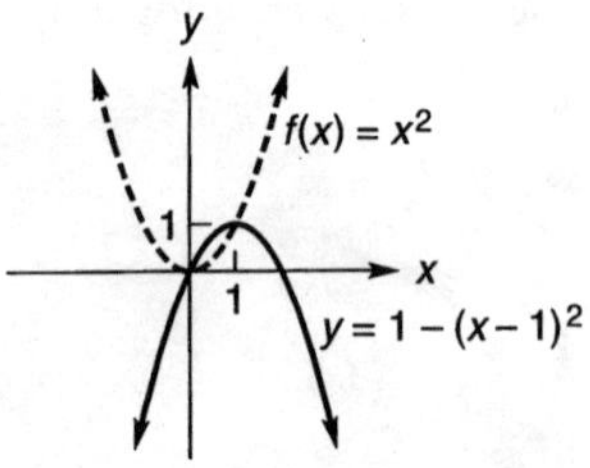

11.

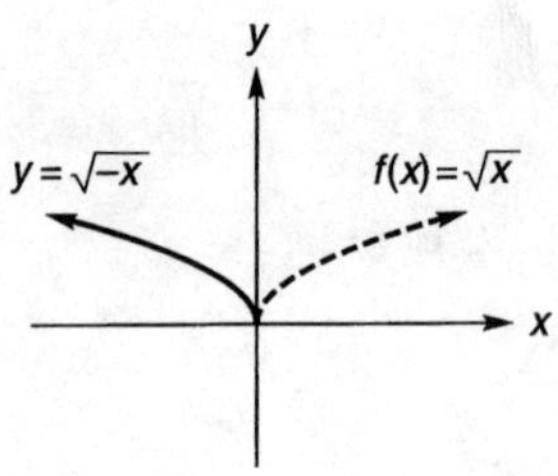

13. Translate 4 units to the right and 3 units upward.
15. Reflect about the y-axis and translate 3 units upward.

REVIEW PROBLEMS CHAPTER 3

1. All real numbers except 1 and 2.
3. All real numbers.
5. All nonnegative real numbers except 1.
7. $7, 46, 62, 3t^2 - 4t + 7$. **9.** $0, 3, \sqrt{t}, \sqrt{x^2 - 1}$.
11. $\frac{3}{5}, 0, \frac{\sqrt{x+4}}{x}, \frac{\sqrt{u}}{u-4}$. **13.** $-8, 4, 4, -92$.
15. **a.** $3 - 7x - 7h$; **b.** -7.
17. **a.** $4x^2 + 8hx + 4h^2 + 2x + 2h - 5$;
b. $8x + 4h + 2$.
19. **a.** $5x + 2$; **b.** 22; **c.** $x - 4$; **d.** $6x^2 + 7x - 3$;
e. 10; **f.** $\frac{3x - 1}{2x + 3}$; **g.** $3(2x + 3) - 1 = 6x + 8$; **h.** 38;
i. $2(3x - 1) + 3 = 6x + 1$.

21. $\frac{1}{x-1}, \frac{1}{x} - 1 = \frac{1-x}{x}$. **23.** $x^3 + 2,\ (x + 2)^3$.
25. $(0, 0), (\pm\sqrt{2/3}, 0)$; sym. about origin.
27. $(0, 9), (\pm 3, 0)$; sym. about y-axis.

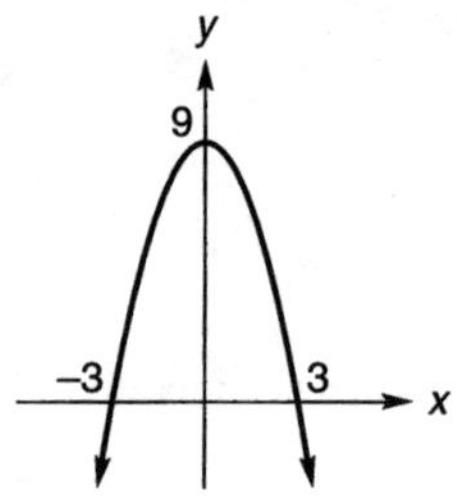

29. $(0, 2), (-4, 0)$; all $u \geq -4$; all real numbers ≥ 0.

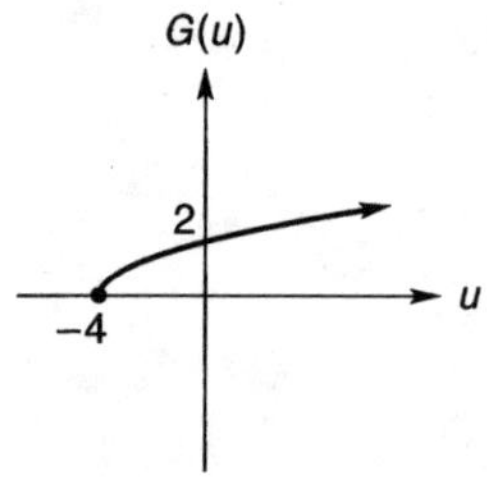

31. $\left(0, -\frac{1}{2}\right)$; all $t \neq 4$; all nonzero real numbers.

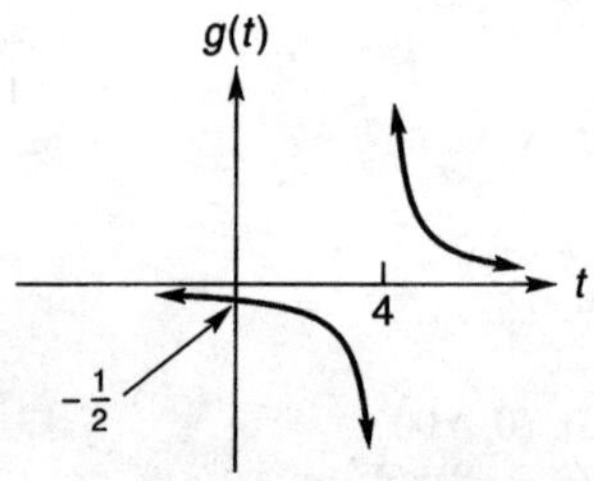

33. All real numbers; all real numbers ≥ 1.

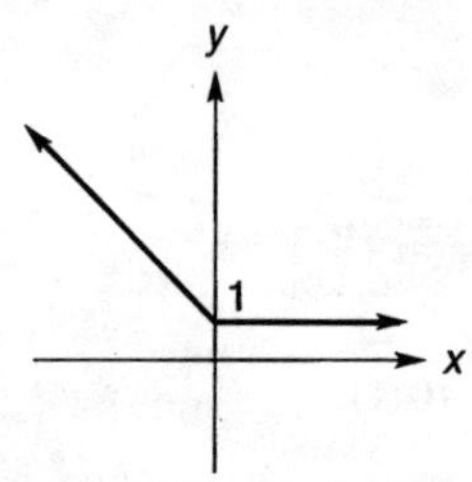

35.

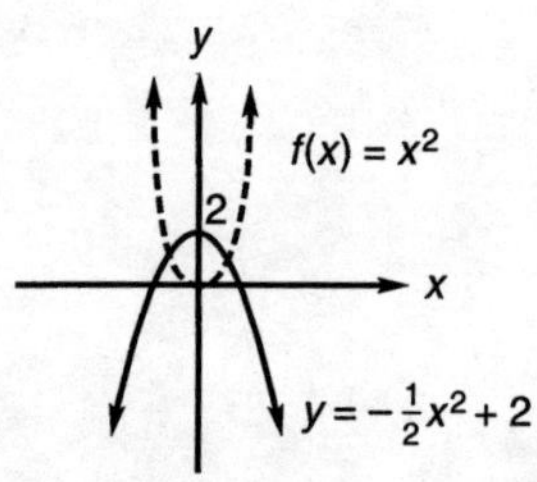

37. a, c.
39. $-0.67, 0.34, 1.73$.

41. −1.50, −0.88, −0.11, 1.09, 1.40.
43. a. $(-\infty, \infty)$; b. (1.92, 0), (0, 7)
45. a. none; b. 1, 3.

MATHEMATICAL SNAPSHOT CHAPTER 3

1. \$22,656. 3. \$85,190.70.

PRINCIPLES IN PRACTICE 4.1

1. −2000; the car depreciated \$2000 per year.
2. $S = 14T + 8$. 3. $F = \frac{9}{5}C + 32$.
4. slope $= \frac{125}{3}$; y-intercept $= \frac{125}{3}$.
5. $9C - 5F + 160 = 0$.
6.

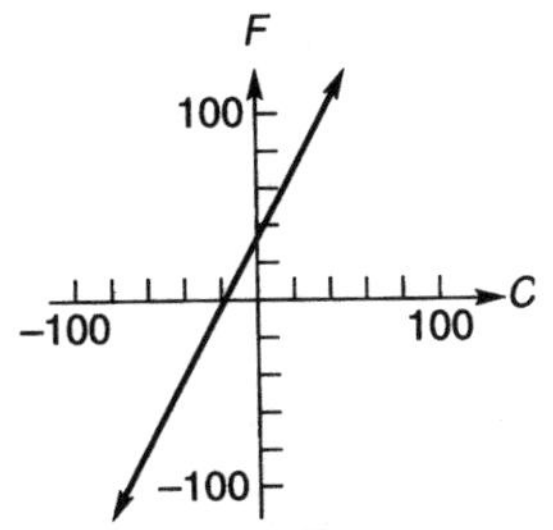

7. The slope of $\overline{AB}$ is 0; the slope of $\overline{BC}$ is 7; the slope of $\overline{CA}$ is 1. None of the slopes are negative reciprocals of each other, so the triangle does not have a right angle. The points do not define a right triangle.

EXERCISE 4.1

1. 3. 3. $-\frac{1}{2}$. 5. Undefined. 7. 0.
9. $6x - y - 4 = 0$. 11. $x + 4y - 18 = 0$.
13. $3x - 7y + 25 = 0$. 15. $8x - 5y - 29 = 0$.
17. $2x - y + 4 = 0$. 19. $x + 2y + 6 = 0$.
21. $y + 2 = 0$. 23. $x - 2 = 0$. 25. 2; −1.
27. $-\frac{1}{2}; \frac{3}{2}$. 29. Slope undefined; no y-intercept.
31. 3; 0. 33. 0; 1.
35. $2x + 3y - 5 = 0$; $y = -\frac{2}{3}x + \frac{5}{3}$.
37. $4x + 9y - 5 = 0$; $y = -\frac{4}{9}x + \frac{5}{9}$.
39. $3x - 2y + 24 = 0$; $y = \frac{3}{2}x + 12$.
41. Parallel. 43. Parallel. 45. Neither.
47. Perpendicular. 49. Perpendicular.
51. $y = 4x + 7$. 53. $y = 1$. 55. $y = -\frac{1}{3}x + 5$.
57. $x = 7$. 59. $y = -\frac{2}{3}x - \frac{29}{3}$. 61. (5, −4).
63. −2; the stock price dropped an average of \$2 per year.
65. $y = 3x + 5$. 67. slope ≈ 0.65; y-intercept ≈ 4.38
69. a. $y = -\frac{1}{3}x + \frac{1}{6}$; b. $y = 3x - \frac{3}{2}$.
71. $y = -x + 3300$; without modification, the approach angle will cause the plane to crash 700 feet short of the airport.
73. $R = 50{,}000T + 80{,}000$.
75. The lines are parallel. This is expected because they each have a slope of 1.5.

PRINCIPLES IN PRACTICE 4.2

1. x = number of skis produced;
y = number of boots produced; $8x + 14y = 1000$.
2. $p = -\frac{3}{8}q + 1025$.
3. Answers may vary, but two possible points are (0, 60) and (2, 140).

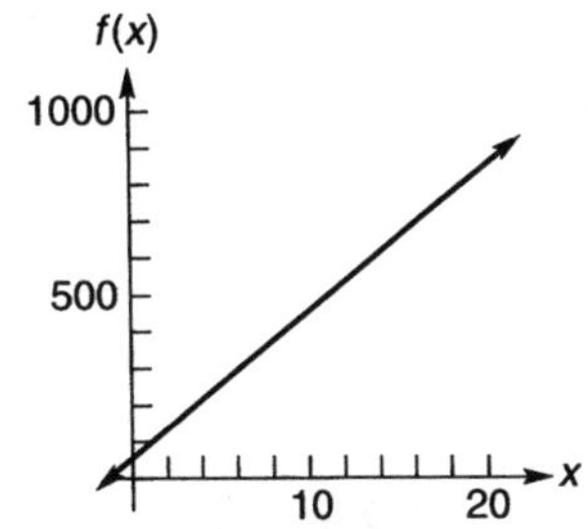

4. $f(t) = 2.3t + 32.2$.
5. $f(x) = 70x + 150$.

EXERCISE 4.2

1. −4; 0.

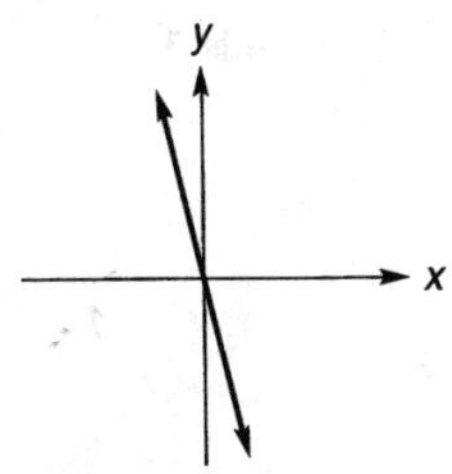

3. 2; −4.

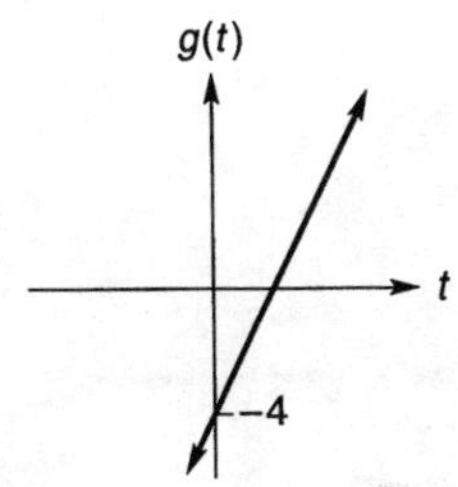

5. $-\frac{1}{2};\frac{7}{2}$.

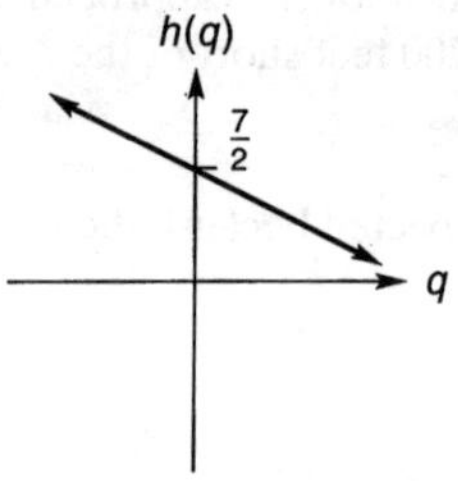

7. $f(x) = 4x$. **9.** $f(x) = -2x + 4$.

11. $f(x) = -\frac{1}{2}x + \frac{15}{4}$. **13.** $f(x) = x + 1$.

15. $p = -\frac{2}{5}q + 28$; \$16. **17.** $p = \frac{5}{2}q + 90$.

19. $c = 3q + 10$; \$115. **21.** $f(x) = 0.115x + 7.95$
23. $v = -800t + 8000$; slope $= -800$.

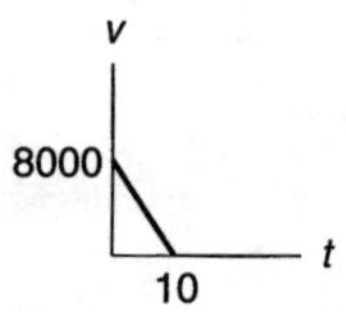

25. $f(x) = 11{,}000x + 198{,}000$. **27.** $f(x) = 65x + 85$.
29. $x + 10y = 100$.
31. a. $y = \frac{5}{11}, x = \frac{600}{11}$; **b.** 12.
33. a. $p = 0.059t + 0.025$; **b.** 0.556.
35. a. $t = \frac{1}{4}c + 37$;
b. Add 37 to the number of chirps in 15 seconds.
37. $P = \frac{T}{4} + 80$. **39. a.** Yes; **b.** 1.8704.

PRINCIPLES IN PRACTICE 4.3

1. Vertex: (1, 400); y-intercept: (0, 399); x-intercepts: (−19, 0), (21, 0).

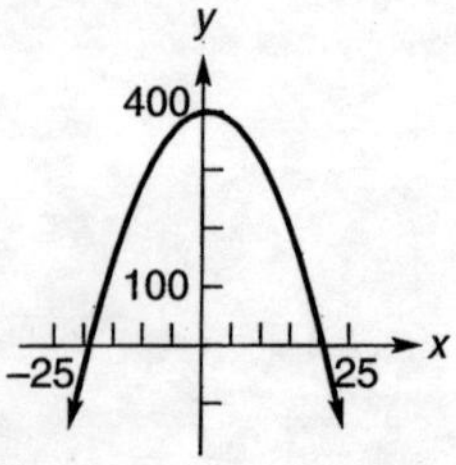

2. Vertex: (1, 24); y-intercept: (0, 8);
x-intercepts: $\left(1 + \frac{\sqrt{6}}{2}, 0\right), \left(1 - \frac{\sqrt{6}}{2}, 0\right)$.

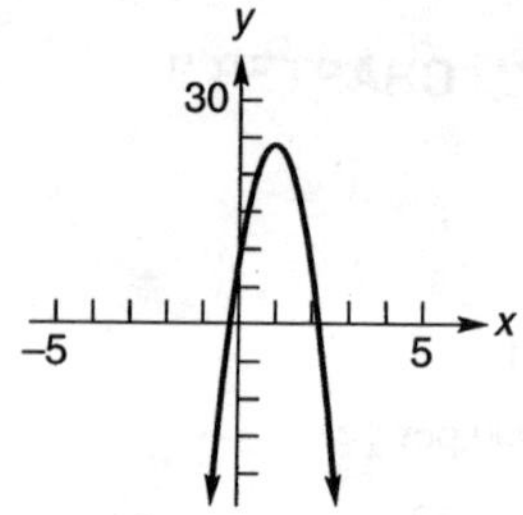

3. 1000 units; \$3000 maximum revenue.

EXERCISE 4.3

1. Quadratic. **3.** Not Quadratic. **5.** Quadratic.
7. Quadratic. **9. a.** (1, 11); **b.** highest.
11. a. −8; **b.** −4, 2; **c.** (−1, −9).
13. Vertex: (3, −4); intercepts: (1, 0), (5, 0), (0, 5); range: all $y \geq -4$.

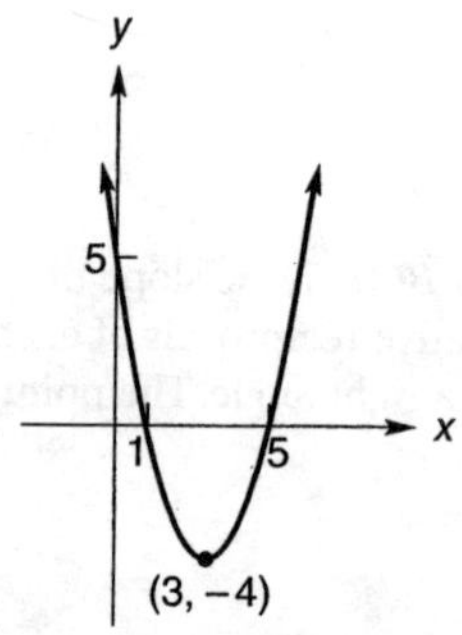

15. Vertex: $\left(-\frac{3}{2}, \frac{9}{2}\right)$; intercepts: (0, 0), (−3, 0); range: all $y \leq \frac{9}{2}$.

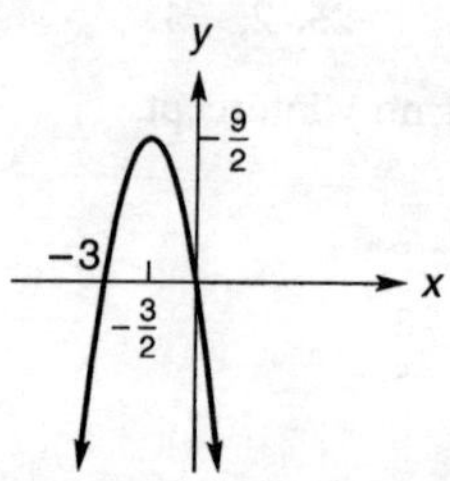

17. Vertex: $(-1, 0)$; intercepts: $(-1, 0)$, $(0, 1)$; range: all $s \geq 0$.

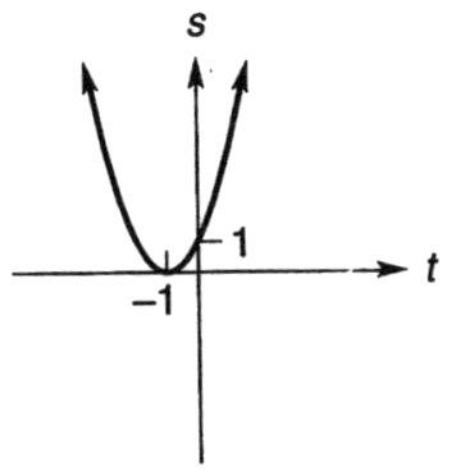

19. Vertex: $(2, -1)$; intercept: $(0, -9)$; range: all $y \leq -1$.

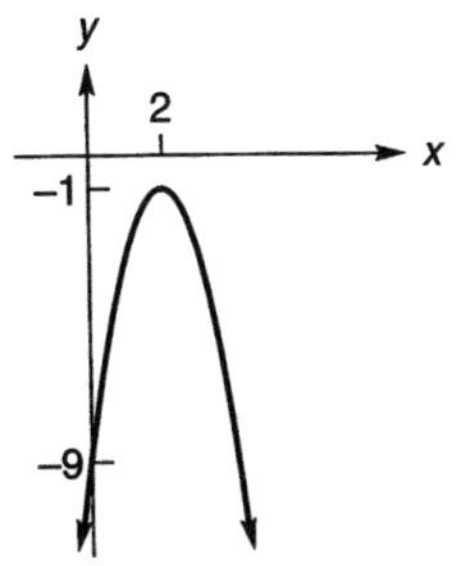

21. Vertex: $(4, -3)$; intercepts: $(4 + \sqrt{3}, 0)$, $(4 - \sqrt{3}, 0)$, $(0, 13)$; range: all $t \geq -3$.

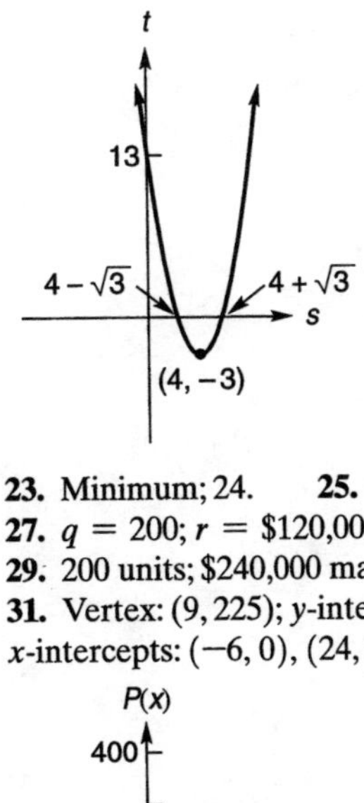

23. Minimum; 24. **25.** Maximum; -10.
27. $q = 200$; $r = \$120{,}000$.
29. 200 units; \$240,000 maximum revenue.
31. Vertex: $(9, 225)$; y-intercept: $(0, 144)$; x-intercepts: $(-6, 0)$, $(24, 0)$.

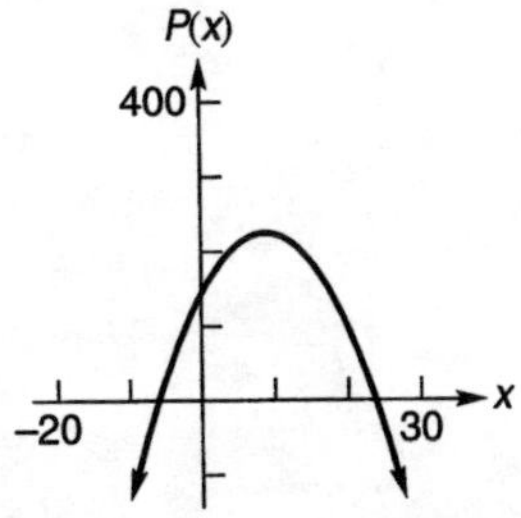

33. 70 grams. **35.** 96 ft; 2 sec.

37. Vertex: $\left(\frac{5}{2}, 116\right)$; y-intercept: $(0, 16)$, x-intercepts: $\left(\frac{5 + \sqrt{29}}{2}, 0\right)$, $\left(\frac{5 - \sqrt{29}}{2}, 0\right)$.

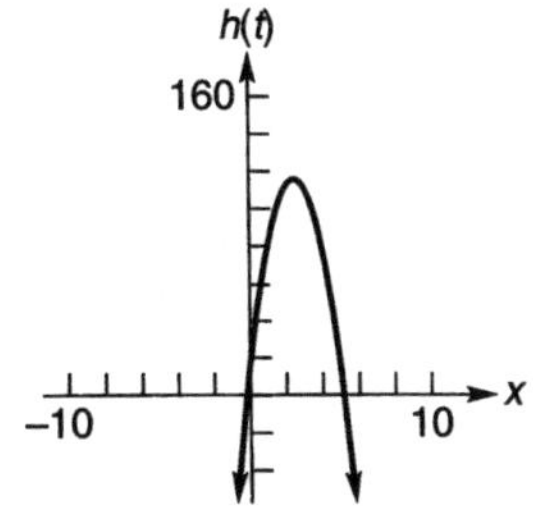

39. **a.** 2.5; **b.** 8.7 m. **41.** **a.** $\frac{l}{2}$; **b.** $\frac{wl^2}{8}$; **c.** 0 and l.
43. 50 ft $\times$ 100 ft. **45.** $(1.11, 2.88)$.
47. **a.** 0; **b.** 1; **c.** 2. **49.** 6.89.

PRINCIPLES IN PRACTICE 4.4

1. \$120,000 at 9% and \$80,000 at 8%.
2. 500 of species A and 1000 of species B.
3. Infinitely many solutions of the form $A = \frac{20{,}000}{3} - \frac{4}{3}r$, $B = r$ where $0 \leq r \leq 5000$.
4. $\frac{1}{6}$ lb of A; $\frac{1}{3}$ lb of B; $\frac{1}{2}$ lb of C.

EXERCISE 4.4

1. $x = -1, y = 1$. **3.** $x = 3, y = -1$.
5. $v = 0, w = 18$. **7.** $x = 2, y = -3$.
9. No solution. **11.** $x = 12, y = -12$.
13. $p = \frac{3}{2} - 3r, q = r$; r is any real number.
15. $x = \frac{1}{2}, y = \frac{1}{2}, z = \frac{1}{4}$. **17.** $x = 1, y = 1, z = 1$
19. $x = 1 + 2r, y = 3 - r, z = r$; r is any real number.
21. $x = -\frac{1}{3}r, y = \frac{5}{3}r, z = r$; r is any real number.
23. $x = \frac{3}{2} - r + \frac{1}{2}s, y = r, z = s$; r and s are any real numbers.
25. 420 gal of 20% solution, 280 gal of 30% solution.
27. 0.5 lb of cotton; 0.25 lb of polyester; 0.25 lb of nylon.
29. 275 mi/h (speed of airplane in still air), 25 mi/h (speed of wind).
31. 240 units (Early American), 200 units (Contemporary).
33. 800 calculators from Exton plant, 700 from Whyton plant.
35. 4% on first \$100,000, 6% on remainder.
37. 60 units of Argon I, 40 units of Argon II.
39. 100 chairs, 100 rockers, 200 chaise lounges.
41. 40 semiskilled workers, 20 skilled workers, 10 shipping clerks.
45. $x = 4, y = 2$. **47.** $x = 8.3, y = 14.0$.

EXERCISE 4.5

1. $x = 4, y = -12; x = -1, y = 3.$
3. $p = -3, q = -5; p = 2, q = 0.$
5. $x = 0, y = 0; x = 1, y = 1.$
7. $x = 4, y = 8; x = -1, y = 3.$
9. $p = 0, q = 0; p = 1, q = 1.$
11. $x = 3\sqrt{2}, y = 2;\ x = -3\sqrt{2}, y = 2;$
$x = \sqrt{15}, y = -1; x = -\sqrt{15}, y = -1.$
13. $x = 21, y = 15.$ **15.** At $(10, 8.1)$ and $(-10, 7.9)$.
17. Three. **19.** $x = -1.3, y = 5.1.$ **21.** $x = 1.76.$
23. $x = -1.46.$

EXERCISE 4.6

1.

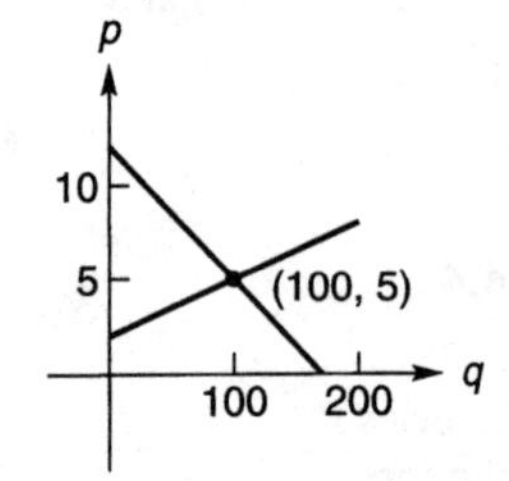

3. $(5, 212.50)$. **5.** $(9, 38)$. **7.** $(15, 5)$.
9.

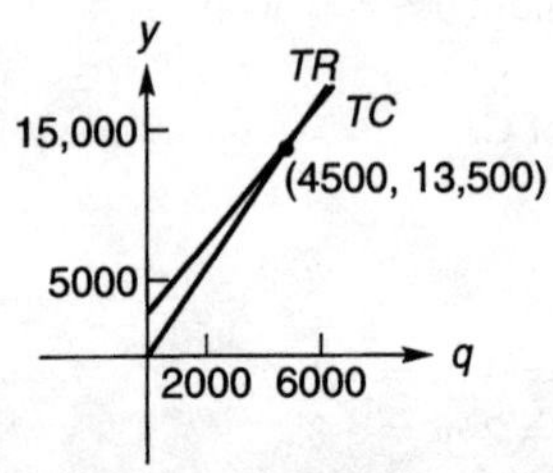

11. Cannot break even at any level of production.
13. 10 units or 40 units. **15. a.** \$12; **b.** \$12.18.
17. 5840 units; 840 units; 1840 units. **19.** \$4.
21. Total cost always exceeds total revenue—no break-even point.
23. Decreases by \$0.70. **25.** $p_A = 5; p_B = 10.$
27. 2.4 and 11.3.

REVIEW PROBLEMS CHAPTER 4

1. 9. **3.** $y = -x + 1; x + y - 1 = 0.$
5. $y = \frac{1}{2}x - 1; x - 2y - 2 = 0.$
7. $y = 4; y - 4 = 0.$
9. $y = \frac{1}{3}x + 2; x - 3y + 6 = 0.$ **11.** Perpendicular.
13. Neither. **15.** Parallel. **17.** $y = \frac{3}{2}x - 2; \frac{3}{2}.$

19. $y = \frac{4}{3}; 0.$ **21.** $-2; (0, 4).$

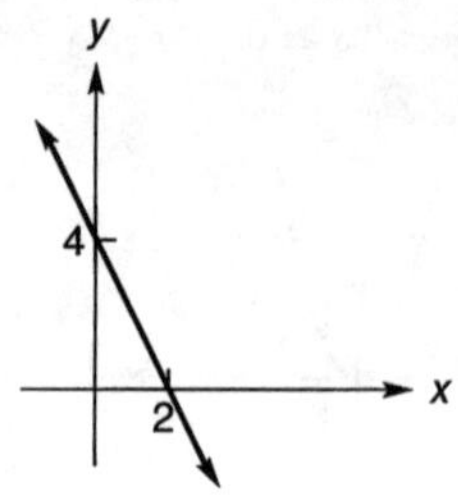

23. $(3, 0), (-3, 0), (0, 9); (0, 9).$

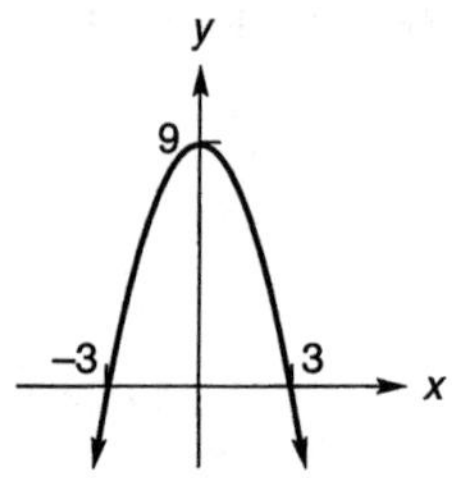

25. $(5, 0), (-1, 0), (0, -5); (2, -9).$

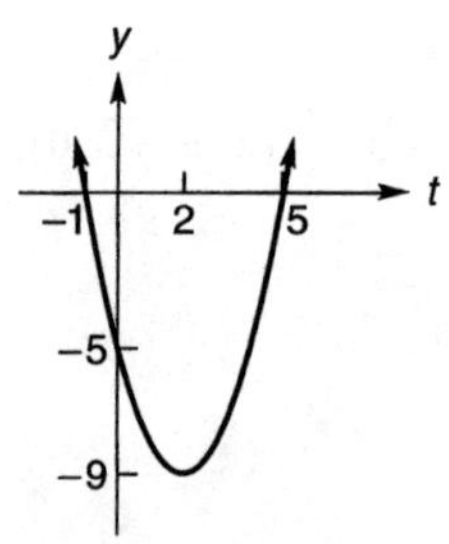

27. 3; $(0, 0)$.

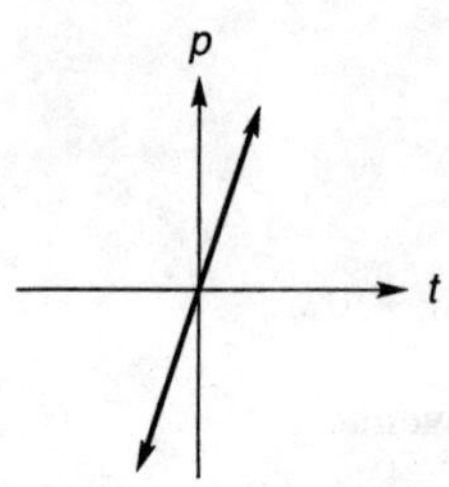

29. $(0, -3); (-1, -2).$

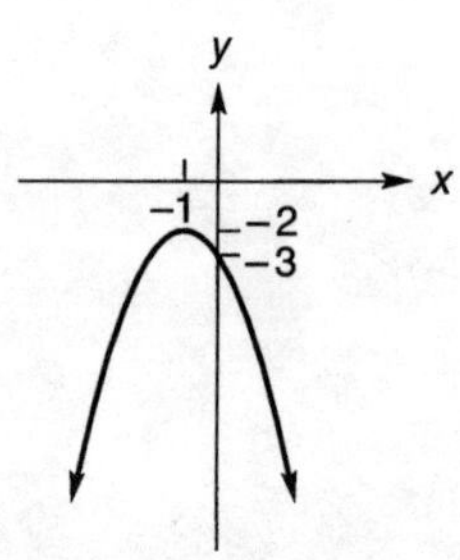

31. $x = \frac{17}{7}, y = -\frac{8}{7}.$ **33.** $x = 2, y = -1.$
35. $x = 8, y = 4.$ **37.** $x = 0, y = 1, z = 0.$
39. $x = -3, y = -4; x = 2, y = 1.$

41. $x = -2 - 2r$, $y = 7 + r$, $z = r$; r is any real number.
43. $x = r$, $y = r$, $z = 0$; r is any real number.
45. $a + b - 3 = 0$; 0. **47.** $f(x) = -\frac{4}{3}x + \frac{19}{3}$.
49. 50 units; \$5000. **51.** 6.
53. a. $R = 75L + 1310$; **b.** 1385 milliseconds;
c. The slope is 75; the time necessary to travel from one level to the next level is 75 milliseconds.
55. 300 ft by 250 ft. **57.** $x = 230$, $y = -130$.
59. $x = 0.75$, $y = 1.43$.

MATHEMATICAL SNAPSHOT CHAPTER 4

1. a. 12; **b.** $F(x) = \begin{cases} 4, & \text{if } x = 2 \\ 6, & \text{if } x = 3 \\ \left\lceil \frac{12}{11}(x-1) + 4.5 \right\rceil & \text{otherwise;} \end{cases}$
c. $F(1) = 4$, $F(2) = 4$, $F(3) = 6$, $F(4) = 7$, $F(5) = 8$, $F(6) = 9$, $F(7) = 11$, $F(8) = 12$, $F(9) = 13$, $F(10) = 14$, $F(11) = 15$, $F(12) = 16$.

Chapter 6
Linear Equations and Straight Lines

1.1
Coordinate Systems and Graphs

1.2
Linear Inequalities

1.3
The Intersection Point of a Pair of Lines

1.4
The Slope of a Straight Line

1.5
The Method of Least Squares

Many applications considered later involve linear equations and their geometric counterparts—straight lines. So let us begin by studying the basic facts about these two important notions.

1.1 Coordinate Systems and Graphs

Often we can display numerical data by using a *Cartesian coordinate system* on either a line or a plane. We construct a Cartesian coordinate system on a line by choosing an arbitrary point O (the *origin*) on the line and a unit of distance along the line. We then assign to each point on the line a number that reflects its directed distance from the origin. Positive numbers refer to points on the right of the origin, negative numbers to points on the left. In Fig. 1 we have drawn a Cartesian coordinate system on the line and have labeled a number of points with their corresponding numbers. Each point on the line corresponds to a number (positive, negative, or zero). Conversely, every number corresponds to a point on the line.

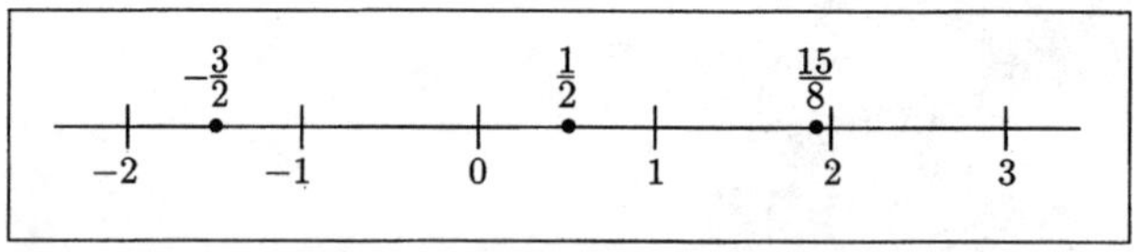

FIGURE 1

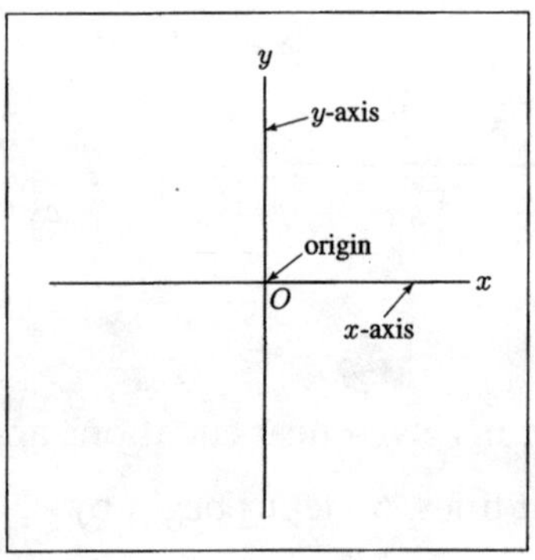

FIGURE 2

A Cartesian coordinate system may be used to numerically describe points on a line. In a similar fashion, we can construct a Cartesian coordinate system to numerically locate points on a plane. Such a system consists of two perpendicular lines called the *coordinate axes.* These lines are usually drawn so that one is horizontal and one is vertical. The horizontal line is called the *x-axis*, the vertical line the *y-axis*. Their point of intersection is called the *origin* (Fig. 2). Each point of the plane is identified by a pair of numbers (a, b). The first number, a, tells the number of units from the point to the y-axis (Fig. 3). When a is positive, the point is to the right of the y-axis; when a is negative, the point is to the left of the y-axis. The second number, b, gives the number of units from the point to the x-axis (Fig. 4). When b is positive, the point is above the x-axis; when b is negative, the point is below. The numbers a and b are called, respectively, the *x- and y-coordinates of the point.* In order to plot the point (a, b), begin at the origin, move a units in the x-direction and then b units in the y-direction.

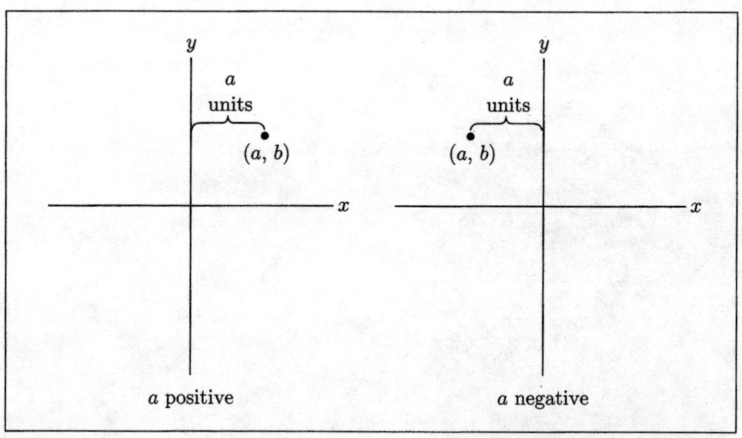

FIGURE 3

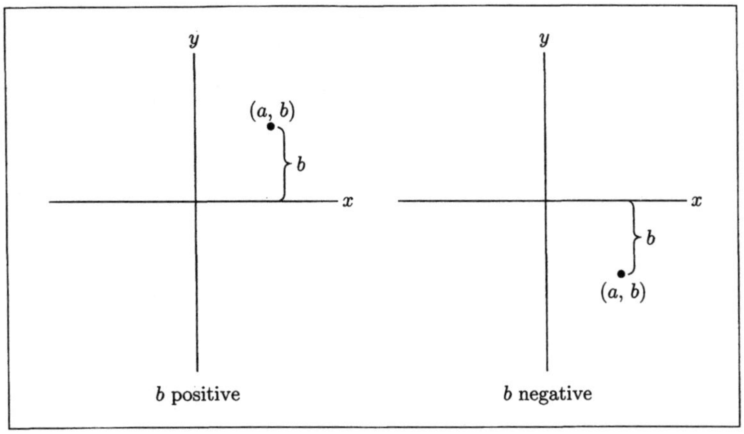

FIGURE 4

EXAMPLE 1 Plot the following points.

(a) $(2, 1)$ (b) $(-1, 3)$ (c) $(-2, -1)$ (d) $(0, -3)$

Solution (a)

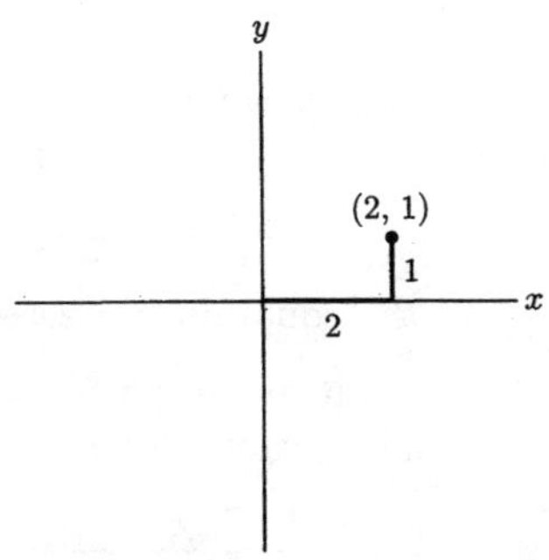

(b)

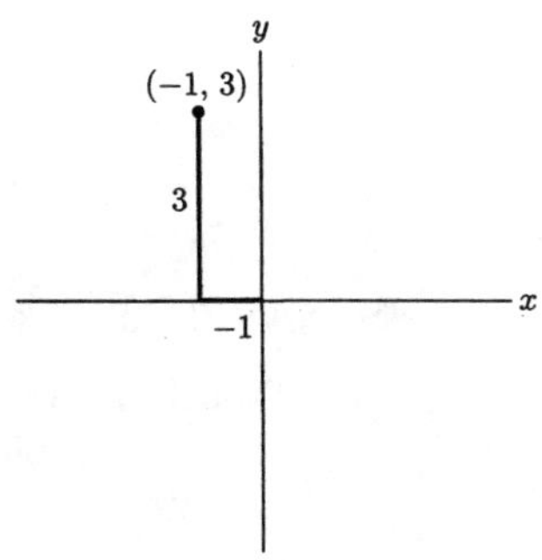

(c)

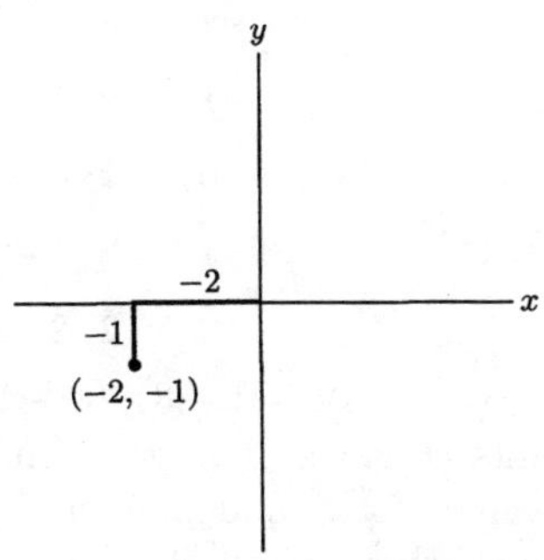

(d)

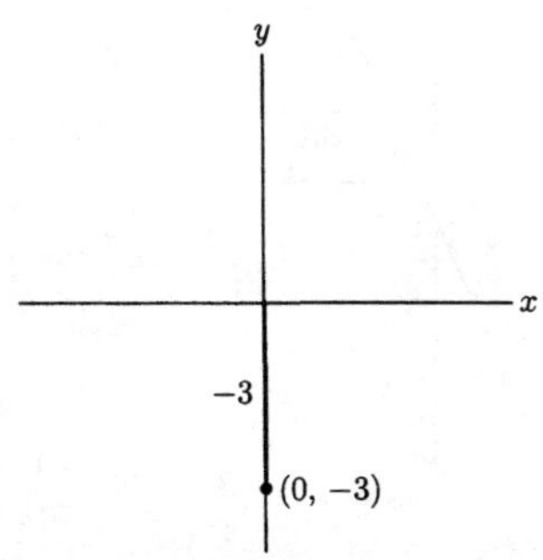

∎

In many applications one encounters equations expressing relationships between variables x and y. Some typical equations are

$$y = 2x - 1$$
$$5x^2 + 3y^3 = 7$$
$$y = 2x^2 - 5x + 8.$$

To any equation in x and y one can associate a certain collection of points in the plane. Namely, the point (a, b) belongs to the collection provided that the equation is satisfied when we substitute a for each occurrence of x and substitute

b for each occurrence of y. This collection of points is usually a curve of some sort and is called the *graph of the equation.*

EXAMPLE 2 Are the following points on the graph of the equation $8x - 4y = 4$?

(a) $(3, 5)$ (b) $(5, 17)$

Solution (a) Substitute 3 for each occurrence of x and 5 for each occurrence of y in the equation.

$$\begin{aligned} 8x - 4y &= 4 \\ 8 \cdot 3 - 4 \cdot 5 &= 4 \\ 24 - 20 &= 4 \\ 4 &= 4. \end{aligned}$$

Since the equation is satisfied, the point $(3, 5)$ is on the graph of the equation.

(b) Replace x by 5 and y by 17 in the equation.

$$\begin{aligned} 8x - 4y &= 4 \\ 8 \cdot 5 - 4 \cdot 17 &= 4 \\ 40 - 68 &= 4 \\ -28 &= 4. \end{aligned}$$

The equation is clearly not satisfied, so the point $(5, 17)$ is *not* on the graph of the equation. ▮

EXAMPLE 3 Sketch the graph of the equation $y = 2x - 1$.

Solution First find some points on the graph by choosing various values for x and determining the corresponding values for y:

x	y
-2	$2 \cdot (-2) - 1 = -5$
-1	$2 \cdot (-1) - 1 = -3$
0	$2 \cdot 0 - 1 = -1$
1	$2 \cdot 1 - 1 = 1$
2	$2 \cdot 2 - 1 = 3$

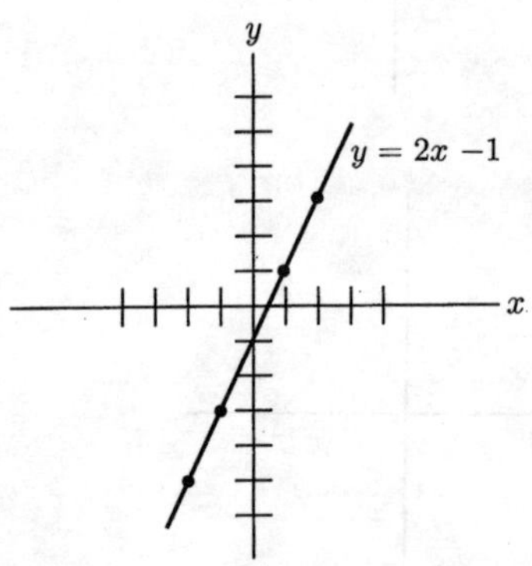

FIGURE 5

Thus the points $(-2, -5)$, $(-1, -3)$, $(0, -1)$, $(1, 1)$, and $(2, 3)$ are all on the graph. Plot these points (Fig. 5). It appears that the points lie on a straight line. By taking more values for x and plotting the corresponding points, it is easy to become convinced that the graph of $y = 2x - 1$ is indeed a straight line. ▮

EXAMPLE 4 Sketch the graph of the equation $x = 3$.

Solution It is clear that the x-coordinate of any point on the graph must be 3. The y-coordinate can be anything. So some points on the graph are $(3, 0)$, $(3, 5)$, $(3, -4)$, $(3, -2)$. Again the graph is a straight line (Fig. 6). ▮

EXAMPLE 5 Sketch the graph of the equation $x = a$, where a is any given number.

Solution The x-coordinate of any point on the graph must be a. Reasoning as in Example 4, the graph is a vertical line a units away from the y-axis (Fig. 7). (Of course, if a is negative, then the line lies to the left of the y-axis.) ▮

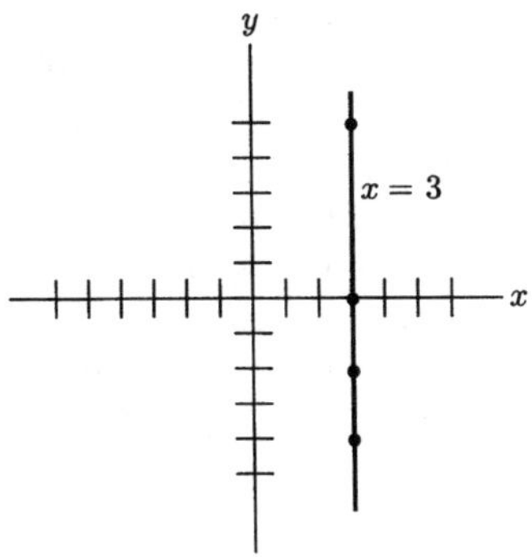

FIGURE 6

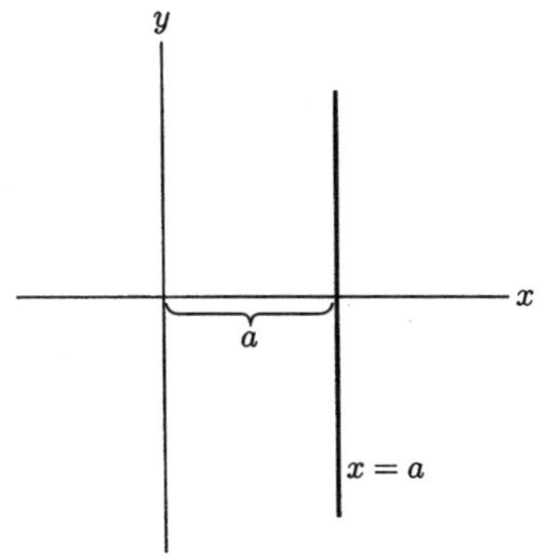

FIGURE 7

The equations in Examples 3 to 5 are all special cases of the general equation

$$cx + dy = e,$$

corresponding to particular choices of the constants c, d, e. Any such equation is called a *linear equation* (in the two variables x and y).

The *standard form* of a linear equation is obtained by solving for y if y appears and for x if y does not appear. In the former case the standard form looks like

$$y = mx + b \qquad (m,\ b \text{ constants}),$$

whereas in the latter it looks like

$$x = a \qquad (a \text{ constant}).$$

EXAMPLE 6 Find the standard form of the following equations.

(a) $8x - 4y = 4$ (b) $2x + 3y = 3$ (c) $2x = 6$

Solution (a) Since y appears, we obtain the standard form by solving for y in terms of x:

$$\begin{aligned} 8x - 4y &= 4 \\ -4y &= -8x + 4 \\ y &= 2x - 1. \end{aligned}$$

Thus the standard form of $8x - 4y = 4$ is $y = 2x - 1$—that is, $y = mx + b$ with $m = 2$, $b = -1$.

(b) Again y occurs, so we solve for y:

$$\begin{aligned} 2x + 3y &= 3 \\ 3y &= -2x + 3 \\ y &= -\tfrac{2}{3}x + 1. \end{aligned}$$

So the standard form is $y = -(2/3)x + 1$. Here $m = -2/3$ and $b = 1$.

(c) Here y does not occur, so we solve for x:

$$\begin{aligned} 2x &= 6 \\ x &= 3. \end{aligned}$$

Thus the standard form of $2x = 6$ is $x = 3$—that is, $x = a$, where a is 3. ∎

We have seen that any linear equation has one of the two standard forms $y = mx + b$ and $x = a$. From Example 5, the graph of $x = a$ is a vertical line, a units from the y-axis. What can be said about the graph of $y = mx + b$? In Example 3, we saw that the graph is a straight line in the special case $y = 2x - 1$. Actually, the graph of $y = mx + b$ is always a straight line. To sketch the graph, we need only locate two points. Two convenient points to locate are the *intercepts*, the points where the line crosses the x- and y-axes. When x is 0, $y = m \cdot 0 + b = b$. Thus $(0, b)$ is on the graph of $y = mx + b$ and is the y-intercept of the line. The x-intercept is found as follows: A point on the x-axis has y-coordinate 0. So the x-coordinate of the x-intercept can be found by setting $y = 0$—that is, $mx + b = 0$—and solving this equation for x.

EXAMPLE 7 Sketch the graph of the equation $y = 2x - 1$.

Solution Here $m = 2$, $b = -1$. The y-intercept is $(0, b) = (0, -1)$. To find the x-intercept, we must solve $2x - 1 = 0$. But then $2x = 1$ and $x = 1/2$. So the x-intercept is $(1/2, 0)$. Plot the two points $(0, -1)$ and $(1/2, 0)$, and draw the straight line through them (Fig. 8). ▮

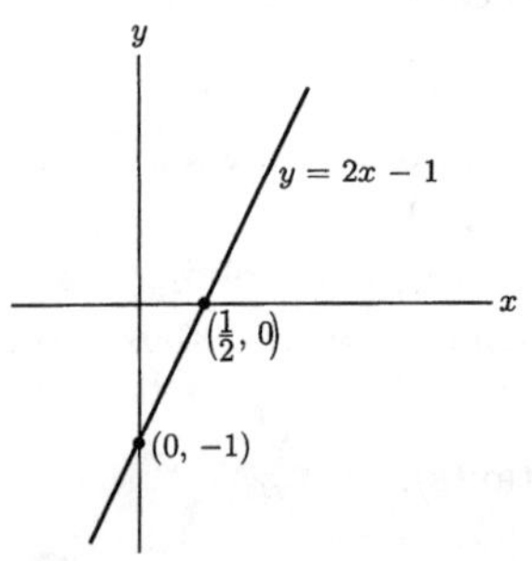

FIGURE 8

Note The line in Example 7 had different x- and y-intercepts. Two other circumstances can occur, however. First, the two intercepts may be the same, as in $y = 3x$. Second, there may be no x-intercept, as in $y = 1$. In both of these circumstances, we must plot some point other than an intercept in order to graph the line.

To summarize, then:

> To graph the equation $y = mx + b$:
>
> 1. Plot the y-intercept $(0, b)$.
> 2. Plot some other point. [The most convenient choice is often the x-intercept $(x, 0)$, where x is determined by solving $mx + b = 0$.]
> 3. Draw a line through the two points.

The next example gives an application of linear equations.

EXAMPLE 8 (*Linear Depreciation*) For tax purposes, businesses must keep track of the current values of each of their assets. A common mathematical model is to assume that the current value y is related to the age x of the asset by a linear equation. A moving company buys a 40-foot van with a useful lifetime of 5 years. After x months of use, the value y of the van is estimated by the linear equation

$$y = 25{,}000 - 400x.$$

(a) Sketch the graph of this linear equation.

(b) What is the value of the van after 5 years?

(c) What economic interpretation can be given to the y-intercept of the graph?

Solution (a) The y-intercept is (0, 25,000). The x-intercept is found from the equation

$$25{,}000 - 400x = 0$$

$$x = \frac{125}{2},$$

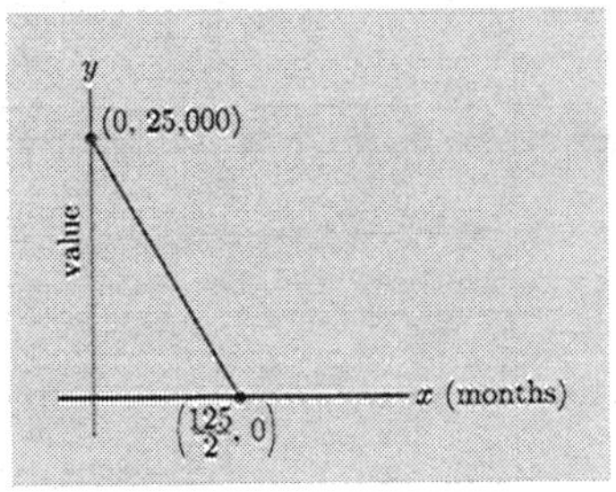

FIGURE 9

so the x-intercept is $(125/2, 0)$. The graph of the linear equation is sketched in Fig. 9. Note how the value decreases as the age of the truck increases. The value of the truck reaches 0 after 125/2 months. Note also that we have sketched only the portion of the graph that has physical meaning, namely the portion for x between 0 and 125/2.

(b) After 5 years (or 60 months), the value of the van is

$$y = 25{,}000 - 400(60) = 25{,}000 - 24{,}000 = 1000.$$

Since the useful life of the van is 5 years, this value represents the *salvage value* of the van.

(c) The y-intercept corresponds to the value of the truck at $x = 0$ months, that is, the initial value of the truck, \$25,000. ▮

GC Accurate graphs of straight lines can be obtained with a *graphing utility*; that is, a graphing calculator or a computer with graphing software. The equation is usually entered in standard form. Figures 10, 11, and 12 show how the equation from Example 7 is entered and graphed on a TI-83 graphing calculator. The window setting in Fig. 11 is sometimes referred to as $[-2, 4]$ *by* $[-2, 2]$. The graphing utility also can be used to find the coordinates of points on the line (see Fig. 13) and to determine the intercepts (see Figs. 14 and 15). Vertical lines can be drawn with the VERTICAL command from the DRAW menu. For instance, to draw the vertical line through the cursor, go to the home screen and press [2nd] [DRAW] **4**.

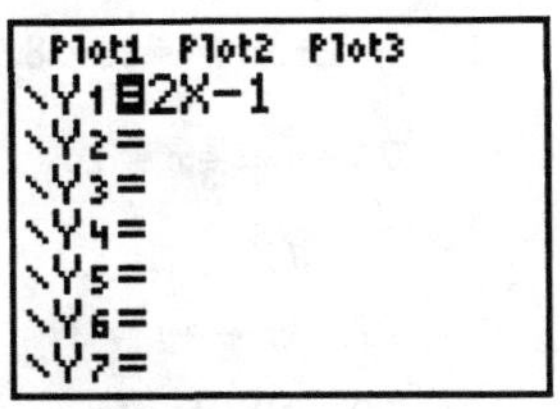

FIGURE 10

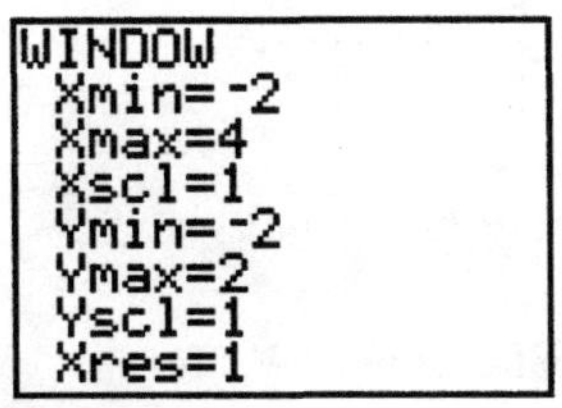

FIGURE 11

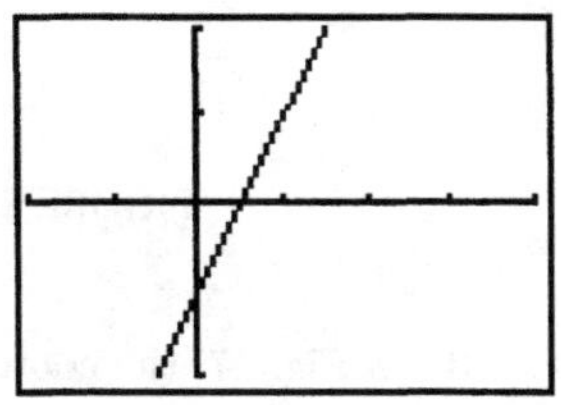

FIGURE 12

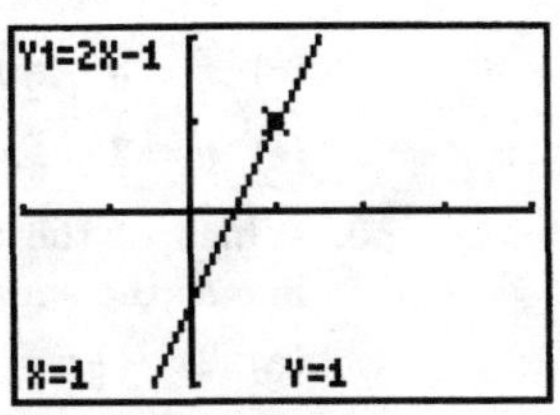

FIGURE 13

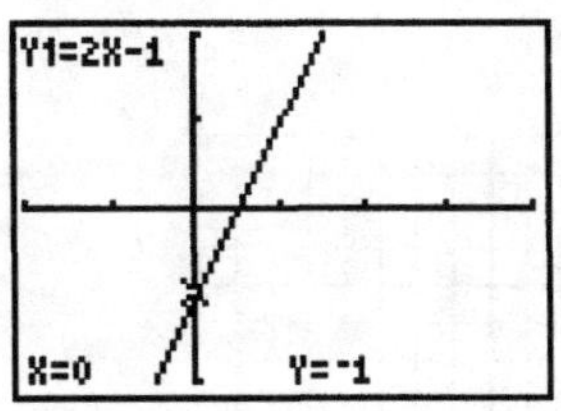

FIGURE 14

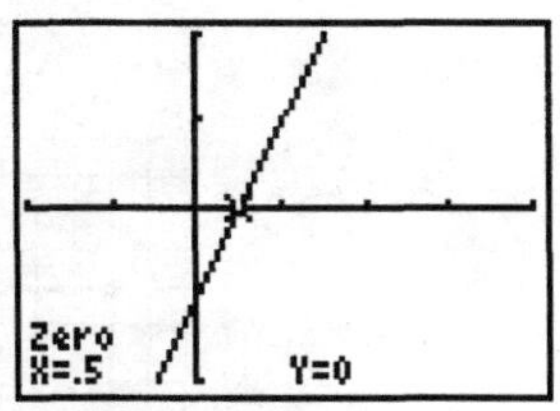

FIGURE 15

Throughout this text we discuss specifics of using the TI-82 and TI-83 graphing calculators. Additional details are provided in Appendix B. For the specific details of using other calculators, consult the guidebook that comes with the calculator.

Practice Problems 1

1. Plot the point (500, 200).
2. Is the point $(4, -7)$ on the graph of the linear equation $2x - 3y = 1$? Is the point $(5, 3)$?
3. Graph the linear equation $5x + y = 10$.
4. Graph the straight line $y = 3x$.

EXERCISES 1

In Exercises 1–8, plot the given point.

1. $(2, 3)$ **2.** $(-1, 4)$
3. $(0, -2)$ **4.** $(2, 0)$
5. $(-2, 1)$ **6.** $\left(-1, -\frac{5}{2}\right)$
7. $(-20, 40)$ **8.** $(25, 30)$

9. ***(From the GMAT exam)*** In Fig. 16, the coordinates of point V are
(a) $(-7, 5)$ (b) $(-5, 7)$ (c) $(5, 7)$
(d) $(7, 5)$ (e) $(7, -5)$.

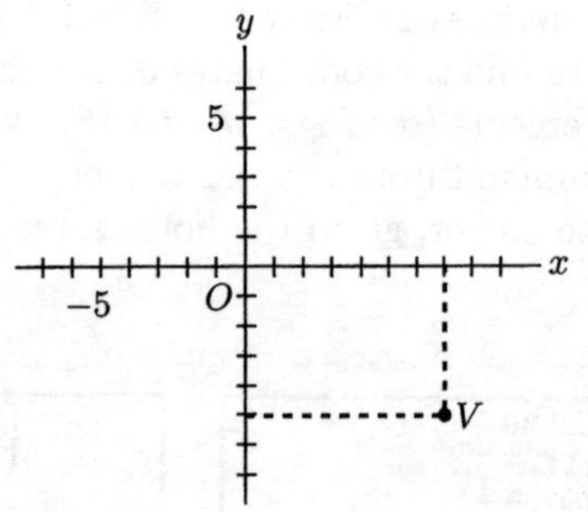

FIGURE 16

10. In Fig. 17, the coordinates of point V are
(a) $(-6, 4)$ (b) $(4, -6)$ (c) $(8, -12)$
(d) $(-12, 8)$ (e) $(6, 4)$.

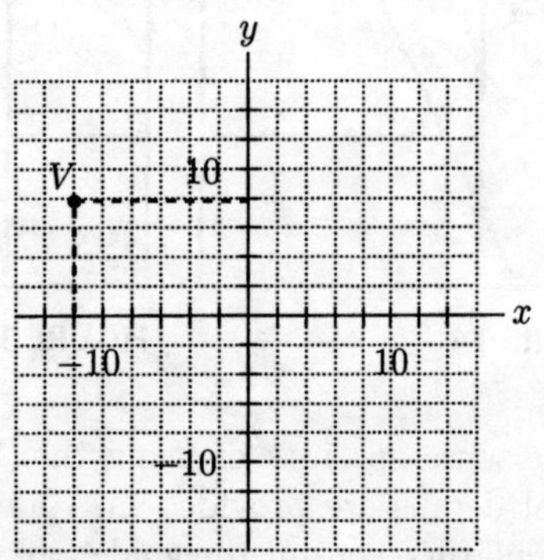

FIGURE 17

In Exercises 11–14, each linear equation is in the standard form $y = mx + b$. Identify m and b.

11. $y = 5x + 8$ **12.** $y = -2x - 6$
13. $y = 3$ **14.** $y = \frac{2}{3}x$

In Exercises 15–18, put the linear equations into standard form.

15. $14x + 7y = 21$ **16.** $x - y = 3$
17. $3x = 5$ **18.** $-\frac{1}{2}x + \frac{2}{3}y = 10$

In Exercises 19–22, find the x-intercept and the y-intercept of each line.

19. $y = -4x + 8$ **20.** $y = 5$
21. $x = 7$ **22.** $y = -8x$

In Exercises 23–28, graph the given linear equation.

23. $y = \frac{1}{3}x - 1$ **24.** $y = 2x$
25. $y = \frac{5}{2}$ **26.** $x = 0$
27. $3x + 4y = 24$ **28.** $x + y = 3$

29. Which of the following equations describe the same line as the equation $2x + 3y = 6$?
(a) $4x + 6y = 12$ (b) $y = -\frac{2}{3}x + 2$
(c) $x = 3 - \frac{3}{2}y$ (d) $6 - 2x - y = 0$
(e) $y = 2 - \frac{2}{3}x$ (f) $x + y = 1$

30. Which of the following equations describe the same line as the equation $\frac{1}{2}x - 5y = 1$?
(a) $2x - \frac{1}{5}y = 1$ (b) $x = 5y + 2$
(c) $2 - 5x + 10y = 0$ (d) $y = .1(x - 2)$
(e) $10y - x = -2$ (f) $1 + .5x = 2 + 5y$

31. Each of the lines L_1, L_2, and L_3 in Fig. 18 is the graph of one of the equations (a), (b), and (c). Match each of the following equations with its corresponding line.
(a) $x + y = 3$ (b) $2x - y = -2$ (c) $x = 3y + 3$

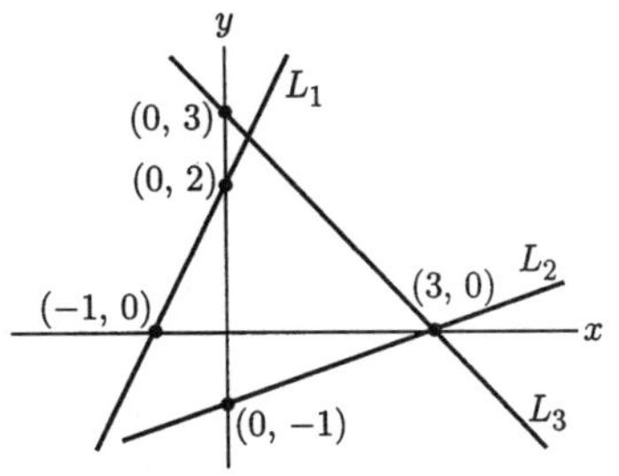

FIGURE 18

32. Which of the following equations is graphed in Fig. 19?
(a) $x + y = 3$ (b) $y = x - 1$ (c) $2y = x + 3$

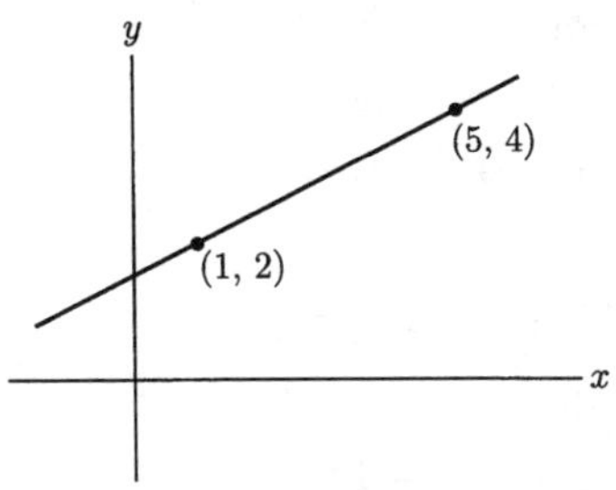

FIGURE 19

33. The temperature of water in a heating tea kettle rises according to the equation $y = 30x + 72$, where y is the temperature (in degrees Fahrenheit) x minutes after the kettle was put on the burner.

(a) After how many minutes will the water boil?

(b) What physical interpretation can be given to the y-intercept of the graph?

(c) Does the x-intercept have a meaningful physical interpretation?

34. The amount of tropical rainforest area in Central America has been decreasing steadily in recent years. The amount y (in thousands of square miles) x years after 1969 is estimated by the linear equation

$$y = \left(-\frac{25}{8}\right) x + 130.$$

(a) Sketch the graph of this linear equation.

(b) What interpretation can be given to the y-intercept of the graph?

(c) When were there 80,000 square miles of tropical rain forests?

(d) If this trend continues, how large will the rain forest be in the year 2001?

35. The worldwide consumption of cigarettes has been increasing steadily in recent years. The number of trillions of cigarettes, y, purchased x years after 1960, is estimated by the linear equation

$$y = .075x + 2.5.$$

(a) Sketch the graph of this linear equation.

(b) What interpretation can be given to the y-intercept of the graph?

(c) When were there 4 trillion cigarettes sold?

(d) If this trend continues, how many cigarettes will be sold in the year 2020?

36. Find an equation of the line having y-intercept $(0, 5)$ and x-intercept $(4, 0)$.

37. Consider an equation of the form $y = mx + b$. When the value of m remains fixed and the value of b changes, the graph is translated vertically. As the value of b increases, does the graph move up or down?

38. Can a line have more than one x-intercept?

39. What is the equation of the x-axis?

40. What is the general form of the equation of a line that is parallel to the x-axis?

41. Does every line have a y-intercept?

42. Does every line have an x-intercept?

Solutions to Practice Problems 1

1. Since the numbers are large, make each hatchmark correspond to 100. Then the point $(500, 200)$ is found by starting at the origin, moving 500 units to the right and then 200 units up (Fig. 20).

2. Substitute the point $(4, -7)$ into the equation $2x - 3y = 1$.

$$2(4) - 3(-7) = 1$$
$$8 + 21 = 1$$
$$29 = 1.$$

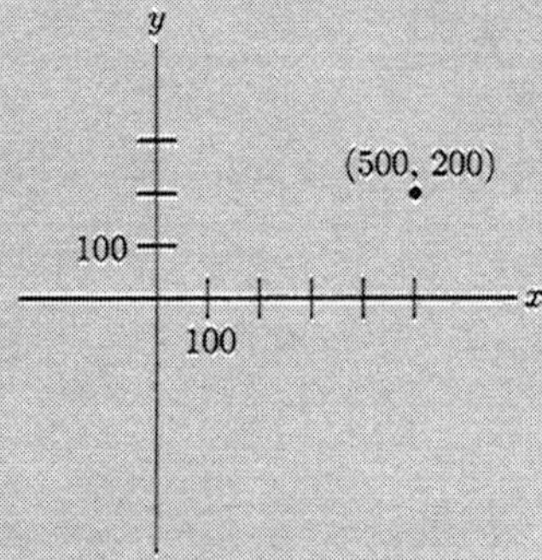

FIGURE 20

The equation does not hold, so $(4, -7)$ is not on the graph. Similarly, if we substitute $(5, 3)$ into the equation, we find that

$$\begin{aligned} 2 \cdot 5 - 3 \cdot 3 &= 1 \\ 10 - 9 &= 1 \\ 1 &= 1. \end{aligned}$$

So the equation holds and $(5, 3)$ is on the graph.

3. The standard form is obtained by solving for y:

$$y = -5x + 10.$$

Thus $m = -5$ and $b = 10$. The y-intercept is $(0, 10)$. To get the x-intercept, set $y = 0$:

$$\begin{aligned} 0 &= -5x + 10 \\ 5x &= 10 \\ x &= 2. \end{aligned}$$

So the x-intercept is $(2, 0)$ (Fig. 21).

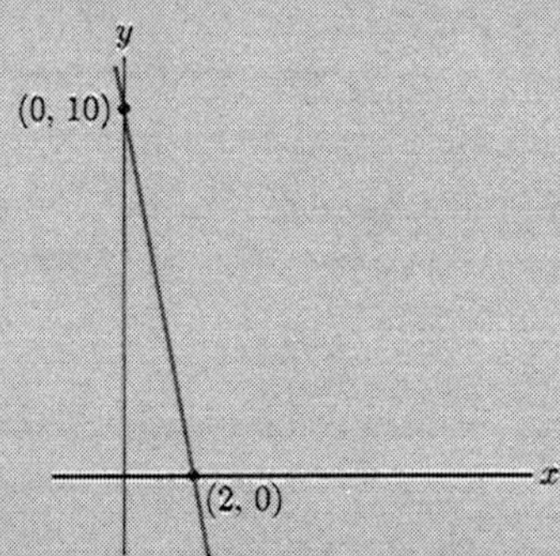

FIGURE 21

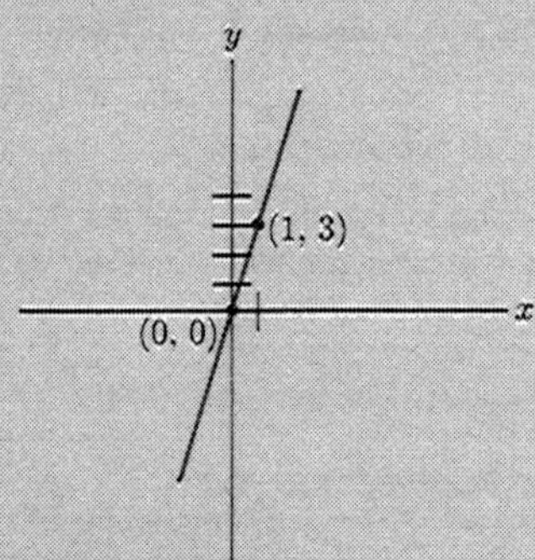

FIGURE 22

4. To find the y-intercept, set $x = 0$; then $y = 3 \cdot 0 = 0$. To find the x-intercept, set $y = 0$; then $3x = 0$ or $x = 0$. The two intercepts are the same point $(0, 0)$. We must therefore plot some other point. Setting $x = 1$, we obtain $y = 3 \cdot 1 = 3$, so another point on the line is $(1, 3)$ (Fig. 22).

1.2 Linear Inequalities

In this section we study the properties of inequalities. Start with a Cartesian coordinate system on the line (Fig. 1). Recall that it is possible to associate with each point of the line a number; and conversely, with each number (positive, negative, or zero) it is possible to associate a point on the line.

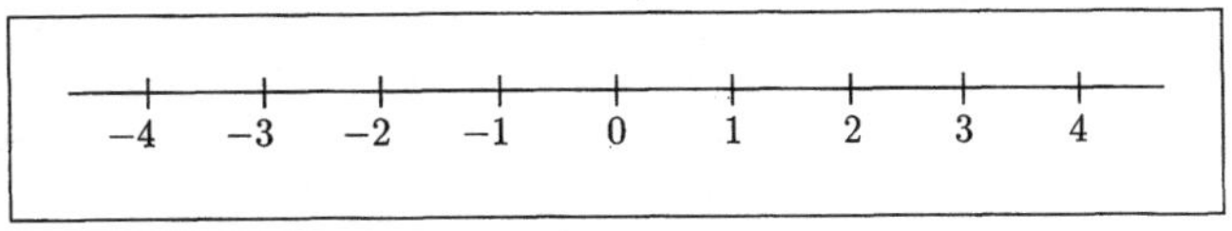

FIGURE 1

Let a and b be any numbers. We say that a *is less than* b if a lies to the left of b on the line. When a is less than b, we write $a < b$ (read: a is less than b). Thus, for example, $2 < 3$, $-1 < 2$, $-3 < -1$, and $0 < 4$. When a is less than b, we also say that b *is greater than* a. Thus, for example, $3 > 2$, $2 > -1$, $-1 > -3$, and $4 > 0$. Sometimes it is convenient to have a notation that means that a is no larger than b. The notation used for this is $a \leq b$ (read: a is less than or equal to b). Similarly, the notation $a \geq b$ (read: a is greater than or equal to b) means that a is no smaller than b. The symbols $>$, $<$, $\geq$, and $\leq$ are called *inequality signs*. It is easiest to remember the meaning of these various symbols by noting that the symbols $>$ and $<$ always *point toward the smaller* of a and b.

An inequality expresses a relationship between the quantities on both of its sides. This relationship is very similar to the relationship expressed by an equation. And just as some problems require solving equations, others require solving inequalities. Our next task will be to state and illustrate the arithmetic operations permissible in dealing with inequalities. These permissible operations form the basis of a technique for solving the inequalities occurring in applications.

> ***Inequality Property 1*** Suppose that $a < b$ and that c is any number. Then $a + c < b + c$ and $a - c < b - c$. In other words, the same number can be added or subtracted from both sides of the inequality.

For example, start with the inequality $2 < 3$ and add 4 to both sides to get

$$2 + 4 < 3 + 4.$$

That is,

$$6 < 7,$$

a correct inequality.

EXAMPLE 1 Solve the inequality $x + 4 > 3$. That is, determine all values of x for which the inequality holds.

Solution We proceed as we would in dealing with an equation. Isolate x on the left by subtracting 4 from both sides, which is permissible by Inequality Property 1:

$$x + 4 > 3$$
$$(x + 4) - 4 > 3 - 4$$
$$x > -1.$$

That is, the values of x for which the inequality holds are exactly those x greater than -1. ▮

In dealing with an equation, both sides may be multiplied or divided by a number. However, multiplying or dividing an inequality by a number requires some care. The result depends on whether the number is positive or negative. More precisely:

> ***Inequality Property 2***
>
> 2A. If $a < b$ and c is positive, then $ac < bc$.
>
> 2B. If $a < b$ and c is negative, then $ac > bc$.

In other words, an inequality may be multiplied by a positive number, just as in the case of equations. But to multiply an inequality by a negative number, it is necessary to reverse the inequality sign. For example, the inequality $-1 < 2$ can be multiplied by 4 to get $-4 < 8$, a correct statement. But if we were to multiply by -4, it would be necessary to reverse the inequality sign, since -4 is negative. In this latter case we would get $4 > -8$, a correct statement.

Note Inequality Properties 1 and 2 are stated using only $<$. However, exactly the same properties hold if $<$ is replaced by $>$, $\leq$, or $\geq$.

EXAMPLE 2 Solve the inequality $-3x + 2 \geq -1$.

Solution Treat the inequality as if it were an equation. The goal is to isolate x on one side. To this end, first subtract 2 from both sides (Property 1). This gives

$$(-3x + 2) - 2 \geq -1 - 2$$
$$-3x \geq -3.$$

Next, we multiply by $-\frac{1}{3}$. (This gives x on the left.) But $-\frac{1}{3}$ is negative, so by Property 2B we must reverse the inequality sign. Thus,

$$-\tfrac{1}{3}(-3x) \leq -\tfrac{1}{3}(-3)$$
$$x \leq 1.$$

Therefore, the values of x satisfying the inequality are precisely those ≤ 1. ▮

The inequalities of greatest interest to us[1] are those in two variables, x and y, and having the form $cx + dy \leq e$ or $cx + dy \geq e$, where c, d, and e are given numbers, with c and d not both 0. We will call such inequalities *linear inequalities.* When $d \neq 0$ (that is, when y actually appears), the inequality can be put into one of the *standard forms* $y \leq mx + b$ or $y \geq mx + b$. When $d = 0$, the inequality can be put into one of the *standard forms* $x \leq a$ or $x \geq a$. The procedure for putting a linear inequality into standard form is analogous to that for putting a linear equation into standard form.

EXAMPLE 3 Put the linear inequality $2x - 3y \geq -9$ into standard form.

Solution Since we want the y-term on the left and all other terms on the right, begin by subtracting $2x$ from both sides:

$$-3y \geq -2x - 9.$$

Next, multiply by $-\frac{1}{3}$, remembering to change the inequality sign, since $-\frac{1}{3}$ is negative:

$$y \leq -\tfrac{1}{3}(-2x - 9)$$
$$y \leq \tfrac{2}{3}x + 3.$$

---

[1]Such inequalities arise in our discussion of linear programming in Chapter 3.

The last inequality is in standard form. ▮

EXAMPLE 4 Find the standard form of the inequality $\frac{1}{2}x \geq 4$.

Solution Note that y does not appear in the inequality. Just as was the case in finding the standard form of a linear equation when y does not appear, solve for x. To do this, multiply by 2 to get

$$x \geq 8,$$

the standard form. ▮

Graphing Linear Inequalities Associated with every linear inequality is a set of points of the plane, the set of all those points that satisfy the inequality. This set of points is called the *graph* of the inequality.

EXAMPLE 5 Determine whether or not the given point satisfies the inequality $y \geq -\frac{2}{3}x + 4$.

(a) $(3, 4)$ (b) $(0, 0)$

Solution Substitute the x-coordinate of the point for x and the y-coordinate for y, and determine if the resulting inequality is correct or not.

(a) $4 \geq -\frac{2}{3}(3) + 4$
$4 \geq -2 + 4$
$4 \geq 2$ (correct)

(b) $0 \geq -\frac{2}{3}(0) + 4$
$0 \geq 0 + 4$
$0 \geq 4$ (not correct)

Therefore, the point $(3, 4)$ satisfies the inequality and the point $(0, 0)$ does not. ▮

It is easiest to determine the graph of a given inequality after it has been written in standard form. Therefore, let us describe the graphs of each of the standard forms. The easiest to handle are the forms $x \geq a$ and $x \leq a$.

A point satisfies the inequality $x \geq a$ if and only if its x-coordinate is greater than or equal to a. The y-coordinate can be anything. Therefore, the graph of $x \geq a$ consists of all points to the right of and on the vertical line $x = a$. We will display the graph by crossing out the portion of the plane to the left of the line (see Fig. 2). Similarly, the graph of $x \leq a$ consists of the points to the left of and on the line $x = a$. This graph is shown in Fig. 3.

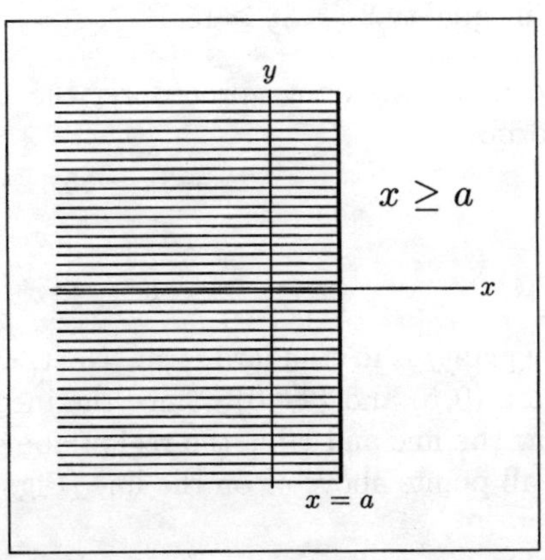

FIGURE 2

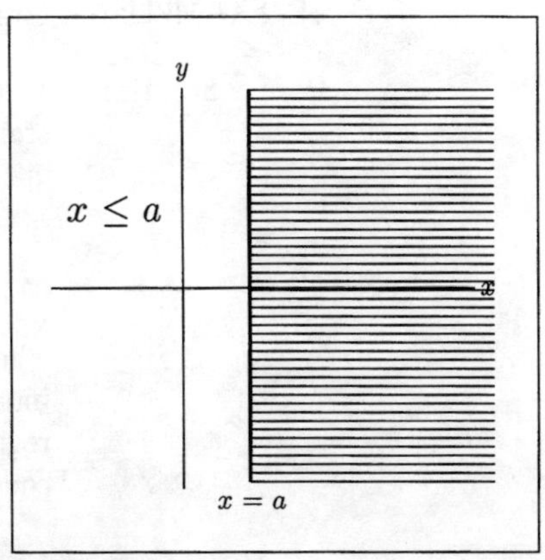

FIGURE 3

Here is a simple procedure for graphing the other two standard forms.

To graph $y \geq mx + b$ or $y \leq mx + b$:

1. Draw the graph of $y = mx + b$.
2. Throw away, that is, "cross out," the portion of the plane not satisfying the inequality. The graph of $y \geq mx + b$ consists of all points above or on the line. The graph of $y \leq mx + b$ consists of all points below or on the line.

The graphs of the inequalities $y \geq mx + b$ and $y \leq mx + b$ are shown in Fig. 4. Some simple reasoning suffices to show why these graphs are correct. Draw the line $y = mx + b$, as in Fig. 5, and pick any point P above the line. Suppose that P has coordinates (x, y). Let Q be the point on the line that lies directly below P. Then Q has the same first coordinate as P, that is, x. Since Q lies on the line, the second coordinate of Q is $mx+b$. Since the second coordinate of P is clearly larger than the second coordinate of Q, we must have $y \geq mx + b$. Similarly, any point below the line satisfies $y \leq mx + b$. Thus the two graphs are as given in Fig. 4.

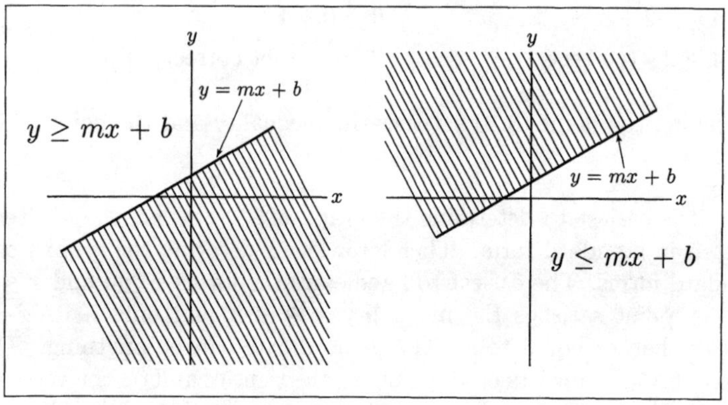

FIGURE 4

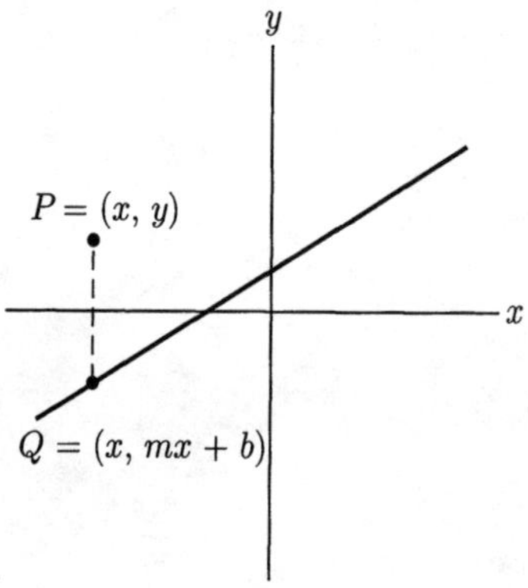

FIGURE 5

EXAMPLE 6 Graph the inequality $2x + 3y \geq 15$.

Solution In order to apply the above procedure, the inequality must first be put into standard form:

$$2x + 3y \geq 15$$
$$3y \geq -2x + 15$$
$$y \geq -\tfrac{2}{3}x + 5.$$

The last inequality is in standard form. Next, we graph the line $y = -\frac{2}{3}x+5$. Its intercepts are $(0, 5)$ and $\left(\frac{15}{2}, 0\right)$. Since the inequality is "$y \geq$" we cross out the region below the line and label the region above with the inequality. The graph consists of all points above or on the line (Fig. 6). ▮

EXAMPLE 7 Graph the inequality $4x - 2y \geq 12$.

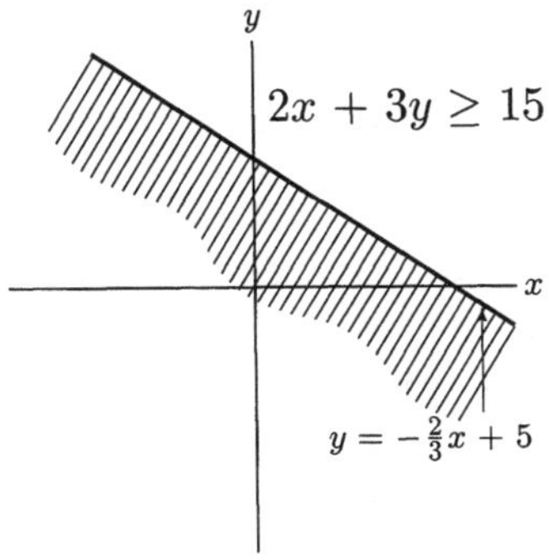

FIGURE 6

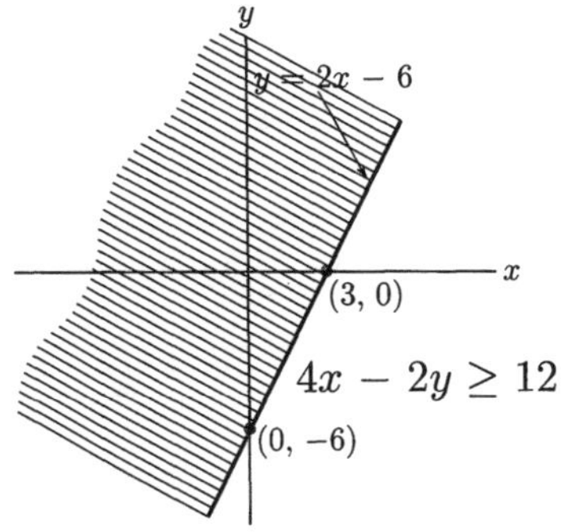

FIGURE 7

Solution First, put the inequality in standard form:

$$\begin{aligned} 4x - 2y &\geq 12 \\ -2y &\geq -4x + 12 \\ y &\leq 2x - 6 \end{aligned}$$

(note the change in the inequality sign!). Next, graph $y = 2x - 6$. The intercepts are $(0, -6)$ and $(3, 0)$. Since the inequality is "$y \leq$" the graph consists of all points below or on the line (Fig. 7). ▮

So far, we have been concerned only with graphing single inequalities. The next example concerns graphing a system of inequalities. That is, it asks us to determine all points of the plane that *simultaneously* satisfy all inequalities of a system.

▮ EXAMPLE 8 Graph the system of inequalities

$$\begin{cases} 2x + 3y \geq 15 \\ 4x - 2y \geq 12 \\ \quad\quad y \geq 0. \end{cases}$$

Solution The first two inequalities have already been graphed in Examples 6 and 7. The graph of $y \geq 0$ consists of all points above or on the x-axis. In Fig. 8, any point that is crossed out is *not* on the graph of at least one inequality. So the points that simultaneously satisfy all three inequalities are those in the remaining clear region and its border. ▮

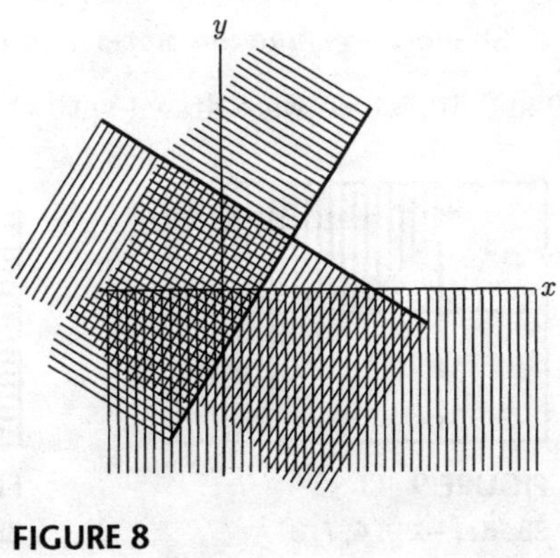

FIGURE 8

Remark At first, our convention of crossing out those points *not* on the graph of an inequality (instead of shading the points *on* the graph) may have seemed odd. However, the real advantage of this convention becomes apparent when graphing a system of inequalities. Imagine trying to find the graph of the system of Example 8 if the points *on* the graph of each inequality had been shaded. It would have been necessary to locate the points that had been shaded three times. This is hard to do.

The graph of a system of inequalities is called a *feasible set.* The feasible set associated to the system of Example 8 is a three-sided, unbounded region.

Given a specific point, we should be able to decide whether or not the point lies in the feasible set. The next example shows how this is done.

EXAMPLE 9 Determine whether the points $(5,3)$ and $(4,2)$ are in the feasible set of the system of inequalities of Example 8.

Solution If we had very accurate measuring devices, we could plot the points in the graph of Fig. 8 and determine whether or not they lie in the feasible set. However, there is a simpler and more reliable algebraic method. Just substitute the coordinates of the points into each of the inequalities of the system and see whether or not *all* of the inequalities are satisfied. So doing, we find that $(5,3)$ is in the feasible set and $(4,2)$ is not.

$$(5,3) \quad \begin{cases} 2(5)+3(3)\geq 15 \\ 4(5)-2(3)\geq 12 \\ (3)\geq 0 \end{cases} \qquad \begin{cases} 19\geq 15 \\ 14\geq 12 \\ 3\geq 0 \end{cases}$$

$$(4,2) \quad \begin{cases} 2(4)+3(2)\geq 15 \\ 4(4)-2(2)\geq 12 \\ (2)\geq 0 \end{cases} \qquad \begin{cases} 14\geq 15 \\ 12\geq 12 \\ 2\geq 0 \end{cases}$$

GC Some graphing calculators will display feasible sets of inequalities. With the TI-82 and TI-83, the **Shade** command (invoked with [2nd] [DRAW] **7** from the home screen) can be used to shade one side of the straight line Y_1. Suppose the window is $[a,b]$ *by* $[c,d]$. The region above the line can be thought of as lying between the lower function Y_1 and the upper function $y=d+1$. The region below the line can be thought of as lying between the lower function $y=c-1$ and the upper function Y_1. These regions can be shaded with the commands

Shade(*lowerfunction*, *upperfunction*, 3) [TI-82]

Shade(*lowerfunction*, *upperfunction*, a, b, 2, 3) [TI-83]

See Figs. 9 and 10, which were drawn with the window $[-5,13]$ *by* $[-6,6]$.

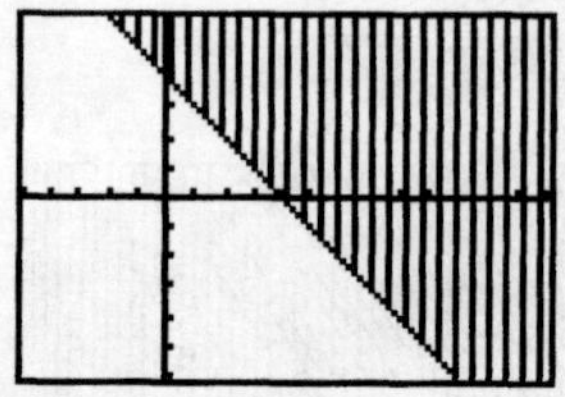

FIGURE 9 TI-82
Shade$(-X+4,7,3)$

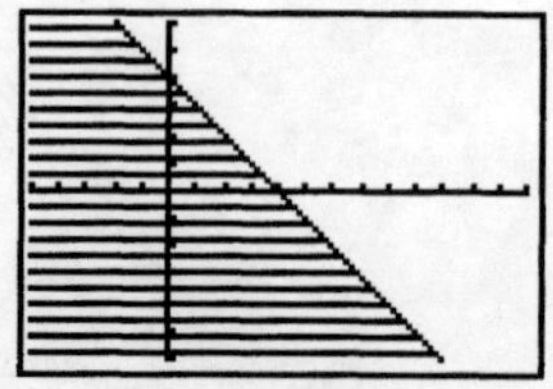

FIGURE 10 TI-83
Shade$(-7,-X+4,-5,13,2,3)$

The regions to the left and right, respectively, of the vertical line $x = k$ can be shaded with the commands

`Shade(c-1,d+1,3,a,k)` and `Shade(c-1,d+1,3,k,b)` [TI-82]

`Shade(c-1,d+1,a,k,2,3)` and `Shade(c-1,d+1,k,b,2,3)` [TI-83]

See Figs. 11 and 12, which were drawn with the window $[-5, 13]$ *by* $[-6, 6]$.

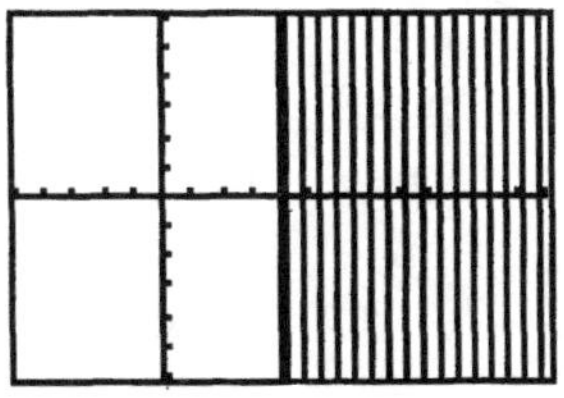

FIGURE 11 TI-82
`Shade(−7,7,3,4,13)`

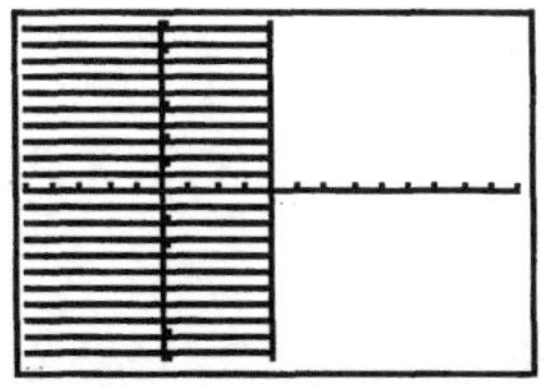

FIGURE 12 TI-83
`Shade(−7,7,−5,4,2,3)`

With the TI-83, there is a style icon (initially a line of four dots) to the left of each function in the `Y=` editor. To change the style icon of a function, move the cursor to the icon and press the ENTER key. Of the seven different icons, the icons ◥ and ◣ specify that the line be graphed and the points above and below the line, respectively, be crossed out. Figure 13 shows the `Y=` editor and the feasible set for Example 8.

(a)

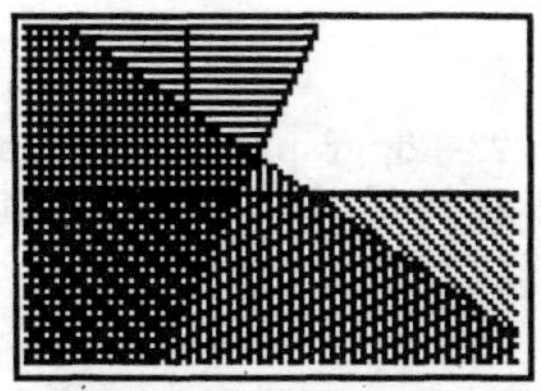

(b) $[-10, 20]$ *by* $[-10, 10]$

FIGURE 13

Practice Problems 2

1. Graph the inequality $3x - y \geq 3$.
2. Graph the feasible set for the system of inequalities

$$\begin{cases} x \geq 0,\ y \geq 0 \\ x + 2y \leq \ \ 4 \\ 4x - 4y \geq -4. \end{cases}$$

Exercises 2

In Exercises 1–4, state whether the inequality is true or false.

1. $2 \leq -3$

2. $-2 \leq 0$

3. $7 \leq 7$

4. $0 \geq \frac{1}{2}$

In Exercises 5–7, solve for x.

5. $2x - 5 \geq 3$

6. $3x - 7 \leq 2$

7. $-5x + 13 \leq -2$

8. Which of the following results from solving $-x+1 \leq 3$ for x?

(a) $x \leq 4$ (b) $x \leq 2$

(c) $x \geq -4$ (d) $x \geq -2$

In Exercises 9–14, put the linear inequality into standard form.

9. $2x + y \leq 5$ **10.** $-3x + y \geq 1$

11. $5x - \frac{1}{3}y \leq 6$ **12.** $\frac{1}{2}x - y \leq -1$

13. $4x \geq -3$ **14.** $-2x \leq 4$

In Exercises 15–22, determine whether or not the given point satisfies the given inequality.

15. $3x + 5y \leq 12$, $(2, 1)$ **16.** $-2x + y \geq 9$, $(3, 15)$

17. $y \geq -2x + 7$, $(3, 0)$ **18.** $y \leq \frac{1}{2}x + 3$, $(4, 6)$

19. $y \leq 3x - 4$, $(3, 5)$ **20.** $y \geq x$, $(-3, -2)$

21. $x \geq 5$, $(7, -2)$ **22.** $x \leq 7$, $(0, 0)$

In Exercises 23–26, graph the given inequality by crossing out (i.e., discarding) the points not satisfying the inequality.

23. $y \leq \frac{1}{3}x + 1$

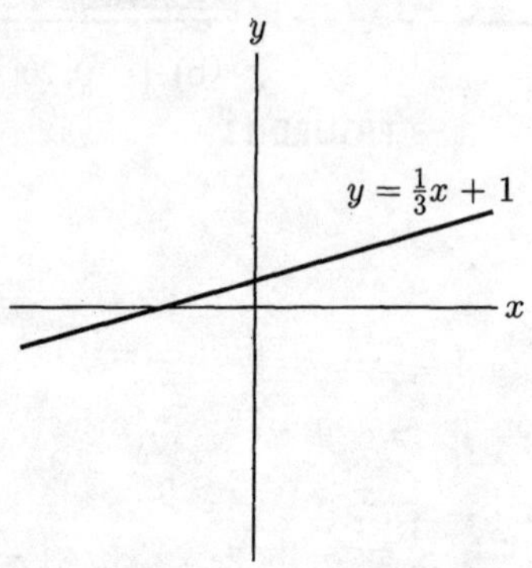

24. $y \geq -x + 1$

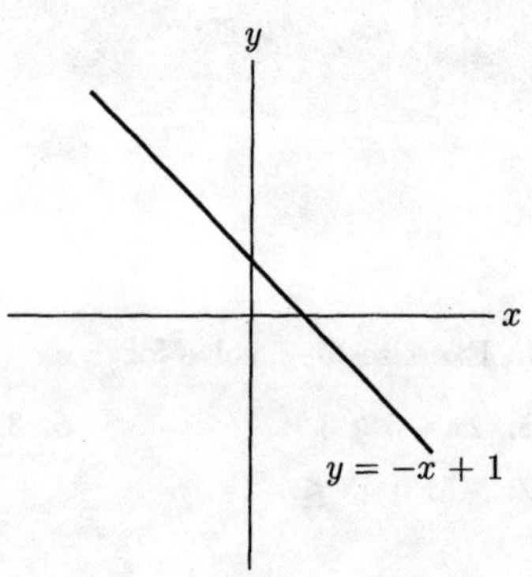

25. $x \geq 4$

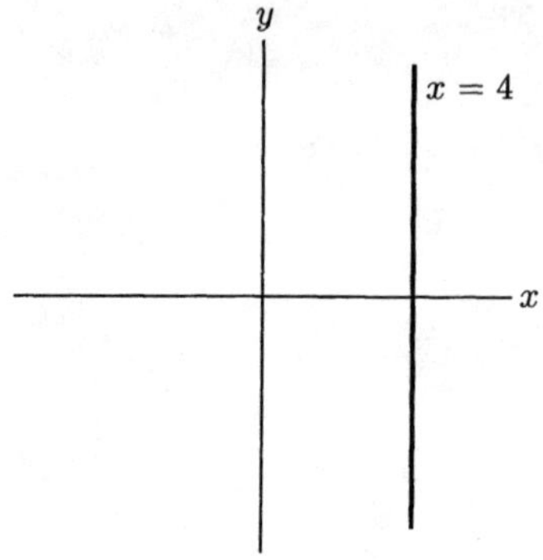

26. $y \leq 2$

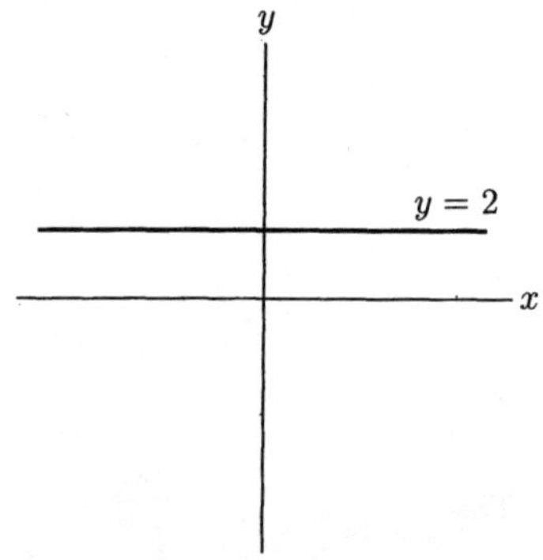

In Exercises 27–36, graph the given inequality.

27. $y \leq 2x + 1$ **28.** $y \geq -3x + 6$

29. $x \geq 2$ **30.** $x \geq 0$

31. $x + 4y \geq 12$ **32.** $4x - 4y \geq 8$

33. $4x - 5y + 25 \geq 0$ **34.** $.2y - x \geq .4$

35. $\frac{1}{2}x - \frac{1}{3}y \leq 1$ **36.** $3y - 3x \leq 2y + x + 1$

In Exercises 37–42, graph the feasible set for the system of inequalities.

37. $\begin{cases} y \leq 2x - 4 \\ y \geq 0 \end{cases}$ **38.** $\begin{cases} y \geq -\frac{1}{3}x + 1 \\ x \geq 0 \end{cases}$

39. $\begin{cases} x + 2y \geq 2 \\ 3x - y \geq 3 \end{cases}$ **40.** $\begin{cases} 3x + 6y \geq 24 \\ 3x + y \geq 6 \end{cases}$

41. $\begin{cases} x + 5y \leq 10 \\ x + y \leq 3 \\ x \geq 0,\ y \geq 0 \end{cases}$ **42.** $\begin{cases} x + 2y \geq 6 \\ x + y \geq 5 \\ x \geq 1 \end{cases}$

In Exercises 43–46, determine whether the given point is in the feasible set of this system of inequalities:

$$\begin{cases} 6x + 3y \leq 96 \\ x + y \leq 18 \\ 2x + 6y \leq 72 \\ x \geq 0,\ y \geq 0. \end{cases}$$

43. $(8, 7)$ **44.** $(14, 3)$

45. $(9, 10)$ **46.** $(16, 0)$

In Exercises 47–50, determine whether the given point is above or below the given line.

47. $y = 2x + 5$, $(3, 9)$ **48.** $3x - y = 4$, $(2, 3)$

49. $7 - 4x + 5y = 0$, $(0, 0)$ **50.** $x = 2y + 5$, $(6, 1)$

51. Give a system of inequalities for which the graph is a region between the pair of lines $8x - 4y - 4 = 0$ and $8x - 4y = 0$.

52. **(*From the GMAT exam*)** In the rectangular coordinate system of Fig. 14, the shaded region is bounded by straight lines. Which of the following is NOT an equation of one of the boundary lines?

(a) $x = 0$ (b) $y = 0$ (c) $x = 1$

(d) $x - y = 0$ (e) $x + 2y = 2$

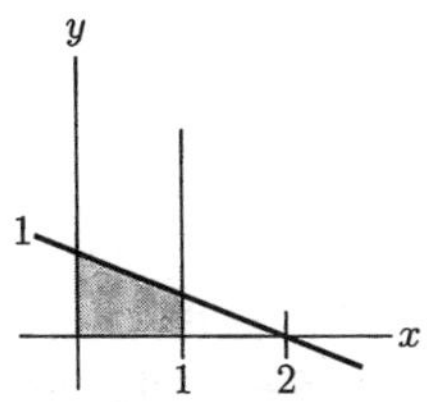

FIGURE 14

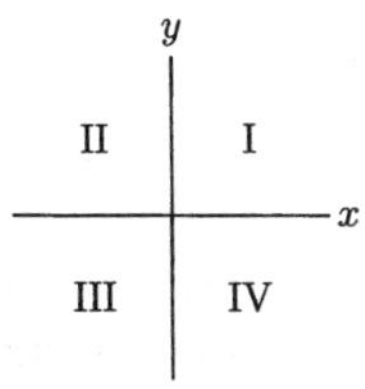

FIGURE 15

53. **(*From the GMAT exam*)** In the rectangular coordinate system of Fig. 15, which quadrant, if any, contains no point (x, y) that satisfies the inequality $2x - 3y \leq -6$?

(a) none (b) I (c) II (d) III (e) IV

Use a graphing calculator or a computer to solve Exercises 54–57.

54. Graph the line $4x - 2y = 7$.

(a) Locate the point on the line with x-coordinate 3.6.

(b) Does the point $(3.6, 3.5)$ lie above or below the line? Explain.

55. Graph the line $x + 2y = 11$.

(a) Locate the point on the line with x-coordinate 6.

(b) Does the point $(6, 2.6)$ lie above or below the line? Explain.

56. If your graphing calculator can shade feasible sets of inequalities, display the feasible set in Exercise 37.

57. If your graphing calculator can shade feasible sets of inequalities, display the feasible set in Exercise 40.

Solutions to Practice Problems 2

1. Linear inequalities are easiest to graph if they are first put into standard form. Subtract $3x$ from both sides and multiply by -1:

$$\begin{aligned} 3x - y &\geq 3 \\ -y &\geq -3x + 3 \\ y &\leq 3x - 3. \end{aligned}$$

Now graph the line $y = 3x - 3$ (Fig. 16). The graph of the inequality is the portion of the plane below and on the line ("$\leq$" corresponds to below), so throw away (that is, cross out) the portion above the line.

2. Begin by putting the linear inequalities into standard form and then graphing them all on the same coordinate system.

$$\begin{cases} x \geq 0,\ y \geq 0 \\ x + 2y \leq 4 \\ 4x - 4y \geq -4 \end{cases} \quad \text{has standard form} \quad \begin{cases} x \geq 0,\ y \geq 0 \\ y \leq -\frac{1}{2}x + 2 \\ y \leq x + 1. \end{cases}$$

A good procedure to follow is to graph all of the linear equations and then cross out the regions to be thrown away one at a time (Fig. 17). The inequalities $x \geq 0$ and $y \geq 0$ arise frequently in applications. The first has the form $x \geq a$, where $a = 0$, and the second has the form $y \geq mx + b$, where $m = 0$ and $b = 0$. To graph them, just cross out all points to the left of the y-axis and all points below the x-axis, respectively.

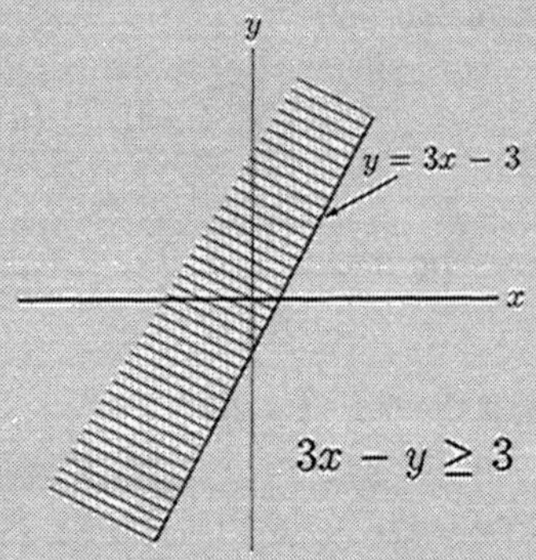

FIGURE 16

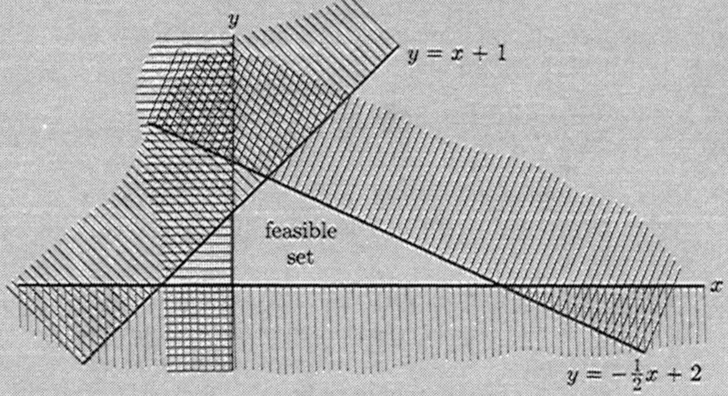

FIGURE 17

1.3 The Intersection Point of a Pair of Lines

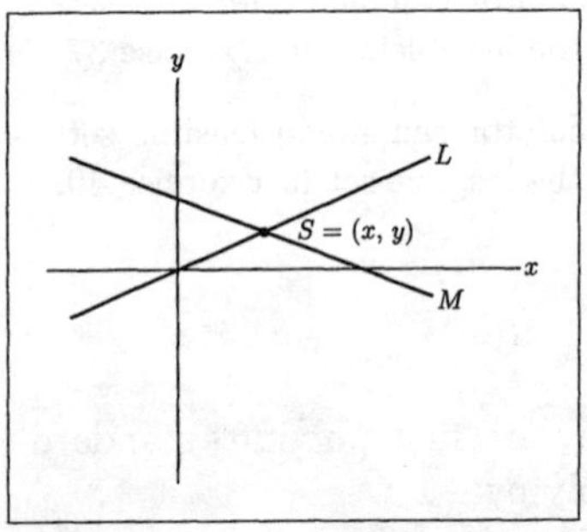

FIGURE 1

Suppose that we are given a pair of intersecting straight lines L and M. Let us consider the problem of determining the coordinates of the point of intersection $S = (x, y)$ (see Fig. 1). We may as well assume that the equations of L and M are given in standard form. First, let us assume that both lines are in the first standard form—that is, that the equations are

$$L\colon\ y = mx + b, \qquad M\colon\ y = nx + c.$$

Since the point S is on both lines, its coordinates satisfy both equations. In particular, we have two expressions for its y-coordinate:

$$y = mx + b = nx + c.$$

The last equality gives an equation from which x can easily be determined. Then the value of y can be determined as $mx + b$. Let us see how this works in a particular example.

EXAMPLE 1 Find the point of intersection of the two lines $y = 2x - 3$, $y = x + 1$.

Solution To find the x-coordinate of the point of intersection, equate the two expressions for y and solve for x:

$$\begin{aligned} 2x - 3 &= x + 1 \\ 2x - x &= 1 + 3 \\ x &= 4. \end{aligned}$$

To find the value of y, set $x = 4$ in either equation, say the first. Then

$$y = 2 \cdot 4 - 3 = 5.$$

So the point of intersection is $(4, 5)$. ▮

EXAMPLE 2 Find the point of intersection of the two lines $x + 2y = 6$, $5x + 2y = 18$.

Solution To use the method described, the equations must be in standard form. Solving both equations for y, we get the standard forms

$$\begin{aligned} y &= -\tfrac{1}{2}x + 3 \\ y &= -\tfrac{5}{2}x + 9. \end{aligned}$$

Equating the expressions for y gives

$$-\tfrac{1}{2}x + 3 = -\tfrac{5}{2}x + 9$$
$$\tfrac{5}{2}x - \tfrac{1}{2}x = 9 - 3$$
$$2x = 6$$
$$x = 3.$$

Setting $x = 3$ in the first equation gives

$$y = -\tfrac{1}{2}(3) + 3 = \tfrac{3}{2}.$$

So the intersection point is $\left(3, \tfrac{3}{2}\right)$. ▮

The above method works when both equations have the first standard form. In case one equation has the standard form $x = a$, things are much simpler. The value of x is then given directly without any work, namely $x = a$. The value of y can be found by substituting a for x in the other equation.

EXAMPLE 3 Find the point of intersection of the lines $y = 2x - 1$, $x = 2$.

Solution The x-coordinate of the intersection point is 2, and the y-coordinate is $y = 2 \cdot 2 - 1 = 3$. Therefore, the intersection point is $(2, 3)$. ▮

The method introduced above may be used to solve systems of two equations in two variables.

EXAMPLE 4 Solve the following system of linear equations:

$$\begin{cases} 2x + 3y = 7 \\ 4x - 2y = 9. \end{cases}$$

Solution First convert the equations to standard form:

$$\begin{aligned} 2x + 3y &= 7 \\ 3y &= -2x + 7 \\ y &= -\tfrac{2}{3}x + \tfrac{7}{3}; \end{aligned} \qquad \begin{aligned} 4x - 2y &= 9 \\ -2y &= -4x + 9 \\ y &= 2x - \tfrac{9}{2}. \end{aligned}$$

Now equate the two expressions for y:

$$2x - \tfrac{9}{2} = -\tfrac{2}{3}x + \tfrac{7}{3}$$
$$\tfrac{8}{3}x = \tfrac{7}{3} + \tfrac{9}{2} = \tfrac{14}{6} + \tfrac{27}{6} = \tfrac{41}{6}$$
$$x = \tfrac{3}{8} \cdot \tfrac{41}{6} = \tfrac{41}{16}$$
$$y = 2x - \tfrac{9}{2} = 2\left(\tfrac{41}{16}\right) - \tfrac{9}{2} = \tfrac{5}{8}.$$

So the solution of the given system is $x = \tfrac{41}{16}$, $y = \tfrac{5}{8}$. ▮

Supply and Demand Curves Let p denote the price of a commodity and q the quantity. Economists study two sorts of graphs that express relationships between p and q. To describe these graphs, let us plot price along the horizontal axis and quantity along the vertical axis. The first graph relating p and q is called a *supply curve* (Fig. 2) and expresses the relationship between p and q from a manufacturer's point of view. For every price p the supply curve specifies the quantity q the manufacturer is willing to produce at the market price p. The higher the price, the more the manufacturer is willing to supply. So supply curves rise when viewed from left to right.

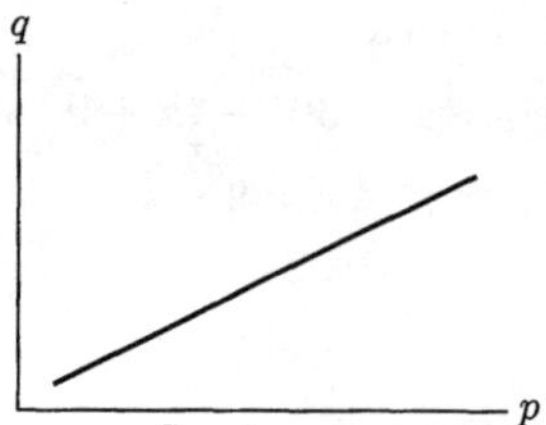

Supply curve

FIGURE 2

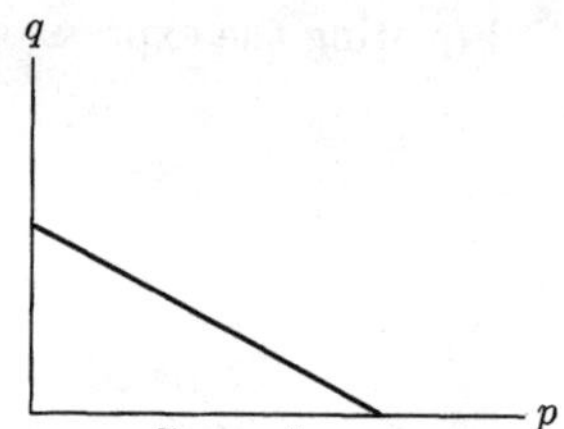

Demand curve

FIGURE 3

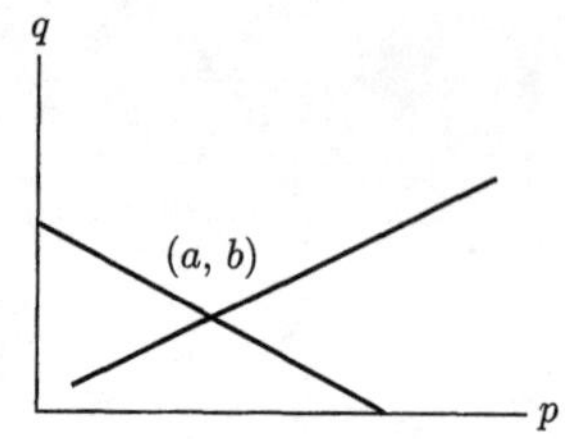

FIGURE 4

The second curve relating p and q is called a *demand curve* (Fig. 3) and expresses the relationship between p and q from the consumer's viewpoint. For each price p, the demand curve expresses the quantity q consumers will buy if the market price is p. The higher the price, the less consumers will buy. So demand curves fall when viewed from left to right.

Suppose that the supply and demand curves for a commodity are drawn on a single coordinate system (Fig. 4). The intersection point (a, b) of the two curves has an economic significance: Prices will stabilize at a dollars per unit and the quantity produced will be b units.

EXAMPLE 5 Suppose that the supply curve for a certain commodity is the straight line whose equation is $q = 5000p - 10{,}000$ (p in dollars). Suppose that the demand curve for the same commodity is the straight line whose equation is $q = -2000p + 11{,}000$. Determine both the price at which the commodity will sell and the quantity produced.

Solution We must solve the system of linear equations

$$\begin{cases} q = 5000p - 10{,}000 \\ q = -2000p + 11{,}000. \end{cases}$$

$$\begin{aligned} 5000p - 10{,}000 &= -2000p + 11{,}000 \\ 7000p &= 21{,}000 \\ p &= 3 \\ q &= 5000(3) - 10{,}000 = 5000. \end{aligned}$$

Thus the commodity will sell for \$3, and 5000 units will be produced. ▮

GC Graphing utilities have commands that find the intersection points of a pair of lines. Figure 5 shows the result of solving Example 4 on the TI-83 with the intersect command of the CALC menu. Since the x-coordinate of the intersection point is assigned to ANS, the x-intercept can be converted to a fraction by pressing [MATH] **1** [ENTER] from the home screen.

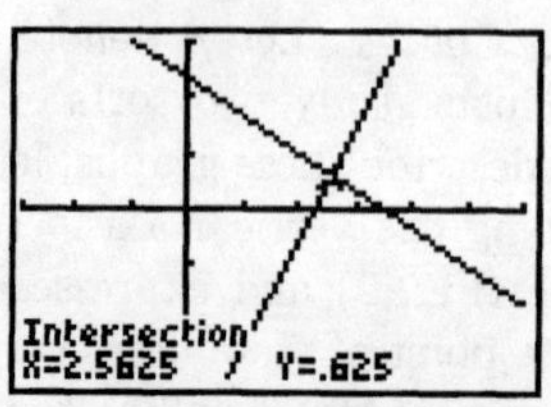

FIGURE 5 $[-3, 6]$ *by* $[-3, 3]$

Practice Problems 3

Figure 6 shows the feasible set of a system of linear inequalities; its four vertices are labeled A, B, C, and D.

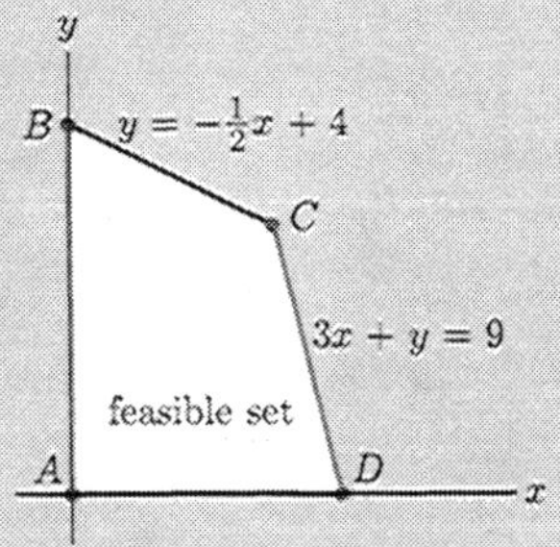

FIGURE 6

1. Use the method of this section to find the coordinates of the point C.
2. Determine the coordinates of the points A and B by inspection.
3. Find the coordinates of the point D.

Exercises 3

In Exercises 1–6, find the point of intersection of the given pair of straight lines.

1. $\begin{cases} y = 4x - 5 \\ y = -2x + 7 \end{cases}$ **2.** $\begin{cases} y = 3x - 15 \\ y = -2x + 10 \end{cases}$

3. $\begin{cases} x - 4y = -2 \\ x + 2y = 4 \end{cases}$ **4.** $\begin{cases} 2x - 3y = 3 \\ y = 3 \end{cases}$

5. $\begin{cases} y = \frac{1}{3}x - 1 \\ x = 12 \end{cases}$ **6.** $\begin{cases} 2x - 3y = 3 \\ x = 6 \end{cases}$

7. Does $(6, 4)$ satisfy the following system of linear equations?

$$\begin{cases} x - 3y = -6 \\ 3x - 2y = 10 \end{cases}$$

8. Does $(12, 4)$ satisfy the following system of linear equations?

$$\begin{cases} y = \frac{1}{3}x - 1 \\ x = 12 \end{cases}$$

In Exercises 9–12, solve the systems of linear equations.

9. $\begin{cases} 2x + y = 7 \\ x - y = 3 \end{cases}$ **10.** $\begin{cases} x + 2y = 4 \\ \frac{1}{2}x + \frac{1}{2}y = 3 \end{cases}$

11. $\begin{cases} 5x - 2y = 1 \\ 2x + y = -4 \end{cases}$ **12.** $\begin{cases} x + 2y = 6 \\ x - \frac{1}{3}y = 4 \end{cases}$

In Exercises 13–16, find the coordinates of the vertices of the feasible sets.

13.

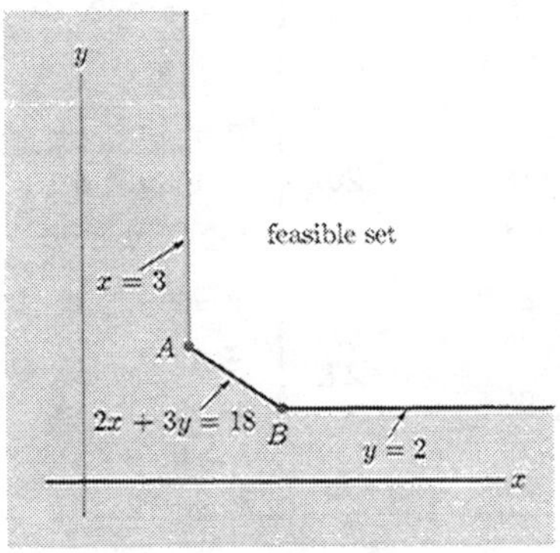

14.

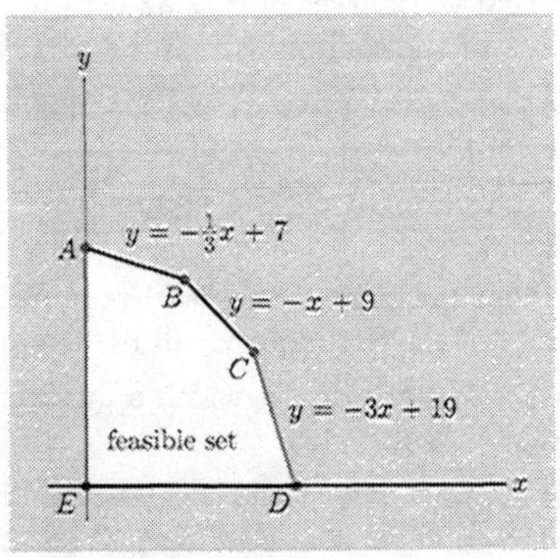

15.

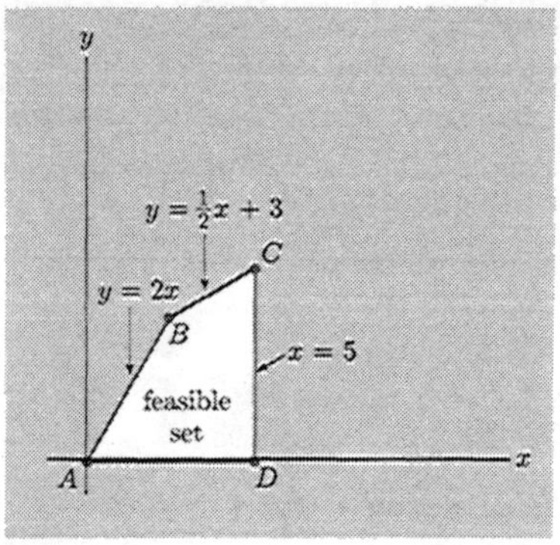

16.

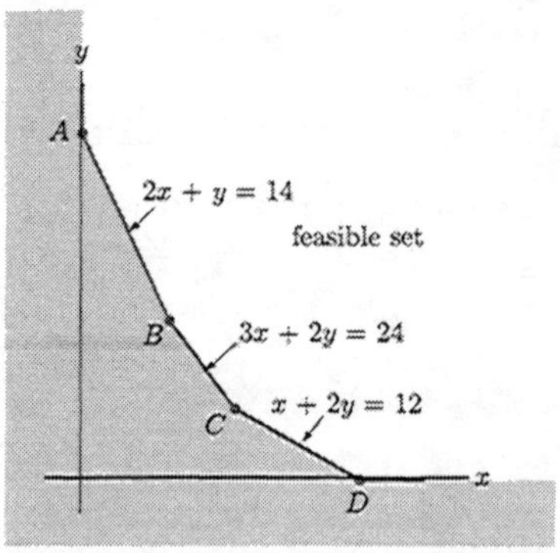

In Exercises 17–22, graph the feasible set for the system of inequalities and find the coordinates of the vertices.

17. $\begin{cases} 2y - x \le 6 \\ x + 2y \ge 10 \\ x \le 6 \end{cases}$ **18.** $\begin{cases} 2x + y \ge 10 \\ x \ge 2 \\ y \ge 2 \end{cases}$

19. $\begin{cases} x + 3y \le 18 \\ 2x + y \le 16 \\ x \ge 0,\ y \ge 0 \end{cases}$ **20.** $\begin{cases} 5x + 2y \ge 14 \\ x + 3y \ge 8 \\ x \ge 0,\ y \ge 0 \end{cases}$

21. $\begin{cases} 4x + y \ge 8 \\ x + y \ge 5 \\ x + 3y \ge 9 \\ x \ge 0,\ y \ge 0 \end{cases}$ **22.** $\begin{cases} x + 4y \le 28 \\ x + y \le 10 \\ 3x + y \le 24 \\ x \ge 0,\ y \ge 0 \end{cases}$

23. The supply curve for a certain commodity is $q = 10{,}000p - 500$.

(a) Suppose that the market price is \$2 per unit. How many units of the commodity will be sold?

(b) What is the highest price at which a quantity of 0 will be supplied?

24. The demand curve for a certain commodity is $q = -1000p + 32{,}500$.

(a) Suppose that the market price is \$1 per unit. How many units of the commodity will be demanded?

(b) What is the lowest price at which a quantity of 0 will be demanded?

25. Suppose that supply and demand are described by the supply and demand curves of Exercises 23 and 24.

(a) Determine the equilibrium selling price of the commodity.

(b) Determine the number of units that will be produced at that price.

26. The x- and y-intercepts of line L_1 are 1 unit to the left and 2 units below the x- and y-intercepts of line L_2, respectively. The equation for L_1 is $y = 3x + 6$. Find the point of intersection of L_1 and L_2.

27. A plant supervisor must apportion her 40-hour work-week between hours working on the assembly line and hours supervising the work of others. She is paid \$12 per hour for working and \$15 per hour for supervising. If her earnings for a certain week are \$504, how much time does she spend on each task?

28. ***(From a GMAT exam)*** An optometrist charges \$150 per pair for soft contact lenses and \$85 per pair for hard contact lenses. Last week she sold five more pairs of soft lenses than hard lenses. If her total sales for pairs of contact lenses last week were \$1690, what was the total number of pairs of contact lenses that she sold?

(a) 11 (b) 13 (c) 15 (d) 17 (e) 19

29. ***(From a GMAT exam)*** In a weight-lifting competition, the total weight of Joe's two lifts was 750 pounds. If twice the weight of his first lift was 300 pounds more than the weight of his second lift, what was the weight, in pounds, of his *first* lift?

(a) 225 (b) 275 (c) 325 (d) 350 (e) 400

Exercises 30–35 require the use of a graphing calculator or a computer.

In Exercises 30–33, graph the lines and use the INTERSECT command to estimate the point of intersection to two decimal places.

30. $\begin{cases} y = -\frac{2}{3}x + 4.5 \\ y = 2x \end{cases}$ **31.** $\begin{cases} y = .25x + 1.3 \\ y = -.5x + 4.1 \end{cases}$

32. $\begin{cases} 2x + 3y = 5 \\ -4x + 5y = 1 \end{cases}$ **33.** $\begin{cases} x - 4y = -5 \\ 3x - 2y = 4.2 \end{cases}$

In Exercises 34 and 35, (a) draw the straight lines corresponding to the inequalities (use a window such as `ZDecimal` that gives nice values of the coordinates of points); (b) find the coordinates of the vertices of the feasible set; (c) locate the given point with the cursor; (d) determine whether or not the given point lies in the feasible set.

34. $\begin{cases} 2x + y \ge 5 \\ x - 2y \le 0 \end{cases}$ (2.2, 1.4)

35. $\begin{cases} -x + 3y \ge 3 \\ .4x + y \ge 3.2 \end{cases}$ (3.2, 2)

Solutions to Practice Problems 3

1. Point C is the point of intersection of the lines with equations $y = -\frac{1}{2}x + 4$ and $3x + y = 9$. To use the method of this section, the second equation must be put into its standard form $y = -3x + 9$. Now equate the two expressions for y and solve.

$$\begin{aligned} -\tfrac{1}{2}x + 4 &= -3x + 9 \\ \tfrac{5}{2}x &= 5 \\ x &= \tfrac{2}{5} \cdot 5 = 2 \\ y &= -\tfrac{1}{2}(2) + 4 = 3. \end{aligned}$$

Therefore, $C = (2, 3)$.

2. $A = (0, 0)$, since the point A is the origin. $B = (0, 4)$, since it is the y-intercept of the line with equation $y = -\frac{1}{2}x + 4$.

3. D is the x-intercept of the line $3x + y = 9$. Its first coordinate is found by setting $y = 0$ and solving for x.

$$\begin{aligned} 3x + (0) &= 9 \\ x &= 3. \end{aligned}$$

Therefore, $D = (3, 0)$.

1.4 The Slope of a Straight Line

As we have seen, any linear equation can be put into one of the two standard forms, $y = mx + b$ or $x = a$. In this section, let us exclude linear equations whose standard form is of the latter type. *Geometrically, this means that we will consider only nonvertical lines.*

Suppose that we are given a nonvertical line L whose equation is $y = mx + b$. The number m is called the *slope of L*. That is, the slope is the coefficient of x in the standard form of the equation of the line.

EXAMPLE 1 Find the slopes of the lines having the following equations.

(a) $y = 2x + 1$ (b) $y = -\frac{3}{4}x + 2$ (c) $y = 3$ (d) $-8x + 2y = 4$

Solution (a) $m = 2$.

(b) $m = -\frac{3}{4}$.

(c) When we write the equation in the form $y = 0 \cdot x + 3$, we see that $m = 0$.

(d) First, we put the equation in standard form:

$$\begin{aligned} -8x + 2y &= 4 \\ 2y &= 8x + 4 \\ y &= 4x + 2. \end{aligned}$$

Thus $m = 4$. ▮

The definition of the slope is given in terms of the standard form of the equation of the line. Let us give an alternative definition.

> ***Geometric Definition of Slope*** Let L be a line passing through the points (x_1, y_1) and (x_2, y_2). Then the slope of L is given by the formula
> $$m = \frac{y_2 - y_1}{x_2 - x_1}. \qquad (1)$$

That is, the slope is the difference in the y-coordinates divided by the difference in the x-coordinates, with both differences formed in the same order.

Before proving this definition equivalent to the first one given, let us show how it can be used.

EXAMPLE 2 Find the slope of the line passing through the points $(1, 3)$ and $(4, 6)$.

Solution We have

$$m = \frac{\text{[difference in } y\text{-coordinates]}}{\text{[difference in } x\text{-coordinates]}} = \frac{6-3}{4-1} = \frac{3}{3} = 1.$$

Thus $m = 1$. [Note that if we reverse the order of the points and use formula (1) to compute the slope, then we get

$$m = \frac{3-6}{1-4} = \frac{-3}{-3} = 1,$$

which is the same answer. The order of the points is immaterial. The important concern is to make sure that the differences in the x- and y-coordinates are formed in the same order.] ∎

The slope of a line does not depend on which pair of points is chosen as (x_1, y_1) and (x_2, y_2). Consider the line $y = 4x - 3$ and two points $(1, 1)$ and $(3, 9)$, which are on the line. Using these two points, we calculate the slope to be

$$m = \frac{9-1}{3-1} = \frac{8}{2} = 4.$$

Now let's choose two other points on the line, say $(2, 5)$ and $(-1, -7)$, and use these points to determine m. We obtain

$$m = \frac{-7-5}{-1-2} = \frac{-12}{-3} = 4.$$

The two pairs of points give the same slope.

Justification of Formula (1) Since (x_1, y_1) and (x_2, y_2) are both on the line, both points satisfy the equation of the line, which has the form $y = mx + b$. Thus

$$y_2 = mx_2 + b$$
$$y_1 = mx_1 + b.$$

Subtracting these two equations gives

$$y_2 - y_1 = mx_2 - mx_1 = m(x_2 - x_1).$$

Dividing by $x_2 - x_1$, we have

$$m = \frac{y_2 - y_1}{x_2 - x_1},$$

which is formula (1). So the two definitions of slope lead to the same number.

Let us now study four of the most important properties of the slope of a straight line. We begin with the *steepness property*, since it provides us with a geometric interpretation for the number m.

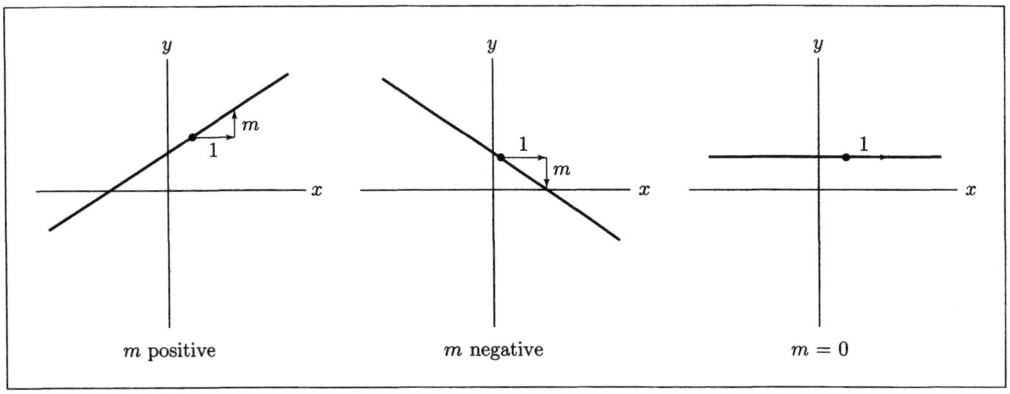

FIGURE 1

> ***Steepness Property*** Let the line L have slope m. If we start at any point on the line and move 1 unit to the right, then we must move m units vertically in order to return to the line (Fig. 1). (Of course, if m is positive, then we move up; and if m is negative, we move down.)

EXAMPLE 3 Illustrate the steepness property for each of the lines.

(a) $y = 2x + 1$ (b) $y = -\frac{3}{4}x + 2$ (c) $y = 3$

Solution (a) Here $m = 2$. So starting from any point on the line, proceeding 1 unit to the right, we must go 2 units up to return to the line (Fig. 2).

(b) Here $m = -\frac{3}{4}$. So starting from any point on the line, proceeding 1 unit to the right, we must go $\frac{3}{4}$ unit down to return to the line (Fig. 3).

(c) Here $m = 0$. So going 1 unit to the right requires going 0 units vertically to return to the line (Fig. 4). ▮

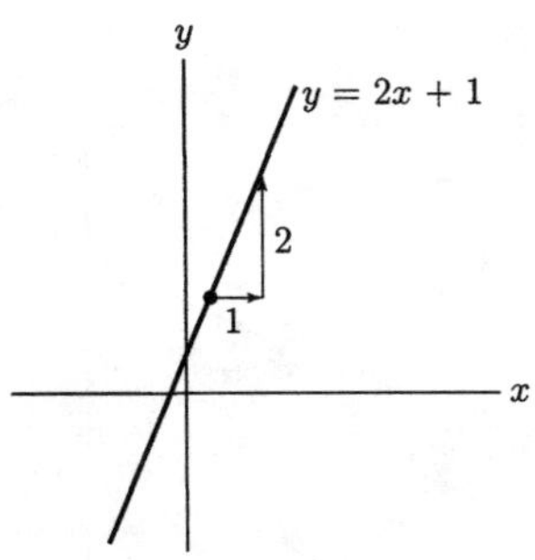

FIGURE 2

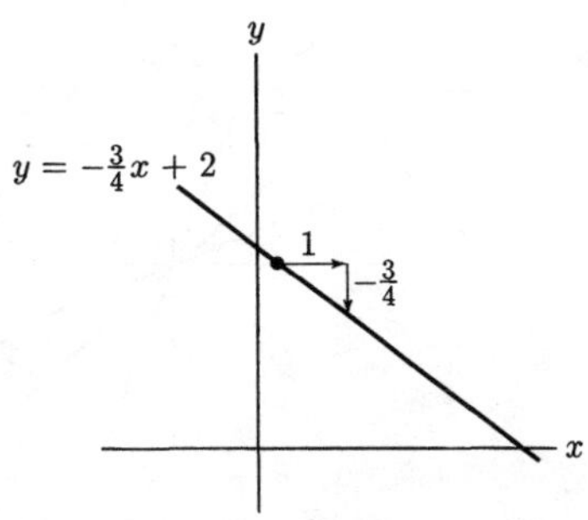

FIGURE 3

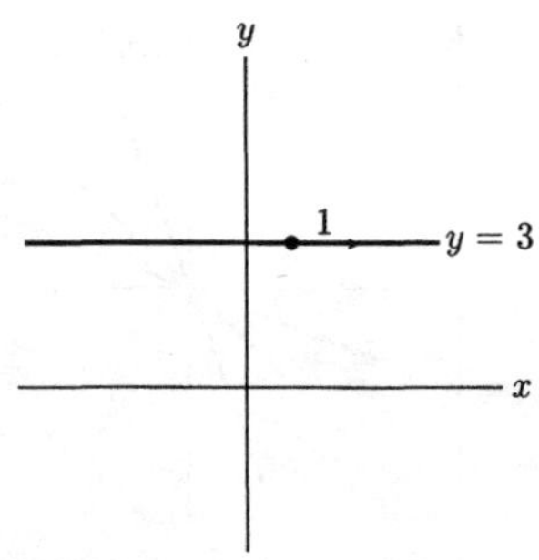

FIGURE 4

In the next example we introduce a new method for graphing a linear equation. This method relies on the steepness property and is much more efficient than finding two points on the line (e.g., the two intercepts).

EXAMPLE 4 Use the steepness property to draw the graph of $y = \frac{1}{2}x + \frac{3}{2}$.

Solution The y-intercept is $\left(0, \frac{3}{2}\right)$, as we read from the equation. We can find another point on the line using the steepness property. Start at $\left(0, \frac{3}{2}\right)$. Go 1 unit to the right. Since the slope is $\frac{1}{2}$, we must move vertically $\frac{1}{2}$ unit to return to the line. But this locates a second point on the line. So we draw the line through the two points. The entire procedure is illustrated in Fig. 5. ▮

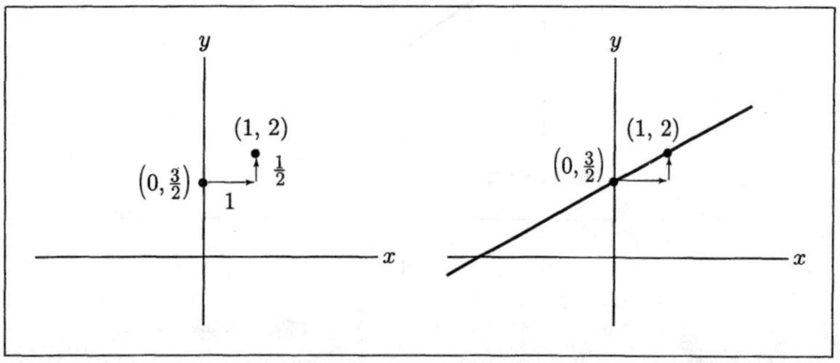

FIGURE 5

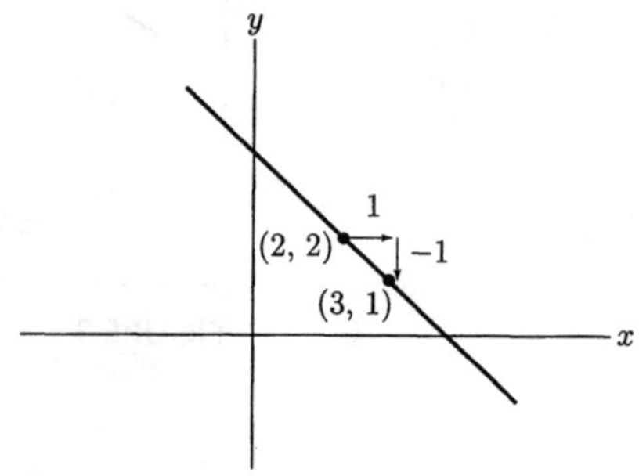

FIGURE 6

Actually, to use the steepness property to graph an equation, all that is needed is the slope plus *any* point (not necessarily the y-intercept).

▮ EXAMPLE 5 Graph the line of slope -1 and passing through the point $(2, 2)$.

Solution Start at $(2, 2)$, move 1 unit to the right and then -1 unit vertically, that is, 1 unit down. The line through $(2, 2)$ and the resulting point is the desired line (see Fig. 6). ▮

Slope measures the *steepness* of a line. Namely, the slope of a line tells whether it is rising or falling, and how fast. Specifically, lines of positive slope rise as we move from left to right. Lines of negative slope fall, and lines of zero slope stay level. The larger the magnitude of the slope, the steeper the ascent or descent. These facts are directly implied by the steepness property (see Fig. 7).

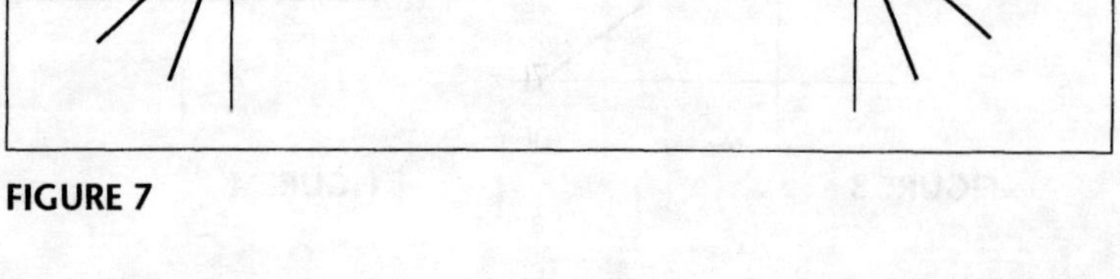

FIGURE 7

FIGURE 8

Justification of the Steepness Property Consider a line with equation $y = mx + b$, and let (x_1, y_1) be any point on the line. If we start from this point and move 1 unit to the right, the first coordinate of the new point is $x_1 + 1$ since the x-coordinate is increased by 1. Now go far enough vertically to return to the line. Denote the y-coordinate of this new point by y_2 (see Fig. 8). We must show that to get y_2, we add m to y_1. That is, $y_2 = y_1 + m$. By equation (1), we can compute m as

$$m = \frac{[\text{difference in } y\text{-coordinates}]}{[\text{difference in } x\text{-coordinates}]} = \frac{y_2 - y_1}{1} = y_2 - y_1.$$

In other words, $y_2 = y_1 + m$, which is what we desired to show. ▮

Often the slopes of the straight lines that occur in applications have interesting and significant interpretations. An application in the field of economics is illustrated in the next example.

▮ EXAMPLE 6 (*Economics*) Suppose a manufacturer finds that the cost y of producing x units of a certain commodity is given by the formula $y = 2x + 5000$. What interpretation can be given to the slope of the graph of this equation?

Solution Suppose that the firm is producing at a certain level and increases production by 1 unit. That is, x is increased by 1 unit. By the steepness property, the value of y then increases by 2, which is the slope of the line whose equation is $y = 2x + 5000$. Thus *each additional unit of production costs \$2.* (The graph of $y = 2x + 5000$ is called a *cost curve.* It relates the size of production to total cost. The graph is a straight line, and economists call its slope the *marginal cost of production.* The y-coordinate of the y-intercept is called the *fixed cost.* In this case the fixed cost is \$5000, and it includes costs such as rent and insurance which are incurred even if no units are produced.) ▮

In applied problems having time as a variable, the letter t is often used in place of the letter x. If so, straight lines have equations of the form $y = mt + b$ and are graphed on a ty-coordinate system.

▮ EXAMPLE 7 (*Depreciation*) The federal government allows an income tax deduction for the decrease in value (or *depreciation*) of capital assets (such as buildings and equipment). One method of calculating the depreciation is to take equal amounts over the expected lifetime of the asset. This method is called *straight-line depreciation.* For tax purposes the value V of an asset t years after purchase is figured according to the equation $V = -100{,}000t + 700{,}000$. The expected life of the asset is 5 years.

(a) How much did the asset originally cost?

(b) What is the annual deduction for depreciation?

(c) What is the salvage value of the asset? (That is, what is the value of the asset after 5 years?)

Solution (a) The original cost is the value of V at $t = 0$, namely

$$V = -100{,}000(0) + 700{,}000 = 700{,}000.$$

That is, the asset originally cost \$700,000.

(b) By the steepness property, each increase of 1 in t causes a decrease in V of 100,000. That is, the value is decreasing by \$100,000 per year. So the depreciation deduction is \$100,000 each year.

(c) After 5 years, the value V is given by

$$V = -100{,}000(5) + 700{,}000 = 200{,}000.$$

The salvage value is \$200,000. ▮

We have see in Example 5 how to sketch a straight line when given its slope and one point on it. Let us now see how to find the equation of the line from this data.

> ***Point-Slope Formula*** The equation of the straight line passing through (x_1, y_1) and having slope m is given by $y - y_1 = m(x - x_1)$.

EXAMPLE 8 Find the equation of the line that passes through $(2, 3)$ and has slope $\frac{1}{2}$.

Solution Here $x_1 = 2$, $y_1 = 3$, and $m = \frac{1}{2}$. So the equation is

$$y - 3 = \tfrac{1}{2}(x - 2)$$

$$y - 3 = \tfrac{1}{2}x - 1$$

$$y = \tfrac{1}{2}x + 2.$$

EXAMPLE 9 Find the equation of the line through the points $(3, 1)$ and $(6, 0)$.

Solution We can compute the slope from equation (1):

$$[\text{slope}] = \frac{[\text{difference in } y\text{-coordinates}]}{[\text{difference in } x\text{-coordinates}]} = \frac{1-0}{3-6} = -\frac{1}{3}.$$

Now we can determine the equation from the point-slope formula with $(x_1, y_1) = (3, 1)$ and $m = -\frac{1}{3}$:

$$y - 1 = -\tfrac{1}{3}(x - 3)$$

$$y = -\tfrac{1}{3}x + 2.$$

[*Question*: What would the equation be if we had chosen $(x_1, y_1) = (6, 0)$?]

EXAMPLE 10 For each dollar in monthly advertising expenditure, a store experiences a 6-dollar increase in sales revenue. Even without advertising, the store has $30,000 in sales revenue per month. Let x be the number of dollars of advertising expenditures per month and let y be the number of dollars in sales revenue per month.

(a) Find the equation of the line that expresses the relationship between x and y.

(b) If the store spends $10,000 in advertising, what will be the sales revenue for the month?

(c) How much would the store have to spend on advertising to attain $150,000 in sales revenue for the month?

Solution (a) The steepness property tells us that the line has slope $m = 6$. Since $x = 0$ (no advertising expenditures) yields $y = \$30{,}000$, the y-intercept of the line is (0, 30,000). Therefore, the standard form of the equation of the line is

$$y = 6x + 30{,}000.$$

(b) If $x = 10{,}000$, then $y = 6(10{,}000) + 30{,}000$. Therefore, the sales revenue for the month will be $90,000.

(c) We are given that $y = 150{,}000$, and we must find the value of x for which

$$150{,}000 = 6x + 30{,}000.$$

Solving for x, we obtain $6x = 120{,}000$, and hence $x = \$20{,}000$. To attain $150,000 in sales revenue, the store should invest $20,000 in advertising.

Verification of the Point-Slope Formula Let (x, y) be any point on the line passing through the point (x_1, y_1) and having slope m. Then, by equation (1), we have

$$m = \frac{y - y_1}{x - x_1}.$$

Multiplying through by $x - x_1$ gives

$$y - y_1 = m(x - x_1). \tag{2}$$

Thus every point (x, y) on the line satisfies equation (2). So (2) gives the equation of the line through (x_1, y_1) and having slope m.

The next property of slope relates the slopes of two perpendicular lines.

> ***Perpendicular Property*** When two lines are perpendicular, their slopes are negative reciprocals of one another. That is, if two lines with slopes m and n are perpendicular to one another, then[1]
>
> $$m = -\frac{1}{n}.$$
>
> Conversely, if two lines have slopes that are negative reciprocals of one another, they are perpendicular.

A proof of the perpendicular property is outlined in Exercise 85. Let us show how it can be used to help find equations of lines.

EXAMPLE 11 Find the equation of the line perpendicular to the graph of $y = 2x - 5$ and passing through $(1, 2)$.

Solution The slope of the graph of $y = 2x - 5$ is 2. By the perpendicular property, the slope of a line perpendicular to it is $-\frac{1}{2}$. If a line has slope $-\frac{1}{2}$ and passes through $(1, 2)$, it has the equation

$$y - 2 = -\tfrac{1}{2}(x - 1) \quad \text{or} \quad y = -\tfrac{1}{2}x + \tfrac{5}{2}$$

(by the point-slope formula). ∎

The final property of slopes gives the relationship between slopes of parallel lines. A proof is outlined in Exercise 84.

> ***Parallel Property*** Parallel lines have the same slope. Conversely, if two lines have the same slope, they are parallel.

EXAMPLE 12 Find the equation of the line through $(2, 0)$ and parallel to the line whose equation is $y = \frac{1}{3}x - 11$.

Solution The slope of the line having equation $y = \frac{1}{3}x - 11$ is $\frac{1}{3}$. Therefore, any line parallel to it also has slope $\frac{1}{3}$. Thus the desired line passes through $(2, 0)$ and has slope $\frac{1}{3}$, so its equation is

$$y - 0 = \tfrac{1}{3}(x - 2) \quad \text{or} \quad y = \tfrac{1}{3}x - \tfrac{2}{3}. \ ∎$$

[1] If $n = 0$, this formula does not say anything, since $1/0$ is undefined. However, in this case, one line is horizontal and one is vertical, the vertical one having an undefined slope.

GC Figure 9 shows the lines $y = x$ and $y = -x$ drawn on a TI calculator with the windows `ZStandard` and `ZSquare`. With the `ZSquare` window, the lines visually bisect the quadrants and appear perpendicular. Also, with the `ZSquare` window, one unit on the x-axis has the same length as one unit on the y-axis. We say that the `ZSquare` window has "true aspect." In general, a window will approximately exhibit true aspect when $(\texttt{Ymax} - \texttt{Ymin}) \approx \frac{2}{3}(\texttt{Xmax} - \texttt{Xmin})$. Only with true aspect do two lines whose slopes are negative reciprocals of each other actually look perpendicular.

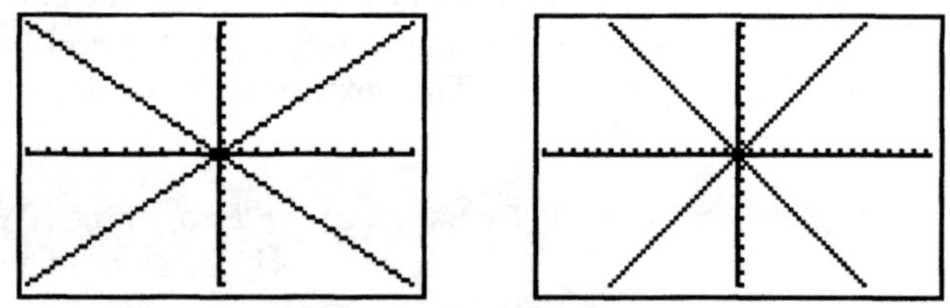

(a) `ZStandard` (b) `ZSquare`

FIGURE 9

Practice Problems 4

Suppose that the revenue y from selling x units of a certain commodity is given by the formula $y = 4x$. (Revenue is the amount of money received from the sale of the commodity.)

1. What interpretation can be given to the slope of the graph of this equation?
2. (See Example 6.) Find the coordinates of the point of intersection of $y = 4x$ and $y = 2x + 5000$.
3. What interpretation can be given to the value of the x-coordinate of the point found in Problem 2?

EXERCISES 4

In Exercises 1–4, find the slope of the line having the given equation.

1. $y = \frac{2}{3}x + 7$ **2.** $y = -4$

3. $y - 3 = 5(x + 4)$ **4.** $7x + 5y = 10$

In Exercises 5–8, plot each pair of points, draw the straight line between them, and find the slope.

5. $(3, 4), (7, 9)$ **6.** $(-2, 1), (3, -3)$

7. $(0, 0), (5, 4)$ **8.** $(4, 17), (-2, 17)$

9. What is the slope of any line parallel to the y-axis?

10. Why doesn't it make sense to talk about the slope of the line between the two points $(2, 3)$ and $(2, -1)$?

In Exercises 11–14, graph the given linear equation by beginning at the y-intercept, moving 1 unit to the right and m units in the y-direction.

11. $y = -2x + 1$ **12.** $y = 4x - 2$

13. $y = 3x$ **14.** $y = -2$

In Exercises 15–22, find the equation of line L.

15.

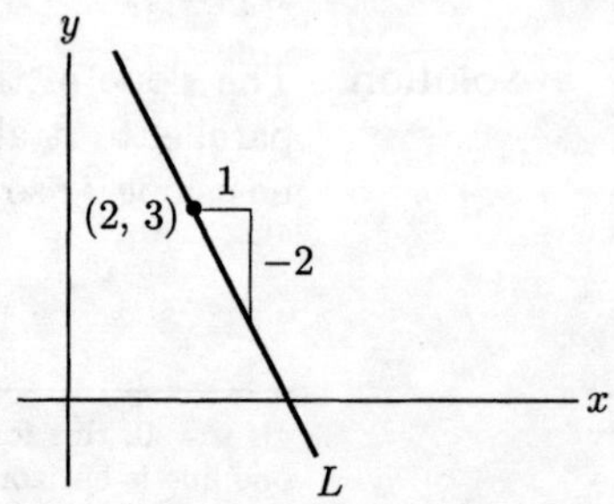

16.

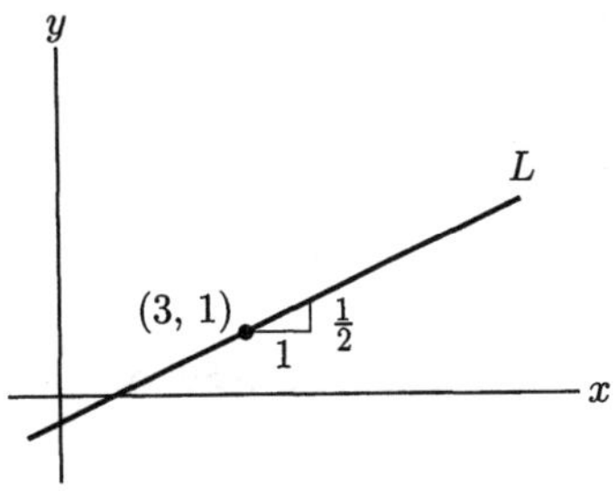

17.

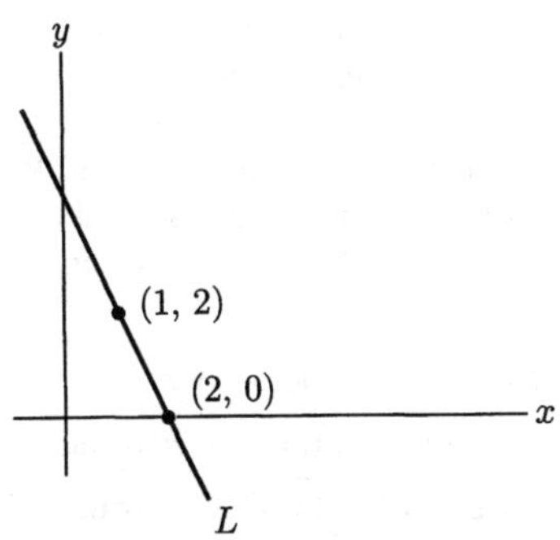

18.

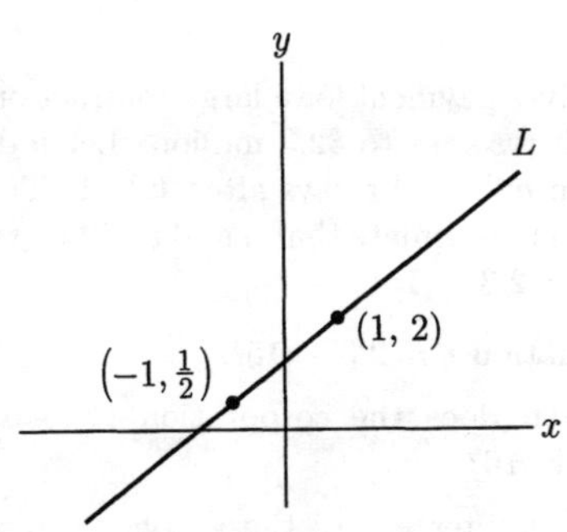

19.

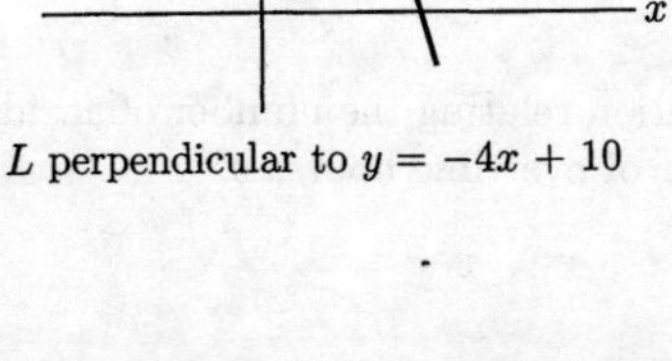

L perpendicular to $y = -4x + 10$

20.

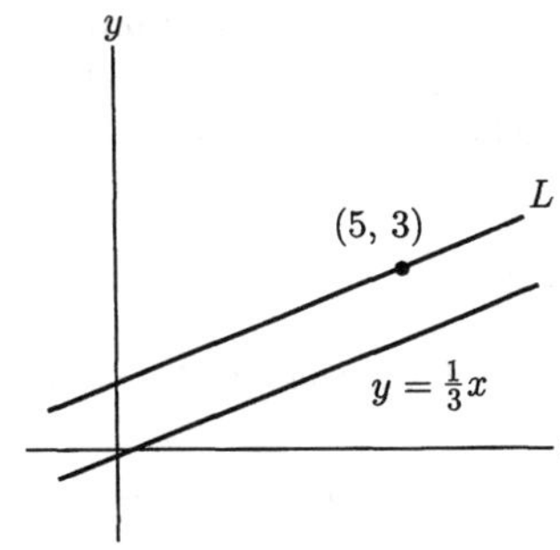

L parallel to $y = \frac{1}{3}x$

21.

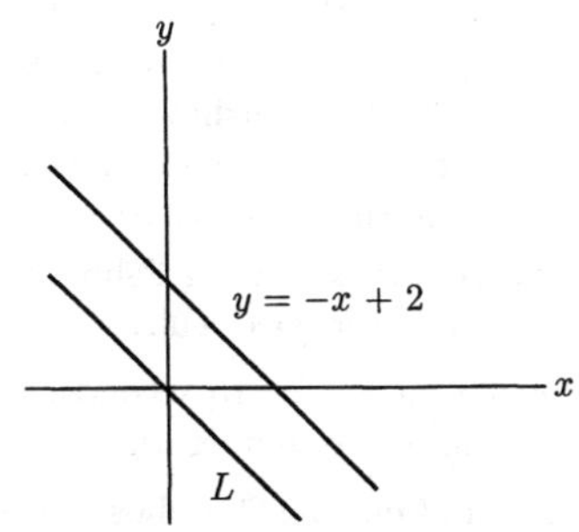

L parallel to $y = -x + 2$

22.

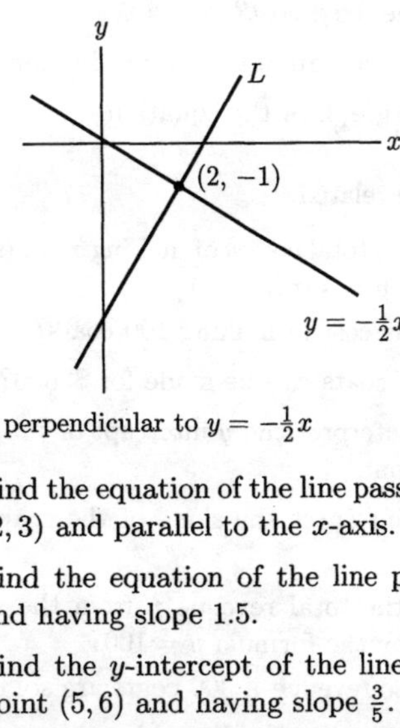

L perpendicular to $y = -\frac{1}{2}x$

23. Find the equation of the line passing through the point $(2, 3)$ and parallel to the x-axis.

24. Find the equation of the line passing through $(0, 0)$ and having slope 1.5.

25. Find the y-intercept of the line passing through the point $(5, 6)$ and having slope $\frac{3}{5}$.

26. Find the slope of the line passing through the point $(1, 4)$ and having y-intercept $(0, 4)$.

27. A salesperson's weekly pay depends on the volume of sales. If she sells x units of goods, then her pay is $y = 5x + 60$ dollars. Give an interpretation to the slope and the y-intercept of this straight line.

28. A manufacturer has fixed costs (such as rent and insurance) of $2000 per month. The cost of producing each unit of goods is $4. Give the linear equation for the cost of producing x units per month.

29. Suppose the price that must be set in order to sell q items is given by the equation $p = -3q + 1200$.
   (a) Find and interpret the p-intercept of the graph of the equation.
   (b) Find and interpret the q-intercept of the graph of the equation.
   (c) Find and interpret the slope of the graph of the equation.
   (d) What price must be set in order to sell 350 items?
   (e) What quantity will be sold if the price is \$300?
   (f) Sketch the graph of the equation.

30. Biologists have found that the number of chirps that crickets of a certain species make per minute is related to the temperature. The relationship is very close to linear. At 68°F those crickets chirp about 124 times a minute. At 80°F they chirp about 172 times a minute.
   (a) Find the linear equation relating Fahrenheit temperature F and the number of chirps c.
   (b) If you only count chirps for 15 seconds, how can you quickly estimate the temperature?

31. Suppose that the cost of making 20 radios is \$6800 and the cost of making 50 radios is \$9500.
   (a) Find the cost equation.
   (b) What is the fixed cost?
   (c) What is the marginal cost of production?
   (d) Sketch the graph of the equation.

Exercises 32–34 are related.

32. Suppose that the total cost y of making x coats is given by the formula $y = 40x + 2400$.
   (a) What is the cost of making 100 coats?
   (b) How many coats can be made for \$3600?
   (c) Find and interpret the y-intercept of the graph of the equation.
   (d) Find and interpret the slope of the graph of the equation.

33. Suppose that the total revenue y from the sale of x coats is given by the formula $y = 100x$.
   (a) What is the revenue if 300 coats are sold?
   (b) How many coats must be sold to have a revenue of \$6000?
   (c) Find and interpret the y-intercept of the graph of the equation.
   (d) Find and interpret the slope of the graph of the equation.

34. Consider a coat factory with the cost and revenue equations given in Exercises 32 and 33.
   (a) Find the equation giving the profit y resulting from making and selling x coats.
   (b) Find and interpret the y-intercept of the graph of the equation.
   (c) Find and interpret the x-intercept of the graph of the equation.
   (d) Find and interpret the slope of the graph of the equation.
   (e) How much profit will be made if 80 coats are sold?
   (f) How many coats must be sold to have a profit of \$6000?
   (g) Sketch the graph of the equation found in part (a).

An apartment complex has a storage tank to hold its heating oil. The tank was filled on January 1, but no more deliveries of oil will be made until sometime in March. Let t denote the number of days after January 1 and let y denote the number of gallons of fuel oil in the tank. Current records show that y and t will be related by the equation $y = 30{,}000 - 400t$.

35. Graph the equation $y = 30{,}000 - 400t$.

36. How much oil will be in the tank on February 1?

37. How much oil will be in the tank on February 15?

38. Determine the y-intercept of the graph. Explain its significance.

39. Determine the t-intercept of the graph. Explain its significance.

A corporation receives payment for a large contract on July 1, bringing its cash reserves to \$2.3 million. Let y denote its cash reserves (in millions) t days after July 1. The corporation's accountants estimate that y and t will be related by the equation $y = 2.3 - .15t$.

40. Graph the equation $y = 2.3 - .15t$.

41. How much cash does the corporation have on the morning of July 16?

42. Determine the y-intercept of the graph. Explain its significance.

43. Determine the t-intercept of the graph. Explain its significance.

44. Determine the cash reserves on July 4.

45. When will the cash reserves be \$.8 million?

In a certain factory, each day the expected number of accidents is related to the number of overtime hours by a linear equation. Suppose that on one day there were 1000 overtime hours logged and 8 accidents reported, and on another day there were 400 overtime hours logged and 5 accidents.

46. Find the equation relating the number of accidents, y, to the number of overtime hours, x.

47. What are the expected number of accidents when no overtime hours are logged?

48. What are the expected number of accidents when 2000 hours of overtime are logged?

49. To what level should the number of overtime hours be restricted if the company feels it cannot tolerate more than 10 accidents a day?

A furniture salesperson earns $160 a week plus 10% commission on her sales. Let x denote her sales and y her income for a week.

50. Express y in terms of x.

51. Determine her week's income if she sells $1000 in merchandise that week.

52. How much must she sell in a week in order to earn $500?

Find the equations of the following lines.

53. Slope is 3; y-intercept is $(0, -1)$

54. Slope is $-\frac{1}{2}$; y-intercept is $(0, 0)$

55. Slope is 1; $(1, 2)$ on line

56. Slope is $-\frac{1}{3}$; $(6, -2)$ on line

57. Slope is -7; $(5, 0)$ on line

58. Slope is $\frac{1}{2}$; $(2, -3)$ on line

59. Slope is 0; $(7, 4)$ on line

60. Slope is $-\frac{2}{5}$; $(0, 5)$ on line

61. $(2, 1)$ and $(4, 2)$ on line

62. $(5, -3)$ and $(-1, 3)$ on line

63. $(0, 0)$ and $(1, -2)$ on line

64. $(2, -1)$ and $(3, -1)$ on line

In each of Exercises 65–68 we specify a line by giving the slope and one point on the line. We give the first coordinate of some points on the line. Without deriving the equation of the line, find the second coordinate of each of the points.

65. Slope is 2, $(1, 3)$ on line; $(2,\ )$; $(0,\ )$; $(-1,\ )$

66. Slope is -3, $(2, 2)$ on line; $(3,\ )$; $(4,\ )$; $(1,\ )$

67. Slope is $-\frac{1}{4}$, $(-1, -1)$ on line; $(0,\ )$; $(1,\ )$; $(-2,\ )$

68. Slope is $\frac{1}{3}$, $(-5, 2)$ on line; $(-4,\ )$; $(-3,\ )$; $(-2,\ )$

69. Each of the lines (A), (B), (C), and (D) in Fig. 10 is the graph of one of the linear equations (a), (b), (c), and (d). Match each line with its equation.

(A)

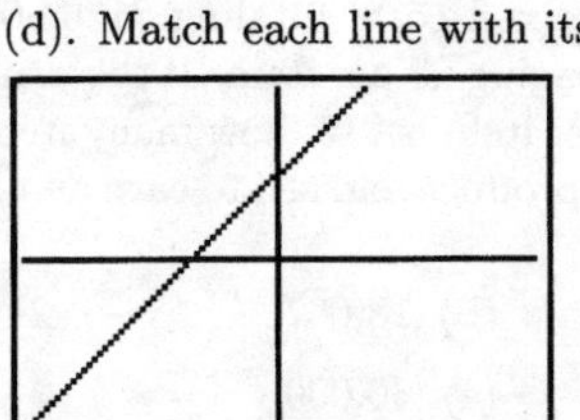

(B) (C) (D)

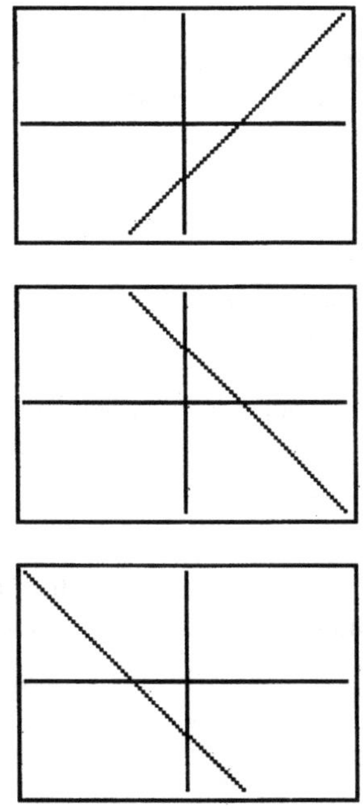

FIGURE 10

(a) $x + y = 1$ (b) $x - y = 1$

(c) $x + y = -1$ (d) $x - y = -1$

70. The following table gives several points on the line $Y_1 = mx + b$. Find m and b.

X	Y1
4.8	3.6
4.9	4.8
5	6
5.1	7.2
5.2	8.4
5.3	9.6
5.4	[illegible]

Y1=10.8

71. Celsius and Fahrenheit temperatures are related by a linear equation. Use the fact that 0°C = 32°F and 100°C = 212°F to find an equation.

72. An archaeologist dates a bone fragment discovered at a depth of 4 feet as approximately 1500 B.C. and dates a pottery shard at a depth of 8 feet as approximately 2100 B.C. Assuming there is a linear relationship between depths and dates at this archeological site, find the equation that relates depth to date. How deep should the archaeologist dig to look for relics from 3000 B.C.?

73. The number of farms in the United States declined from 6 million in 1920 to 2 million in 1980. Assuming that the number of farms declined linearly with respect to time, find the equation that relates the number of farms, y, to the number of years after 1920, x. How many farms were there in 1965?

74. The average size of a farm in the United States increased from 100 acres in 1920 to 700 acres in 1980. Assuming that farm size increased linearly with respect to time, find the equation that relates the average size of a farm, y, to the number of years after 1920, x. When was the average size of a farm 600 acres?

75. Consider a linear equation of the form $y = mx + b$. When the value of b remains fixed and the value of m changes, the graph rotates about the point $(0, b)$. As the value of m increases, will the graph rotate in a clockwise or counterclockwise direction?

76. Write an inequality whose graph consists of the points on or below the straight line passing through the two points $(2, 5)$ and $(4, 9)$.

77. Write an inequality whose graph consists of the points on or above the line with slope 4 and y-intercept $(0, 3)$.

78. Find a system of inequalities having the feasible set in Fig. 11.

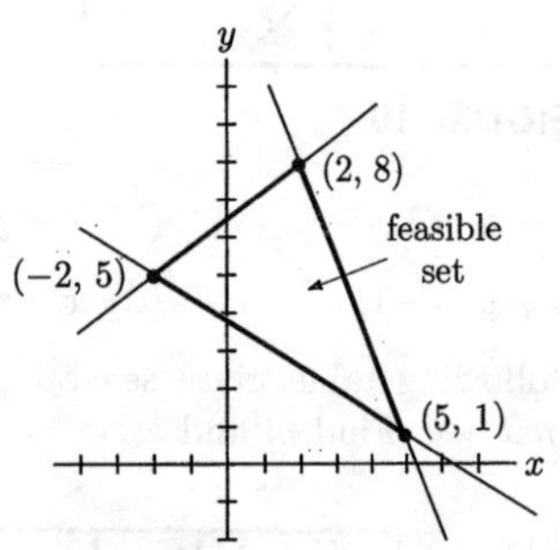

FIGURE 11

79. Find a system of inequalities having the feasible set in Fig. 12.

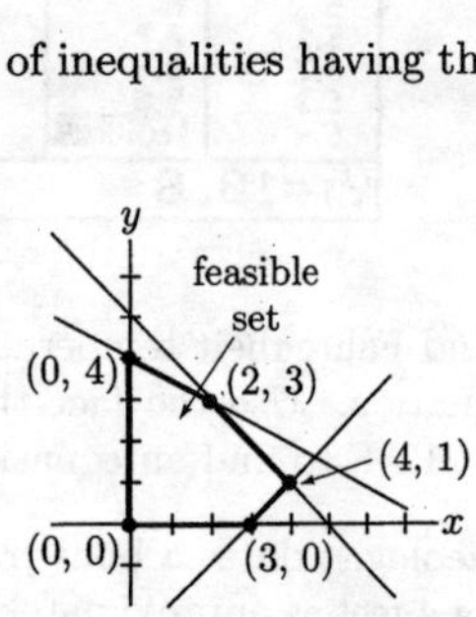

FIGURE 12

80. Show that the points $(1, 3)$, $(2, 4)$, and $(3, -1)$ are *not* on the same line.

81. For what value of k will the three points $(1, 5)$, $(2, 7)$, and $(3, k)$ be on the same line?

82. Find the value of a for which the line through the points $(a, 1)$ and $(2, -3.1)$ is parallel to the line through the points $(-1, 0)$ and $(3.8, 2.4)$.

83. Rework Exercise 82, where the word "parallel" is replaced by the word "perpendicular."

84. Prove the parallel property. [*Hint*: If $y = mx + b$ and $y = m'x + b'$ are the equations of two lines, then the two lines have a point in common if and only if the equation $mx + b = m'x + b'$ has a solution x.]

85. Prove the perpendicular property. [*Hint*: Without loss of generality, assume that both lines pass through the origin. Use the point-slope formula, the Pythagorean theorem, and Fig. 13.]

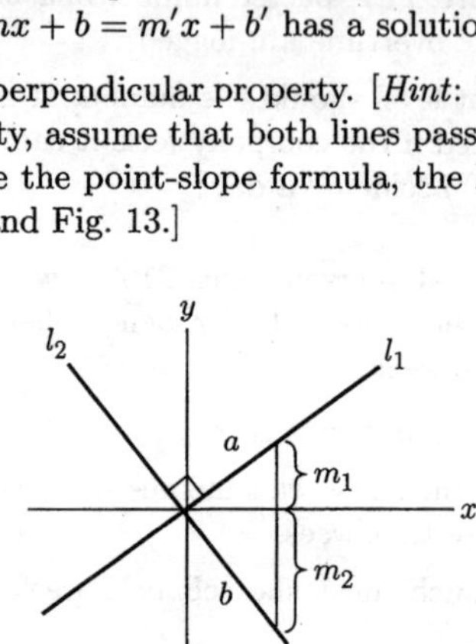

FIGURE 13

86. What is the difference between having no slope and zero slope?

87. ***(From CPA exam Nov. 95, ARE, #39)*** Sender Inc. estimates parcel mailing costs using data shown in Fig. 14. What is Sender's estimated cost for mailing 12,000 parcels?

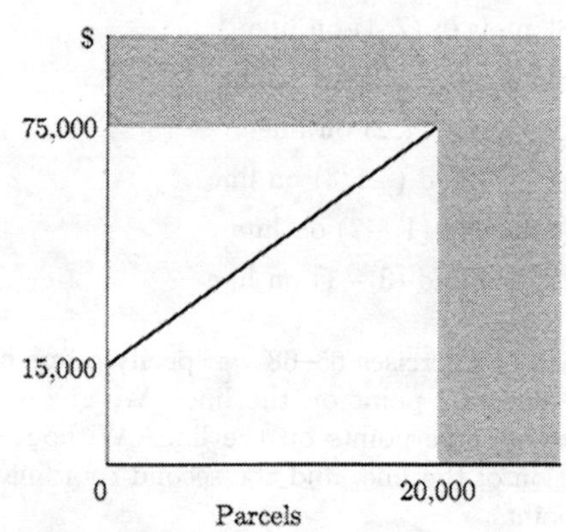

FIGURE 14

(a) \$36,000 (b) \$45,000
(c) \$51,000 (d) \$60,000

88. ***(From a GMAT exam)*** A certain manufacturer produces items for which the production costs consist of annual fixed costs totaling \$130,000 and variable costs averaging \$8 per item. If the manufacturer's selling price per item is \$15, how many items must the manufacturer produce and sell to earn an annual profit of \$150,000?

(a) 2,858 (b) 18,667 (c) 21,429
(d) 35,000 (e) 40,000

89. (***From a GMAT exam***) If the number n of calculators sold per week varies with the price p in dollars according to the equation $n = 300 - 20p$, what would be the total weekly revenue from the sale of \$10 calculators?

(a) \$100 (b) \$300 (c) \$1000
(d) \$2800 (e) \$3000

90. (***From a CPA exam May 94, TI, #46***) During 1995 Thor Lab supplied hospitals with a comprehensive diagnostic kit for \$120. At a volume of 80,000 kits, Thor had fixed costs of \$1,000,000 and a profit before income taxes of \$200,000. Due to an adverse legal decision, Thor's 1996 liability insurance increased by \$1,200,000 over 1995. Assuming the volume and other costs are unchanged, what should the 1996 price be if Thor is to make the same \$200,000 profit before income taxes?

(a) \$120 (b) \$135 (c) \$150 (d) \$240

Exercises 91–94 require the use of a graphing calculator or a computer.

91. Graph the two lines $y = .5x + 1$ and $y = -2x + 9$ in the standard window $[-10, 10]$ *by* $[-10, 10]$. Do they appear perpendicular? If not, change the window to one having true aspect, such as $[-15, 15]$ *by* $[-10, 10]$, and look at the graphs.

92. Graph the three lines $y = 2x - 3$, $y = 2x$, and $y = 2x+3$ together and then identify each line without using TRACE.

93. Graph the three lines $y = 2x + 1$, $y = x + 1$, and $y = .5x + 1$ together and then identify each line without using TRACE.

94. Graph the line $y = -.5x + 2$ with the window `ZDecimal`. Without pressing [TRACE], move the cursor to a point on the line. Then move the cursor one unit to the right and down $\frac{1}{2}$ unit to return to the line. If you start at a point on the line and move 2 units to the right, how many units down will you have to move the cursor to return to the line? Test your answer.

95. Repeat Exercise 94 for the line $y = .75x - 2$, using "up" instead of "down."

Solutions to Practice Problems 4

1. By the steepness property, whenever x is increased by 1 unit, the value of y is increased by 4 units. Therefore, *each additional unit of production brings in \$4 of revenue.* (The graph of $y = 4x$ is called a *revenue curve* and its slope is called the *marginal revenue of production.*)

2. $\begin{cases} y = 4x \\ y = 2x + 5000 \end{cases}$

   Equate expressions for y:

$$4x = 2x + 5000$$
$$2x = 5000$$
$$x = 2500$$
$$y = 4(2500) = 10{,}000.$$

3. When producing 2500 units, the revenue equals the cost. This value of x is called the *break-even point.* Since (profit) = (revenue) − (cost), the company will make a profit only if its level of production is greater than the break-even point (Fig. 15).

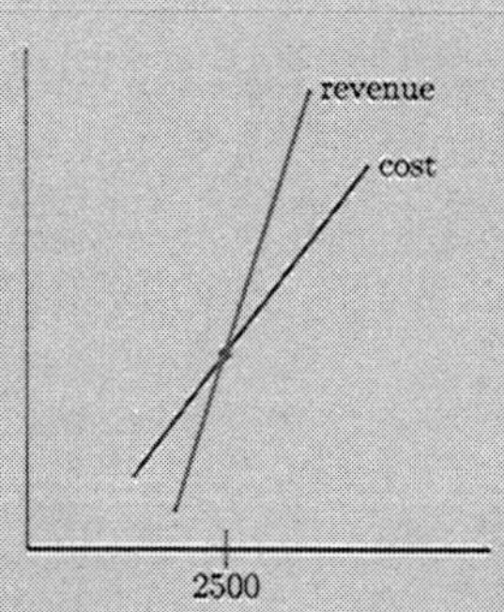

FIGURE 15

1.5 The Method of Least Squares

Modern people compile graphs of literally thousands of different quantities: the purchasing value of the dollar as a function of time; the pressure of a fixed volume of air as a function of temperature; the average income of people as a function of their years of formal education; or the incidence of strokes as a function of blood pressure. The observed points on such graphs tend to be irregularly distributed due to the complicated nature of the phenomena underlying them as well as to errors made in observation (for example, a given procedure for measuring average income may not count certain groups). In spite of the imperfect nature of the data, we are often faced with the problem of making assessments and predictions based on them. Roughly speaking, this problem amounts to filtering the sources of errors in the data and isolating the basic underlying trend. Frequently, on the basis of a suspicion or a working hypothesis, we may suspect that the underlying trend is linear—that is, the data should lie on a straight line. But which straight line? This is the problem that the *method of least squares* attempts to answer. To be more specific, let us consider the following problem:

> ***Problem*** Given observed data points $(x_1, y_1), (x_2, y_2), \dots, (x_N, y_N)$ on a graph, find the straight line that "best" fits these points.

In order to completely understand the statement of the problem being considered, we must define what it means for a line to "best" fit a set of points. If (x_i, y_i) is one of our observed points, then we will measure how far it is from a given line $y = ax + b$ by the vertical distance, E_i, from the point to the line. (See Fig. 1.)

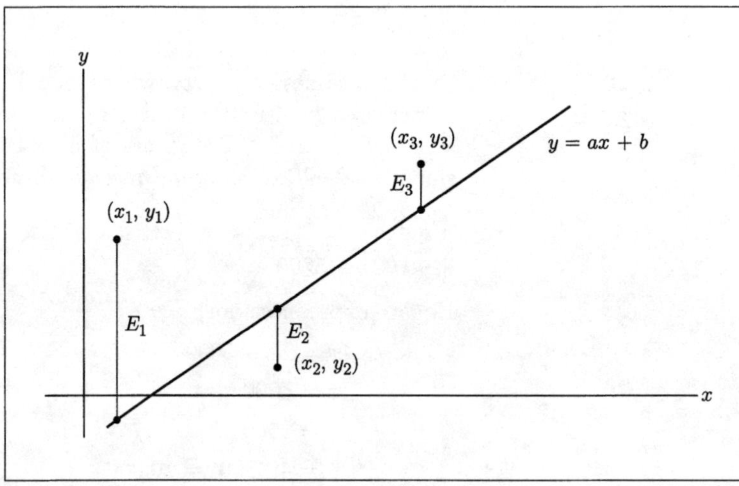

FIGURE 1 Fitting a line to data points

Statisticians prefer to work with the square of the vertical distance E_i. The total error in approximating the data points $(x_1, y_1), \dots, (x_N, y_N)$ by the line $y = ax + b$ is usually measured by the sum E of the squares of the vertical distances from the points to the line,

$$E = E_1^2 + E_2^2 + \cdots + E_N^2.$$

E is called the *least-squares error* of the observed points with respect to the line. If all the observed points lie on the line $y = ax + b$, then all the E_i are zero and the error E is zero. If a given observed point is far away from the line, the corresponding E_i^2 is large and hence makes a large contribution to the error E.

EXAMPLE 1 Determine the least-squares error when the line $y = 1.5x + 3$ is used to approximate the data points $(1, 6)$, $(4, 5)$, and $(6, 14)$.

Solution Figure 2 shows the line, the points, and the vertical distances. The vertical distance of a point from the line is determined by finding the second coordinate of

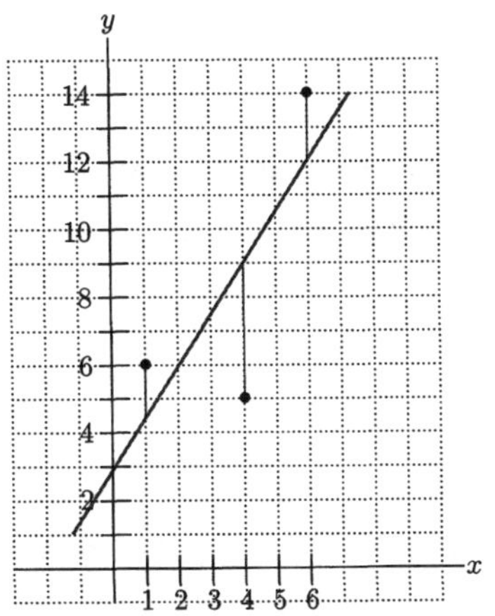

FIGURE 2

TABLE 1 Vertical distances from the line $y = 1.5x + 3$

Data Point	*Point on Line*	*Vertical Distance*
$(1, 6)$	$(1, 4.5)$	1.5
$(4, 5)$	$(4, 9)$	4
$(6, 14)$	$(6, 12)$	2

the point on the line having the same x-coordinate as the point. For instance, for the data point $(1, 6)$, the point on the line with x-coordinate 1 has y-coordinate $y = 1.5(1) + 3 = 4.5$ and therefore vertical distance $6 - 4.5 = 1.5$. Table 1 summarizes the vertical distances. The table shows that the least-squares error is $1.5^2 + 4^2 + 2^2 = 2.25 + 16 + 4 = 22.25$. ▮

In general, we cannot expect to find a line $y = mx + b$ that fits the observed points so well that the error E is zero. Actually, this situation will occur only if the observed points lie on a straight line. However, we can rephrase our original problem as follows:

Problem Given observed data points $(x_1, y_1), (x_2, y_2), \ldots, (x_N, y_N)$ on a graph, find the straight line $y = mx + b$ for which the error E is as small as possible.

This line, called the *least-squares line* or the *regression line*, can be found from the data points with the formulas

$$m = \frac{N \cdot \sum xy - \sum x \cdot \sum y}{N \cdot \sum x^2 - \left(\sum x\right)^2}$$

$$b = \frac{\sum y - m \cdot \sum x}{N},$$

where

$$\sum x = \text{sum of the } x\text{-coordinates of the data points}$$
$$\sum y = \text{sum of the } y\text{-coordinates of the data points}$$
$$\sum xy = \text{sum of the products of the coordinates of the data points}$$
$$\sum x^2 = \text{sum of the squares of the } x\text{-coordinates of the data points}$$
$$N = \text{number of data points.}$$

That is,

$$\sum x = x_1 + x_2 + \cdots + x_N$$
$$\sum y = y_1 + y_2 + \cdots + y_N$$

$$\sum xy = x_1 \cdot y_1 + x_2 \cdot y_2 + \cdots + x_N \cdot y_N$$
$$\sum x^2 = x_1^2 + x_2^2 + \cdots + x_N^2.$$

EXAMPLE 2 Find the least-squares line for the data points of Example 1.

Solution The sums are calculated in Table 2 and then used to determine the values of m and b.

TABLE 2

x	y	xy	x^2
1	6	6	1
4	5	20	16
6	14	84	36
$\sum x = 11$	$\sum y = 25$	$\sum xy = 110$	$\sum x^2 = 53$

$$m = \frac{3 \cdot 110 - 11 \cdot 25}{3 \cdot 53 - 11^2} = \frac{55}{38} \approx 1.45$$

$$b = \frac{25 - \frac{55}{38} \cdot 11}{3} = \frac{38 \cdot 25 - 55 \cdot 11}{38 \cdot 3} = \frac{345}{114} = \frac{115}{38} \approx 3.03$$

Therefore, the equation of the least-squares line is $y = \frac{55}{38}x + \frac{115}{38}$. With this line, the least-squares error can be shown to be about 22.13. ▮

Obtaining the Least-Squares Line with Technology Least-squares lines can be obtained with graphing calculators, spreadsheets, or mathematics software. For instance, on the TI-83 graphing calculator screens in Fig. 3, the data points are entered into lists, the least-squares line is calculated with the item `LinReg(ax + b)` of the STAT/CALC menu, and the data points and line are plotted with [STAT PLOT] and [GRAPH]. The last part of this section contains the details for obtaining least-squares lines with TI-82 and TI-83 graphing calculators. For other graphing calculators, see the guidebook for that particular calculator.

FIGURE 3 Obtaining a least-squares line with a TI-83

	A	B
1	1	6
2	4	5
3	6	14
4	Slope	1.447368
5	Intercept	3.026316

FIGURE 4 Using Excel to calculate a least-squares line

Spreadsheet programs, such as Excel and Lotus 1-2-3, have special functions that calculate the slope and y-intercept of the least-squares line for a collection of data points. Figure 4 shows how the least-squares line of Example 2 is calculated in Excel. For those familiar with Excel, cell B4 contains the formula `"=SLOPE(B1:B3,A1:A3)"` and cell B5 contains the formula `"=INTERCEPT(B1:B3,A1:A3)"`.

The next example obtains a least-squares line and uses the line to make projections.

EXAMPLE 3 Table 3 gives the U.S. per capita health care expenditures[2] for several years.

(a) Find the least-squares line for this data.

(b) Use the least-squares line to predict the per capita health care expenditures for the year 2000.

(c) Use the least-squares line to predict when per capita health care expenditures will reach $6000.

TABLE 3 U.S. per Capita Health Care Expenditures

Year	1990	1991	1992	1993	1994
Dollars	2688	2902	3144	3331	3510

Solution (a) The numbers can be made more manageable by counting years beginning with 1990 and measuring dollars in thousands. See Table 4.

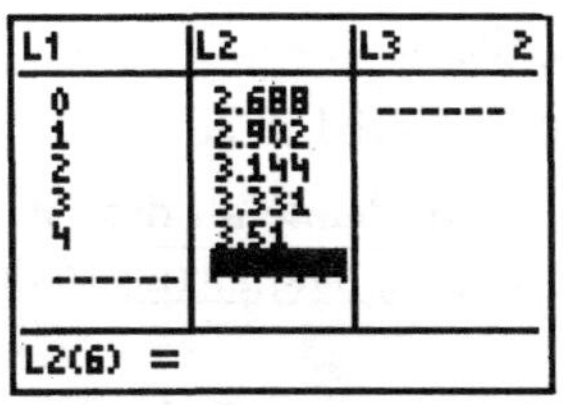

TABLE 4 U.S. per Capita Health Care Expenditures

Years (after 1990)	0	1	2	3	4
Dollars (in thousands)	2.688	2.902	3.144	3.331	3.510

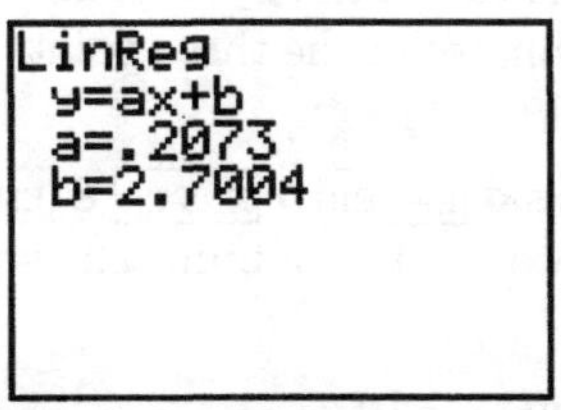

The slope and y-intercept of the least-squares line can be found with the formulas involving sums or with technology. The sums are calculated in Table 5 and then used to determine the values of m and b. The screens in the margin show the results of calculating m and b with a graphing calculator.

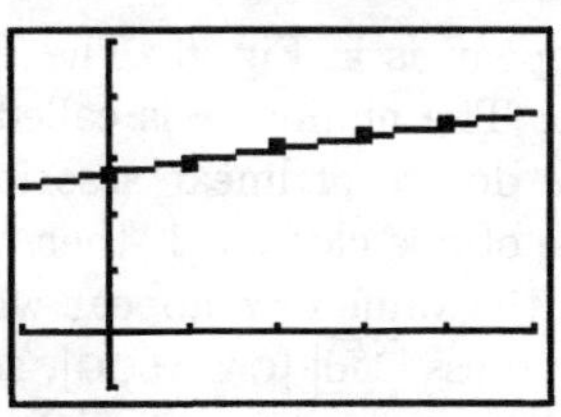

TABLE 5

x	y	xy	x^2
0	2.688	0	0
1	2.902	2.902	1
2	3.144	6.288	4
3	3.331	9.993	9
4	3.510	14.04	16
$\sum x = 10$	$\sum y = 15.575$	$\sum xy = 33.223$	$\sum x^2 = 30$

$$m = \frac{5 \cdot 33.223 - 10 \cdot 15.575}{5 \cdot 30 - 10^2} = \frac{10.365}{50} = .2073$$

$$b = \frac{15.575 - .2073 \cdot 10}{5} = \frac{13.502}{5} = 2.7004$$

Therefore, the least-squares line is $y = .2073x + 2.7004$.

[2]U.S. Health Care Financing Administration, *Health Care Financing Review*, Spring 1996.

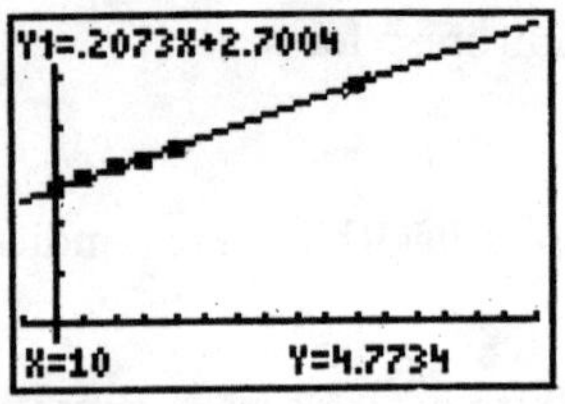

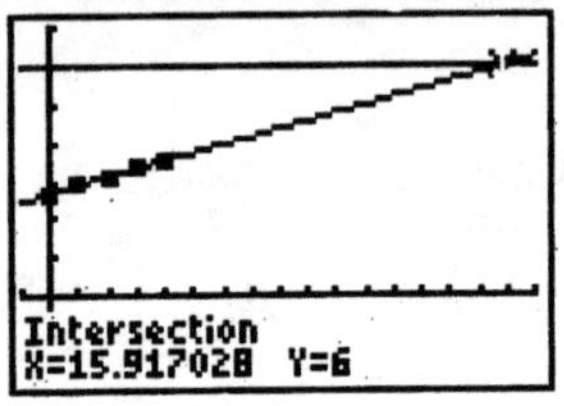

(b) The year 2000 corresponds to $x = 10$. The value of y is

$$y = .2073(10) + 2.7004 = 4.7734.$$

Therefore, an estimate of per capita health care expenditures in the year 2000 is \$4773.40.

(c) Set the value of y equal to 6 and solve for x.

$$6 = .2073x + 2.7004$$

$$x = \frac{6 - 2.7004}{.2073} \approx 15.9$$

Therefore, expenditures are projected to reach \$6000 in 15.9 years after 1990; that is, near the end of the year 2005. ▮

Obtaining the Least-Squares Line with a TI-82 or TI-83 The following steps find the straight line that minimizes the least-squares error for the points $(1, 4)$, $(2, 5)$, and $(3, 8)$.

1. Press `STAT` **1** to obtain a table used for entering the data.
2. If there are no data in columns labeled L_1 and L_2, proceed to step 4.
3. Move the cursor up to L_1 and press `CLEAR` `ENTER` to delete all data in L_1's column. Move the cursor right and up to L_2 and press `CLEAR` `ENTER` to delete all data in L_2's column.
4. If necessary, move the cursor left to the first blank row of the L_1 column. Press **1** `ENTER` **2** `ENTER` **3** `ENTER` to place the x-coordinates of the three points into the L_1 column.
5. Move the cursor right to the L_2 column and press **4** `ENTER` **5** `ENTER` **8** `ENTER` to place the y-coordinates of the three points into the L_2 column. The screen should now appear as in Fig. 5.

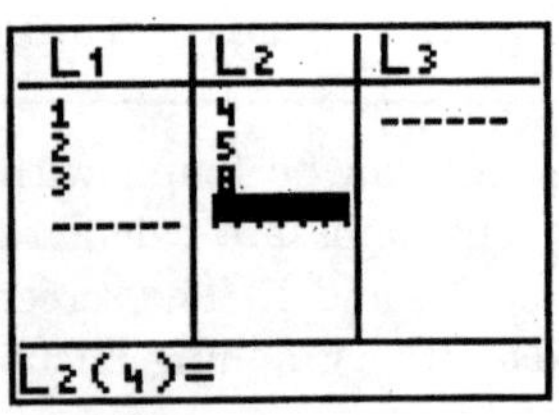

FIGURE 5

6. Press `STAT` `▶` and press the number for `LinReg(ax + b)`.
7. Press `ENTER`. The TI-82 screen should now appear as in Fig. 6. The TI-83 screen normally is missing the last line. (*Note*: The number r is called the *correlation coefficient* and is a measure of the degree of linear association between the two variables. If the absolute value of r is close to 1, then there is a high degree of linear association. To have the value of r appear with a TI-83, create a blank line on the home screen, press `2nd` [CATALOG], select DiagnosticOn, and press `ENTER` twice.) The least-squares line is $y = 2x + \frac{5}{3}$. (*Note*: $\frac{5}{3} \approx 1.666666667$.)

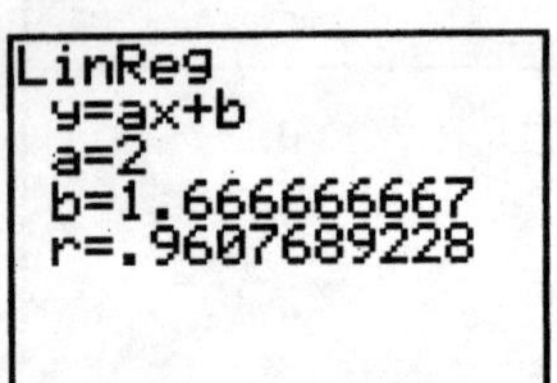

FIGURE 6

8. If desired, the linear function can be assigned to Y_1 with the following steps:
   (a) Press `Y=` `CLEAR` to erase the current expression in Y_1.
   (b) Press `VARS` **5** `▶` `▶` and the number for `RegEQ` to assign the linear function to Y_1.
9. The original points can be easily plotted along with the least-squares line. Assume that the linear function has been assigned to Y_1, that all other functions have been cleared or deselected, and that the window has been set to $[-4, 4]$ *by* $[-4, 9]$. Press `2nd` [STAT PLOT] `ENTER` `ENTER` `GRAPH` to see the display in Fig. 7. (*Note 1*: To turn off the point-plotting feature, press `2nd` [STAT PLOT] `ENTER` `▶` `ENTER`. *Note 2*: With the TI-83, the point-plotting feature can be toggled from the function-declaration screen by moving the cursor to the word `"Plot1"` on the top line and pressing `ENTER`.)

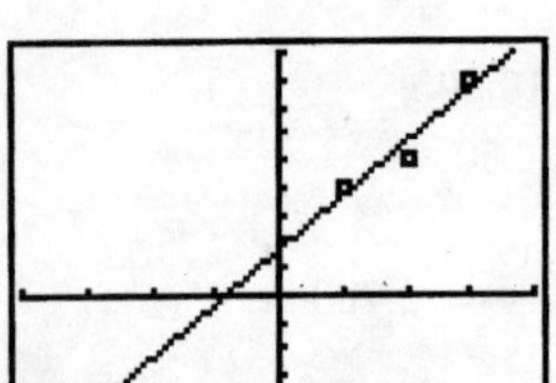

FIGURE 7

Practice Problems 5

1. Can a vertical distance be negative?
2. Under what condition will a vertical distance be zero?

EXERCISES 5

1. Suppose the line $y = 3x + 1$ is used to fit the four data points in Table 6. Complete the table and determine the least-squares error E.

TABLE 6

Data Point	*Point on Line*	*Vertical Distance*
$(1,3)$		
$(2,6)$		
$(3,11)$		
$(4,12)$		

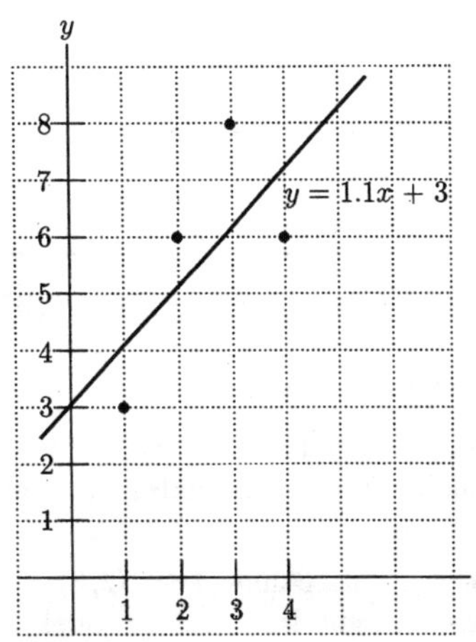

FIGURE 8

2. Suppose the line $y = -2x + 12$ is used to fit the four data points in Table 7. Complete the table and determine the least-squares error E.

TABLE 7

Data Point	*Point on Line*	*Vertical Distance*
$(1,11)$		
$(2,7)$		
$(3,5)$		
$(4,5)$		

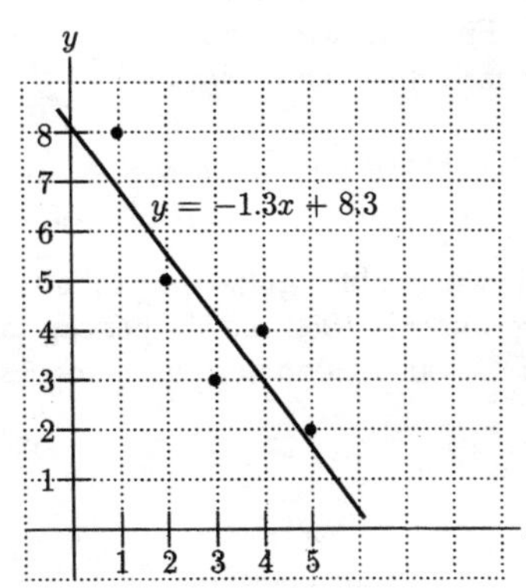

FIGURE 9

3. Find the least-squares error E for the least-squares line fit to the four points in Fig. 8.
4. Find the least-squares error E for the least-squares line fit to the five points in Fig. 9.
5. Complete Table 8 and find the values of m and b for the straight line that provides the best least-squares fit to the data.

TABLE 8

x	y	xy	x^2
1	7		
2	6		
3	4		
4	3		
$\sum x =$	$\sum y =$	$\sum xy =$	$\sum x^2 =$

6. Complete Table 9 and find the values of m and b for the straight line that provides the best least-squares fit to the data.

TABLE 9

x	y	xy	x^2
1	2		
2	4		
3	7		
4	9		
5	12		
$\sum x =$	$\sum y =$	$\sum xy =$	$\sum x^2 =$

7. Consider the data points $(1, 2)$, $(2, 5)$, and $(3, 11)$. Find the straight line that provides the best least-squares fit to these data.

8. Consider the data points $(1, 8)$, $(2, 4)$, and $(4, 3)$. Find the straight line that provides the best least-squares fit to these data.

9. Consider the data points $(1, 9)$, $(2, 8)$, $(3, 6)$, and $(4, 3)$. Find the straight line that provides the best least-squares fit to these data.

10. Consider the data points $(1, 5)$, $(2, 7)$, $(3, 6)$, and $(4, 10)$. Find the straight line that provides the best least-squares fit to these data.

Since Exercises 11–16 use real data, they are best answered with a graphing calculator or spreadsheet.

11. The following table[3] gives the crude male death rate for lung cancer in 1950 and the per capita consumption of cigarettes in 1930 in various countries.

Country	*Cigarette Consumption (per Capita)*	*Lung Cancer Deaths (per Million Males)*
Norway	250	95
Sweden	300	120
Denmark	350	165
Australia	470	170

(a) Use the method of least-squares to obtain the straight line that best fits these data.

(b) In 1930 the per capita cigarette consumption in Finland was 1100. Use the straight line found in part (a) to estimate the male lung cancer death rate in Finland in 1950.

12. The accompanying table shows the 1994 price of a gallon (in U.S. dollars) of fuel and the average miles driven per automobile for several countries.[4]

Country	*Price per Gallon*	*Average Miles per Auto*
Canada	\$1.57	10,371
England	\$2.86	10,186
France	\$3.31	8740
Germany	\$3.34	7674
Sweden	\$3.44	7456
United States	\$1.24	11,099

(a) Find the straight line that provides the best least-squares fit to these data.

(b) In 1994, the price of gas in Japan was \$4.14 per gallon. Use the straight line of part (a) to estimate the average number of miles automobiles were driven in Japan.

13. The following table gives the percent of persons 25 years and over who have completed four or more years of college.[5]

Year	*Percent*
1970	10.7
1975	13.9
1980	16.2
1985	19.4
1990	21.3
1995	20.0

(a) Use the method of least-squares to obtain the straight line that best fits these data.

(b) Estimate the percent for the year 1993.

(c) If the trend determined by the straight line in part (a) continues, when will the percent reach 27.1?

[3]These data were obtained from *Smoking and Health*, Report of the Advisory Committee to the Surgeon General of the Public Health Service, U.S. Department of Health, Education, and Welfare, Washington, D.C., Public Health Service Publication No. 1103, p. 176.
[4]Source: Energy Information Administration, *International Energy Annual*. U.S. Highway Administration, *Highway Statistics*, 1994.
[5]U.S. Bureau of the Census

14. The following table gives the number (in millions) of cars in use in the United States[6] for certain years.

Year	*Cars*
1980	104.6
1985	114.7
1989	122.8
1990	123.3
1991	123.3
1992	120.3
1993	121.1
1994	122.0

(a) Use the method of least squares to obtain the straight line that best fits these data.

(b) Estimate the number of cars in use in 1983.

(c) If the trend determined by the straight line in part (a) continues, when will the number of cars in use reach 130 million?

15. The following table gives the 1996 life expectancy at birth for several countries.[7]

Country	*Men*	*Women*
Finland	70.9	78.9
United States	72.0	78.8
Australia	73.9	80.0
Japan	75.9	81.8

(a) Use the method of least-squares to obtain the straight line that best fits these data.

(b) In Switzerland the life expectancy for men is 74 years. Use the least-squares line to estimate the life expectancy for women.

(c) In Denmark the life expectancy for women is 77.7 years. Use the least-squares line to estimate the life expectancy for men.

16. Two Harvard economists studied countries' relationships between the independence of banks and inflation rates from 1955 to 1990.[8] The independence of banks was rated on a scale of -1.5 to 2.5, with -1.5, 0, and 2.5 corresponding to least, average, and most independence, respectively. The table below gives the values for various countries.

(a) Use the method of least squares to obtain the straight line that best fits these data.

Country	*Independence Rating*	*Inflation Rate (%)*
New Zealand	−1.4	7.6
Italy	−.75	7.2
Belgium	.3	4.0
France	.4	6.0
Canada	.9	4.5
United States	1.6	4.0
Switzerland	2.2	3.1

(b) What relationship between independence of banks and inflation is indicated by the least-squares line?

(c) Japan has a .6 independence rating. Use the least-squares line to estimate Japan's inflation rate.

(d) The inflation rate for Britain is 6.8. Use the least-squares line to estimate Britain's independence rating.

17. Suppose you have found the line of best least-squares fit to a collection of points and that you edit the data by adding a point on the line to the data. Will the expanded data have the same least-squares line? Explain the rationale for your conclusion, and then experiment to test whether your conclusion is correct.

18. Find the best least-squares fit to the points $(5,4)$ and $(7,3)$. Show that it is the straight line through the two points.

Solutions to Practice Problems 5

1. No. The word "distance" implies a nonnegative number. It is the absolute value of the difference between the y-coordinate of the data point and the y-coordinate of the point on the line.

2. The vertical distance will be zero when the point actually lies on the least-squares line.

[6] American Automobile Manufacturers Association, Inc., Detroit, MI. *Motor Vehicle Facts and Figures.*

[7] U.S. Bureau of the Census

[8] J. Bradford DeLong (Harvard) and L.H. Summers (World Bank)

Chapter 1 Review of Fundamental Concepts

1. How do you determine the coordinates of a point in the plane?
2. What is meant by the graph of an equation in x and y?
3. What is the general form of a linear equation in x and y?
4. What is the standard form of a linear equation in x and y?
5. What is the y-intercept of a line? How do you find the y-intercept from an equation of a line?
6. What is the x-intercept of a line? How do you find the x-intercept from an equation of a line?
7. Give a method for graphing the equation $y = mx + b$.
8. State the inequality properties for addition, subtraction, and multiplication.
9. What are the general and standard forms of a linear inequality in x and y?
10. Explain how to obtain the graph of a linear inequality.
11. What is meant by the *feasible set* of a system of linear inequalities?
12. Describe how to obtain the point of intersection of two lines.
13. Define the slope of a line, and give a physical description.
14. Suppose you know the slope and the coordinates of a point on a line. How could you draw the line without first finding its equation?
15. What is the point-slope form of the equation of a line?
16. Describe how to find the equation for a line when you know the coordinates of two points on the line.
17. What can you say about the slopes of perpendicular lines?
18. What can you say about the slopes of parallel lines?
19. What is the least-squares line approximation to a set of data points?

CHAPTER 1 SUPPLEMENTARY EXERCISES

1. What is the equation of the y-axis?
2. Graph the linear equation $y = -\frac{1}{2}x$.
3. Find the point of intersection of the pair of straight lines $x - 5y = 6$ and $3x = 6$.
4. Find the slope of the line having the equation $3x - 4y = 8$.
5. Find the equation of the line having y-intercept $(0, 5)$ and x-intercept $(10, 0)$.
6. Graph the linear inequality $x - 3y \geq 12$.
7. Does the point $(1, 2)$ satisfy the linear inequality $3x + 4y \geq 11$?
8. Find the point of intersection of the pair of straight lines $2x - y = 1$ and $x + 2y = 13$.
9. Find the equation of the straight line passing through the point $(15, 16)$ and parallel to the line $2x - 10y = 7$.
10. Find the y-coordinate of the point having x-coordinate 1 and lying on the line $y = 3x + 7$.
11. Find the x-intercept of the straight line with equation $x = 5$.
12. Graph the linear inequality $y \leq 6$.
13. Solve the following system of linear equations:

$$\begin{cases} 3x - 2y = 1 \\ 2x + y = 24. \end{cases}$$

14. Graph the feasible set for the following system of inequalities:

$$\begin{cases} 2y + 7x \geq 30 \\ 4y - x \geq 0 \\ y \leq 8. \end{cases}$$

15. Find the y-intercept of the line passing through the point $(4, 9)$ and having slope $\frac{1}{2}$.

16. The fee charged by a local moving company depends on the amount of time required for the move. If t hours are required, then the fee is $y = 35t + 20$ dollars. Give an interpretation of the slope and y-intercept of this line.

17. Are the points $(1, 2)$, $(2, 0)$, and $(3, 1)$ on the same line?

18. Write an equation of the line with x-intercept $(3, 0)$ and y-intercept $(0, -2)$.

19. ***(From the GMAT exam)*** If $x + 5y = 16$ and $x = -3y$, then $y =$

(a) -24 (b) -8 (c) -2 (d) 2 (e) 8.

20. Write the inequality whose graph is the half-plane below the line with slope $\frac{2}{3}$ and y-intercept $\left(0, \frac{3}{2}\right)$.

21. Write the inequality whose graph is the half-plane above the line through $(2, -1)$ and $(6, 8.6)$.

22. Solve the system of linear equations

$$\begin{cases} 1.2x + 2.4y = .6 \\ 4.8y - 1.6x = 2.4. \end{cases}$$

23. Find the equation of the line through $(1, 1)$ and the intersection of the lines $y = -x + 1$ and $y = 2x + 3$.

24. Find all numbers x such that $2x + 3(x - 2) \geq 0$.

25. Graph the equation $x + \frac{1}{2}y = 4$, and give the slope and both intercepts.

26. Do the three graphs of the linear equations $2x - 3y = 1$, $5x + 2y = 0$, and $x + y = 1$ contain a common point?

27. Show that the lines with equations $2x - 3y = 1$ and $3x + 2y = 4$ are perpendicular.

28. Each of the half-planes (A), (B), (C), and (D) is the graph of one of the linear inequalities (a), (b), (c) and (d). Match each half-plane with its inequality.

(A)

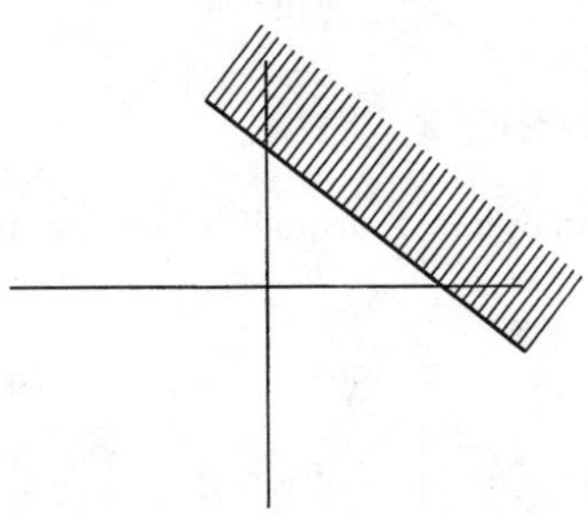

(B)

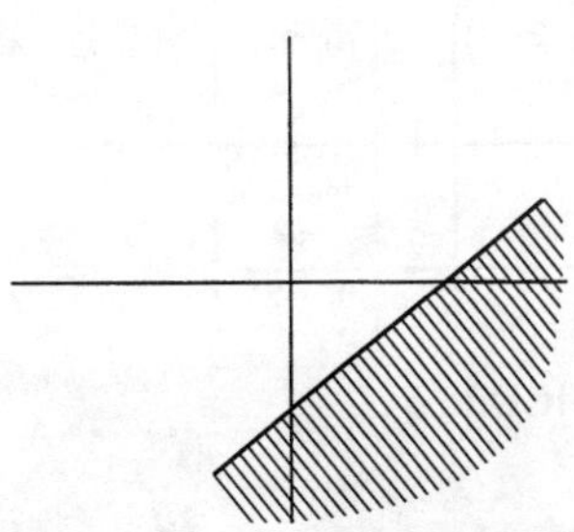

(C)

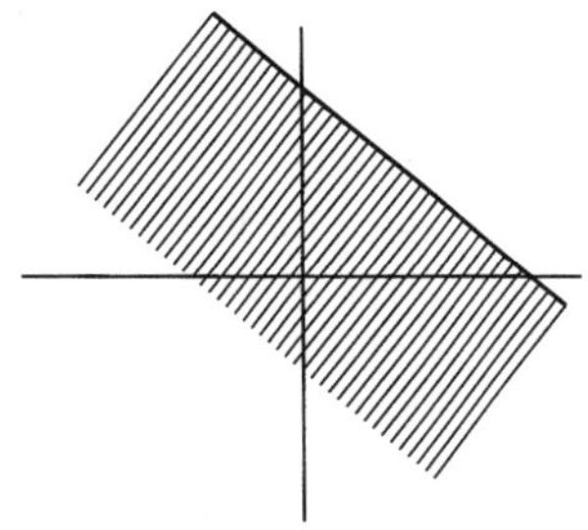

(D)

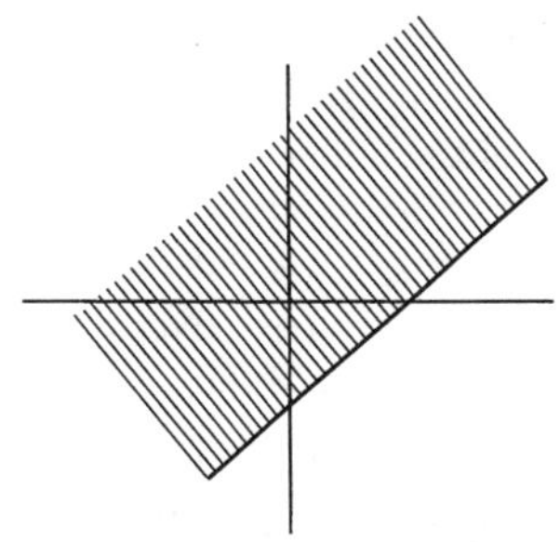

(a) $x + y \geq 1$ (b) $x + y \leq 1$

(c) $x - y \leq 1$ (d) $y - x \leq -1$

29. Each of the lines L_1, L_2, and L_3 in Fig. 1 is the graph of one of the equations (a), (b), and (c). Match each of the following equations with its corresponding line.

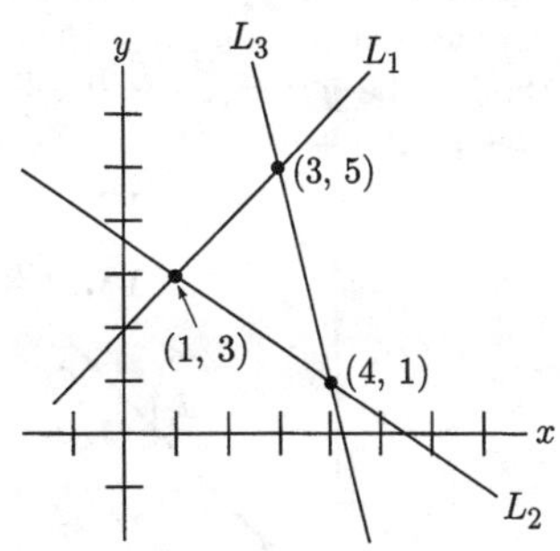

FIGURE 1

(a) $4x + y = 17$ (b) $y = x + 2$

(c) $2x + 3y = 11$

30. Find a system of inequalities having the feasible set in Fig. 2 and find the coordinates of the unspecified vertex. [*Note:* There is a right angle at the vertex $\left(4, \frac{3}{2}\right)$.]

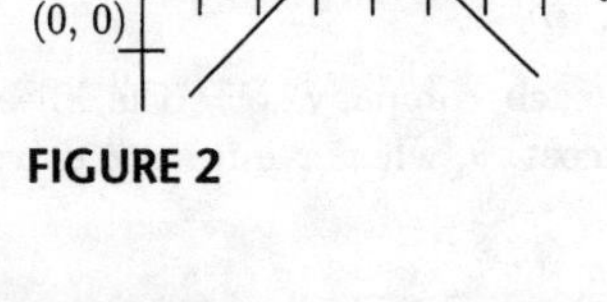

FIGURE 2

31. Consider the following four equations:

$$q = 100p + 500$$
$$q = 150p - 100$$
$$q = -75p - 750$$
$$q = -75p + 500.$$

One is the equation of a supply curve and another is the equation of a demand curve. Identify the two equations and then find the intersection point of those two curves.

32. Find the vertices of the following feasible set.

$$\begin{cases} x \geq 0 \\ y \geq 0 \\ 5x + y \leq 50 \\ 2x + 3y \leq 33 \\ x - 2y \geq -8 \end{cases}$$

33. It is not possible to draw a straight line through all three of the points $(2,4)$, $(5,8)$, and $(7,9)$. However, it *is* possible for a straight line to *miss* all three points by the same amount; that is, there is a line that makes the vertical distances d_1, d_2, and d_3 in Fig. 3 equal. Find the equation of this line. [*Hint*: Let the equation of the line be $y = mx + b$. Then the point P has coordinates $(2, 2m + b)$.]

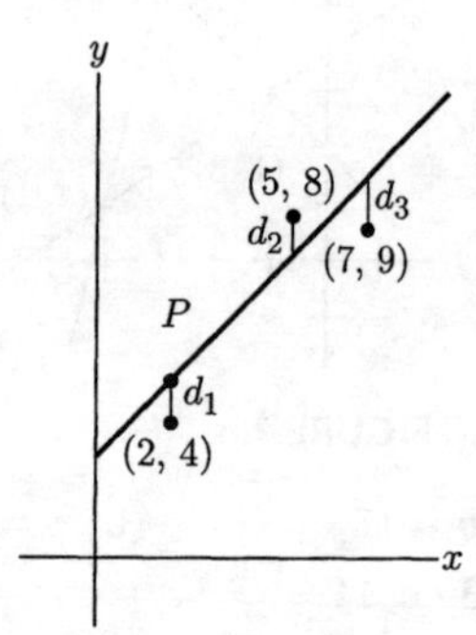

FIGURE 3

34. For a certain manufacturer, the production and sale of each additional unit yields an additional profit of \$10. The sale of 1000 units yields a profit of \$4000.

(a) Find the profit equation.

(b) Find the y- and x-intercepts of the graph of the profit equation.

(c) Sketch the graph of the profit equation.

35. One-day car rentals cost \$50 plus 10 cents per mile from company A and \$40 plus 20 cents per mile from company B.

(a) For each company, give the linear equations for the cost, y, when x miles are driven.

(b) Which company offers the best value when the car is driven for 80 miles?

(c) Which company offers the best value when the car is driven for 160 miles?

(d) For what mileage do the two companies offer the same value?

36. (***Linear Depreciation***) For accounting purposes, the value of certain items is depreciated linearly over time from the purchase price to the salvage value. Suppose that a computer was purchased in 1986 for \$5000 and sold in 1994 for \$1000.

(a) Find the linear equation giving its value, y, after x years.

(b) How much was the computer worth in 1990?

(c) When was the computer valued at \$2000?

37. Graph the linear inequality $x \leq 3y + 2$.

38. A furniture store offers its new employees a weekly salary of \$200 plus a 3% commission on sales. After one year, employees receive \$100 per week plus a 5% sales commission. For what weekly sales level will the two scales produce the same salary?

39. Find a system of linear inequalities having the feasible set of Fig. 4

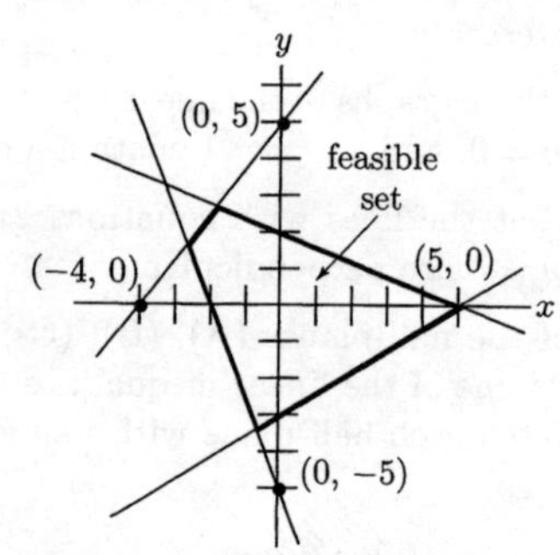

FIGURE 4

40. Find a system of linear inequalities having the feasible set of Fig. 5.

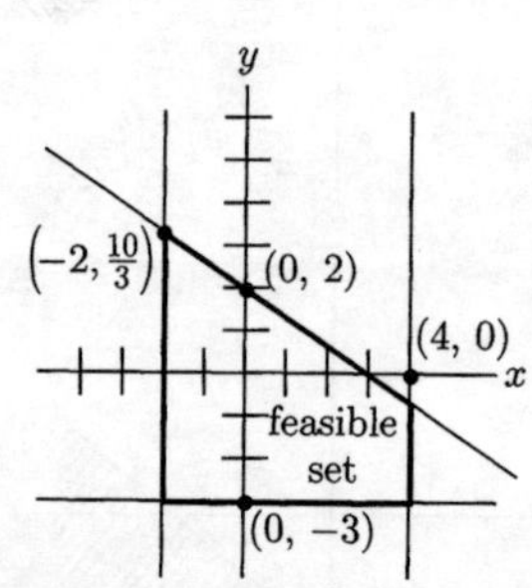

FIGURE 5

41. The following table gives the number (in thousands) of bachelor's degrees in business and administration earned in the United States during several years.[9]

Year	*Number of Degrees*
1980	185.9
1985	232.6
1990	248.7
1993	256.8

(a) Use the method of least squares to obtain the straight line that best fits these data.

(b) Estimate the number of degrees awarded in 1995.

(c) If the trend determined by the straight line in part (a) continues, when will the number of degrees awarded reach 300,000?

42. The following table[10] gives the (age-adjusted) death rate per 100,000 women from breast cancer and the daily dietary fat intake (in grams per day) for various countries.

Country	*Fat Intake*	*Death Rate*
Japan	41	4
Poland	90	10
Finland	118	13
United States	148	21

(a) Use the method of least squares to obtain the straight line that best fits these data.

(b) In Denmark women consume an average of 160 grams of fat per day. Use the least-squares line to estimate the breast cancer death rate.

(c) In New Zealand the breast cancer death rate is 22 women per 100,000. Use the least-squares line to estimate the daily fat intake in New Zealand.

[9]U.S. National Center for Educational Statistics, *Digest of Educational Statistics*.

[10]B.S. Reddy, et al.: "Nutrition and Its Relationship to Cancer," *Advances in Cancer Research* 32:237, 1980.

Chapter 7
Matrices

2.1 Solving Systems of Linear Equations, I

2.2 Solving Systems of Linear Equations, II

2.3 Arithmetic Operations on Matrices

2.4 The Inverse of a Matrix

2.5 The Gauss–Jordan Method for Calculating Inverses

2.6 Input–Output Analysis

We begin this chapter by developing a method for solving systems of linear equations in any number of variables. Our discussion of this method will lead naturally into the study of mathematical objects called *matrices*. The arithmetic and applications of matrices are the main topics of the chapter. We discuss in detail the application of matrix arithmetic to input–output analysis, which can be (and is) used to make production decisions for large businesses and entire economies.

2.1 Solving Systems of Linear Equations, I

In Chapter 1 we presented a method for solving systems of linear equations in two variables. The method of Chapter 1 is very efficient for determining the solutions. Unfortunately, it works only for systems of linear equations having *two* variables. In many applications we meet systems having more than two variables, as the following example illustrates.

EXAMPLE 1 The Upside Down company specializes in making down jackets, ski vests, and comforters. The requirements for down and labor and the profits earned are given in the following chart.

	Down (*pounds*)	*Time* (*labor-hours*)	*Profit* ($)
Jacket	3	2	6
Vest	2	1	6
Comforter	4	1	2

Each week the company has available 600 pounds of down and 275 labor-hours. It wants to earn a weekly profit of $1150. How many of each item should the company make each week?

Solution The requirements and earnings can be expressed by a system of equations. Let x be the number of jackets, y the number of vests, and z the number of comforters. If 600 pounds of down are used, then

$$\begin{aligned} [\text{down in jackets}] + [\text{down in vests}] + [\text{down in comforters}] &= 600. \\ 3\,[\text{no. jackets}] \;+\; 2\,[\text{no. vests}] \;+\; 4\,[\text{no. comforters}] &= 600. \end{aligned}$$

That is,

$$3x + 2y + 4z = 600.$$

Similarly, the equation for labor is

$$2x + y + z = 275,$$

and the equation for the profit is

$$6x + 6y + 2z = 1150.$$

The numbers x, y, and z must simultaneously satisfy a system of three linear equations in three variables.

$$\left\{\begin{aligned} 3x + 2y + 4z &= 600 \\ 2x + y + z &= 275 \\ 6x + 6y + 2z &= 1150 \end{aligned}\right. \tag{1}$$

Later we present a method for determining the solution to this system. This method yields the solution $x = 50$, $y = 125$, and $z = 50$. It is easy to confirm that these values of x, y, and z satisfy all three equations:

$$\begin{aligned} 3(50) + 2(125) + 4(50) &= 600 \\ 2(50) + (125) + (50) &= 275 \\ 6(50) + 6(125) + 2(50) &= 1150. \end{aligned}$$

Thus, the Upside Down company can make a profit of $1150 by producing 50 jackets, 125 vests, and 50 comforters. ▮

In this section we develop a step-by-step procedure for solving systems of linear equations such as (1). The procedure, called the *Gaussian elimination method*, consists of repeatedly simplifying the system, using so-called elementary row operations, until the solution stares us in the face!

In the system of linear equations (1) the equations have been written in such a way that the x-terms, the y-terms, and the z-terms lie in different columns. We shall always be careful to display systems of equations with separate columns for each variable. One of the key ideas of the Gaussian elimination method is to think of the solution as a system of linear equations in its own right. For example, we can write the solution of system (1) as:

$$\begin{cases} x \qquad\quad = 50 \\ \quad y \qquad = 125 \\ \qquad\quad z = 50. \end{cases} \tag{2}$$

This is just a system of linear equations in which the coefficients of most terms are zero! Since the only terms with nonzero coefficients are arranged on a diagonal, such a system is said to be in *diagonal form.*

Our method for solving a system of linear equations consists of repeatedly using three operations that alter the system but do not change the solutions. The operations are used to transform the system into a system in diagonal form. Since the operations involve only elementary arithmetic and are applied to entire equations (i.e., rows of the system), they are called *elementary row operations.* Let us begin our study of the Gaussian elimination method by introducing these operations.

Elementary Row Operation 1

Rearrange the equations in any order.

This operation is harmless enough. It certainly does not change the solutions of the system.

Elementary Row Operation 2

Multiply an equation by a nonzero number.

For example, if we are given the system of linear equations

$$\begin{cases} 2x - 3y + 4z = 11 \\ 4x - 19y + z = 31 \\ 5x + 7y - z = 12, \end{cases}$$

then we may replace it by a new system obtained by leaving the last two equations unchanged and multiplying the first equation by 3. To accomplish this, multiply each term of the first equation by 3. The transformed system is

$$\begin{cases} 6x - 9y + 12z = 33 \\ 4x - 19y + z = 31 \\ 5x + 7y - z = 12. \end{cases}$$

The operation of multiplying an equation by a nonzero number does not change the solutions of the system. For if a particular set of values of the variables satisfies the original equation, it satisfies the resulting equation, and vice versa.

Elementary row operation 2 may be used to make the coefficient of a particular variable 1.

EXAMPLE 2 Replace the system

$$\begin{cases} -5x + 10y + 20z = 4 \\ x \qquad\quad - 12z = 1 \\ x + \quad y + \quad z = 0 \end{cases}$$

by an equivalent system in which the coefficient of x in the first equation is 1.

Solution The coefficient of x in the first equation is -5, so we use elementary row operation 2 to multiply the first equation by $-\frac{1}{5}$. Multiplying each term of the first equation by $-\frac{1}{5}$ gives

$$\begin{cases} x - 2y - \ 4z = -\frac{4}{5} \\ x \qquad - 12z = \ \ 1 \\ x + \ y + \quad z = \ \ 0. \end{cases}$$ ∎

Another operation that can be performed on a system without changing its solutions is to replace one equation by its sum with some other equation. For example, consider this system of equations:

$$\text{A:} \quad \begin{cases} x + \ y - \ 2z = \ 3 \\ x + 2y - \ 5z = \ 4 \\ 5x + 8y - 18z = 14. \end{cases}$$

We can replace the second equation by the sum of the first and the second. Since

$$\begin{array}{r} x + \ y - 2z = 3 \\ + \quad x + 2y - 5z = 4 \\ \hline 2x + 3y - 7z = 7, \end{array}$$

the resulting system is

$$\text{B:} \quad \begin{cases} x + \ y - \ 2z = \ 3 \\ 2x + 3y - \ 7z = \ 7 \\ 5x + 8y - 18z = 14. \end{cases}$$

If a particular choice of x, y, and z satisfies system A, it also satisfies system B. This is because system B results from adding equations. Similarly, system A can be derived from system B by subtracting equations. So any particular solution of system A is a solution of system B, and vice versa.

The operation of adding equations is usually used in conjunction with elementary row operation 2. That is, an equation is changed by adding to it a nonzero multiple of another equation. For example, consider the system

$$\begin{cases} x + \ y - \ 2z = \ 3 \\ x + 2y - \ 5z = \ 4 \\ 5x + 8y - 18z = 14. \end{cases}$$

Let us change the second equation by adding to it twice the first. Since

$$\begin{array}{rr} 2(\text{first}) & 2x + 2y - 4z = \ 6 \\ +\ (\text{second}) & x + 2y - 5z = \ 4 \\ \hline & 3x + 4y - 9z = 10, \end{array}$$

the new second equation is

$$3x + 4y - 9z = 10$$

and the transformed system is

$$\begin{cases} x + y - 2z = 3 \\ 3x + 4y - 9z = 10 \\ 5x + 8y - 18z = 14. \end{cases}$$

Since addition of equations and elementary row operation 2 are often used together, let us define a third elementary row operation to be the combination:

> ***Elementary Row Operation 3***
>
> Change an equation by adding to it a multiple of another equation.

For reference, let us summarize the elementary row operations we have just defined.

> ***Elementary Row Operations***
>
> 1. Rearrange the equations in any order.
> 2. Multiply an equation by a nonzero number.
> 3. Change an equation by adding to it a multiple of another equation.

The idea of the Gaussian elimination method is to transform an arbitrary system of linear equations into diagonal form by repeated applications of the three elementary row operations. To see how the method works, consider the following example.

EXAMPLE 3 Solve the following system by the Gaussian elimination method:

$$\begin{cases} x - 3y = 7 \\ -3x + 4y = -1. \end{cases}$$

Solution Let us transform this system into diagonal form by examining one column at a time, starting from the left. Examine the first column:

$$\begin{matrix} x \\ -3x \end{matrix}$$

The coefficient of the top x is 1, which is exactly what it should be for the system to be in diagonal form. So we do nothing to this term. Now examine the next term in the column, $-3x$. In diagonal form this term must be absent. In order to accomplish this, we add a multiple of the first equation to the second. Since the coefficient of x in the second is -3, we add three times the first equation to the second equation in order to cancel the x-term. (Abbreviation: $[2]+3[1]$. The $[2]$ means that we are changing equation 2. The expression $[2]+3[1]$ means that we are replacing equation 2 by the original equation plus three times equation 1.)

$$\begin{cases} x - 3y = 7 \\ -3x + 4y = -1 \end{cases} \xrightarrow{[2]+3[1]} \begin{cases} x - 3y = 7 \\ - 5y = 20 \end{cases}$$

The first column now has the proper form, so we proceed to the second column. In diagonal form that column will have one nonzero term, namely the second, and the coefficient of y in that term must be 1. To bring this about, multiply the second equation by $-\frac{1}{5}$ (abbreviation: $-\frac{1}{5}[2]$):

$$\begin{cases} x - 3y = 7 \\ \quad - 5y = 20 \end{cases} \xrightarrow{-\frac{1}{5}[2]} \begin{cases} x - 3y = 7 \\ \quad y = -4. \end{cases}$$

The second column still does not have the correct form. We must get rid of the $-3y$-term in the first equation. We do this by adding a multiple of the second equation to the first. Since the coefficient of the term to be canceled is -3, we add three times the second equation to the first:

$$\begin{cases} x - 3y = 7 \\ \quad y = -4 \end{cases} \xrightarrow{[1]+3[2]} \begin{cases} x \quad = -5 \\ \quad y = -4. \end{cases}$$

The system is now in diagonal form and the solution can be read off: $x = -5$, $y = -4$. ▮

▮ EXAMPLE 4 Use the Gaussian elimination method to solve the system

$$\begin{cases} 2x - 6y = -8 \\ -5x + 13y = 1. \end{cases}$$

Solution We can perform the calculations in a mechanical way, proceeding column by column from the left:

$$\begin{cases} 2x - 6y = -8 \\ -5x + 13y = 1 \end{cases} \xrightarrow{\frac{1}{2}[1]} \begin{cases} x - 3y = -4 \\ -5x + 13y = 1 \end{cases}$$

$$\xrightarrow{[2]+5[1]} \begin{cases} x - 3y = -4 \\ \quad - 2y = -19 \end{cases}$$

$$\xrightarrow{-\frac{1}{2}[2]} \begin{cases} x - 3y = -4 \\ \quad y = \frac{19}{2} \end{cases}$$

$$\xrightarrow{[1]+3[2]} \begin{cases} x \quad = \frac{49}{2} \\ \quad y = \frac{19}{2}. \end{cases}$$

So the solution of the system is $x = \frac{49}{2}$, $y = \frac{19}{2}$. ▮

The calculation becomes easier to follow if we omit writing down the variables at each stage and work only with the coefficients. At each stage of the computation, the system is represented by a rectangular array of numbers. For instance, the original system is written[1]

$$\left[\begin{array}{rr|r} 2 & -6 & -8 \\ -5 & 13 & 1 \end{array}\right].$$

The elementary row operations are performed on the rows of this rectangular array just as if the variables were there. So, for example, the first step above is

[1]The vertical line between the second and third columns is a placemarker that separates the data obtained from the right- and left-hand sides of the equations. It is inserted for visual convenience.

to multiply the first equation by $\frac{1}{2}$. This corresponds to multiplying the first row of the array by $\frac{1}{2}$ to get

$$\left[\begin{array}{cc|c} 1 & -3 & -4 \\ -5 & 13 & 1 \end{array}\right].$$

The diagonal form just corresponds to the array

$$\left[\begin{array}{cc|c} 1 & 0 & \frac{49}{2} \\ 0 & 1 & \frac{19}{2} \end{array}\right].$$

Note that this array has ones down the diagonal and zeros everywhere else on the left. The solution of the system appears on the right.

A rectangular array of numbers is called a *matrix* (plural: *matrices*). In the next example, we use matrices to carry out the Gaussian elimination method.

EXAMPLE 5 Use the Gaussian elimination method to solve the system

$$\begin{cases} 3x - 6y + 9z = 0 \\ 4x - 6y + 8z = -4 \\ -2x - y + z = 7. \end{cases}$$

Solution The initial array corresponding to the system is

$$\left[\begin{array}{ccc|c} 3 & -6 & 9 & 0 \\ 4 & -6 & 8 & -4 \\ -2 & -1 & 1 & 7 \end{array}\right].$$

We must use elementary row operations to transform this array into diagonal form—that is, with ones down the diagonal and zeros everywhere else on the left:

$$\left[\begin{array}{ccc|c} 1 & 0 & 0 & * \\ 0 & 1 & 0 & * \\ 0 & 0 & 1 & * \end{array}\right].$$

We proceed one column at a time.

$$\left[\begin{array}{ccc|c} 3 & -6 & 9 & 0 \\ 4 & -6 & 8 & -4 \\ -2 & -1 & 1 & 7 \end{array}\right] \xrightarrow{\frac{1}{3}[1]} \left[\begin{array}{ccc|c} 1 & -2 & 3 & 0 \\ 4 & -6 & 8 & -4 \\ -2 & -1 & 1 & 7 \end{array}\right] \xrightarrow{[2]+(-4)[1]}$$

$$\left[\begin{array}{ccc|c} 1 & -2 & 3 & 0 \\ 0 & 2 & -4 & -4 \\ -2 & -1 & 1 & 7 \end{array}\right] \xrightarrow{[3]+2[1]} \left[\begin{array}{ccc|c} 1 & -2 & 3 & 0 \\ 0 & 2 & -4 & -4 \\ 0 & -5 & 7 & 7 \end{array}\right] \xrightarrow{\frac{1}{2}[2]}$$

$$\left[\begin{array}{ccc|c} 1 & -2 & 3 & 0 \\ 0 & 1 & -2 & -2 \\ 0 & -5 & 7 & 7 \end{array}\right] \xrightarrow{[1]+2[2]} \left[\begin{array}{ccc|c} 1 & 0 & -1 & -4 \\ 0 & 1 & -2 & -2 \\ 0 & -5 & 7 & 7 \end{array}\right] \xrightarrow{[3]+5[2]}$$

$$\left[\begin{array}{ccc|c} 1 & 0 & -1 & -4 \\ 0 & 1 & -2 & -2 \\ 0 & 0 & -3 & -3 \end{array}\right] \xrightarrow{(-\frac{1}{3})[3]} \left[\begin{array}{ccc|c} 1 & 0 & -1 & -4 \\ 0 & 1 & -2 & -2 \\ 0 & 0 & 1 & 1 \end{array}\right] \xrightarrow{[1]+1[3]}$$

$$\left[\begin{array}{ccc|c} 1 & 0 & 0 & -3 \\ 0 & 1 & -2 & -2 \\ 0 & 0 & 1 & 1 \end{array}\right] \xrightarrow{[2]+2[3]} \left[\begin{array}{ccc|c} 1 & 0 & 0 & -3 \\ 0 & 1 & 0 & 0 \\ 0 & 0 & 1 & 1 \end{array}\right].$$

The last array is in diagonal form, so we just put back the variables and read off the solution:

$$x = -3, \qquad y = 0, \qquad z = 1.$$

Because so much arithmetic has been performed, it is a good idea to check the solution by substituting the values for x, y, and z into each of the equations of the original system. This will uncover any arithmetic errors that may have occurred.

$$\begin{cases} 3x - 6y + 9z = 0 \\ 4x - 6y + 8z = -4 \\ -2x - y + z = 7 \end{cases} \qquad \begin{cases} 3(-3) - 6(0) + 9(1) = 0 \\ 4(-3) - 6(0) + 8(1) = -4 \\ -2(-3) - (0) + (1) = 7 \end{cases}$$

$$\begin{cases} -9 - 0 + 9 = 0 \\ -12 - 0 + 8 = -4 \\ 6 - 0 + 1 = 7 \end{cases}$$

$$\begin{cases} 0 = 0 \\ -4 = -4 \\ 7 = 7. \end{cases}$$

So we have indeed found a solution of the system. ▮

Remark Note that so far we have not had to use elementary row operation 1, which allows interchange of equations. But in some examples it is definitely needed.

Consider this system:

$$\begin{cases} y + z = 0 \\ 3x - y + z = 6 \\ 6x - z = 3. \end{cases}$$

The first step of the Gaussian elimination method consists of making the x-coefficient 1 in the first equation. But we cannot do this, since the first equation does not involve x. To remedy this difficulty, just interchange the first two equations to guarantee that the first equation involves x. Now proceed as before. Of course, in terms of the matrix of coefficients, interchanging equations corresponds to interchanging rows of the matrix.

GC Graphing utilities can perform many operations with matrices, including elementary row operations. With a TI-82 or TI-83, pressing MATRX presents three menus. See Fig. 1. You define a new matrix or alter an existing matrix with the EDIT menu. You place the name of an existing matrix on the home screen with the NAMES menu. You perform operations on existing matrices with the MATH menu.

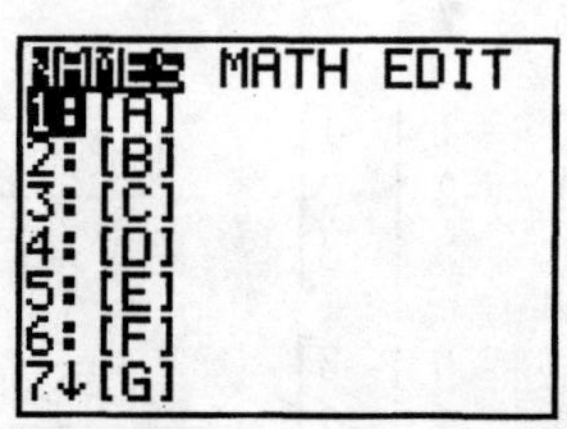

FIGURE 1

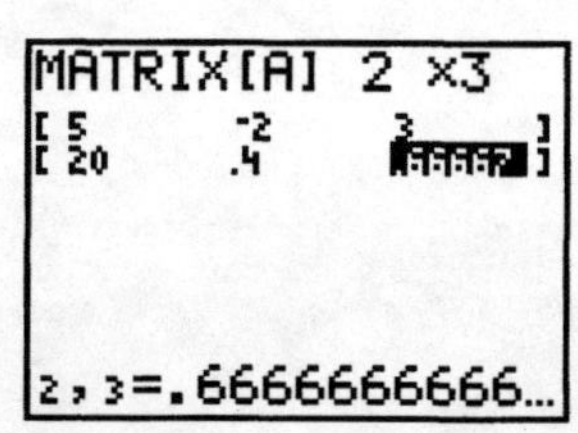

FIGURE 2

With the EDIT menu, you first select one of the names `[A]`, `[B]`, `[C]`, ... for the matrix and then specify the size and fill in the entries as shown in Fig. 2. The three elementary row operations are carried out with commands from the MATRX/MATH menu of the form

`rowSwap(`*matrix, rowA, rowB*`)`	Interchange *rowA* and *rowB* of *matrix*
`*row(`*value, matrix, row*`)`	Multiply *row* of *matrix* by *value*
`*row+(`*value, matrix, rowA, rowB*`)`	Add *value*∗*rowA* to *rowB* of *matrix*

When one of these commands is carried out, the resulting matrix is displayed but the stored matrix is not changed. Therefore, when a sequence of commands is executed to carry out the Gaussian elimination method, each command should be followed with → *matrix* to change the stored matrix. (*Note*: To display the name of a matrix, press MATRX, cursor down to the matrix, and press ENTER. To display an arrow, press STO ▶.) For instance, if in Example 5 the original matrix is named [A], then the first three row operations are carried out with

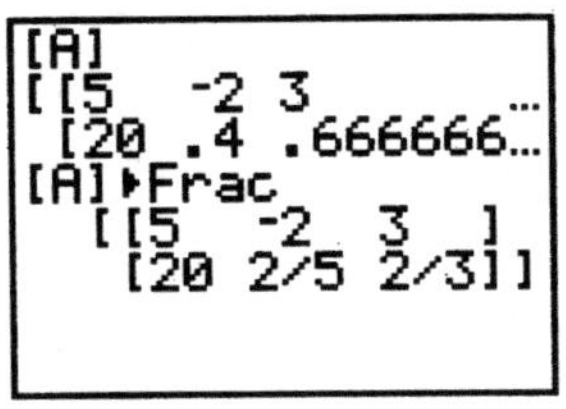

FIGURE 3

`*row(1/3,[A],1)` STO ▶ `[A]`

`*row+(-4,[A],1,2)` STO ▶ `[A]`

`*row+(2,[A],1,3)` STO ▶ `[A]`

Matrix entries are normally displayed on the calculator as decimals. From the home screen you can display the entries as fractions with a command such as `[A] ▶ Frac` or `Ans ▶ Frac`. (`▶ Frac` is displayed by pressing MATH **1**.) See Fig. 3.

Practice Problems 1

1. Determine whether the following systems of linear equations are in diagonal form.

(a) $\begin{cases} x \quad + z = 3 \\ \quad y \quad = 2 \\ \qquad z = 7 \end{cases}$ (b) $\begin{cases} x \qquad = 3 \\ \qquad y = 5 \\ \quad z \quad = 7 \end{cases}$ (c) $\begin{cases} x \qquad = -1 \\ \quad y \quad = 0 \\ \qquad 3z = 4 \end{cases}$

2. Perform the indicated elementary row operation.

(a) $\begin{cases} x - 3y = 2 \\ 2x + 3y = 5 \end{cases} \xrightarrow{[2]+(-2)[1]}$ (b) $\begin{cases} x + y = 3 \\ -x + 2y = 5 \end{cases} \xrightarrow{[2]+(1)[1]}$

3. State the next elementary row operation that should be performed when applying the Gaussian elimination method.

(a) $\left[\begin{array}{ccc|c} 0 & 2 & 4 & 1 \\ 0 & 3 & -7 & 0 \\ 3 & 6 & -3 & 3 \end{array}\right]$ (b) $\left[\begin{array}{ccc|c} 1 & -3 & 4 & 5 \\ 0 & 2 & 3 & 4 \\ -6 & 5 & -7 & 0 \end{array}\right]$

EXERCISES 1

In Exercises 1–8, perform the indicated elementary row operations and give their abbreviations.

1. Operation 2: Multiply the first equation by 2.

$$\begin{cases} \frac{1}{2}x - 3y = 2 \\ 5x + 4y = 1 \end{cases}$$

2. Operation 2: Multiply the second equation by −1.

$$\begin{cases} x + 4y = 6 \\ \quad -y = 2 \end{cases}$$

3. Operation 3: Change the second equation by adding to it 5 times the first equation.

$$\begin{cases} x + 2y = 3 \\ -5x + 4y = 1 \end{cases}$$

4. Operation 3: Change the second equation by adding to it $\left(-\frac{1}{2}\right)$ times the first equation.

$$\begin{cases} x - 6y = 4 \\ \frac{1}{2}x + 2y = 1 \end{cases}$$

5. Operation 3: Change the third equation by adding to it (-4) times the first equation.

$$\begin{cases} x - 2y + z = 0 \\ y - 2z = 4 \\ 4x + y + 3z = 5 \end{cases}$$

6. Operation 3: Change the third equation by adding to it 3 times the second equation.

$$\begin{cases} x + 6y - 4z = 1 \\ y + 3z = 1 \\ -3y + 7z = 2 \end{cases}$$

7. Operation 3: Change the first row by adding to it $\frac{1}{2}$ times the second row.

$$\left[\begin{array}{cc|c} 1 & -\frac{1}{2} & 3 \\ 0 & 1 & 4 \end{array}\right]$$

8. Operation 3: Change the third row by adding to it (-4) times the second row.

$$\left[\begin{array}{ccc|c} 1 & 0 & 7 & 9 \\ 0 & 1 & -2 & 3 \\ 0 & 4 & 8 & 5 \end{array}\right]$$

In Exercises 9–16, state the next elementary row operation that should be performed in order to put the matrix into diagonal form. Do not perform the operation.

9. $\left[\begin{array}{cc|c} 1 & -5 & 1 \\ -2 & 4 & 6 \end{array}\right]$

10. $\left[\begin{array}{cc|c} 1 & 3 & 4 \\ 0 & 2 & 6 \end{array}\right]$

11. $\left[\begin{array}{cc|c} 1 & 2 & 3 \\ 0 & 1 & 4 \end{array}\right]$

12. $\left[\begin{array}{ccc|c} 1 & -2 & 5 & 7 \\ 0 & -3 & 6 & 9 \\ 4 & 5 & -6 & 7 \end{array}\right]$

13. $\left[\begin{array}{ccc|c} 0 & 5 & -3 & 6 \\ 2 & -3 & 4 & 5 \\ 4 & 1 & -7 & 8 \end{array}\right]$

14. $\left[\begin{array}{ccc|c} 1 & 4 & -2 & 5 \\ 0 & -3 & 6 & 9 \\ 0 & 4 & 3 & 1 \end{array}\right]$

15. $\left[\begin{array}{ccc|c} 1 & 0 & 3 & 4 \\ 0 & 1 & 2 & 5 \\ 0 & 0 & 1 & 6 \end{array}\right]$

16. $\left[\begin{array}{ccc|c} 1 & 2 & 4 & 5 \\ 0 & 0 & 3 & 6 \\ 0 & 1 & 1 & 7 \end{array}\right]$

In Exercises 17–30, solve the linear system by using the Gaussian elimination method.

17. $\begin{cases} 3x + 9y = 6 \\ 2x + 8y = 6 \end{cases}$

18. $\begin{cases} \frac{1}{3}x + 2y = 1 \\ -2x - 4y = 6 \end{cases}$

19. $\begin{cases} x - 3y + 4z = 1 \\ 4x - 10y + 10z = 4 \\ -3x + 9y - 5z = -6 \end{cases}$

20. $\begin{cases} \frac{1}{2}x + y = 4 \\ -4x - 7y + 3z = -31 \\ 6x + 14y + 7z = 50 \end{cases}$

21. $\begin{cases} 2x - 2y + 4 = 0 \\ 3x + 4y - 1 = 0 \end{cases}$

22. $\begin{cases} 2x + 3y = 4 \\ -x + 2y = -2 \end{cases}$

23. $\begin{cases} 4x - 4y + 4z = -8 \\ x - 2y - 2z = -1 \\ 2x + y + 3z = 1 \end{cases}$

24. $\begin{cases} x + 2y + 2z - 11 = 0 \\ x - y - z + 4 = 0 \\ 2x + 5y + 9z - 39 = 0 \end{cases}$

25. $\begin{cases} .2x + .3y = 4 \\ .6x + 1.1y = 15 \end{cases}$

26. $\begin{cases} \frac{3}{2}x + 6y = 9 \\ \frac{1}{2}x - \frac{2}{3}y = 11 \end{cases}$

27. $\begin{cases} x + y + 4z = 3 \\ 4x + y - 2z = -6 \\ -3x + 2z = 1 \end{cases}$

28. $\begin{cases} -2x - 3y + 2z = -2 \\ x + y = 3 \\ -x - 3y + 5z = 8 \end{cases}$

29. $\begin{cases} -x + y = -1 \\ x + z = 4 \\ 6x - 3y + 2z = 10 \end{cases}$

30. $\begin{cases} x + 2z = 9 \\ y + z = 1 \\ 3x - 2y = 9 \end{cases}$

31. **(*From a GMAT exam*)** A rabbit on a controlled diet is fed daily 300 grams of a mixture of two foods, food X and food Y. Food X contains 10% protein, and food Y contains 15% protein. If the rabbit's diet provides exactly 38 grams of protein daily, how many grams of food X are in the mixture?

(a) 100 (b) 140 (c) 150 (d) 160 (e) 200

32. **(*From a GMAT exam*)** Elena purchased brand X pens for \$4.00 apiece and brand Y pens for \$2.80 apiece. If Elena purchased a total of 12 of these pens for \$42.00, how many brand X pens did she purchase?

(a) 4 (b) 5 (c) 6 (d) 7 (e) 8

33. A 600-seat movie theater charges \$5.50 admission for adults and \$2.50 for children. If the theater is full and \$1911 is collected, how many adults and how many children are in the audience?

34. A baseball player's batting average is determined by dividing the number of hits by the number of times at bat and multiplying by 1000. (Batting averages are usually, but not necessarily, rounded to the nearest whole number.) For instance, if a player gets 2 hits in 5 times at bat, his batting average is 400: $\left(\frac{2}{5} \times 1000 = 400\right)$. Partway through the season, a player thinks to himself: "If I get a hit in my next time at bat, my average will go up to 250; if I don't get a hit, it will drop to 187.5." How many times has this player batted, how many hits has he had, and what is his current batting average?

35. A bank wishes to invest a \$100,000 trust fund in three sources: bonds paying 8%; certificates of deposit paying 7%; and first mortgages paying 10%. The bank wishes to realize an \$8000 annual income from the investment. A condition of the trust is that the total amount invested in bonds and certificates of deposit must be triple the amount invested in mortgages. How much should the bank invest in each possible category? Let x, y, and z, respectively, be the amounts invested in bonds, certificates of deposit, and first mortgages. Solve the system of equations by the Gaussian elimination method.

36. A dietitian wishes to plan a meal around three foods. Each ounce of food I contains 10% of the daily requirements for carbohydrates, 10% for protein, and 15% for vitamin C. Each ounce of food II contains 10% of the daily requirements for carbohydrates, 5% for protein, and 0% for vitamin C. Each ounce of food III contains 10% of the daily requirements for carbohydrates, 25% for protein, and 10% for vitamin C. How many ounces of each food should be served in order to supply exactly the daily requirements for each nutrient? Let x, y, and z, respectively, be the number of ounces of foods I, II, and III.

Exercises 37–40 require the use of a graphing calculator or a computer. Most graphing calculators are able to perform an elementary row operation with a single instruction.

37. Enter the matrix corresponding to the equations in Exercise 3, and change the second row by adding to it 5 times the first row.

38. Enter the matrix from Exercise 5, and change the third row by adding to it (-4) times the first row.

39. Enter the matrix in Exercise 7, and interchange the rows.

40. Enter the matrix corresponding to the linear system in Exercise 25, and multiply the first row by 5.

In Exercises 41 and 42, solve the system of linear equations by the Gaussian elimination method.

41. $$\begin{cases} 2x + 2y + 2z + 4w = -12 \\ y + z + w = -5 \\ z + 2w = -6 \\ x + y + z + 4w = -14 \end{cases}$$

42. $$\begin{cases} 2x - z + 5w = 19 \\ 2x + 3y - z + 5w = 28 \\ - z + 5w = 21 \\ 6x - 3z = -3 \end{cases}$$

Solutions to Practice Problems 1

1. (a) Not in diagonal form, since the first equation contains both x and z.
   (b) Not in diagonal form, since the variables are not arranged in diagonal fashion.
   (c) Not in diagonal form, since the coefficient of z is not 1.
2. (a) Change the system into another system in which the second equation is altered by having (-2)(first equation) added to it. The new system is

$$\begin{cases} x - 3y = 2 \\ 9y = 1. \end{cases}$$

The equation $9y = 1$ was obtained as follows:

$$\begin{array}{lr} (-2)(\text{first equation}) & -2x + 6y = -4 \\ +\ (\text{second equation}) & \underline{2x + 3y = 5} \\ & 9y = 1. \end{array}$$

(b) Replace the second equation by the first equation multiplied by 1 and added to the second. This is the same thing as adding the first equation to the second. The result is

$$\begin{cases} x + \ \ y = 3 \\ \quad\ \ 3y = 8. \end{cases}$$

3. (a) The first row should contain a nonzero number as its first entry. This can be accomplished by interchanging the first and third rows.

(b) The first column can be put into proper form by eliminating the −6. To accomplish this, multiply the first row by 6 and add this product to the third row. The notation for this operation is

$$\xrightarrow{[3]+6[1]}$$

2.2 Solving Systems of Linear Equations, II

In this section we introduce the operation of pivoting and consider systems of linear equations that do not have unique solutions.

Roughly speaking, the Gaussian elimination method applied to a matrix proceeds as follows: Consider the columns one at a time, from left to right. For each column use the elementary row operations to transform the appropriate entry to a one and the remaining entries in the column to zeros. (The "appropriate" entry is the first entry in the first column, the second entry in the second column, and so forth.) This sequence of elementary row operations performed for each column is called *pivoting*. More precisely:

> *Method* To pivot a matrix about a given nonzero entry:
>
> 1. Transform the given entry into a one.
> 2. Transform all other entries in the same column into zeros.

Pivoting is used in solving problems other than systems of linear equations. As we shall see in Chapter 4, it is the basis for the simplex method of solving linear programming problems.

EXAMPLE 1 Pivot the matrix about the circled element.

$$\left[\begin{array}{cc|c} 18 & \textcircled{-6} & 15 \\ 5 & -2 & 4 \end{array}\right]$$

Solution The first step is to transform the −6 to a 1. We do this by multiplying the first row by $-\frac{1}{6}$:

$$\left[\begin{array}{cc|c} 18 & -6 & 15 \\ 5 & -2 & 4 \end{array}\right] \xrightarrow{-\frac{1}{6}[1]} \left[\begin{array}{cc|c} -3 & 1 & -\frac{5}{2} \\ 5 & -2 & 4 \end{array}\right].$$

Next, we transform the −2 (the only remaining entry in column 2) into a 0:

$$\left[\begin{array}{cc|c} -3 & 1 & -\frac{5}{2} \\ 5 & -2 & 4 \end{array}\right] \xrightarrow{[2]+2[1]} \left[\begin{array}{cc|c} -3 & 1 & -\frac{5}{2} \\ -1 & 0 & -1 \end{array}\right].$$

The last matrix is the result of pivoting the original matrix about the circled entry. ▮

In terms of pivoting, we can give the following summary of the Gaussian elimination method.

Gaussian Elimination Method to Transform a System of Linear Equations into Diagonal Form

1. Write down the matrix corresponding to the linear system.
2. Make sure that the first entry in the first column is nonzero. Do this by interchanging the first row with one of the rows below it, if necessary.
3. Pivot the matrix about the first entry in the first column.
4. Make sure that the second entry in the second column is nonzero. Do this by interchanging the second row with one of the rows below it, if necessary.
5. Pivot the matrix about the second entry in the second column.
6. Continue in this manner.

All the systems considered in the preceding section had only a single solution. In this case we say that the solution is *unique.* Let us now use the Gaussian elimination method to study the various possibilities other than a unique solution. We first experiment with an example.

▮ EXAMPLE 2 Determine all solutions of the system

$$\begin{cases} 2x + 2y + 4z = 8 \\ x - y + 2z = 2 \\ -x + 5y - 2z = 2. \end{cases}$$

Solution We set up the matrix corresponding to the system and perform the appropriate pivoting operations. (The elements pivoted about are circled.)

$$\left[\begin{array}{ccc|c} \textcircled{2} & 2 & 4 & 8 \\ 1 & -1 & 2 & 2 \\ -1 & 5 & -2 & 2 \end{array}\right] \xrightarrow[\substack{[3]+(1)[1]}]{\substack{\frac{1}{2}[1] \\ [2]+(-1)[1]}} \left[\begin{array}{ccc|c} 1 & 1 & 2 & 4 \\ 0 & \textcircled{-2} & 0 & -2 \\ 0 & 6 & 0 & 6 \end{array}\right]$$

$$\xrightarrow[\substack{[3]+(-6)[2]}]{\substack{\left(-\frac{1}{2}\right)[2] \\ [1]+(-1)[2]}} \left[\begin{array}{ccc|c} 1 & 0 & 2 & 3 \\ 0 & 1 & 0 & 1 \\ 0 & 0 & 0 & 0 \end{array}\right].$$

Note that our method must terminate here, since there is no way to transform the third entry in the third column into a 1 without disturbing the columns already in appropriate form. The equations corresponding to the last matrix read

$$\begin{cases} x + 2z = 3 \\ y = 1 \\ 0 = 0. \end{cases}$$

The last equation does not involve any of the variables and so may be omitted. This leaves the two equations

$$\begin{cases} x \quad\;\; + 2z = 3 \\ \quad y \qquad\;\; = 1. \end{cases}$$

Now, taking the $2z$-term in the first equation to the right side, we can write the equations

$$\begin{cases} x = 3 - 2z \\ y = 1. \end{cases}$$

The value of y is given: $y = 1$. The value of x is given in terms of z. To find a solution to this system, assign any value to z. Then the first equation gives a value for x and thereby a specific solution to the system. For example, if we take $z = 1$, then the corresponding specific solution is

$$\begin{aligned} z &= 1 \\ x &= 3 - 2(1) = 1 \\ y &= 1. \end{aligned}$$

If we take $z = -3$, the corresponding specific solution is

$$\begin{aligned} z &= -3 \\ x &= 3 - 2(-3) = 9 \\ y &= 1. \end{aligned}$$

Thus, we see that the original system has infinitely many specific solutions, corresponding to the infinitely many possible different choices for z.

We say that the *general solution* of the system is

$$\begin{aligned} z &= \text{any value} \\ x &= 3 - 2z \\ y &= 1. \end{aligned}$$ ▮

When a linear system cannot be *completely* diagonalized:

1. Apply the Gaussian elimination method to as many columns as possible. Proceed from left to right, but do not disturb columns that have already been put into proper form.
2. Variables corresponding to columns not in proper form can assume any value.
3. The other variables can be expressed in terms of the variables of step 2.

▮ EXAMPLE 3 Find all solutions of the linear system

$$\begin{cases} x + 2y - \;\; z + 3w = 5 \\ \qquad\; y + 2z + \;\; w = 7. \end{cases}$$

Solution The Gaussian elimination method proceeds as follows:

$$\left[\begin{array}{cccc|c} 1 & 2 & -1 & 3 & 5 \\ 0 & \textcircled{1} & 2 & 1 & 7 \end{array}\right] \quad \text{(The first column is already in proper form.)}$$

$$\xrightarrow{[1]+(-2)[2]} \left[\begin{array}{cccc|c} 1 & 0 & -5 & 1 & -9 \\ 0 & 1 & 2 & 1 & 7 \end{array}\right].$$

We cannot do anything further with the last two columns (without disturbing the first two columns), so the corresponding variables, z and w, can assume any values. Writing down the equations corresponding to the last matrix yields

$$\begin{cases} x \phantom{{}+y} - 5z + w = -9 \\ \phantom{x+{}} y + 2z + w = 7, \end{cases}$$

or

$$\begin{aligned} z &= \text{any value} \\ w &= \text{any value} \\ x &= -9 + 5z - w \\ y &= 7 - 2z - w. \end{aligned}$$

To determine a specific solution, let $z = 1$, $w = 2$. Then a specific solution of the original system is

$$\begin{aligned} z &= 1 \\ w &= 2 \\ x &= -9 + 5(1) - (2) = -6 \\ y &= 7 - 2(1) - (2) = 3. \end{aligned}$$ ▮

▮ EXAMPLE 4 Find all solutions of the system of equations

$$\begin{cases} x - 7y + z = 3 \\ 2x - 14y + 3z = 4. \end{cases}$$

Solution The first pivot operation is routine:

$$\left[\begin{array}{ccc|c} \textcircled{1} & -7 & 1 & 3 \\ 2 & -14 & 3 & 4 \end{array}\right] \xrightarrow{[2]+(-2)[1]} \left[\begin{array}{ccc|c} 1 & -7 & 1 & 3 \\ 0 & 0 & 1 & -2 \end{array}\right].$$

However, it is impossible to pivot about the zero in the second column. So skip the second column and pivot about the second entry in the third column to get

$$\left[\begin{array}{ccc|c} 1 & -7 & 0 & 5 \\ 0 & 0 & 1 & -2 \end{array}\right].$$

This is as far as we can go. The variable corresponding to the second column, namely y, can assume any value, and the general solution of the system is obtained from the equations

$$\begin{cases} x - 7y \phantom{{}+z} = 5 \\ \phantom{x - 7y + {}} z = -2. \end{cases}$$

Therefore, the general solution of the system is

$$\begin{aligned} y &= \text{any value} \\ x &= 5 + 7y \\ z &= -2. \ \blacksquare \end{aligned}$$

We have seen that a linear system may have a unique solution or it may have infinitely many solutions. But another phenomenon can occur: A system may have no solutions at all, as the next example shows.

▮ EXAMPLE 5 Find all solutions of the system

$$\begin{cases} x - \ \ y + \ \ z = 3 \\ x + \ \ y - \ \ z = 5 \\ -2x + 4y - 4z = 1. \end{cases}$$

Solution We apply the Gaussian elimination method to the matrix of the system.

$$\left[\begin{array}{rrr|r} \textcircled{1} & -1 & 1 & 3 \\ 1 & 1 & -1 & 5 \\ -2 & 4 & -4 & 1 \end{array}\right] \xrightarrow[{[3]+2[1]}]{{[2]+(-1)[1]}} \left[\begin{array}{rrr|r} 1 & -1 & 1 & 3 \\ 0 & \textcircled{2} & -2 & 2 \\ 0 & 2 & -2 & 7 \end{array}\right]$$

$$\xrightarrow[{[3]+(-2)[2]}]{\substack{\frac{1}{2}[2] \\ [1]+(1)[2]}} \left[\begin{array}{rrr|r} 1 & 0 & 0 & 4 \\ 0 & 1 & -1 & 1 \\ 0 & 0 & 0 & 5 \end{array}\right]$$

We cannot pivot about the last zero in the third column, so we have carried the method as far as we can. Let us write out the equations corresponding to the last matrix:

$$\begin{cases} x \qquad\quad = 4 \\ \quad y - z = 1 \\ 0 \qquad\quad = 5. \end{cases}$$

Note that the last equation is a built-in contradiction. In mathematical terms, the last equation is *inconsistent.* So the last equation can never be satisfied, no matter what the values of x, y, and z. Thus, the original system has no solutions. Systems with no solutions can always be detected by the presence of inconsistent equations in the last matrix resulting from the Gaussian elimination method. ▮

At first it might seem strange that some systems have no solutions, some have one, and yet others have infinitely many. The reason for the difference can be explained geometrically. For simplicity, consider the case of systems of two equations in two variables. Each equation in this case has a graph in the xy-plane, and the graph is a straight line. As we have seen, solving the system corresponds to finding the points lying on both lines. There are three possibilities. First, the two lines may intersect. In this case the solution is unique. Second, the two lines may be parallel. Then the two lines do not intersect, and the system has no solutions. Finally, the two equations may represent the same line, as, for example, the equations $2x + 3y = 1$, $4x + 6y = 2$ do. In this case every point on the line is a solution of the system; that is, there are infinitely many solutions (Fig. 1).

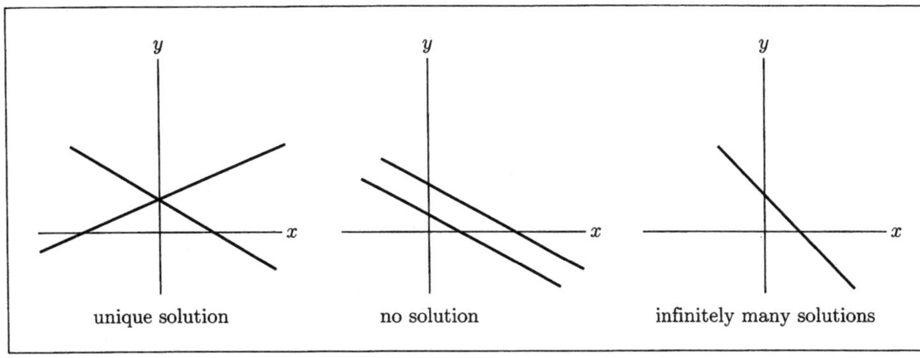

FIGURE 1

```
[A]
   [[1  -7   1 3]
    [2  -14  3 4]]
rref([A])
   [[1  -7  0  5 ]
    [0  0   1  -2]]
```

FIGURE 2

GC With the TI-83, the complete Gaussian elimination method can be carried out in one step as shown in Fig. 2 for the matrix of Example 4. To enter the symbols `rref(`, press MATRX ▶ to obtain the MATRX/MATH menu, cursor down to `B:rref(`, and press ENTER. (*Note*: `rref` stands for "row-reduced echelon form," the name given to the final form of a matrix that has been completely row reduced.)

Practice Problems 2

1. Find a specific solution to a system of linear equations whose general solution is

$$\begin{aligned} w &= \text{any value} \\ y &= \text{any value} \\ z &= 7 + 6w \\ x &= 26 - 2y + 14w. \end{aligned}$$

2. Find all solutions of this system of linear equations.

$$\begin{cases} 2x + 4y - 4z - 4w = 24 \\ -3x - 6y + 10z - 18w = -8 \\ -x - 2y + 4z - 10w = 2 \end{cases}$$

EXERCISES 2

In Exercises 1–8, pivot each matrix about the circled element.

1. $\begin{bmatrix} \textcircled{2} & -4 & 6 \\ 3 & 7 & 1 \end{bmatrix}$

2. $\begin{bmatrix} 1 & 2 & 3 \\ 4 & \textcircled{8} & -12 \end{bmatrix}$

3. $\begin{bmatrix} 7 & 1 & 4 & 5 \\ -1 & 1 & \textcircled{2} & 6 \\ 4 & 0 & 2 & 3 \end{bmatrix}$

4. $\begin{bmatrix} 5 & 10 & -10 & 12 \\ 4 & 3 & 6 & 12 \\ 4 & \textcircled{-4} & 4 & -16 \end{bmatrix}$

5. $\begin{bmatrix} \textcircled{2} & 3 \\ 6 & 0 \\ 1 & 5 \end{bmatrix}$

6. $\begin{bmatrix} 2 & 1 \\ \textcircled{-1} & 0 \end{bmatrix}$

7. $\begin{bmatrix} 4 & 3 & 0 \\ \frac{2}{3} & 0 & -2 \\ 1 & 3 & \textcircled{6} \end{bmatrix}$

8. $\begin{bmatrix} 1 & 0 & 2 \\ -1 & 1 & \textcircled{-2} \\ 1 & 2 & 6 \end{bmatrix}$

In Exercises 9–22, use the Gaussian elimination method to find all solutions of the systems of linear equations.

9. $\begin{cases} 2x - 4y = 6 \\ -x + 2y = -3 \end{cases}$

10. $\begin{cases} -\frac{1}{2}x + y = \frac{3}{2} \\ -3x + 6y = 10 \end{cases}$

11. $\begin{cases} x + 2y = 5 \\ 3x - y = 1 \\ -x + 3y = 5 \end{cases}$

12. $\begin{cases} x - 6y = 12 \\ -\frac{1}{2}x + 3y = -6 \\ \frac{1}{3}x - 2y = 4 \end{cases}$

13. $\begin{cases} x - y + 3z = 3 \\ -2x + 3y - 11z = -4 \\ x - 2y + 8z = 6 \end{cases}$

14. $\begin{cases} x - 3y + z = 5 \\ -2x + 7y - 6z = -9 \\ x - 2y - 3z = 6 \end{cases}$

15. $\begin{cases} x + y + z = -1 \\ 2x + 3y + 2z = 3 \\ 2x + y + 2z = -7 \end{cases}$

16. $\begin{cases} x - 3y + 2z = 10 \\ -x + 3y - z = -6 \\ -x + 3y + 2z = 6 \end{cases}$

17. $\begin{cases} x + 2y + 3z = 4 \\ 5x + 6y + 7z = 8 \\ x + 2y + 3z = 5 \end{cases}$

18. $\begin{cases} x + 3y = 7 \\ x + 2y = 5 \\ -x + y = 2 \end{cases}$

19. $\begin{cases} x + y - 2z + 2w = 5 \\ 2x + y - 4z + w = 5 \\ 3x + 4y - 6z + 9w = 20 \\ 4x + 4y - 8z + 8w = 20 \end{cases}$

20. $\begin{cases} 2y + z - w = 1 \\ x - y + z + w = 14 \\ -x - 9y - z + 4w = 11 \\ x + y + z = 9 \end{cases}$

21. $\begin{cases} 6x - 4y = 2 \\ -3x + 3y = 6 \\ 5x + 2y = 39 \end{cases}$

22. $\begin{cases} 3x + 2y = 5 \\ -x + 3y = 2 \\ 5x + 2y = 6 \\ 6x + y = 39 \end{cases}$

In Exercises 22–25, find three solutions to the systems of equations.

23. $\begin{cases} x + 2y + z = 5 \\ y + 3z = 9 \end{cases}$

24. $\begin{cases} x + 5y + 3z = 9 \\ 2x + 9y + 7z = 5 \end{cases}$

25. $\begin{cases} x + 7y - 3z = 8 \\ z = 5 \end{cases}$

26. An out-of-shape athlete runs 6 miles per hour, swims 1 mile per hour, and bikes 10 miles per hour. He entered a triathalon, which requires all three events, and finished 5 hours and 40 minutes later. A friend who runs 8 miles per hour, swims 2 miles per hour, and bikes 15 miles per hour finished the same course in 3 hours and 35 minutes. The total course was 32 miles long. How many miles was each segment (running, swimming, and biking)?

27. In a laboratory experiment, a researcher wants to provide a rabbit with exactly 1000 units of vitamin A, exactly 1600 units of vitamin C, and exactly 2400 units of vitamin E. The rabbit is fed a mixture of three foods. Each gram of food 1 contains 2 units of vitamin A, 3 units of vitamin C, and 5 units of vitamin E. Each gram of food 2 contains 4 units of vitamin A, 7 units of vitamin C, and 9 units of vitamin E. Each gram of food 3 contains 6 units of vitamin A, 10 units of vitamin C, and 14 units of vitamin E. How many grams of each food should the rabbit be fed?

28. Rework Exercise 27 with the requirement for vitamin E changed to 2000 units.

29. Granny's Custom Quilts receives an order for a patchwork quilt made from square patches of three types: solid green, solid blue, and floral. The quilt is to be 8 squares by 12 squares, and there must be 15 times as many solid squares as floral squares. If Granny's charges \$3 per solid square and \$5 per floral square, and if the customer wishes to spend exactly \$300, how many of each type of square may be used in the quilt?

30. Amanda is decorating her new home and wants to buy some house plants. She is interested in three types of plants costing \$7, \$10, and \$13. If she has budgeted exactly \$150 for the plants and wants to buy exactly 15 of them, what are her options?

31. Find all solutions to the following system of equations.

$$\begin{cases} x^2 + y^2 + z^2 = 14 \\ x^2 - y^2 + 2z^2 = 15 \\ x^2 + 2y^2 + 3z^2 = 36 \end{cases}$$

32. For what value of k will the following system of linear equations have a solution?

$$\begin{cases} 2x + 6y = 4 \\ x + 7y = 10 \\ kx + 8y = 4 \end{cases}$$

33. For what values(s) of k will the following system of linear equations have no solution? Infinitely many solutions?

$$\begin{cases} 2x - 3y = 4 \\ -6x + 9y = k \end{cases}$$

Exercises 34–42 require the use of a graphing calculator or a computer.

In Exercises 34–37, graph the three equations together, and determine the number of solutions (exactly one, none, or infinitely many). If there is exactly one solution, estimate the solution.

34. $\begin{cases} x + y = 10 \\ 2x - 3y = 5 \\ -x + 3y = 2 \end{cases}$

35. $\begin{cases} 2x + 3y = 5 \\ -3x + 5y = 22 \\ 2x + y = -1 \end{cases}$

36. $\begin{cases} 2x + y = 12 \\ 3x - y = 2 \\ x + 2y = 16 \end{cases}$

37. $\begin{cases} 3x - 2y = 3 \\ -2x + 4y = 14 \\ x + y = 11 \end{cases}$

In Exercises 38–41, solve the system of linear equations by the Gaussian elimination method.

38. $\begin{cases} 4x - 3y + 2z = 3 \\ -7x + 5y = 2 \\ -10x + 7y + 2z = 4 \end{cases}$

39. $\begin{cases} 2x + 3y + 6z = 4 \\ 4x + 7y = 2 \\ 3x + 5y + 3z = 3 \end{cases}$

40. $\begin{cases} 2x - 5y - 3z = -3 \\ -5x + 12y + 3z = 11 \\ -3x + 7y = 8 \end{cases}$

41. $\begin{cases} 8x + 3y - 2z = 5 \\ 12x + 5y + 2z = 3 \\ 5x + 2y = 7 \end{cases}$

42. If your calculator or computer has `rref` or an analogous command, apply the command to the matrix in Example 5. How does the final matrix differ from the one appearing in the text?

Solutions to Practice Problems 2

1. Since w and y can each assume any value, select any numbers, say $w = 1$ and $y = 2$. Then $z = 7 + 6(1) = 13$ and $x = 26 - 2(2) + 14(1) = 36$. So $x = 36$, $y = 2$, $z = 13$, $w = 1$ is a specific solution. There are infinitely many different specific solutions since there are infinitely many different choices for w and y.

2. We apply the Gaussian elimination method to the matrix of the system.

$$\left[\begin{array}{cccc|c} \textcircled{2} & 4 & -4 & -4 & 24 \\ -3 & -6 & 10 & -18 & -8 \\ -1 & -2 & 4 & -10 & 2 \end{array}\right] \xrightarrow[\substack{[2]+3[1] \\ [3]+1[1]}]{\frac{1}{2}[1]} \left[\begin{array}{cccc|c} 1 & 2 & -2 & -2 & 12 \\ 0 & 0 & \textcircled{4} & -24 & 28 \\ 0 & 0 & 2 & -12 & 14 \end{array}\right]$$

$$\xrightarrow[\substack{[1]+2[2] \\ [3]+(-2)[2]}]{\frac{1}{4}[2]} \left[\begin{array}{cccc|c} 1 & 2 & 0 & -14 & 26 \\ 0 & 0 & 1 & -6 & 7 \\ 0 & 0 & 0 & 0 & 0 \end{array}\right]$$

The corresponding system of equations is

$$\begin{cases} x + 2y \quad - 14w = 26 \\ \qquad z - 6w = 7. \end{cases}$$

The general solution is

$$\begin{aligned} w &= \text{any value} \\ y &= \text{any value} \\ z &= 7 + 6w \\ x &= 26 - 2y + 14w. \end{aligned}$$

2.3 Arithmetic Operations on Matrices

We introduced matrices in Sections 2.1 and 2.2 to display the coefficients of a system of linear equations. For example, the linear system

$$\begin{cases} 5x - 3y = \frac{1}{2} \\ 4x + 2y = -1 \end{cases}$$

is represented by the matrix

$$\left[\begin{array}{cc|c} 5 & -3 & \frac{1}{2} \\ 4 & 2 & -1 \end{array}\right].$$

After we have become accustomed to using such matrices in solving linear systems, we may omit the vertical line that separates the left and right sides of the equations. We need only remember that the right side of the equations is recorded in the right column. So, for example, we would write the matrix above in the form

$$\begin{bmatrix} 5 & -3 & \frac{1}{2} \\ 4 & 2 & -1 \end{bmatrix}.$$

A matrix is *any* rectangular array of numbers and may be of any size. Here are some examples of matrices of various sizes:

$$\begin{bmatrix} 3 & 7 \\ 0 & -1 \end{bmatrix}, \quad \begin{bmatrix} 1 \\ 2 \end{bmatrix}, \quad \begin{bmatrix} 2 & 1 \end{bmatrix}, \quad \begin{bmatrix} 6 \end{bmatrix}, \quad \begin{bmatrix} 5 & 7 & -1 \\ 0 & 3 & 5 \\ 6 & 0 & 5 \end{bmatrix}.$$

Examples of matrices abound in everyday life. For example, the newspaper stock market report is a large matrix with several thousand rows, one for each listed stock. The columns of the matrix give the various data about each stock, such as opening and closing price, number of shares traded, and so on. Another example of a matrix is a mileage chart on a road map. The rows and columns are labeled with the names of cities. The number at a given row and column gives the distance between the corresponding cities.

In these everyday examples, matrices are used only to display data. However, the most important applications involve arithmetic operations on matrices—namely, addition, subtraction, and multiplication of matrices. The major goal of this section is to discuss these operations. Before we can do so, however, we need some vocabulary with which to describe matrices.

A matrix is described by the number of rows and columns it contains. For example, the matrix

$$\begin{bmatrix} 7 & 5 \\ \frac{1}{2} & -2 \\ 2 & -11 \end{bmatrix}$$

has three rows and two columns and is referred to as a 3×2 (read: "three-by-two") *matrix*. The matrix $\begin{bmatrix} 4 & 5 & 0 \end{bmatrix}$ has one row and three columns and is a 1×3 *matrix*. A matrix with only one row is often called a *row matrix* (sometimes also called a *row vector*). A matrix, such as

$$\begin{bmatrix} 2 \\ 7 \end{bmatrix},$$

which has only one column is called a *column matrix* or *column vector*. If a matrix has the same number of rows and columns, it is called a *square matrix*. Here are some square matrices of various sizes:

$$\begin{bmatrix} 5 \end{bmatrix}, \qquad \begin{bmatrix} 1 & 2 \\ 3 & 4 \end{bmatrix}, \qquad \begin{bmatrix} 2 & -1 & 0 \\ 3 & 5 & 4 \\ 0 & 3 & -7 \end{bmatrix}.$$

The rows of a matrix are numbered from the top down, and the columns are numbered from left to right. For example, the first row of the matrix

$$\begin{bmatrix} 1 & -1 & 0 \\ 2 & 1 & 7 \\ -3 & 2 & 4 \end{bmatrix}$$

is $\begin{bmatrix} 1 & -1 & 0 \end{bmatrix}$, and its third column is

$$\begin{bmatrix} 0 \\ 7 \\ 4 \end{bmatrix}.$$

The numbers in a matrix, called *entries*, may be identified in terms of the row and column containing the entry in question. For example, the entry in the first row, third column of the following matrix is 0:

$$\begin{bmatrix} 1 & -1 & \boxed{0} \\ 2 & 1 & 7 \\ -3 & 2 & 4 \end{bmatrix};$$

the entry in the second row, first column is 2:

$$\begin{bmatrix} 1 & -1 & 0 \\ \boxed{2} & 1 & 7 \\ -3 & 2 & 4 \end{bmatrix};$$

and the entry in the third row, third column is 4:

$$\begin{bmatrix} 1 & -1 & 0 \\ 2 & 1 & 7 \\ -3 & 2 & \boxed{4} \end{bmatrix}.$$

We use double-subscripted letters to indicate the locations of the entries of a matrix. We denote the entry in the ith row, jth column by a_{ij}. For instance, in the above matrix we have $a_{13} = 0$, $a_{21} = 2$, and $a_{33} = 4$.

We say that two matrices A and B are *equal*, denoted $A = B$, provided that they have the same size and that all their corresponding entries are equal.

Addition and Subtraction of Matrices We define the sum $A+B$ of two matrices A and B only if A and B are two matrices of the same size—that is, if A and B have the same number of rows and the same number of columns. In this case $A+B$ is the matrix formed by adding the corresponding entries of A and B. For example,

$$\begin{bmatrix} 2 & 0 \\ 1 & 1 \\ 5 & 3 \end{bmatrix} + \begin{bmatrix} 5 & 4 \\ 0 & 2 \\ 2 & 6 \end{bmatrix} = \begin{bmatrix} 2+5 & 0+4 \\ 1+0 & 1+2 \\ 5+2 & 3+6 \end{bmatrix} = \begin{bmatrix} 7 & 4 \\ 1 & 3 \\ 7 & 9 \end{bmatrix}.$$

We subtract matrices of the same size by subtracting corresponding entries. Thus, we have

$$\begin{bmatrix} 7 \\ 1 \end{bmatrix} - \begin{bmatrix} 3 \\ 2 \end{bmatrix} = \begin{bmatrix} 7-3 \\ 1-2 \end{bmatrix} = \begin{bmatrix} 4 \\ -1 \end{bmatrix}.$$

Multiplication of Matrices It might seem that to define the product of two matrices, one would start with two matrices of like size and multiply the corresponding entries. But this definition is not useful, since the calculations that arise in applications require a somewhat more complex multiplication. In the interests of simplicity, we start by defining the product of a row matrix times a column matrix.

If A is a row matrix and B is a column matrix, then we can form the product $A \cdot B$ provided that the two matrices have the same length. The product $A \cdot B$ is the 1×1 matrix obtained by multiplying corresponding entries of A and B and then forming the sum.

We may put this definition into algebraic terms as follows. Suppose that A is the row matrix

$$A = \begin{bmatrix} a_1 & a_2 & \cdots & a_n \end{bmatrix},$$

and B is the column matrix

$$B = \begin{bmatrix} b_1 \\ b_2 \\ \vdots \\ b_n \end{bmatrix}.$$

Note that A and B are both of the same length, namely n. Then

$$A \cdot B = \begin{bmatrix} a_1 & a_2 & \cdots & a_n \end{bmatrix} \cdot \begin{bmatrix} b_1 \\ b_2 \\ \vdots \\ b_n \end{bmatrix}$$

is calculated by multiplying corresponding entries of A and B and forming the sum; that is,

$$A \cdot B = \begin{bmatrix} a_1b_1 + a_2b_2 + \cdots + a_nb_n \end{bmatrix}.$$

Notice that the product is a 1×1 matrix, namely a single number in brackets.

Here are some examples of the product of a row matrix times a column matrix:

$$\begin{bmatrix} 3 & \frac{1}{2} \end{bmatrix} \cdot \begin{bmatrix} 1 \\ 4 \end{bmatrix} = \begin{bmatrix} 3 \cdot 1 + \frac{1}{2} \cdot 4 \end{bmatrix} = \begin{bmatrix} 5 \end{bmatrix};$$

$$\begin{bmatrix} 2 & 0 & -1 \end{bmatrix} \cdot \begin{bmatrix} 6 \\ 5 \\ 3 \end{bmatrix} = \begin{bmatrix} 2 \cdot 6 + 0 \cdot 5 + (-1) \cdot 3 \end{bmatrix} = \begin{bmatrix} 9 \end{bmatrix}.$$

In multiplying a row matrix times a column matrix, it helps to use both of your hands. Use your left index finger to point to an element of the row matrix and your right to point to the corresponding element of the column. Multiply the elements you are pointing to and keep a running total of the products in your head. After each multiplication move your fingers to the next elements of each matrix. With a little practice you should be able to multiply a row times a column quickly and accurately.

The definition of multiplication above may seem strange. But products of this sort occur in many down-to-earth problems. Consider, for instance, the next example.

EXAMPLE 1 A dairy farm produces three items—milk, eggs, and cheese. The prices of the three items are \$1.50 per gallon, \$0.80 per dozen, and \$2.00 per pound, respectively. In a certain week the dairy farm sells 30,000 gallons of milk, 2000 dozen eggs, and 5000 pounds of cheese. Represent its total revenue as a matrix product.

Solution The total revenue equals

$$(1.50)(30{,}000) + (.80)(2000) + (2)(5000).$$

This suggests that we define two matrices: The first displays the prices of the various produce:

$$\begin{bmatrix} 1.50 & .80 & 2 \end{bmatrix}.$$

The second represents the production:

$$\begin{bmatrix} 30{,}000 \\ 2{,}000 \\ 5{,}000 \end{bmatrix}.$$

Then the revenue for the week, when placed in a 1×1 matrix, equals

$$\begin{bmatrix} 1.50 & .80 & 2 \end{bmatrix} \begin{bmatrix} 30{,}000 \\ 2{,}000 \\ 5{,}000 \end{bmatrix} = \begin{bmatrix} 56{,}600 \end{bmatrix}. \blacksquare$$

The principle behind Example 1 is this: Any sum of products of the form $a_1b_1 + a_2b_2 + \cdots + a_nb_n$, when placed in a 1×1 matrix, can be written as the matrix product

$$\begin{bmatrix} a_1b_1 + a_2b_2 + \cdots + a_nb_n \end{bmatrix} = \begin{bmatrix} a_1 & a_2 & \cdots & a_n \end{bmatrix} \cdot \begin{bmatrix} b_1 \\ b_2 \\ \vdots \\ b_n \end{bmatrix}.$$

Let us illustrate the procedure for multiplying more general matrices by working out a typical product:

$$\begin{bmatrix} 2 & 1 \\ 0 & 1 \\ 1 & 0 \end{bmatrix} \cdot \begin{bmatrix} 1 & 1 \\ 4 & 2 \end{bmatrix}.$$

To obtain the entries of the product, we multiply the rows of the left matrix by the columns of the right matrix, taking care to arrange the products in a specific way to yield a matrix, as follows. Start with the first row on the left, $\begin{bmatrix} 2 & 1 \end{bmatrix}$, and the first column on the right, $\begin{bmatrix} 1 \\ 4 \end{bmatrix}$. Their product is $\begin{bmatrix} 6 \end{bmatrix}$, so we enter 6 as the element in the first row, first column of the product:

$$\begin{bmatrix} 2 & 1 \\ 0 & 1 \\ 1 & 0 \end{bmatrix} \cdot \begin{bmatrix} 1 & 1 \\ 4 & 2 \end{bmatrix} = \begin{bmatrix} 6 & \\ & \\ & \end{bmatrix}.$$

The product of the first row of the left matrix and the second column of the right matrix is $\begin{bmatrix} 4 \end{bmatrix}$, so we put a 4 in the first row, second column of the product:

$$\begin{bmatrix} 2 & 1 \\ 0 & 1 \\ 1 & 0 \end{bmatrix} \cdot \begin{bmatrix} 1 & 1 \\ 4 & 2 \end{bmatrix} = \begin{bmatrix} 6 & 4 \\ & \\ & \end{bmatrix}.$$

There are no more columns that can be multiplied by the first row, so let us move to the second row and shift back to the first column. Correspondingly, we move down one row in the product:

$$\begin{bmatrix} 2 & 1 \\ 0 & 1 \\ 1 & 0 \end{bmatrix} \cdot \begin{bmatrix} 1 & 1 \\ 4 & 2 \end{bmatrix} = \begin{bmatrix} 6 & 4 \\ 4 & \\ & \end{bmatrix};$$

$$\begin{bmatrix} 2 & 1 \\ 0 & 1 \\ 1 & 0 \end{bmatrix} \cdot \begin{bmatrix} 1 & 1 \\ 4 & 2 \end{bmatrix} = \begin{bmatrix} 6 & 4 \\ 4 & 2 \\ & \end{bmatrix}.$$

We have now exhausted the second row of the left matrix, so we shift to the third row and correspondingly move down one row in the product:

$$\begin{bmatrix} 2 & 1 \\ 0 & 1 \\ 1 & 0 \end{bmatrix} \cdot \begin{bmatrix} 1 & 1 \\ 4 & 2 \end{bmatrix} = \begin{bmatrix} 6 & 4 \\ 4 & 2 \\ 1 & \end{bmatrix};$$

$$\begin{bmatrix} 2 & 1 \\ 0 & 1 \\ 1 & 0 \end{bmatrix} \cdot \begin{bmatrix} 1 & 1 \\ 4 & 2 \end{bmatrix} = \begin{bmatrix} 6 & 4 \\ 4 & 2 \\ 1 & 1 \end{bmatrix}.$$

Note that we have now multiplied every row of the left matrix by every column of the right matrix. This completes the computation of the product:

$$\begin{bmatrix} 2 & 1 \\ 0 & 1 \\ 1 & 0 \end{bmatrix} \cdot \begin{bmatrix} 1 & 1 \\ 4 & 2 \end{bmatrix} = \begin{bmatrix} 6 & 4 \\ 4 & 2 \\ 1 & 1 \end{bmatrix}.$$

EXAMPLE 2 Calculate the following product:

$$\begin{bmatrix} 1 & 5 \\ 3 & 2 \end{bmatrix} \cdot \begin{bmatrix} 1 & 2 \\ 1 & 0 \end{bmatrix}.$$

Solution

$$\begin{bmatrix} 1 & 5 \\ 3 & 2 \end{bmatrix} \cdot \begin{bmatrix} 1 & 2 \\ 1 & 0 \end{bmatrix} = \begin{bmatrix} 6 & \\ & \end{bmatrix}$$

$$\begin{bmatrix} 1 & 5 \\ 3 & 2 \end{bmatrix} \cdot \begin{bmatrix} 1 & 2 \\ 1 & 0 \end{bmatrix} = \begin{bmatrix} 6 & 2 \\ & \end{bmatrix}$$

$$\begin{bmatrix} 1 & 5 \\ 3 & 2 \end{bmatrix} \cdot \begin{bmatrix} 1 & 2 \\ 1 & 0 \end{bmatrix} = \begin{bmatrix} 6 & 2 \\ 5 & \end{bmatrix}$$

$$\begin{bmatrix} 1 & 5 \\ 3 & 2 \end{bmatrix} \cdot \begin{bmatrix} 1 & 2 \\ 1 & 0 \end{bmatrix} = \begin{bmatrix} 6 & 2 \\ 5 & 6 \end{bmatrix}.$$

Thus,

$$\begin{bmatrix} 1 & 5 \\ 3 & 2 \end{bmatrix} \cdot \begin{bmatrix} 1 & 2 \\ 1 & 0 \end{bmatrix} = \begin{bmatrix} 6 & 2 \\ 5 & 6 \end{bmatrix}. \quad \blacksquare$$

Notice that we cannot use the method above to compute the product $A \cdot B$ of *any* matrices A and B. For the procedure to work, it is crucial that the number of entries of each row of A be the same as the number of entries of each column of B. (Or, to put it another way, the number of columns of the left matrix must equal the number of rows of the right matrix.) Therefore, in order for us to form the product $A \cdot B$, the sizes of A and B must match up in a special way. If A is $m \times n$ and B is $p \times q$, then the product $A \cdot B$ is defined only in case the "inner" dimensions n and p are equal. In that case, the size of the product is determined by the "outer" dimensions m and q. It is an $m \times q$ matrix:

$$\underset{m \times n}{A} \cdot \underset{p \times q}{B} = \underset{m \times q}{C}.$$

(n and p: equal)

So, for example,

$$\underset{3 \times 4}{\begin{bmatrix} \quad & \quad & \quad & \quad \end{bmatrix}} \underset{4 \times 2}{\begin{bmatrix} \quad & \quad \end{bmatrix}} = \underset{3 \times 2}{\begin{bmatrix} \quad & \quad \end{bmatrix}}$$

$$\underset{2 \times 2}{\begin{bmatrix} \quad & \quad \end{bmatrix}} \underset{2 \times 1}{\begin{bmatrix} \quad \end{bmatrix}} = \underset{2 \times 1}{\begin{bmatrix} \quad \end{bmatrix}}.$$

If the sizes of A and B do not match up in the way just described, the product $A \cdot B$ is not defined.

EXAMPLE 3 Calculate the following products, if defined.

(a) $\begin{bmatrix} 3 & -1 \\ 2 & 0 \\ 1 & 5 \end{bmatrix} \begin{bmatrix} 1 & 0 \\ 5 & -4 \\ 2 & -1 \end{bmatrix}$ (b) $\begin{bmatrix} 3 & -1 \\ 2 & 0 \\ 1 & 5 \end{bmatrix} \begin{bmatrix} 5 & 4 \\ -2 & 3 \end{bmatrix}$

Solution (a) The matrices to be multiplied are 3×2 and 3×2. The inner dimensions do not match, so the product is undefined.

(b) We are asked to multiply a 3×2 matrix times a 2×2 matrix. The inner dimensions match, so the product is defined and has size determined by the outer dimensions, that is, 3×2.

$$\begin{bmatrix} 3 & -1 \\ 2 & 0 \\ 1 & 5 \end{bmatrix} \cdot \begin{bmatrix} 5 & 4 \\ -2 & 3 \end{bmatrix} = \begin{bmatrix} 3 \cdot 5 + (-1) \cdot (-2) & 3 \cdot 4 + (-1) \cdot 3 \\ 2 \cdot 5 + 0 \cdot (-2) & 2 \cdot 4 + 0 \cdot 3 \\ 1 \cdot 5 + 5 \cdot (-2) & 1 \cdot 4 + 5 \cdot 3 \end{bmatrix}$$

$$= \begin{bmatrix} 17 & 9 \\ 10 & 8 \\ -5 & 19 \end{bmatrix}. \quad \blacksquare$$

Multiplication of matrices has many properties in common with multiplication of ordinary numbers. However, there is at least one important difference. With matrix multiplication, the order of the factors is usually important. For example, the product of a 2×3 matrix times a 3×2 matrix is defined: The product is a 2×2 matrix. If the order is reversed to a 3×2 matrix times a 2×3 matrix, the product is a 3×3 matrix. So reversing the order may change the size of the product. Even when it does not, reversing the order may still change the entries in the product, as the following two products demonstrate:

$$\begin{bmatrix} 1 & 5 \\ 3 & 2 \end{bmatrix}\begin{bmatrix} 1 & 2 \\ 1 & 0 \end{bmatrix} = \begin{bmatrix} 6 & 2 \\ 5 & 6 \end{bmatrix}; \qquad \begin{bmatrix} 1 & 2 \\ 1 & 0 \end{bmatrix}\begin{bmatrix} 1 & 5 \\ 3 & 2 \end{bmatrix} = \begin{bmatrix} 7 & 9 \\ 1 & 5 \end{bmatrix}.$$

EXAMPLE 4 An investment trust has investments in three states. Its deposits in each state are divided among bonds, mortgages, and consumer loans. On January 1 the amount (in millions of dollars) of money invested in each category by state is given by the matrix

	Bonds	Mortgages	Consumer loans
State A	10	5	20
State B	30	12	10
State C	15	6	25

The current average yields are 7% for bonds, 9% for mortgages, and 15% for consumer loans. Determine the earnings of the trust from its investments in each state.

Solution Define the matrix of investment yields by

$$\begin{bmatrix} .07 \\ .09 \\ .15 \end{bmatrix} \begin{matrix} \text{Bonds} \\ \text{Mortgages} \\ \text{Consumer loans.} \end{matrix}$$

The amount earned in state A, for instance, is

$$\begin{aligned} &[\text{amount of bonds}][\text{yield of bonds}] \\ &\quad + [\text{amount of mortgages}][\text{yield of mortgages}] \\ &\quad + [\text{amount of consumer loans}][\text{yield of consumer loans}] \\ &= (10)(.07) + (5)(.09) + (20)(.15). \end{aligned}$$

And this is just the first entry of the product:

$$\begin{bmatrix} 10 & 5 & 20 \\ 30 & 12 & 10 \\ 15 & 6 & 25 \end{bmatrix}\begin{bmatrix} .07 \\ .09 \\ .15 \end{bmatrix}.$$

Similarly, the earnings for the other states are the second and third entries of the product. Carrying out the arithmetic, we find that

$$\begin{bmatrix} 10 & 5 & 20 \\ 30 & 12 & 10 \\ 15 & 6 & 25 \end{bmatrix}\begin{bmatrix} .07 \\ .09 \\ .15 \end{bmatrix} = \begin{bmatrix} 4.15 \\ 4.68 \\ 5.34 \end{bmatrix}.$$

Therefore, the trust earns \$4.15 million in state A, \$4.68 million in state B, and \$5.34 million in state C. ▮

EXAMPLE 5 A clothing manufacturer has factories in Los Angeles, San Antonio, and Newark. Sales (in thousands) during the first quarter of last year are summarized in the production matrix

$$\begin{array}{l} \\ \text{Coats} \\ \text{Shirts} \\ \text{Sweaters} \\ \text{Ties} \end{array} \begin{array}{c} \begin{array}{ccc} \text{Los Angeles} & \text{San Antonio} & \text{Newark} \end{array} \\ \begin{bmatrix} 12 & 13 & 38 \\ 25 & 5 & 26 \\ 11 & 8 & 8 \\ 5 & 0 & 12 \end{bmatrix} \end{array}.$$

During this period the selling price of a coat was \$100, of a shirt \$10, of a sweater \$25, and of a tie \$5.

(a) Use a matrix calculation to determine the total revenue produced by each of the factories.

(b) Suppose that the prices had been \$110, \$8, \$20, and \$10, respectively. How would this have affected the revenue of each factory?

Solution (a) For each factory, we wish to multiply the price of each item by the number produced to arrive at revenue. Since the production figures for the various items of clothing are arranged down the columns, we arrange the prices in a row matrix, ready for multiplication. The price matrix is

$$\begin{bmatrix} 100 & 10 & 25 & 5 \end{bmatrix}.$$

The revenues of the various factories are then the entries of the product

$$\begin{bmatrix} 100 & 10 & 25 & 5 \end{bmatrix} \begin{bmatrix} 12 & 13 & 38 \\ 25 & 5 & 26 \\ 11 & 8 & 8 \\ 5 & 0 & 12 \end{bmatrix} = \begin{array}{c} \begin{array}{ccc} \text{Los Angeles} & \text{San Antonio} & \text{Newark} \end{array} \\ \begin{bmatrix} 1750 & 1550 & 4320 \end{bmatrix} \end{array}.$$

Since the production figures are in thousands, the revenue figures are in thousands of dollars. That is, the Los Angeles factory has revenues of \$1,750,000, the San Antonio factory \$1,550,000, and the Newark factory \$4,320,000.

(b) In a similar way, we determine the revenue of each factory if the price matrix had been $\begin{bmatrix} 110 & 8 & 20 & 10 \end{bmatrix}$.

$$\begin{bmatrix} 110 & 8 & 20 & 10 \end{bmatrix} \begin{bmatrix} 12 & 13 & 38 \\ 25 & 5 & 26 \\ 11 & 8 & 8 \\ 5 & 0 & 12 \end{bmatrix} = \begin{array}{c} \begin{array}{ccc} \text{Los Angeles} & \text{San Antonio} & \text{Newark} \end{array} \\ \begin{bmatrix} 1790 & 1630 & 4668 \end{bmatrix} \end{array}.$$

The change in revenue at each factory can be read from the difference of the revenue matrices:

$$\begin{bmatrix} 1790 & 1630 & 4668 \end{bmatrix} - \begin{bmatrix} 1750 & 1550 & 4320 \end{bmatrix} = \begin{bmatrix} 40 & 80 & 348 \end{bmatrix}.$$

If prices had been as given in (b), revenues of the Los Angeles factory would have increased by \$40,000, revenues at San Antonio would have increased by \$80,000, and revenues at Newark would have increased by \$348,000. ∎

There are special matrices analogous to the number 1. Such matrices are called *identity matrices.* The identity matrix I_n of size n is the $n \times n$ square matrix with all zeros except for ones down the upper-left-to-lower-right diagonal. Here are the identity matrices of sizes 2, 3, and 4:

$$I_2 = \begin{bmatrix} 1 & 0 \\ 0 & 1 \end{bmatrix}; \qquad I_3 = \begin{bmatrix} 1 & 0 & 0 \\ 0 & 1 & 0 \\ 0 & 0 & 1 \end{bmatrix}; \qquad I_4 = \begin{bmatrix} 1 & 0 & 0 & 0 \\ 0 & 1 & 0 & 0 \\ 0 & 0 & 1 & 0 \\ 0 & 0 & 0 & 1 \end{bmatrix}.$$

The characteristic property of an identity matrix is that it plays the role of the number 1; that is,

$$I_n \cdot A = A \cdot I_n = A$$

for all $n \times n$ matrices A.

One of the principal uses of matrices is in dealing with systems of linear equations. Matrices provide a compact way of writing systems, as the next example shows.

EXAMPLE 6 Write the system of linear equations

$$\begin{cases} -2x + 4y = 2 \\ -3x + 7y = 7 \end{cases}$$

as a matrix equation.

Solution The system of equations can be written in the form

$$\begin{bmatrix} -2x + 4y \\ -3x + 7y \end{bmatrix} = \begin{bmatrix} 2 \\ 7 \end{bmatrix}.$$

So consider the matrices

$$A = \begin{bmatrix} -2 & 4 \\ -3 & 7 \end{bmatrix}, \qquad X = \begin{bmatrix} x \\ y \end{bmatrix}, \qquad B = \begin{bmatrix} 2 \\ 7 \end{bmatrix}.$$

Notice that

$$AX = \begin{bmatrix} -2 & 4 \\ -3 & 7 \end{bmatrix} \begin{bmatrix} x \\ y \end{bmatrix} = \begin{bmatrix} -2x + 4y \\ -3x + 7y \end{bmatrix}.$$

Thus, AX is a 2×1 column matrix whose entries correspond to the left side of the given system of linear equations. Since the entries of B correspond to the right side of the system of equations, we can rewrite the given system in the form

$$AX = B$$

—that is,

$$\begin{bmatrix} -2 & 4 \\ -3 & 7 \end{bmatrix} \begin{bmatrix} x \\ y \end{bmatrix} = \begin{bmatrix} 2 \\ 7 \end{bmatrix}.$$ ▮

The matrix A of the preceding example displays the coefficients of the variables x and y, and so it is called the *coefficient matrix* of the system.

GC The arithmetic operations +, −, and ∗ can be applied to matrices in much the same way as to numbers, as Figs. 1 and 2 show. (*Note*: `[A]*[B]` also can be written as `[A][B]`.) Identity matrices are placed on the home screen with the command `identity(` from the MATRX/MATH menu, as shown in Fig. 3. The value in the ith row, jth column of the matrix `[A]` can be displayed with `[A](i,j)`. See Fig. 4.

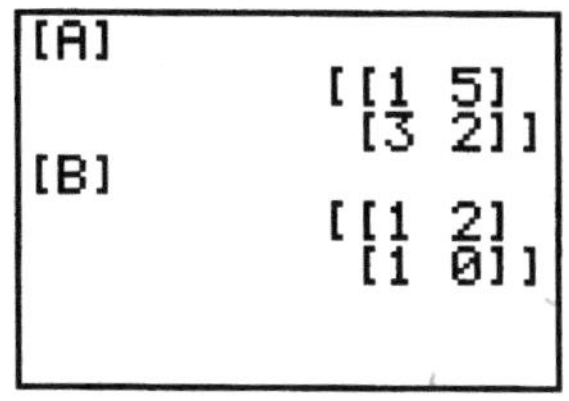

FIGURE 1

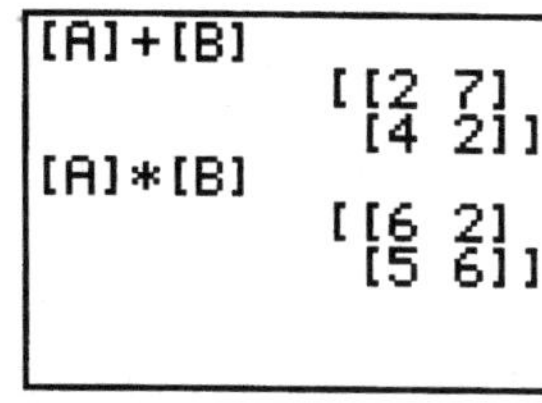

FIGURE 2

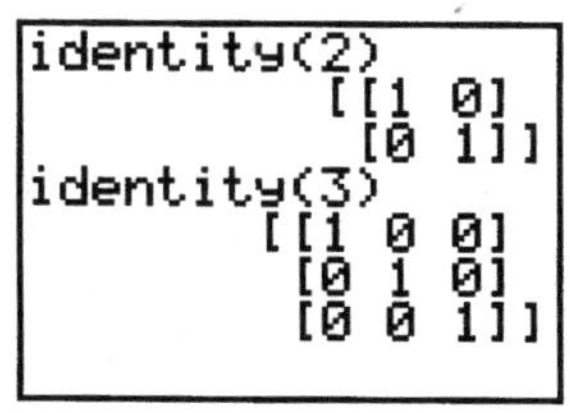

FIGURE 3

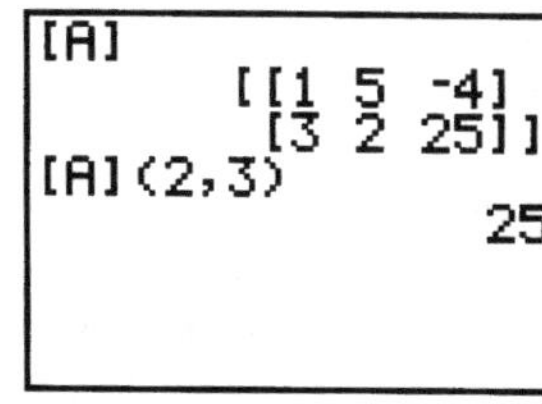

FIGURE 4

Practice Problems 3

1. Compute

$$\begin{bmatrix} 3 & 1 & 2 \\ -1 & 0 & \frac{1}{2} \\ 0 & 4 & 1 \end{bmatrix} \begin{bmatrix} 7 & -1 & 0 \\ 5 & 4 & 2 \\ -6 & 0 & 4 \end{bmatrix}.$$

2. Give the system of linear equations that is equivalent to the matrix equation

$$\begin{bmatrix} 3 & -6 \\ 2 & 1 \end{bmatrix} \begin{bmatrix} x \\ y \end{bmatrix} = \begin{bmatrix} 5 \\ 0 \end{bmatrix}.$$

3. Give a matrix equation equivalent to this system of equations:

$$\begin{cases} 8x + 3y = 7 \\ 9x - 2y = -5. \end{cases}$$

EXERCISES 3

In Exercises 1–6, give the size and special characteristics of the given matrix (such as square, column, row, identity).

1. $\begin{bmatrix} 3 & 2 & 4 \\ \frac{1}{2} & 0 & 6 \end{bmatrix}$ **2.** $\begin{bmatrix} 3 \\ -1 \end{bmatrix}$

3. $\begin{bmatrix} 2 & \frac{1}{3} & 0 \end{bmatrix}$ **4.** $\begin{bmatrix} 1 & 0 \\ 0 & 1 \end{bmatrix}$

5. $\begin{bmatrix} -2 \end{bmatrix}$ **6.** $\begin{bmatrix} 0 & 0 & 0 & 0 \\ 0 & 0 & 0 & 0 \end{bmatrix}$

In Exercises 7–14, perform the indicated matrix calculations.

7. $\begin{bmatrix} 4 & -2 \\ 3 & 0 \end{bmatrix} + \begin{bmatrix} 5 & 5 \\ 4 & -1 \end{bmatrix}$ **8.** $\begin{bmatrix} 8 \\ -3 \end{bmatrix} + \begin{bmatrix} 5 \\ 6 \end{bmatrix}$

9. $\begin{bmatrix} 2 & 8 \\ \frac{4}{3} & 4 \\ 1 & -2 \end{bmatrix} - \begin{bmatrix} 1 & 5 \\ \frac{1}{3} & 2 \\ -3 & 0 \end{bmatrix}$

10. $\begin{bmatrix} 1 & 0 \\ 0 & 1 \end{bmatrix} - \begin{bmatrix} .8 & .5 \\ .2 & .5 \end{bmatrix}$

11. $\begin{bmatrix} 5 & 3 \end{bmatrix} \begin{bmatrix} 1 \\ 2 \end{bmatrix}$ **12.** $\begin{bmatrix} 1 & 0 & 0 \end{bmatrix} \begin{bmatrix} \frac{1}{2} \\ 6 \\ 2 \end{bmatrix}$

13. $\begin{bmatrix} 6 & 1 & 5 \end{bmatrix} \begin{bmatrix} \frac{1}{2} \\ -3 \\ 2 \end{bmatrix}$ 14. $\begin{bmatrix} 0 & 0 \end{bmatrix} \begin{bmatrix} 5 \\ -3 \end{bmatrix}$

In Exercises 15–20, the sizes of two matrices are given. Tell whether or not the product AB is defined. If so, give its size.

15. A, 3×4; B, 4×5
16. A, 3×3; B, 3×4
17. A, 3×2; B, 3×2
18. A, 1×1; B, 1×1
19. A, 3×3; B, 3×1
20. A, 4×2; B, 3×4

In Exercises 21–30, perform the multiplication.

21. $\begin{bmatrix} 3 & 1 \\ 0 & 2 \end{bmatrix} \begin{bmatrix} 1 & 4 \\ 3 & 5 \end{bmatrix}$ 22. $\begin{bmatrix} 4 & -1 \\ 2 & \frac{1}{2} \end{bmatrix} \begin{bmatrix} 3 \\ 2 \end{bmatrix}$

23. $\begin{bmatrix} 4 & 1 & 0 \\ -2 & 0 & 3 \\ 1 & 5 & -1 \end{bmatrix} \begin{bmatrix} 5 \\ 1 \\ 2 \end{bmatrix}$ 24. $\begin{bmatrix} 0 & 0 \\ 0 & 0 \\ 0 & 0 \end{bmatrix} \begin{bmatrix} 1 & 2 \\ 3 & 4 \end{bmatrix}$

25. $\begin{bmatrix} 1 & 0 \\ 0 & 1 \end{bmatrix} \begin{bmatrix} 5 & 6 \\ 7 & 8 \end{bmatrix}$ 26. $\begin{bmatrix} 1 & 2 \\ 1 & 3 \end{bmatrix} \begin{bmatrix} 3 & -2 \\ -1 & 1 \end{bmatrix}$

27. $\begin{bmatrix} .6 & .3 \\ .4 & .7 \end{bmatrix} \begin{bmatrix} .6 & .3 \\ .4 & .7 \end{bmatrix}$

28. $\begin{bmatrix} 0 & 1 & 2 \\ -1 & 4 & \frac{1}{2} \\ 1 & 3 & 0 \end{bmatrix} \begin{bmatrix} 3 & -1 & 5 \\ 0 & 2 & 2 \\ 4 & -6 & 0 \end{bmatrix}$

29. $\begin{bmatrix} 2 & -1 & 4 \\ 0 & 1 & 0 \\ \frac{1}{2} & 3 & -2 \end{bmatrix} \begin{bmatrix} 4 & 8 & 0 \\ 3 & -1 & 2 \\ 5 & 0 & 1 \end{bmatrix}$

30. $\begin{bmatrix} 1 & 0 & 0 \\ 0 & 1 & 0 \\ 0 & 0 & 1 \end{bmatrix} \begin{bmatrix} 1 \\ 2 \\ 3 \end{bmatrix}$

In Exercises 31–34, give the system of linear equations that is equivalent to the matrix equation. Do not solve.

31. $\begin{bmatrix} 2 & 3 \\ 4 & 5 \end{bmatrix} \begin{bmatrix} x \\ y \end{bmatrix} = \begin{bmatrix} 6 \\ 7 \end{bmatrix}$ 32. $\begin{bmatrix} -3 & 4 \\ 0 & 1 \end{bmatrix} \begin{bmatrix} x \\ y \end{bmatrix} = \begin{bmatrix} 1 \\ 1 \end{bmatrix}$

33. $\begin{bmatrix} 1 & 2 & 3 \\ 4 & 5 & 6 \\ 7 & 8 & 9 \end{bmatrix} \begin{bmatrix} x \\ y \\ z \end{bmatrix} = \begin{bmatrix} 10 \\ 11 \\ 12 \end{bmatrix}$

34. $\begin{bmatrix} 1 & 0 & 0 \\ 0 & 1 & 0 \\ 0 & 0 & 1 \end{bmatrix} \begin{bmatrix} x \\ y \\ z \end{bmatrix} = \begin{bmatrix} 1 \\ 2 \\ 3 \end{bmatrix}$

In Exercises 35–38, write the given system of linear equations in matrix form.

35. $\begin{cases} 3x + 2y = -1 \\ 7x - y = 2 \end{cases}$ 36. $\begin{cases} 5x - 2y = 6 \\ -2x + 4y = 0 \end{cases}$

37. $\begin{cases} x - 2y + 3z = 5 \\ y + z = 6 \\ z = 2 \end{cases}$ 38. $\begin{cases} -2x + 4y - z = 5 \\ x + 6y + 3z = -1 \\ 7x + 4z = 8 \end{cases}$

The distributive law says that $(A+B)C = AC+BC$. That is, adding A and B and then multiplying on the right by C gives the same result as first multiplying each of A and B on the right by C and then adding. In Exercises 39 and 40, verify the distributive law for the given matrices.

39. $A = \begin{bmatrix} 1 & 2 \\ 0 & 3 \end{bmatrix}$, $B = \begin{bmatrix} 3 & -2 \\ 4 & 5 \end{bmatrix}$, $C = \begin{bmatrix} 1 & 6 \\ 2 & 0 \end{bmatrix}$

40. $A = \begin{bmatrix} 1 & 0 & 0 \\ 0 & 1 & 0 \\ 0 & 0 & 1 \end{bmatrix}$, $B = \begin{bmatrix} 2 & 1 & 3 \\ 0 & 5 & -1 \\ 3 & 6 & 0 \end{bmatrix}$, C= $\begin{bmatrix} 0 \\ 3 \\ -4 \end{bmatrix}$

Two $n \times n$ matrices A and B are called *inverses* (of one another) if both products AB and BA equal I_n. Check that the pairs of matrices in Exercises 41 and 42 are inverses.

41. $\begin{bmatrix} 3 & -1 \\ -1 & \frac{1}{2} \end{bmatrix}$, $\begin{bmatrix} 1 & 2 \\ 2 & 6 \end{bmatrix}$

42. $\begin{bmatrix} 2 & 8 & -11 \\ -1 & -5 & 7 \\ 1 & 2 & -3 \end{bmatrix}$, $\begin{bmatrix} 1 & 2 & 1 \\ 4 & 5 & -3 \\ 3 & 4 & -2 \end{bmatrix}$

43. The quantities of pants, shirts, and jackets owned by Mike and Don are given by the matrix A, and the costs of these items are given by matrix B.

$$\begin{array}{r} \\ \text{Mike} \\ \text{Don} \end{array} \begin{array}{c} \text{Pants} \quad \text{Shirts} \quad \text{Jackets} \\ \begin{bmatrix} 6 & \quad 8 \quad & 2 \\ 2 & \quad 5 \quad & 3 \end{bmatrix} \end{array} = A$$

$$\begin{array}{l} \text{Pants} \\ \text{Shirts} \\ \text{Jackets} \end{array} \begin{bmatrix} 20 \\ 15 \\ 50 \end{bmatrix} = B.$$

(a) Calculate the matrix AB.

(b) Interpret the entries of the matrix AB.

44. A company has three appliance stores that sell washers, dryers, and stoves. Matrices A and B give the wholesale and retail prices of these items, respectively. Matrices C and D give the quantities of these items

sold by the three stores in September and October, respectively.

$$A = \begin{bmatrix} 300 & 250 & 450 \end{bmatrix}$$ (columns: Washers, Dryers, Stoves)

$$B = \begin{bmatrix} 500 & 450 & 750 \end{bmatrix}$$ (columns: Washers, Dryers, Stoves)

	Store 1	Store 2	Store 3	
$C =$	30	40	20	Washers
	20	30	10	Dryers
	10	5	35	Stoves

	Store 1	Store 2	Store 3	
$D =$	20	50	30	Washers
	30	10	20	Dryers
	10	20	30	Stoves

Determine and interpret the following matrices.

(a) AC (b) AD

(c) BC (d) BD

(e) $B - A$ (f) $(B - A)C$

(g) $(B - A)D$ (h) $C + D$

(i) $(B - A)(C + D)$

45. Three professors teaching the same course have entirely different grading policies. The percentage of each grade given by the professors is summarized in the matrix

	Grade				
	A	B	C	D	F
Prof. I	25	35	30	10	0
Prof. II	10	20	40	20	10
Prof. III	5	10	20	40	25

(a) The point values of the grades are A = 4, B = 3, C = 2, D = 1, and F = 0. Use matrix multiplication to determine the average grade given by each professor.

(b) Professor I has 240 students, professor II has 120 students, and professor III has 40 students. Use matrix multiplication to determine the numbers of As, Bs, Cs, Ds, and Fs given.

46. A professor bases semester grades on four 100-point items: homework, quizzes, a midterm exam, and a final exam. Students may choose one of three schemes summarized in the matrix below for weighting the points from the four items. Use matrix multiplication to determine the most advantageous weighting scheme for a student who earned 97 points on homework, 72 points on the quizzes, 83 points on the midterm exam, and 75 points on the final exam.

	Items			
	Hw	Qu	ME	FE
Scheme I	.10	.10	.30	.50
Scheme II	.10	.20	.30	.40
Scheme III	.15	.15	.35	.35

47. In a certain town the proportions of voters voting Democratic and Republican by various age groups is summarized by this matrix:

	Dem.	Rep.
Under 30	.65	.35
30–50	.55	.45
Over 50	.45	.55

$= A.$

The population of voters in the town by age group is given by the matrix

$$B = \begin{bmatrix} \underbrace{6000}_{\text{Under 30}} & \underbrace{8000}_{30\text{–}50} & \underbrace{4000}_{\text{Over 50}} \end{bmatrix}.$$

Interpret the entries of the matrix product BA.

48. Refer to Exercise 47.

(a) Using the given data, which party would win and what would be the percentage of the winning vote?

(b) Suppose that the population of the town shifted toward older residents as reflected in the population matrix $B = \begin{bmatrix} 2000 & 4000 & 12{,}000 \end{bmatrix}$. What would be the result of the election now?

49. Suppose that a contractor employs carpenters, bricklayers, and plumbers, working three shifts per day. The number of labor-hours employed in each of the shifts is summarized in the matrix

	Shift		
	1	2	3
Carpenters	50	20	10
Bricklayers	30	30	15
Plumbers	20	20	5

Labor in shift 1 costs \$10 per hour, in shift 2 \$15 per hour, and in shift 3 \$20 per hour. Use matrix multiplication to compute the amount spent on each type of labor.

50. A flu epidemic hits a large city. Each resident of the city is either sick, well, or a carrier. The proportion of people in each of the categories is expressed by the matrix

	Age		
	0–10	10–30	Over 30
Well	.70	.70	.60
Sick	.10	.20	.30
Carrier	.20	.10	.10

$= A.$

The population of the city is distributed by age and sex as follows:

$$\text{Age}\;\begin{array}{|l} 0\text{–}10 \\ 10\text{–}30 \\ \text{Over } 30 \end{array}\;\begin{array}{c} \begin{array}{cc} \text{Male} & \text{Female} \end{array} \\ \begin{bmatrix} 60{,}000 & 65{,}000 \\ 100{,}000 & 110{,}000 \\ 200{,}000 & 230{,}000 \end{bmatrix} \end{array} = B.$$

(a) Compute AB.

(b) How many sick males are there?

(c) How many female carriers are there?

51. Mikey's diet consists of food X and food Y. The matrix N represents the number of units of nutrients 1, 2, and 3 per ounce for each of the foods.

$$N = \begin{array}{c} \begin{array}{ccc} 1 & 2 & 3 \end{array} \\ \begin{bmatrix} 60 & 50 & 38 \\ 42 & 50 & 67 \end{bmatrix} \end{array} \begin{array}{l} X \\ Y \end{array}$$

The matrices B, L, and D represent the number of ounces of each food that Mikey eats each day for breakfast, lunch, and dinner, respectively.

$$B = \overset{\begin{array}{cc} X & Y \end{array}}{\begin{bmatrix} 2 & 1 \end{bmatrix}} \quad L = \overset{\begin{array}{cc} X & Y \end{array}}{\begin{bmatrix} 1 & 3 \end{bmatrix}} \quad D = \overset{\begin{array}{cc} X & Y \end{array}}{\begin{bmatrix} 2 & 4 \end{bmatrix}}$$

Calculate and interpret the following.

(a) BN (b) LN (c) DN

(d) $B+L+D$ (e) $(B+L+D)N$

52. A bakery makes three types of cookies, I, II, and III. Each type of cookie is made from the four ingredients A, B, C, and D. The number of units of each ingredient used in each type of cookie is given by the matrix M. The cost per unit of each of the four ingredients (in cents) is given by the matrix N. The selling price for each of the cookies (in cents) is given by the matrix S. The baker receives an order for 10 type I cookies, 20 type II cookies, and 15 type III cookies, as represented by the matrix R.

$$M = \begin{array}{c} \begin{array}{cccc} A & B & C & D \end{array} \\ \begin{bmatrix} 1 & 0 & 2 & 4 \\ 3 & 2 & 1 & 1 \\ 2 & 5 & 3 & 1 \end{bmatrix} \end{array} \begin{array}{l} \text{I} \\ \text{II} \\ \text{III} \end{array} \qquad N = \begin{bmatrix} 10 \\ 20 \\ 15 \\ 17 \end{bmatrix} \begin{array}{l} A \\ B \\ C \\ D \end{array}$$

$$S = \begin{bmatrix} 175 \\ 150 \\ 225 \end{bmatrix} \begin{array}{l} \text{I} \\ \text{II} \\ \text{III} \end{array} \qquad R = \overset{\begin{array}{ccc} \text{I} & \text{II} & \text{III} \end{array}}{\begin{bmatrix} 10 & 20 & 15 \end{bmatrix}}$$

Calculate and interpret the following

(a) RM (b) MN (c) RMN

(d) $S-MN$ (e) $R(S-MN)$ (f) RS

53. A community fitness center has a pool and a weight room. The admission prices (in dollars) for residents and nonresidents are given by the matrix

$$P = \begin{bmatrix} 4.50 \\ 5.00 \end{bmatrix} \begin{array}{l} \text{Residents} \\ \text{Nonresidents} \end{array}.$$

The average daily numbers of customers for the fitness center are given by the matrix

$$A = \begin{array}{c} \begin{array}{cc} \text{Residents} & \text{Nonresidents} \end{array} \\ \begin{bmatrix} 90 & 63 \\ 78 & 59 \end{bmatrix} \end{array} \begin{array}{l} \text{Pool} \\ \text{Weight room} \end{array}$$

(a) Compute AP.

(b) What is the average amount of money taken in by the pool each day?

54. A company makes radios and TV sets. Each radio requires 3 hours of assembly and $\frac{1}{2}$ hour of packaging, while each TV set requires 5 hours of assembly and 1 hour of packaging.

(a) Write a matrix T representing the required time for assembly and packaging of radios and TV sets.

(b) The company receives an order from a retail outlet for 30 radios and 20 TV sets. Find a matrix S so that either ST or TS gives the total assembly time and the total packaging time required to fill the order. What is the total assembly time? What is the total packaging time?

55. The matrix equation $AX = B$, where A is a 3×3 matrix, has the solution $x = 5$, $y = 4$, $z = 3$. What will be the new solution to this equation if the first and third columns of A are interchanged?

56. Find the values of a and b for which $A \cdot B = I_3$, where

$$A = \begin{bmatrix} 3 & 2 & 0 \\ 1 & 1 & 0 \\ 0 & 0 & 1 \end{bmatrix} \text{ and } B = \begin{bmatrix} a & b & 0 \\ -1 & 3 & 0 \\ 0 & 0 & 1 \end{bmatrix}.$$

In Exercises 57 and 58, determine the matrix B based on the screen shown.

57.

```
[A]
      [[1 5 -4]
       [3 2 25]]
[A]+[B]
      [[4  3 -3]
       [-2 8 32]]
```

58.

```
[A]
      [[1 5 -4]
       [3 2 25]]
[A]-[B]
      [[-4 1 -1]
       [3  3 23]]
```

Exercises 59–68 require the use of a graphing calculator or a computer.

In Exercises 59–67, calculate the given expression, where

$$A = \begin{bmatrix} .4 & 7 & -3 \\ 19 & .5 & 1.6 \\ -9 & 11 & 2 \end{bmatrix}, \qquad B = \begin{bmatrix} 6 & -9 & .3 \\ 1.5 & 22 & -4 \\ -5 & 6.6 & 14 \end{bmatrix},$$

and

$$C = \begin{bmatrix} 2.4 & 8 & -.2 \\ -11 & .3 & 6 \\ 7 & -4 & 5.1 \end{bmatrix}.$$

59. AB

60. BA

61. $A(B+C)$

62. $AB+AC$

63. Display a_{23} on the home screen.

64. Display a_{31} on the home screen.

65. A^2 (that is, AA)

66. A^3 (that is, AAA)

67. Calculate `[A]^2` and `[A]^3`. Compare your results with your answers in Exercises 65 and 66.

68. Try multiplying matrices A and B, where the number of columns of A differs from the number of rows of B. How does your calculator or computer respond?

Solutions to Practice Problems 3

1. Answer:

$$\begin{bmatrix} 3 & 1 & 2 \\ -1 & 0 & \frac{1}{2} \\ 0 & 4 & 1 \end{bmatrix} \begin{bmatrix} 7 & -1 & 0 \\ 5 & 4 & 2 \\ -6 & 0 & 4 \end{bmatrix} = \begin{bmatrix} 14 & 1 & 10 \\ -10 & 1 & 2 \\ 14 & 16 & 12 \end{bmatrix}.$$

The systematic steps to be taken are:

(a) Determine the size of the product matrix. Since we have a

$$\textcircled{3} \times 3 \text{ times a } 3 \times \textcircled{3},$$
└ outer dimensions ┘

the size of the product is given by the outer dimensions or 3×3. Begin by drawing a 3×3 rectangular array.

(b) Find the entries one at a time. To find the entry in the first row, first column of the product, look at the first row of the left given matrix and the first column of the right given matrix and form their product.

$$\begin{bmatrix} 3 & 1 & 2 \\ -1 & 0 & \frac{1}{2} \\ 0 & 4 & 1 \end{bmatrix} \begin{bmatrix} 7 & -1 & 0 \\ 5 & 4 & 2 \\ -6 & 0 & 4 \end{bmatrix} = \begin{bmatrix} 14 & & \\ & & \\ & & \end{bmatrix}$$

since $3 \cdot 7 + 1 \cdot 5 + 2(-6) = 14$. In general, to find the entry in the ith row, jth column of the product, put one finger on the ith row of the left given matrix and another finger on the jth column of the right given matrix. Then multiply the row matrix times the column matrix to get the desired entry.

2. Denote the three matrices by A, X, and B, respectively. Since b_{11} (the entry of the first row, first column of B) is 5, this means that

$$[\text{first row of } A]\begin{bmatrix}\text{first}\\ \text{column}\\ \text{of } X\end{bmatrix} = [b_{11}].$$

That is,

$$\begin{bmatrix}3 & -6\end{bmatrix}\begin{bmatrix}x\\ y\end{bmatrix} = [5] \quad \text{or} \quad 3x - 6y = 5.$$

Similarly, $b_{21} = 0$ says that $2x + y = 0$. Therefore, the corresponding system of linear equations is

$$\begin{cases} 3x - 6y = 5 \\ 2x + \ \ y = 0. \end{cases}$$

3. The coefficient matrix is

$$\begin{bmatrix}8 & 3\\ 9 & -2\end{bmatrix}.$$

So the system is equivalent to the matrix equation

$$\begin{bmatrix}8 & 3\\ 9 & -2\end{bmatrix}\begin{bmatrix}x\\ y\end{bmatrix} = \begin{bmatrix}7\\ -5\end{bmatrix}.$$

2.4 The Inverse of a Matrix

In Section 2.3 we introduced the operations of addition, subtraction, and multiplication of matrices. In this section let us pursue the algebra of matrices a bit further and consider equations involving matrices. Specifically, we consider equations of the form

$$AX = B, \tag{1}$$

where A and B are given matrices and X is an unknown matrix whose entries are to be determined. Such equations among matrices are intimately bound up with the theory of systems of linear equations. Indeed, we described the connection in a special case in Example 6 of Section 2.3. In that example we wrote the system of linear equations

$$\begin{cases} -2x + 4y = 2 \\ -3x + 7y = 7 \end{cases}$$

as a matrix equation of the form (1), where

$$A = \begin{bmatrix}-2 & 4\\ -3 & 7\end{bmatrix}, \qquad B = \begin{bmatrix}2\\ 7\end{bmatrix}, \qquad X = \begin{bmatrix}x\\ y\end{bmatrix}.$$

Note that by determining the entries (x and y) of the unknown matrix X, we solve the system of linear equations. We will return to this example after we have made a complete study of the matrix equation (1).

As motivation for our solution of equation (1), let us consider the analogous equation among numbers:

$$ax = b,$$

where a and b are given numbers[1] and x is to be determined. Let us examine its solution in great detail. Multiply both sides by $1/a$. (Note that $1/a$ makes sense, since $a \neq 0$.)

$$\begin{aligned} \left(\frac{1}{a}\right) \cdot (ax) &= \frac{1}{a} \cdot b \\ \left(\frac{1}{a} \cdot a\right) \cdot x &= \frac{1}{a} \cdot b \\ 1 \cdot x &= \frac{1}{a} \cdot b \\ x &= \frac{1}{a} \cdot b. \end{aligned}$$

Let us model our solution of equation (1) on the calculation above. To do so, we need to multiply both sides of the equation by a matrix that plays the same role in matrix arithmetic as $1/a$ plays in ordinary arithmetic. Our first task then will be to introduce this matrix and study its properties.

The number $1/a$ has the following relationship to the number a:

$$\frac{1}{a} \cdot a = a \cdot \frac{1}{a} = 1. \tag{2}$$

The matrix analog of the number 1 is an identity matrix I. This prompts us to generalize equation (2) to matrices as follows. Suppose that we are given a square matrix A. Then the *inverse* of A, denoted A^{-1}, is a square matrix with the property

$$A^{-1}A = I \quad \text{and} \quad AA^{-1} = I, \tag{3}$$

where I is an identity matrix of the same size as A. The matrix A^{-1} is the matrix analog of the number $1/a$. It can be shown that a matrix A has at most one inverse. (However, A may not have an inverse at all; see below.)

If we are given a matrix A, then it is easy to determine whether or not a given matrix is its inverse. Merely check equation (3) with the given matrix substituted for A^{-1}. For example, if

$$A = \begin{bmatrix} -2 & 4 \\ -3 & 7 \end{bmatrix}, \quad \text{then} \quad A^{-1} = \begin{bmatrix} -\frac{7}{2} & 2 \\ -\frac{3}{2} & 1 \end{bmatrix}.$$

Indeed, we have

$$\underset{A^{-1}}{\begin{bmatrix} -\frac{7}{2} & 2 \\ -\frac{3}{2} & 1 \end{bmatrix}} \underset{A}{\begin{bmatrix} -2 & 4 \\ -3 & 7 \end{bmatrix}} = \begin{bmatrix} 7-6 & -14+14 \\ 3-3 & -6+7 \end{bmatrix} = \underset{I_2}{\begin{bmatrix} 1 & 0 \\ 0 & 1 \end{bmatrix}}$$

and

$$\underset{A}{\begin{bmatrix} -2 & 4 \\ -3 & 7 \end{bmatrix}} \underset{A^{-1}}{\begin{bmatrix} -\frac{7}{2} & 2 \\ -\frac{3}{2} & 1 \end{bmatrix}} = \begin{bmatrix} 7-6 & -4+4 \\ \frac{21}{2}-\frac{21}{2} & -6+7 \end{bmatrix} = \underset{I_2}{\begin{bmatrix} 1 & 0 \\ 0 & 1 \end{bmatrix}}.$$

The inverse of a matrix can be calculated using Gaussian elimination, as the next example illustrates.

---

[1] We may as well assume that $a \neq 0$. Otherwise, x does not occur.

EXAMPLE 1 Let $A = \begin{bmatrix} 3 & 1 \\ 5 & 2 \end{bmatrix}$. Determine A^{-1}.

Solution Since A is a 2×2 matrix, A^{-1} is also a 2×2 matrix and satisfies

$$AA^{-1} = I_2 \quad \text{and} \quad A^{-1}A = I_2, \tag{4}$$

where $I_2 = \begin{bmatrix} 1 & 0 \\ 0 & 1 \end{bmatrix}$ is a 2×2 identity matrix. Suppose that

$$A^{-1} = \begin{bmatrix} x & y \\ z & w \end{bmatrix}.$$

Then the first equation of (4) reads

$$\begin{bmatrix} 3 & 1 \\ 5 & 2 \end{bmatrix} \begin{bmatrix} x & y \\ z & w \end{bmatrix} = \begin{bmatrix} 1 & 0 \\ 0 & 1 \end{bmatrix}.$$

Multiplying out the matrices on the left gives

$$\begin{bmatrix} 3x + z & 3y + w \\ 5x + 2z & 5y + 2w \end{bmatrix} = \begin{bmatrix} 1 & 0 \\ 0 & 1 \end{bmatrix}.$$

Now equate corresponding elements in the two matrices to obtain the equations

$$\begin{cases} 3x + z = 1 \\ 5x + 2z = 0, \end{cases} \qquad \begin{cases} 3y + w = 0 \\ 5y + 2w = 1. \end{cases}$$

Notice that the equations break up into two pairs of linear equations, each pair involving only two variables. Solving these two systems of linear equations yields $x = 2$, $z = -5$, $y = -1$, and $w = 3$. Therefore,

$$A^{-1} = \begin{bmatrix} 2 & -1 \\ -5 & 3 \end{bmatrix}.$$

Indeed, we may readily verify that

$$\begin{bmatrix} 3 & 1 \\ 5 & 2 \end{bmatrix} \begin{bmatrix} 2 & -1 \\ -5 & 3 \end{bmatrix} = \begin{bmatrix} 1 & 0 \\ 0 & 1 \end{bmatrix}$$

$$\begin{bmatrix} 2 & -1 \\ -5 & 3 \end{bmatrix} \begin{bmatrix} 3 & 1 \\ 5 & 2 \end{bmatrix} = \begin{bmatrix} 1 & 0 \\ 0 & 1 \end{bmatrix}. \quad \blacksquare$$

The above method can be used to calculate the inverse of matrices of any size, although it involves considerable calculation. We provide a rather efficient computational method for calculating A^{-1} in the next section. For now, however, let us be content with the above method. Using it, we can derive a general formula for A^{-1} in case A is a 2×2 matrix.

To determine the inverse of a 2×2 matrix, let

$$A = \begin{bmatrix} a & b \\ c & d \end{bmatrix}.$$

Let $\Delta = ad - bc$, and assume that $\Delta \neq 0$. Then A^{-1} is given by the formula

$$A^{-1} = \begin{bmatrix} \dfrac{d}{\Delta} & -\dfrac{b}{\Delta} \\[2ex] -\dfrac{c}{\Delta} & \dfrac{a}{\Delta} \end{bmatrix}. \tag{5}$$

We will omit the derivation of this formula. It proceeds along lines similar to those of Example 1. Notice that formula (5) involves division by Δ. Since division by 0 is not permissible, it is necessary that $\Delta \neq 0$ for formula (5) to be applied. We discuss the case $\Delta = 0$ in Example 3 ahead.

Equation (5) can be reduced to a simple step-by-step procedure.

To determine the inverse of $\begin{bmatrix} a & b \\ c & d \end{bmatrix}$ if $\Delta = ad - bc \neq 0$:

1. Interchange a and d to get $\begin{bmatrix} d & b \\ c & a \end{bmatrix}$.
2. Change the signs of b and c to get $\begin{bmatrix} d & -b \\ -c & a \end{bmatrix}$.
3. Divide all entries by Δ to get $\begin{bmatrix} \frac{d}{\Delta} & -\frac{b}{\Delta} \\ -\frac{c}{\Delta} & \frac{a}{\Delta} \end{bmatrix}$.

EXAMPLE 2 Calculate the inverse of $\begin{bmatrix} -2 & 4 \\ -3 & 7 \end{bmatrix}$.

Solution $\Delta = (-2) \cdot 7 - 4 \cdot (-3) = -2$, so $\Delta \neq 0$, and we may use the computation above.

1. Interchange a and d:

$$\begin{bmatrix} 7 & 4 \\ -3 & -2 \end{bmatrix}.$$

2. Change the signs of b and c:

$$\begin{bmatrix} 7 & -4 \\ 3 & -2 \end{bmatrix}.$$

3. Divide all entries by $\Delta = -2$:

$$\begin{bmatrix} -\frac{7}{2} & 2 \\ -\frac{3}{2} & 1 \end{bmatrix}.$$

Thus

$$\begin{bmatrix} -2 & 4 \\ -3 & 7 \end{bmatrix}^{-1} = \begin{bmatrix} -\frac{7}{2} & 2 \\ -\frac{3}{2} & 1 \end{bmatrix}.$$

Not every square matrix has an inverse. Indeed, it may be impossible to satisfy equation (3) for any choice of A^{-1}. This phenomenon can even occur in the case of 2×2 matrices. Here one can show that *if* $\Delta = 0$, *then the matrix does not have an inverse.* The next example illustrates this phenomenon in a special case.

EXAMPLE 3 Show that $\begin{bmatrix} 1 & 1 \\ 1 & 1 \end{bmatrix}$ does not have an inverse.

Solution Note first that $\Delta = 1 \cdot 1 - 1 \cdot 1 = 0$, so the inverse cannot be computed via equation (5). Suppose that the given matrix did have an inverse, say

$$\begin{bmatrix} s & t \\ u & v \end{bmatrix}.$$

Then the following equation would hold:

$$\begin{bmatrix} s & t \\ u & v \end{bmatrix} \begin{bmatrix} 1 & 1 \\ 1 & 1 \end{bmatrix} = \begin{bmatrix} 1 & 0 \\ 0 & 1 \end{bmatrix}.$$

On multiplying out the two matrices on the left, we get the equation

$$\begin{bmatrix} s+t & s+t \\ u+v & u+v \end{bmatrix} = \begin{bmatrix} 1 & 0 \\ 0 & 1 \end{bmatrix},$$

or, on equating entries in the first row:

$$s+t=1, \qquad s+t=0.$$

But $s+t$ cannot equal both 1 and 0. So we reach a contradiction, and therefore the original matrix cannot have an inverse. ▮

We were led to introduce the inverse of a matrix from a discussion of the matrix equation $AX = B$. Let us now return to that discussion. Suppose that A and B are given matrices and that we wish to solve the matrix equation

$$AX = B$$

for the unknown matrix X. Suppose further that A has an inverse A^{-1}. Multiply both sides of the equation on the left by A^{-1} to obtain

$$A^{-1} \cdot AX = A^{-1}B.$$

Because $A^{-1} \cdot A = I$, we have

$$IX = A^{-1}B$$
$$X = A^{-1}B.$$

Thus the matrix X is found by simply multiplying B on the left by A^{-1}, and we can summarize our findings as follows:

> ***Solving a Matrix Equation*** If the matrix A has an inverse, then the solution of the matrix equation
>
> $$AX = B \quad \text{is given by} \quad X = A^{-1}B.$$

Matrix equations can be used to solve systems of linear equations, as illustrated in the next example.

▮ EXAMPLE 4 Use a matrix equation to solve the system of linear equations

$$\begin{cases} -2x + 4y = 2 \\ -3x + 7y = 7. \end{cases}$$

Solution In Example 6 of Section 2.3 we saw that the system could be written as a matrix equation:

$$\underset{A}{\begin{bmatrix} -2 & 4 \\ -3 & 7 \end{bmatrix}} \underset{X}{\begin{bmatrix} x \\ y \end{bmatrix}} = \underset{B}{\begin{bmatrix} 2 \\ 7 \end{bmatrix}}.$$

We happen to know A^{-1} from Example 2, namely

$$A^{-1} = \begin{bmatrix} -\frac{7}{2} & 2 \\ -\frac{3}{2} & 1 \end{bmatrix}.$$

So we may compute the matrix $X = A^{-1}B$:

$$X = \begin{bmatrix} x \\ y \end{bmatrix} = \begin{bmatrix} -\frac{7}{2} & 2 \\ -\frac{3}{2} & 1 \end{bmatrix} \begin{bmatrix} 2 \\ 7 \end{bmatrix} = \begin{bmatrix} 7 \\ 4 \end{bmatrix}.$$

Thus, the solution of the system is $x = 7$, $y = 4$. ▮

▮ EXAMPLE 5 Let x and y denote the number of married and single adults in a certain town as of January 1. Let m and s denote the corresponding numbers for the following year. A statistical survey shows that x, y, m, and s are related by the equations

$$\begin{aligned} .9x + .2y &= m \\ .1x + .8y &= s. \end{aligned}$$

In a given year there were found to be 490,000 married adults and 147,000 single adults.

(a) How many married adults were there in the preceding year?

(b) How many married adults were there two years ago?

Solution (a) The given equations can be written in the matrix form

$$AX = B,$$

where

$$A = \begin{bmatrix} .9 & .2 \\ .1 & .8 \end{bmatrix}, \quad X = \begin{bmatrix} x \\ y \end{bmatrix}, \quad B = \begin{bmatrix} m \\ s \end{bmatrix}.$$

We are given that $B = \begin{bmatrix} 490{,}000 \\ 147{,}000 \end{bmatrix}$. So, since

$$X = A^{-1}B \quad \text{and} \quad A^{-1} = \begin{bmatrix} \frac{8}{7} & -\frac{2}{7} \\ -\frac{1}{7} & \frac{9}{7} \end{bmatrix},$$

we have

$$X = \begin{bmatrix} \frac{8}{7} & -\frac{2}{7} \\ -\frac{1}{7} & \frac{9}{7} \end{bmatrix} \begin{bmatrix} 490{,}000 \\ 147{,}000 \end{bmatrix} = \begin{bmatrix} 518{,}000 \\ 119{,}000 \end{bmatrix}.$$

Thus last year there were 518,000 married adults and 119,000 single adults.

(b) We deduce x and y for two years ago from the values of m and s for last year, namely $m = 518{,}000$, $s = 119{,}000$.

$$X = A^{-1}B = \begin{bmatrix} \frac{8}{7} & -\frac{2}{7} \\ -\frac{1}{7} & \frac{9}{7} \end{bmatrix} \begin{bmatrix} 518{,}000 \\ 119{,}000 \end{bmatrix} = \begin{bmatrix} 558{,}000 \\ 79{,}000 \end{bmatrix}.$$

That is, two years ago there were 558,000 married adults and 79,000 single adults. ▮

EXAMPLE 6 In Section 2.5 we will show that if

$$A = \begin{bmatrix} 4 & -2 & 3 \\ 8 & -3 & 5 \\ 7 & -2 & 4 \end{bmatrix}, \quad \text{then} \quad A^{-1} = \begin{bmatrix} -2 & 2 & -1 \\ 3 & -5 & 4 \\ 5 & -6 & 4 \end{bmatrix}.$$

(a) Use this fact to solve the system of linear equations

$$\begin{cases} 4x - 2y + 3z = 1 \\ 8x - 3y + 5z = 4 \\ 7x - 2y + 4z = 5. \end{cases}$$

(b) Solve the system of equations

$$\begin{cases} 4x - 2y + 3z = 4 \\ 8x - 3y + 5z = 7 \\ 7x - 2y + 4z = 6. \end{cases}$$

Solution (a) The system can be written in the matrix form

$$\underset{A}{\begin{bmatrix} 4 & -2 & 3 \\ 8 & -3 & 5 \\ 7 & -2 & 4 \end{bmatrix}} \underset{X}{\begin{bmatrix} x \\ y \\ z \end{bmatrix}} = \underset{B}{\begin{bmatrix} 1 \\ 4 \\ 5 \end{bmatrix}}.$$

The solution of this matrix equation is $X = A^{-1}B$, or

$$\begin{bmatrix} x \\ y \\ z \end{bmatrix} = \begin{bmatrix} -2 & 2 & -1 \\ 3 & -5 & 4 \\ 5 & -6 & 4 \end{bmatrix} \begin{bmatrix} 1 \\ 4 \\ 5 \end{bmatrix} = \begin{bmatrix} 1 \\ 3 \\ 1 \end{bmatrix}.$$

Thus the solution of the system is $x = 1$, $y = 3$, $z = 1$.

(b) This system has the same left-hand side as the preceding system, so its solution is

$$\begin{bmatrix} x \\ y \\ z \end{bmatrix} = \begin{bmatrix} -2 & 2 & -1 \\ 3 & -5 & 4 \\ 5 & -6 & 4 \end{bmatrix} \begin{bmatrix} 4 \\ 7 \\ 6 \end{bmatrix} = \begin{bmatrix} 0 \\ 1 \\ 2 \end{bmatrix}.$$

That is, the solution of the system is $x = 0$, $y = 1$, $z = 2$. ▮

Using the method of matrix equations to solve a system of linear equations is especially efficient if one wishes to solve a number of systems all having the same left-hand sides but different right-hand sides. For then A^{-1} must be computed only once for all the systems under consideration. (This point is useful in Exercises 19–22.)

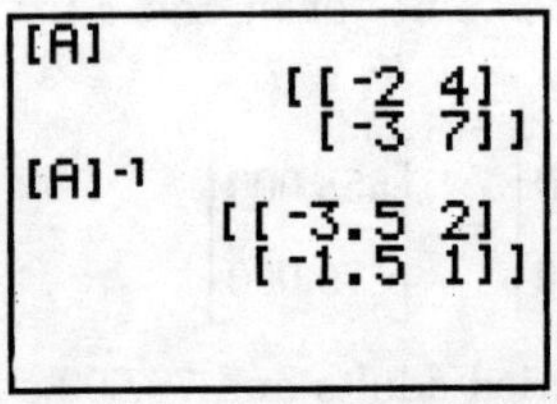

FIGURE 1

GC The inverse of a square matrix can be obtained directly with the inverse key $\boxed{x^{-1}}$. See Fig. 1. (*Note*: Do not use `[A]^-1`.)

Practice Problems 4

1. Show that the inverse of

$$\begin{bmatrix} -4 & 1 & 2 \\ 7 & -1 & -4 \\ -\frac{1}{2} & 0 & \frac{1}{2} \end{bmatrix} \text{ is } \begin{bmatrix} 1 & 1 & 4 \\ 3 & 2 & 4 \\ 1 & 1 & 6 \end{bmatrix}.$$

2. Use the method of this section to solve the system of linear equations

$$\begin{cases} .8x + .6y = 5 \\ .2x + .4y = 2. \end{cases}$$

Exercises 4

In Exercises 1 and 2, use the fact that

$$\begin{bmatrix} 2 & 2 \\ \frac{1}{2} & 1 \end{bmatrix}^{-1} = \begin{bmatrix} 1 & -2 \\ -\frac{1}{2} & 2 \end{bmatrix}.$$

1. Solve $\begin{cases} 2x + 2y = 4 \\ \frac{1}{2}x + \ \ y = 1. \end{cases}$

2. Solve $\begin{cases} 2x + 2y = 14 \\ \frac{1}{2}x + \ \ y = \ 4. \end{cases}$

In Exercises 3–10, find the inverse of the given matrix.

3. $\begin{bmatrix} 7 & 2 \\ 3 & 1 \end{bmatrix}$ **4.** $\begin{bmatrix} 2 & 3 \\ 5 & 7 \end{bmatrix}$

5. $\begin{bmatrix} 6 & 2 \\ 5 & 2 \end{bmatrix}$ **6.** $\begin{bmatrix} 1 & .5 \\ 0 & .5 \end{bmatrix}$

7. $\begin{bmatrix} .7 & .2 \\ .3 & .8 \end{bmatrix}$ **8.** $\begin{bmatrix} 0 & 1 \\ 1 & 0 \end{bmatrix}$

9. $\begin{bmatrix} 3 \end{bmatrix}$ **10.** $\begin{bmatrix} .2 \end{bmatrix}$

In Exercises 11–14, use the method of this section to solve the system of linear equations.

11. $\begin{cases} x + 2y = 3 \\ 2x + 6y = 5 \end{cases}$ **12.** $\begin{cases} 5x + 3y = 1 \\ 7x + 4y = 2 \end{cases}$

13. $\begin{cases} \frac{1}{2}x + \ 2y = 4 \\ 3x + 16y = 0 \end{cases}$ **14.** $\begin{cases} .8x + .6y = 2 \\ .2x + .4y = 1 \end{cases}$

15. It is found that the number of married and single adults in a certain town are subject to the following statistics. Suppose that x and y denote the number of married and single adults, respectively, in a given year (say as of January 1) and let m, s denote the corresponding numbers for the following year. Then

$$.8x + .3y = m$$
$$.2x + .7y = \ s.$$

(a) Write this system of equations in matrix form.

(b) Solve the resulting matrix equation for $X = \begin{bmatrix} x \\ y \end{bmatrix}$.

(c) Suppose that in a given year there were found to be 100,000 married adults and 50,000 single adults. How many married (respectively, single) adults were there the preceding year?

(d) How many married (respectively, single) adults were there two years ago?

16. A flu epidemic is spreading through a town of 48,000 people. It is found that if x and y denote the numbers of people sick and well in a given week, respectively, and if s and w denote the corresponding numbers for the following week, then

$$\tfrac{1}{3}x + \tfrac{1}{4}y = s$$
$$\tfrac{2}{3}x + \tfrac{3}{4}y = w.$$

(a) Write this system of equations in matrix form.

(b) Solve the resulting matrix equation for $X = \begin{bmatrix} x \\ y \end{bmatrix}$.

(c) Suppose that 13,000 people are sick in a given week. How many were sick the preceding week?

(d) Same question as part (c), except assume that 14,000 are sick.

17. Statistics show that at a certain university, 70% of the students who live on campus during a given semester will remain on campus the following semester, and 90% of students living off campus during a given semester will remain off campus the following semester. Let x and y denote the number of students who live on and off campus this semester, and let u and v be the corresponding numbers for the next semester. Then

$$.7x + .1y = u$$
$$.3x + .9y = v.$$

(a) Write this system of equations in matrix form.

(b) Solve the resulting matrix equation for $\begin{bmatrix} x \\ y \end{bmatrix}$.

(c) Suppose that out of a group of 9000 students, 6000 currently live on campus and 3000 live off campus. How many lived on campus last semester? How many will live off campus next semester?

18. A teacher estimates that of the students who pass a test, 80% will pass the next test, while of the students who fail a test, 50% will pass the next test. Let x and y denote the number of students who pass and fail a given test, and let u and v be the corresponding numbers for the following test.

(a) Write a matrix equation relating $\begin{bmatrix} x \\ y \end{bmatrix}$ to $\begin{bmatrix} u \\ v \end{bmatrix}$.

(b) Suppose that 25 of the teacher's students pass the third test and 8 fail the third test. How many students will pass the fourth test? Approximately how many passed the second test?

In Exercises 19 and 20, use the fact that

$$\begin{bmatrix} 1 & 2 & 2 \\ 1 & 3 & 2 \\ 1 & 2 & 3 \end{bmatrix}^{-1} = \begin{bmatrix} 5 & -2 & -2 \\ -1 & 1 & 0 \\ -1 & 0 & 1 \end{bmatrix}.$$

19. Solve $\begin{cases} x + 2y + 2z = 1 \\ x + 3y + 2z = -1 \\ x + 2y + 3z = -1. \end{cases}$

20. Solve $\begin{cases} x + 2y + 2z = 1 \\ x + 3y + 2z = 0 \\ x + 2y + 3z = 0. \end{cases}$

In Exercises 21 and 22, use the fact that

$$\begin{bmatrix} 9 & 0 & 2 & 0 \\ -20 & -9 & -5 & 5 \\ 4 & 0 & 1 & 0 \\ -4 & -2 & -1 & 1 \end{bmatrix}^{-1} = \begin{bmatrix} 1 & 0 & -2 & 0 \\ 0 & 1 & 0 & -5 \\ -4 & 0 & 9 & 0 \\ 0 & 2 & 1 & -9 \end{bmatrix}.$$

21. Solve $\begin{cases} 9x + 2z = 1 \\ -20x - 9y - 5z + 5w = 0 \\ 4x + z = 0 \\ -4x - 2y - z + w = -1. \end{cases}$

22. Solve $\begin{cases} 9x + 2z = 2 \\ -20x - 9y - 5z + 5w = 1 \\ 4x + z = 3 \\ -4x - 2y - z + w = 0. \end{cases}$

23. Without computing Δ, show that the matrix $\begin{bmatrix} 6 & 3 \\ 2 & 1 \end{bmatrix}$ does not have an inverse.

24. If $A^{-1} = \begin{bmatrix} 2 & 7 \\ 1 & -3 \end{bmatrix}$, what is the matrix A?

25. There are two age groups for a particular species of organism. Group I consists of all organisms aged under 1 year, while group II consists of all organisms aged from 1 to 2 years. No organism survives more than 2 years. The average number of offspring per year born to each member of group I is 1, while the average number of organisms per year born to each member of group II is 2. Nine-tenths of group I survive to enter group II each year.

(a) Let x and y represent the initial number of organisms in groups I and II, respectively. Let a and b represent the number of organisms in groups I and II, respectively, after one year. Write a matrix equation relating $\begin{bmatrix} x \\ y \end{bmatrix}$ to $\begin{bmatrix} a \\ b \end{bmatrix}$.

(b) If there are initially 450,000 organisms in group I and 360,000 organisms in group II, calculate the number of organisms in each of the groups after 1 year and after 2 years.

(c) Suppose that at a certain time there were 810,000 organisms in group I and 630,000 organisms in group II. Determine the population of each group 1 year earlier.

26. If $A^2 = \begin{bmatrix} -2 & -1 \\ 2 & -1 \end{bmatrix}$ and $A^3 = \begin{bmatrix} -2 & 1 \\ -2 & -3 \end{bmatrix}$, what is A?

Exercises 27–35 require the use of a graphing calculator or a computer.

In Exercises 27–30, use the inverse key to find the inverse of the given matrix. Display the answer as a fraction.

27. $\begin{bmatrix} .2 & 3 \\ 4 & 1.6 \end{bmatrix}$

28. $\begin{bmatrix} -12 & 3.3 \\ 6 & .4 \end{bmatrix}$

29. $\begin{bmatrix} .6 & 3 & -7 \\ 2.5 & -1 & 4 \\ -2 & .3 & 9 \end{bmatrix}$

30. $\begin{bmatrix} 5 & 2.3 & 6 \\ 1.2 & 5 & -7 \\ -3 & -4 & 6.5 \end{bmatrix}$

In Exercises 31–34, calculate the answer using $[\mathtt{A}]^{-1} * [\mathtt{B}]$ and give the answer using fractions.

31. $\begin{cases} 2x - 4y + 7z = 11 \\ x + 3y - 5z = -9 \\ 3x - y + 3z = 7 \end{cases}$

32. $\begin{cases} 5x + 2y - 3z = 1 \\ 4x - y + z = 22 \\ -x + 5y - 6z = 4 \end{cases}$

33. $\begin{cases} 2x + 7z + 5w = 10 \\ 5x - y + 3z = -2 \\ x + 2y - 2w = 0 \\ 3x - 4y + 2z - 5w = -18 \end{cases}$

34. $\begin{cases} x + 4y - z + 2w = 9 \\ 3x - 8y + 2z + 4w = -2 \\ -x - 3y + 7z - 6w = 10 \\ 4x + 2y - 3z + w = -6 \end{cases}$

35. Try finding the inverse of a matrix that does not have an inverse. How does your calculator or computer respond?

Solutions to Practice Problems 4

1. To see if this matrix is indeed the inverse, multiply it by the original matrix and find out if the products are identity matrices.

$$\begin{bmatrix} 1 & 1 & 4 \\ 3 & 2 & 4 \\ 1 & 1 & 6 \end{bmatrix} \begin{bmatrix} -4 & 1 & 2 \\ 7 & -1 & -4 \\ -\frac{1}{2} & 0 & \frac{1}{2} \end{bmatrix} = \begin{bmatrix} 1 & 0 & 0 \\ 0 & 1 & 0 \\ 0 & 0 & 1 \end{bmatrix}, \quad \text{an identity matrix.}$$

$$\begin{bmatrix} -4 & 1 & 2 \\ 7 & -1 & -4 \\ -\frac{1}{2} & 0 & \frac{1}{2} \end{bmatrix} \begin{bmatrix} 1 & 1 & 4 \\ 3 & 2 & 4 \\ 1 & 1 & 6 \end{bmatrix} = \begin{bmatrix} 1 & 0 & 0 \\ 0 & 1 & 0 \\ 0 & 0 & 1 \end{bmatrix}.$$

2. The matrix form of this system is

$$\begin{bmatrix} .8 & .6 \\ .2 & .4 \end{bmatrix} \begin{bmatrix} x \\ y \end{bmatrix} = \begin{bmatrix} 5 \\ 2 \end{bmatrix}.$$

Therefore, the solution is

$$\begin{bmatrix} x \\ y \end{bmatrix} = \begin{bmatrix} .8 & .6 \\ .2 & .4 \end{bmatrix}^{-1} \begin{bmatrix} 5 \\ 2 \end{bmatrix}.$$

To compute the inverse of the 2×2 matrix, first compute Δ.

$$\Delta = ad - bc = (.8)(.4) - (.6)(.2) = .32 - .12 = .2.$$

Thus,

$$\begin{bmatrix} .8 & .6 \\ .2 & .4 \end{bmatrix}^{-1} = \begin{bmatrix} .4/.2 & -.6/.2 \\ -.2/.2 & .8/.2 \end{bmatrix} = \begin{bmatrix} 2 & -3 \\ -1 & 4 \end{bmatrix}.$$

Therefore,

$$\begin{bmatrix} x \\ y \end{bmatrix} = \begin{bmatrix} 2 & -3 \\ -1 & 4 \end{bmatrix} \begin{bmatrix} 5 \\ 2 \end{bmatrix} = \begin{bmatrix} 4 \\ 3 \end{bmatrix}.$$

so the solution is $x = 4$, $y = 3$.

2.5 The Gauss–Jordan Method for Calculating Inverses

Of the several popular methods for finding the inverse of a matrix, the Gauss–Jordan method is probably the easiest to describe. It can be used on square matrices of any size. Also, the mechanical nature of the computations allows this method to be programmed for a computer with relative ease. We shall illustrate the procedure with a 2×2 matrix, whose inverse can also be calculated using the method of the previous section. Let

$$A = \begin{bmatrix} \frac{1}{2} & 1 \\ 1 & 3 \end{bmatrix}.$$

It is simple to check that

$$A^{-1} = \begin{bmatrix} 6 & -2 \\ -2 & 1 \end{bmatrix}.$$

Let us now derive this result using the Gauss–Jordan method.

Step 1 Write down the matrix A, and on its right write an identity matrix of the same size.

This is most conveniently done by placing I_2 beside A in a single matrix.

$$\underbrace{\begin{bmatrix} \frac{1}{2} & 1 \\ 1 & 3 \end{bmatrix}}_{A}\ \underbrace{\begin{bmatrix} 1 & 0 \\ 0 & 1 \end{bmatrix}}_{I_2}$$

Step 2 Perform elementary row operations on the left-hand matrix so as to transform it into an identity matrix. Each operation performed on the left-hand matrix is also performed on the right-hand matrix.

This step proceeds exactly like the Gaussian elimination method and may be most conveniently expressed in terms of pivoting.

$$\left[\begin{array}{cc|cc} \textcircled{\tfrac{1}{2}} & 1 & 1 & 0 \\ 1 & 3 & 0 & 1 \end{array}\right], \quad \left[\begin{array}{cc|cc} 1 & 2 & 2 & 0 \\ 0 & \textcircled{1} & -2 & 1 \end{array}\right], \quad \left[\begin{array}{cc|cc} 1 & 0 & 6 & -2 \\ 0 & 1 & -2 & 1 \end{array}\right].$$

Step 3 When the matrix on the left becomes an identity matrix, the matrix on the right is the desired inverse.

So, from the last matrix of our calculation above, we have

$$A^{-1} = \begin{bmatrix} 6 & -2 \\ -2 & 1 \end{bmatrix}.$$

This is the same result obtained earlier.

We will demonstrate why the method above works after some further examples.

EXAMPLE 1 Find the inverse of the matrix

$$A = \begin{bmatrix} 4 & -2 & 3 \\ 8 & -3 & 5 \\ 7 & -2 & 4 \end{bmatrix}.$$

Solution

$$\left[\begin{array}{ccc|ccc} \textcircled{4} & -2 & 3 & 1 & 0 & 0 \\ 8 & -3 & 5 & 0 & 1 & 0 \\ 7 & -2 & 4 & 0 & 0 & 1 \end{array}\right]$$

$$\left[\begin{array}{ccc|ccc} 1 & -\frac{1}{2} & \frac{3}{4} & \frac{1}{4} & 0 & 0 \\ 0 & \textcircled{1} & -1 & -2 & 1 & 0 \\ 0 & \frac{3}{2} & -\frac{5}{4} & -\frac{7}{4} & 0 & 1 \end{array}\right]$$

$$\left[\begin{array}{ccc|ccc} 1 & 0 & \frac{1}{4} & -\frac{3}{4} & \frac{1}{2} & 0 \\ 0 & 1 & -1 & -2 & 1 & 0 \\ 0 & 0 & \textcircled{\frac{1}{4}} & \frac{5}{4} & -\frac{3}{2} & 1 \end{array}\right]$$

$$\left[\begin{array}{ccc|ccc} 1 & 0 & 0 & -2 & 2 & -1 \\ 0 & 1 & 0 & 3 & -5 & 4 \\ 0 & 0 & 1 & 5 & -6 & 4 \end{array}\right]$$

Therefore,

$$A^{-1} = \begin{bmatrix} -2 & 2 & -1 \\ 3 & -5 & 4 \\ 5 & -6 & 4 \end{bmatrix}.$$ ∎

Not all square matrices have inverses. If a matrix does not have an inverse, this will become apparent when applying the Gauss–Jordan method. At some point there will be no way to continue transforming the left-hand matrix into an identity matrix. This is illustrated in the next example.

EXAMPLE 2 Find the inverse of the matrix

$$A = \begin{bmatrix} 1 & 3 & 2 \\ 0 & 1 & 4 \\ 1 & 5 & 10 \end{bmatrix}.$$

Solution

$$\left[\begin{array}{ccc|ccc} \textcircled{1} & 3 & 2 & 1 & 0 & 0 \\ 0 & 1 & 4 & 0 & 1 & 0 \\ 1 & 5 & 10 & 0 & 0 & 1 \end{array}\right]$$

$$\left[\begin{array}{ccc|ccc} 1 & 3 & 2 & 1 & 0 & 0 \\ 0 & \textcircled{1} & 4 & 0 & 1 & 0 \\ 0 & 2 & 8 & -1 & 0 & 1 \end{array}\right]$$

$$\left[\begin{array}{ccc|ccc} 1 & 0 & -10 & 1 & -3 & 0 \\ 0 & 1 & 4 & 0 & 1 & 0 \\ 0 & 0 & 0 & -1 & -2 & 1 \end{array}\right]$$

Since the third row of the left-hand matrix has only zero entries, it is impossible to complete the Gauss–Jordan method. Therefore, the matrix A has no inverse matrix. ∎

Verification of the Gauss–Jordan Method for Calculating Inverses In Section 2.4 we showed how to calculate the inverse by solving several systems of linear equations. Actually, the Gauss–Jordan method is just an organized way of going about the calculation. To see why, let us consider a concrete example:

$$A = \begin{bmatrix} 4 & -2 & 3 \\ 8 & -3 & 5 \\ 7 & -2 & 4 \end{bmatrix}.$$

We wish to determine A^{-1}, so regard it as a matrix of unknowns:

$$A^{-1} = \begin{bmatrix} x_1 & x_2 & x_3 \\ y_1 & y_2 & y_3 \\ z_1 & z_2 & z_3 \end{bmatrix}.$$

The statement $AA^{-1} = I_3$ is

$$\begin{bmatrix} 4 & -2 & 3 \\ 8 & -3 & 5 \\ 7 & -2 & 4 \end{bmatrix} \begin{bmatrix} x_1 & x_2 & x_3 \\ y_1 & y_2 & y_3 \\ z_1 & z_2 & z_3 \end{bmatrix} = \begin{bmatrix} 1 & 0 & 0 \\ 0 & 1 & 0 \\ 0 & 0 & 1 \end{bmatrix}.$$

Multiplying out the matrices on the left and comparing the result with the matrix on the right give us nine equations, namely

$$\begin{cases} 4x_1 - 2y_1 + 3z_1 = 1 \\ 8x_1 - 3y_1 + 5z_1 = 0 \\ 7x_1 - 2y_1 + 4z_1 = 0 \end{cases}$$
$$\begin{cases} 4x_2 - 2y_2 + 3z_2 = 0 \\ 8x_2 - 3y_2 + 5z_2 = 1 \\ 7x_2 - 2y_2 + 4z_2 = 0 \end{cases}$$
$$\begin{cases} 4x_3 - 2y_3 + 3z_3 = 0 \\ 8x_3 - 3y_3 + 5z_3 = 0 \\ 7x_3 - 2y_3 + 4z_3 = 1. \end{cases}$$

Notice that each system of equations corresponds to one column of unknowns in A^{-1}. More precisely, if we set

$$X_1 = \begin{bmatrix} x_1 \\ y_1 \\ z_1 \end{bmatrix}, \qquad X_2 = \begin{bmatrix} x_2 \\ y_2 \\ z_2 \end{bmatrix}, \qquad X_3 = \begin{bmatrix} x_3 \\ y_3 \\ z_3 \end{bmatrix},$$

then the three systems above have the respective matrix forms

$$AX_1 = \begin{bmatrix} 1 \\ 0 \\ 0 \end{bmatrix}, \qquad AX_2 = \begin{bmatrix} 0 \\ 1 \\ 0 \end{bmatrix}, \qquad AX_3 = \begin{bmatrix} 0 \\ 0 \\ 1 \end{bmatrix}.$$

Now imagine the process of applying Gaussian elimination to solve these three systems. We apply elementary row operations to the matrices

$$\left[\begin{array}{c|c} A & \begin{matrix} 1 \\ 0 \\ 0 \end{matrix} \end{array}\right], \qquad \left[\begin{array}{c|c} A & \begin{matrix} 0 \\ 1 \\ 0 \end{matrix} \end{array}\right], \qquad \left[\begin{array}{c|c} A & \begin{matrix} 0 \\ 0 \\ 1 \end{matrix} \end{array}\right].$$

The process ends when we convert A into the identity matrix, at which point the solutions may be read off the right column. So the procedure ends with the matrices

$$\left[I_3 \mid X_1\right], \qquad \left[I_3 \mid X_2\right], \qquad \left[I_3 \mid X_3\right].$$

Realize, however, that at each step of the three Gaussian eliminations we are performing the same operations, since all three start with the matrix A on the left. So, in order to save calculations, perform the three Gaussian eliminations simultaneously by performing the row operations on the composite matrix

$$\left[\begin{array}{c|ccc} & 1 & 0 & 0 \\ A & 0 & 1 & 0 \\ & 0 & 0 & 1 \end{array}\right] = \left[A \mid I_3\right].$$

The procedure ends when this matrix is converted into

$$\left[I_3 \mid X_1 \quad X_2 \quad X_3\right].$$

That is, since $A^{-1} = \left[X_1 \quad X_2 \quad X_3\right]$, the procedure ends with A^{-1} on the right. This is the reasoning behind the Gauss–Jordan method of calculating inverses.

Practice Problems 5

1. Use the Gauss–Jordan method to calculate the inverse of the matrix
$$\begin{bmatrix} 1 & 0 & 2 \\ 0 & 1 & -4 \\ 0 & 0 & 2 \end{bmatrix}.$$

2. Solve the system of linear equations
$$\begin{cases} x + 2z = 4 \\ y - 4z = 6 \\ 2z = 9. \end{cases}$$

EXERCISES 5

In Exercises 1–12, use the Gauss–Jordan method to compute the inverse of the matrix.

1. $\begin{bmatrix} 7 & 3 \\ 5 & 2 \end{bmatrix}$

2. $\begin{bmatrix} 5 & -2 \\ 6 & 2 \end{bmatrix}$

3. $\begin{bmatrix} 10 & 12 \\ 3 & -4 \end{bmatrix}$

4. $\begin{bmatrix} 1 & -3 \\ 0 & 1 \end{bmatrix}$

5. $\begin{bmatrix} 2 & -4 \\ -1 & 2 \end{bmatrix}$

6. $\begin{bmatrix} 1 & 3 & 1 \\ -1 & 2 & 0 \\ 2 & 11 & 3 \end{bmatrix}$

7. $\begin{bmatrix} 1 & 2 & -2 \\ 1 & 1 & 1 \\ 0 & 0 & 1 \end{bmatrix}$

8. $\begin{bmatrix} 2 & 2 & 0 \\ 0 & -2 & 0 \\ 3 & 0 & 1 \end{bmatrix}$

9. $\begin{bmatrix} -2 & 5 & 2 \\ 1 & -3 & -1 \\ -1 & 2 & 1 \end{bmatrix}$

10. $\begin{bmatrix} 1 & 0 & 0 \\ 2 & 1 & -2 \\ -1 & 2 & 1 \end{bmatrix}$

11. $\begin{bmatrix} 1 & 6 & 0 & 0 \\ 1 & 5 & 0 & 0 \\ 0 & 0 & 4 & 2 \\ 0 & 0 & 50 & 2 \end{bmatrix}$

12. $\begin{bmatrix} 6 & 0 & 2 & 0 \\ -6 & 1 & 0 & 1 \\ 1 & 0 & 1 & 0 \\ -9 & 0 & -1 & 1 \end{bmatrix}$

In Exercises 13–16, use matrix inversion to solve the system of linear equations.

13. $\begin{cases} x + y + 2z = 3 \\ 3x + 2y + 2z = 4 \\ x + y + 3z = 5 \end{cases}$

14. $\begin{cases} x + 2y + 3z = 4 \\ 3x + 5y + 5z = 3 \\ 2x + 4y + 2z = 4 \end{cases}$

15. $\begin{cases} x - 2z - 2w = 0 \\ y - 5w = 1 \\ -4x + 9z + 9w = 2 \\ 2y + z - 8w = 3 \end{cases}$

16. $\begin{cases} y + 2z = 1 \\ 2x + y + 3z = 2 \\ x + y + 2z = 3 \end{cases}$

17. Suppose that we try to solve the matrix equation $AX = B$ by using matrix inversion but find that even though the matrix A is a square matrix, it has no inverse. What can be said about the outcome from solving the associated system of linear equations by the Gaussian elimination method?

18. Find the 2×2 matrix C for which $A \cdot C = B$, where
$$A = \begin{bmatrix} 3 & 2 \\ 4 & 3 \end{bmatrix} \quad \text{and} \quad B = \begin{bmatrix} 5 & 6 \\ 1 & 7 \end{bmatrix}.$$

19. Find a 2×2 matrix A for which

$$A \cdot \begin{bmatrix} 2 \\ 1 \end{bmatrix} = \begin{bmatrix} -1 \\ 4 \end{bmatrix} \quad \text{and} \quad A \cdot \begin{bmatrix} 5 \\ 3 \end{bmatrix} = \begin{bmatrix} 0 \\ 2 \end{bmatrix}.$$

20. Let

$$A = \begin{bmatrix} 7 & 4 \\ 3 & 2 \end{bmatrix} \quad \text{and} \quad B = \begin{bmatrix} 1 & 5 \\ -3 & 4 \end{bmatrix}.$$

Each of the equations $AX = B$ and $XA = B$ has a 2×2 matrix X as a solution. Find X in each case and explain why the two answers are different.

Exercises 21–24 require the use of a graphing calculator or a computer.

In Exercises 21–24, use the Gauss–Jordan method to compute the inverse of the matrix.

21. $\begin{bmatrix} 4 & 1 \\ 7 & 2 \end{bmatrix}$

22. $\begin{bmatrix} 8 & 3 \\ 13 & 5 \end{bmatrix}$

23. $\begin{bmatrix} 2 & 5 & 3 \\ 1 & 3 & 0 \\ 2 & 3 & 4 \end{bmatrix}$

24. $\begin{bmatrix} 4 & -2 & -8 \\ -3 & 1 & 3 \\ 5 & 0 & 6 \end{bmatrix}$

Solutions to Practice Problems 5

1. First write the given matrix beside an identity matrix of the same size

$$\left[\begin{array}{ccc|ccc} 1 & 0 & 2 & 1 & 0 & 0 \\ 0 & 1 & -4 & 0 & 1 & 0 \\ 0 & 0 & 2 & 0 & 0 & 1 \end{array}\right].$$

The object is to use elementary row operations to transform the 3×3 matrix on the left into the identity matrix. The first two columns are already in the correct form.

$$\left[\begin{array}{ccc|ccc} 1 & 0 & 2 & 1 & 0 & 0 \\ 0 & 1 & -4 & 0 & 1 & 0 \\ 0 & 0 & 2 & 0 & 0 & 1 \end{array}\right] \xrightarrow{\frac{1}{2}[3]} \left[\begin{array}{ccc|ccc} 1 & 0 & 2 & 1 & 0 & 0 \\ 0 & 1 & -4 & 0 & 1 & 0 \\ 0 & 0 & 1 & 0 & 0 & \frac{1}{2} \end{array}\right]$$

$$\xrightarrow{[1]+(-2)[3]} \left[\begin{array}{ccc|ccc} 1 & 0 & 0 & 1 & 0 & -1 \\ 0 & 1 & -4 & 0 & 1 & 0 \\ 0 & 0 & 1 & 0 & 0 & \frac{1}{2} \end{array}\right]$$

$$\xrightarrow{[2]+(4)[3]} \left[\begin{array}{ccc|ccc} 1 & 0 & 0 & 1 & 0 & -1 \\ 0 & 1 & 0 & 0 & 1 & 2 \\ 0 & 0 & 1 & 0 & 0 & \frac{1}{2} \end{array}\right]$$

Thus the inverse of the given matrix is

$$\begin{bmatrix} 1 & 0 & -1 \\ 0 & 1 & 2 \\ 0 & 0 & \frac{1}{2} \end{bmatrix}.$$

2. The matrix form of this system of equations is $AX = B$, where A is the matrix whose inverse was found in Problem 1, and

$$B = \begin{bmatrix} 4 \\ 6 \\ 9 \end{bmatrix}.$$

Therefore, $X = A^{-1}B$, so that

$$\begin{bmatrix} x \\ y \\ z \end{bmatrix} = \begin{bmatrix} 1 & 0 & -1 \\ 0 & 1 & 2 \\ 0 & 0 & \frac{1}{2} \end{bmatrix} \begin{bmatrix} 4 \\ 6 \\ 9 \end{bmatrix} = \begin{bmatrix} -5 \\ 24 \\ \frac{9}{2} \end{bmatrix}.$$

So the solution of the system is $x = -5$, $y = 24$, $z = \frac{9}{2}$.

2.6 Input–Output Analysis

In recent years matrix arithmetic has played an ever-increasing role in economics, especially in that branch of economics called *input–output analysis.* Pioneered by the Harvard economist Vassily Leontieff, input–output analysis is used to analyze an economy in order to meet given consumption and export demands. As we shall see, such analysis leads into matrix calculations and in particular to inverses. Input–output analysis has been of such great significance that Leontieff was awarded the 1973 Nobel Prize in economics for his fundamental work in the subject.

Suppose that we divide an economy up into a number of industries—transportation, agriculture, steel, and so on. Each industry produces a certain output using certain raw materials (or input). The input of each industry is made up in part by the outputs of other industries. For example, in order to produce food, agriculture uses as input the output of many industries, such as transportation (tractors and trucks) and oil (gasoline and fertilizers). This interdependence among the industries of the economy is summarized in a matrix—an *input–output matrix.* There is one column for each industry's input requirements. The entries in the column reflect the amount of input required from each of the industries. A typical input–output matrix looks like this:

		Input requirements of:			
		Industry 1	Industry 2	Industry 3	...
From	Industry 1				
	Industry 2				
	Industry 3				
	⋮				

It is most convenient to express the entries of this matrix in monetary terms. That is, each column gives the dollar values of the various inputs needed by an industry in order to produce \$1 worth of output.

There are consumers (other than the industries themselves) who want to purchase some of the output of these industries. The quantity of goods that these consumers want (or demand) is called the *final demand* on the economy. The final demand can be represented by a column matrix, with one entry for each industry, indicating the amount of consumable output demanded from the industry:

$$[\text{final demand}] = \begin{bmatrix} \text{amount from industry 1} \\ \text{amount from industry 2} \\ \vdots \end{bmatrix}.$$

We shall consider the situation in which the final-demand matrix is given and it is necessary to determine how much output should be produced by each industry in order to provide the needed inputs of the various industries and also to satisfy the final demand. The proper level of output can be computed using matrix calculations, as illustrated in the next example.

EXAMPLE 1 Suppose that an economy is composed of only three industries—coal, steel, and electricity. Each of these industries depends on the others for some of its raw materials. Suppose that to make \$1 of coal, it takes no coal, but \$.02 of steel and \$.01 of electricity; to make \$1 of steel, it takes \$.15 of coal, \$.03 of steel, and \$.08 of electricity; and to make \$1 of electricity, it takes \$.43 of coal, \$.20 of steel, and \$.05 of electricity. How much should each industry produce to allow for consumption (not used for production) at these levels: \$2 billion coal, \$1 billion steel, \$3 billion electricity?

Solution Put all the data indicating the interdependence of the industries in a matrix. In each industry's column, put the amount of input from each of the industries needed to produce \$1 of output in that particular industry:

$$\begin{array}{l} \\ \text{Coal} \\ \text{Steel} \\ \text{Electricity} \end{array} \begin{array}{c} \begin{array}{ccc} \text{Coal} & \text{Steel} & \text{Electricity} \end{array} \\ \begin{bmatrix} 0 & .15 & .43 \\ .02 & .03 & .20 \\ .01 & .08 & .05 \end{bmatrix} \end{array} = A.$$

This matrix is the *input–output matrix* corresponding to the economy. Let D denote the final-demand matrix. Then, letting the numbers in D stand for billions of dollars, we have

$$D = \begin{bmatrix} 2 \\ 1 \\ 3 \end{bmatrix}.$$

Suppose that the coal industry produces x billion dollars of output, the steel industry y billion dollars, and the electrical industry z billion dollars. Our problem is to determine the x, y, and z that yield the desired amounts left over from the production process. As an example, consider coal. The amount of coal that can be consumed or exported is just

$$x - [\text{amount of coal used in production}].$$

To determine the amount of coal used in production, refer to the input–output matrix. Production of x billion dollars of coal takes $0 \cdot x$ billion dollars of coal; production of y billion dollars of steel takes $.15y$ billion dollars of coal; and production of z billion dollars of electricity takes $.43z$ billion dollars of coal. Thus,

$$[\text{amount of coal used in production}] = 0 \cdot x + .15y + .43z.$$

This quantity should be recognized as the first entry of a matrix product. Namely, if we let

$$X = \begin{bmatrix} x \\ y \\ z \end{bmatrix},$$

then

$$\begin{bmatrix} \text{coal} \\ \text{steel} \\ \text{electricity} \end{bmatrix}_{\text{used in production}} = \begin{bmatrix} 0 & .15 & .43 \\ .02 & .03 & .20 \\ .01 & .08 & .05 \end{bmatrix} \begin{bmatrix} x \\ y \\ z \end{bmatrix} = AX.$$

But then the amount of each output available for purposes other than production is $X - AX$. That is, we have the matrix equation

$$X - AX = D.$$

To solve this equation for X, proceed as follows. Since $IX = X$, write the equation in the form

$$IX - AX = D$$

$$(I - A)X = D$$

$$X = (I - A)^{-1}D. \tag{1}$$

So, in other words, X may be found by multiplying D on the left by $(I - A)^{-1}$. Let us now do the arithmetic.

$$I - A = \begin{bmatrix} 1 & 0 & 0 \\ 0 & 1 & 0 \\ 0 & 0 & 1 \end{bmatrix} - \begin{bmatrix} 0 & .15 & .43 \\ .02 & .03 & .20 \\ .01 & .08 & .05 \end{bmatrix} = \begin{bmatrix} 1 & -.15 & -.43 \\ -.02 & .97 & -.20 \\ -.01 & -.08 & .95 \end{bmatrix}.$$

Applying the Gauss–Jordan method, we find that

$$(I - A)^{-1} = \begin{bmatrix} 1.01 & .20 & .50 \\ .02 & 1.05 & .23 \\ .01 & .09 & 1.08 \end{bmatrix},$$

where all figures are carried to two decimal places (Exercise 5). Therefore,

$$X = (I - A)^{-1}D = \begin{bmatrix} 1.01 & .20 & .50 \\ .02 & 1.05 & .23 \\ .01 & .09 & 1.08 \end{bmatrix} \begin{bmatrix} 2 \\ 1 \\ 3 \end{bmatrix} = \begin{bmatrix} 3.72 \\ 1.78 \\ 3.35 \end{bmatrix}.$$

In other words, coal should produce \$3.72 billion worth of output, steel \$1.78 billion, and electricity \$3.35 billion. This output will meet the required final demands from each industry. ▮

The analysis above is useful in studying not only entire economies but also segments of economies and even individual companies.

▮ EXAMPLE 2 A conglomerate has three divisions, which produce computers, semiconductors, and business forms. For each \$1 of output, the computer division needs \$.02 worth of computers, \$.20 worth of semiconductors, and \$.10 worth of business forms. For each \$1 of output, the semiconductor division needs \$.02 worth of computers, \$.01 worth of semiconductors, and \$.02 worth of business forms. For each \$1 of output, the business forms division requires \$.10 worth of computers and \$.01 worth of business forms. The conglomerate estimates the sales demand to be \$300,000,000 for the computer division, \$100,000,000 for the semiconductor division, and \$200,000,000 for the business forms division. At what level should each division produce in order to satisfy this demand?

Solution The conglomerate can be viewed as a miniature economy and its sales as the final demand. The input–output matrix for this "economy" is

	Computers	Semiconductors	Business forms
Computers	.02	.02	.10
Semiconductors	.20	.01	0
Business forms	.10	.02	.01

$$= A.$$

The final-demand matrix is

$$D = \begin{bmatrix} 3 \\ 1 \\ 2 \end{bmatrix},$$

where the demand is expressed in hundreds of millions of dollars. By equation (1) the matrix X, giving the desired levels of production for the various divisions, is given by

$$X = (I - A)^{-1}D.$$

But

$$I - A = \begin{bmatrix} .98 & -.02 & -.10 \\ -.20 & .99 & 0 \\ -.10 & -.02 & .99 \end{bmatrix},$$

so that (Exercise 6)

$$(I - A)^{-1} = \begin{bmatrix} 1.04 & .02 & .10 \\ .21 & 1.01 & .02 \\ .11 & .02 & 1.02 \end{bmatrix} \quad \text{and} \quad (I - A)^{-1}D = \begin{bmatrix} 3.34 \\ 1.68 \\ 2.39 \end{bmatrix}.$$

Therefore,

$$X = \begin{bmatrix} 3.34 \\ 1.68 \\ 2.39 \end{bmatrix}.$$

That is, the computer division should produce \$334,000,000, the semiconductor division \$168,000,000, and the business forms division \$239,000,000. ∎

Input–output analysis is usually applied to the entire economy of a country having hundreds of industries. The resulting matrix equation $(I-A)X = D$ could be solved by the Gaussian elimination method. However, it is best to find the inverse of $I - A$ and solve for X as we have done in the examples of this section. Over a short period, D might change but A is unlikely to change. Therefore, the proper outputs to satisfy the new demand can easily be determined by using the already computed inverse of $I - A$.

The Closed Leontieff Model The foregoing description of an economy is usually called the *Leontieff open model* since it views exports as an activity that takes place external to the economy. However, it is possible to consider export as but another industry in the economy. Instead of describing exports by a demand column D, we describe it by a column in the input–output matrix. That is, the export column describes how each dollar of exports is divided among the various industries. Since exports are now regarded as another industry, each of the original columns has an additional entry, namely the amount of output from the export industry (that is, imports) used to produce \$1 of goods (of the industry corresponding to the column). If A denotes the expanded input–output matrix and X the production matrix (as before), then AX is the matrix describing the total demand experienced by each of the industries. In order for the economy to function efficiently, the total amount demanded by the various industries should equal the amount produced. That is, the production matrix must satisfy the equation

$$AX = X.$$

By studying the solutions to this equation, it is possible to determined the equilibrium states of the economy—that is, the production matrices X for which the amounts produced exactly equal the amounts needed by the various industries. The model just described is called *Leontieff's closed model.*

We may expand the Leontieff closed model to include the effects of labor and monetary phenomena by considering labor and banking as yet further industries to be incorporated in the input–output matrix.

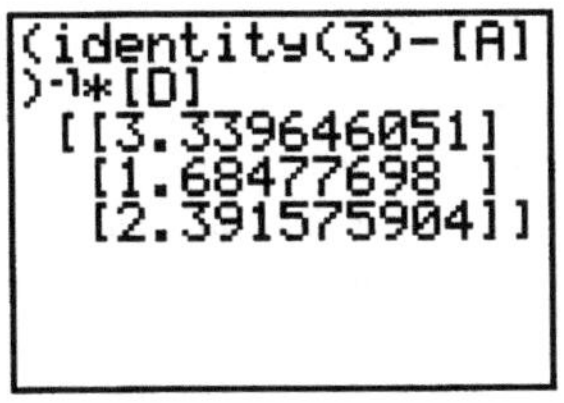

FIGURE 1

GC With a graphing calculator, the matrix $(I - A)^{-1}D$ can be calculated with a single press of the ENTER key. Figure 1 shows the calculation for the matrices of Example 2.

Practice Problems 6

1. Let

$$I = \begin{bmatrix} 1 & 0 & 0 \\ 0 & 1 & 0 \\ 0 & 0 & 1 \end{bmatrix}, \quad A = \begin{bmatrix} .1 & 0 & .1 \\ .2 & .1 & .1 \\ .1 & .2 & 0 \end{bmatrix}, \quad X = \begin{bmatrix} x \\ y \\ z \end{bmatrix}, \quad D = \begin{bmatrix} 100 \\ 200 \\ 50 \end{bmatrix}.$$

Solve the matrix equation

$$(I - A)X = D.$$

2. Let I, A, and X be as in Problem 1, but let

$$D = \begin{bmatrix} 300 \\ 100 \\ 100 \end{bmatrix}.$$

Solve the matrix equation $(I - A)X = D$.

EXERCISES 6

1. Suppose that in the economy of Example 1 the demand for electricity triples and the demand for coal doubles, whereas the demand for steel increases only by 50%. At what levels should the various industries produce in order to satisfy the new demand?

2. Suppose that the conglomerate of Example 2 is faced with an increase of 50% in demand for computers, a doubling in demand for semiconductors, and a decrease of 50% in demand for business forms. At what levels should the various divisions produce in order to satisfy the new demand?

3. Suppose that the conglomerate of Example 2 experiences a doubling in the demand for business forms. At what levels should the computer and semiconductor divisions produce?

4. A multinational corporation does business in the United States, Canada, and England. Its branches in one country purchase goods from the branches in other countries according to the matrix

		Branch in:		
		United States	Canada	England
Purchase from:	United States	.02	0	.02
	Canada	.01	.03	.01
	England	.03	0	.01

where the entries in the matrix represent proportions of total sales by the respective branch. The external sales by each of the offices are \$800,000,000 for the U.S. branch, \$300,000,000 for the Canadian branch, and \$1,400,000,000 for the English branch. At what level should each of the branches produce in order to satisfy the total demand?

5. Show that to two decimal places

$$\begin{bmatrix} 1 & -.15 & -.43 \\ -.02 & .97 & -.20 \\ -.01 & -.08 & .95 \end{bmatrix}^{-1} = \begin{bmatrix} 1.01 & .20 & .50 \\ .02 & 1.05 & .23 \\ .01 & .09 & 1.08 \end{bmatrix}.$$

6. Show that to two decimal places

$$\begin{bmatrix} .98 & -.02 & -.10 \\ -.2 & .99 & 0 \\ -.10 & -.02 & .99 \end{bmatrix}^{-1} = \begin{bmatrix} 1.04 & .02 & .10 \\ .21 & 1.01 & .02 \\ .11 & .02 & 1.02 \end{bmatrix}.$$

7. A corporation has a plastics division and an industrial equipment division. For each \$1 worth of output, the plastics division needs \$.02 worth of plastics and \$.10 worth of equipment. For each \$1 worth of output, the industrial equipment division needs \$.01 worth of plastics and \$.05 worth of equipment. At what level should the divisions produce to meet a demand for \$930,000 worth of plastics and \$465,000 worth of industrial equipment?

8. Rework Exercise 7 under the condition that the demand for plastics is \$1,860,000 and the demand for industrial equipment is \$2,790,000.

9. An industrial system involves manufacturing, transportation, and agriculture. The interdependence of the three industries is given by the input–output matrix

$$A = \begin{array}{c} \begin{matrix} M & T & A \end{matrix} \\ \begin{bmatrix} .4 & .3 & .1 \\ .2 & .2 & .2 \\ .1 & .1 & .4 \end{bmatrix} \end{array} \begin{matrix} M \\ T \\ A \end{matrix}$$

At what levels must the industries produce to satisfy a demand for \$100 million worth of manufactured goods, \$80 million of transportation, and \$200 million worth of agricultural products?

10. A town has a merchant, a baker, and a farmer. To produce \$1 worth of output, the merchant requires \$.30 worth of baked goods and \$.40 worth of the farmer's products. To produce \$1 worth of output, the baker requires \$.50 worth of the merchant's goods, \$.10 worth of his own goods, and \$.30 worth of the farmer's goods. To produce \$1 worth of output, the farmer requires \$.30 worth of the merchant's goods, \$.20 worth of baked goods, and \$.30 worth of his own products. How much should the merchant, baker, and farmer produce to meet a demand for \$20,000 worth of output from the merchant, \$15,000 worth of output from the baker, and \$18,000 worth of output from the farmer?

Exercises 11 and 12 require the use of a graphing calculator or a computer.

In Exercises 11 and 12, use the input–output matrix A and the final-demand matrix D to find the production matrix X for the Leontieff open model. Round your answers to two decimal places.

11. $A = \begin{bmatrix} .1 & .2 & .4 \\ .05 & .3 & .25 \\ .15 & .1 & .2 \end{bmatrix}, \ D = \begin{bmatrix} 3 \\ 7 \\ 4 \end{bmatrix}.$

12. $A = \begin{bmatrix} .2 & .35 & .15 & .05 \\ .1 & .1 & .3 & .2 \\ .075 & .2 & .05 & .1 \\ .3 & .04 & .1 & .15 \end{bmatrix}, \ D = \begin{bmatrix} 5 \\ 2 \\ 1 \\ 7 \end{bmatrix}$

Solutions to Practice Problems 6

1. The equation $(I - A)X = D$ has the form $CX = D$, where C is the matrix $I - A$. From Section 2.4 we know that $X = C^{-1}D$. That is, $X = (I - A)^{-1}D$. Now

$$I - A = \begin{bmatrix} 1 & 0 & 0 \\ 0 & 1 & 0 \\ 0 & 0 & 1 \end{bmatrix} - \begin{bmatrix} .1 & 0 & .1 \\ .2 & .1 & .1 \\ .1 & .2 & 0 \end{bmatrix} = \begin{bmatrix} .9 & 0 & -.1 \\ -.2 & .9 & -.1 \\ -.1 & -.2 & 1 \end{bmatrix}.$$

Using the Gauss–Jordan method to find the inverse of this matrix, we have (to two decimal places)

$$(I - A)^{-1} = \begin{bmatrix} 1.13 & .03 & .12 \\ .27 & 1.14 & .14 \\ .17 & .23 & 1.04 \end{bmatrix}.$$

Therefore, rounding to the nearest integer, we have

$$X = (I - A)^{-1}D = \begin{bmatrix} 1.13 & .03 & .12 \\ .27 & 1.14 & .14 \\ .17 & .23 & 1.04 \end{bmatrix} \begin{bmatrix} 100 \\ 200 \\ 50 \end{bmatrix} = \begin{bmatrix} 125 \\ 262 \\ 115 \end{bmatrix}.$$

2. We have $X = (I - A)^{-1}D$, where $(I - A)^{-1}$ is as computed in Problem 1. So

$$X = (I - A)^{-1}D = \begin{bmatrix} 1.13 & .03 & .12 \\ .27 & 1.14 & .14 \\ .17 & .23 & 1.04 \end{bmatrix} \begin{bmatrix} 300 \\ 100 \\ 100 \end{bmatrix} = \begin{bmatrix} 354 \\ 209 \\ 178 \end{bmatrix}.$$

Chapter 2 Review of Fundamental Concepts

1. What is meant by a solution to a system of linear equations?
2. What is a matrix?
3. State three elementary row operations on equations or matrices.
4. What does it mean for a system of equations or a matrix to be in diagonal form?
5. What is meant by pivoting a matrix about a nonzero entry?
6. State the Gaussian elimination method for transforming a system of linear equations into diagonal form.
7. What is a row matrix? Column matrix? Square matrix? Identity matrix, I_n?
8. What is meant by a_{ij}, the ijth entry of a matrix?
9. Define the sum and difference of two matrices.
10. Define the product of two matrices.
11. Define the inverse of a matrix, A^{-1}.
12. Give the formula for the inverse of a 2×2 matrix.
13. Explain how to use the inverse of a matrix to solve a system of linear equations.
14. Describe the steps of the Gauss–Jordan method for calculating the inverse of a matrix.
15. What is an input–output matrix and a final–demand matrix?
16. Explain how to solve an input–output analysis problem.

CHAPTER 2 SUPPLEMENTARY EXERCISES

In Exercises 1 and 2, pivot each matrix around the circled element.

1. $\begin{bmatrix} \textcircled{3} & -6 & 1 \\ 2 & 4 & 6 \end{bmatrix}$

2. $\begin{bmatrix} -5 & -3 & 1 \\ 4 & \textcircled{2} & 0 \\ 0 & 6 & 7 \end{bmatrix}$

In Exercises 3–8, use the Gaussian elimination method to find all solutions of the system of linear equations.

3. $\begin{cases} \frac{1}{2}x - y = -3 \\ 4x - 5y = -9 \end{cases}$

4. $\begin{cases} 3x + 9z = 42 \\ 2x + y + 6z = 30 \\ -x + 3y - 2z = -20 \end{cases}$

5. $\begin{cases} 3x - 6y + 6z = -5 \\ -2x + 3y - 5z = \frac{7}{3} \\ x + y + 10z = 3 \end{cases}$

6. $\begin{cases} 3x + 6y - 9z = 1 \\ 2x + 4y - 6z = 1 \\ 3x + 4y + 5z = 0 \end{cases}$

7. $\begin{cases} x + 2y - 5z + 3w = 16 \\ -5x - 7y + 13z - 9w = -50 \\ -x + y - 7z + 2w = 9 \\ 3x + 4y - 7z + 6w = 33 \end{cases}$

8. $\begin{cases} 5x - 10y = 5 \\ 3x - 8y = -3 \\ -3x + 7y = 0 \end{cases}$

In Exercises 9 and 10, perform the indicated matrix operations.

9. $\begin{bmatrix} 2 \\ -1 \\ 0 \end{bmatrix} + \begin{bmatrix} 3 \\ 4 \\ 7 \end{bmatrix}$

10. $\begin{bmatrix} 1 & 3 & -2 \\ 4 & 0 & -1 \end{bmatrix} \cdot \begin{bmatrix} 3 & 5 \\ 1 & 0 \\ 0 & -6 \end{bmatrix}$

11. Find the inverse of the appropriate matrix, and use it to solve the system of equations

$$\begin{cases} 3x + 2y = 0 \\ 5x + 4y = 2. \end{cases}$$

12. The matrices

$$\begin{bmatrix} 4 & -2 & 3 \\ 8 & -3 & 5 \\ 7 & -2 & 4 \end{bmatrix} \quad \text{and} \quad \begin{bmatrix} -2 & 2 & -1 \\ 3 & -5 & 4 \\ 5 & -6 & 4 \end{bmatrix}$$

are inverses of each other. Use these matrices to solve the following systems of linear equations.

(a) $\begin{cases} -2x + 2y - z = 1 \\ 3x - 5y + 4z = 0 \\ 5x - 6y + 4z = 3 \end{cases}$

(b) $\begin{cases} 4x - 2y + 3z = 0 \\ 8x - 3y + 5z = -1 \\ 7x - 2y + 4z = 2 \end{cases}$

In Exercises 13 and 14, use the Gauss–Jordan method to calculate the inverses of the following matrices.

13. $\begin{bmatrix} 2 & 6 \\ 1 & 2 \end{bmatrix}$ **14.** $\begin{bmatrix} 1 & 1 & 1 \\ 3 & 4 & 3 \\ 1 & 1 & 2 \end{bmatrix}$

15. Farmer Brown has 1000 acres of land on which he plans to grow corn, wheat, and soybeans. The cost of cultivating these crops is \$28 per acre for corn, \$40 per acre for wheat, and \$32 per acre for soybeans. If Farmer Brown wishes to use all his available land and his entire budget of \$30,000, and if he wishes to plant the same number of acres of corn as wheat and soybeans combined, how many acres of each crop can he grow?

16. In an arms race between two superpowers, each nation takes stock of its own and its enemy's nuclear arsenal each year. Each nation has the policy of dismantling a certain percentage of its stockpile each year and adding that same percentage of its competitor's stockpile. Nation A uses 20% and nation B uses 10%. Suppose that the current stockpiles of nations A and B are 10,000 and 7000 weapons, respectively.

(a) What will the stockpiles be in each of the next two years?

(b) What were the stockpiles in each of the preceding two years?

(c) Show that the "missile gap" between the superpowers decreases by 30% each year under these policies. Show that the total number of weapons decreases each year if nation A begins with the most weapons, and the total number of weapons increases if nation B begins with the most weapons.

17. **(*From a GMAT exam*)** Marion rented a car for \$18.00 plus \$0.10 per mile driven. Craig rented a car for \$25.00 plus \$0.05 per mile driven. If each drove d miles and each was charged exactly the same amount for the rental, then d equals

(a) 100 (b) 120 (c) 135 (d) 140 (e) 150

18. **(*From a GMAT exam*)** Three types of pencils, J, K, and L, cost \$0.05, \$0.10, and \$0.25 each, respectively. If a box of 32 of these pencils costs a total of \$3.40 and if there are twice as many K pencils as L pencils in the box, how many J pencils are in the box?

(a) 6 (b) 12 (c) 14 (d) 18 (e) 20

19. The economy of a small country can be regarded as consisting of two industries, I and II, whose input–output matrix is

$$A = \begin{bmatrix} .4 & .2 \\ .1 & .3 \end{bmatrix}.$$

How many units should be produced by each industry in order to meet a demand for 8 units from industry I and 12 units from industry II?

Chapter 8
Linear Programming, A Geometric Approach

3.1
A Linear Programming Problem

3.2
Linear Programming I

3.3
Linear Programming II

Linear programming is a method for solving problems in which a linear function (representing cost, profit, distance, weight, or the like) is to be maximized or minimized. Such problems are called *optimization problems*. As we shall see, these problems, when translated into mathematical language, involve systems of linear inequalities, systems of linear equations, and eventually (in Chapter 4) matrices.

3.1 A Linear Programming Problem

Let us begin with a detailed discussion of a typical problem that can be solved by linear programming.

Furniture Manufacturing Problem A furniture manufacturer makes two types of furniture—chairs and sofas. For simplicity, divide the production process into three distinct operations—carpentry, finishing, and upholstery. The amount of labor required for each operation varies. Manufacture of a chair requires 6 hours of carpentry, 1 hour of finishing, and 2 hours of upholstery. Manufacture of a sofa requires 3 hours of carpentry, 1 hour of finishing, and 6 hours of upholstery. Due to limited availability of skilled labor as well as of tools and equipment, the factory has available each day 96 labor-hours for carpentry, 18 labor-hours for finishing, and 72 labor-hours for upholstery. The profit per chair is \$80 and the profit per sofa \$70. How many chairs and how many sofas should be produced each day to maximize the profit?

It is often helpful to tabulate data given in verbal problems. Our first step, then, is to construct a chart.

	Chair	*Sofa*	*Available time*
Carpentry	6 hours	3 hours	96 labor-hours
Finishing	1 hour	1 hour	18 labor-hours
Upholstery	2 hours	6 hours	72 labor-hours
Profit	\$80	\$70	

The next step is to translate the problem into mathematical language. As you know, this is done by identifying what is unknown and denoting the unknown quantities by letters. Since the problem asks for the optimal number of chairs and sofas to be produced each day, there are two unknowns—the number of chairs produced each day and the number of sofas produced each day. Let x denote the former and y the latter.

To achieve a large profit, one need only manufacture a large number of chairs and sofas. But due to restricted availability of tools and labor, the factory cannot manufacture an unlimited quantity of furniture. Let us translate the restrictions into mathematical language. Each row of the chart gives one restriction. The first row says that the amount of carpentry required is 6 hours for each chair and 3 hours for each sofa. Also, there are available only 96 labor-hours of carpentry per day. We can compute the total number of labor-hours of carpentry required per day to produce x chairs and y sofas as follows:

$$\begin{aligned}&[\text{number of labor-hours per day of carpentry}]\\ &\quad = (\text{number of hours carpentry per chair}) \cdot (\text{number of chairs per day})\\ &\qquad + (\text{number of hours carpentry per sofa}) \cdot (\text{number of sofas per day})\\ &\quad = 6 \cdot x + 3 \cdot y.\end{aligned}$$

The requirement that at most 96 labor-hours of carpentry be used per day means that x and y must satisfy the inequality

$$6x + 3y \leq 96. \tag{1}$$

The second row of the chart gives a restriction imposed by finishing. Since 1 hour of finishing is required for each chair and sofa, and since at most 18 labor-hours of finishing are available per day, the same reasoning as used to derive inequality (1) yields

$$x + y \leq 18. \tag{2}$$

Similarly, the third row of the chart gives the restriction due to upholstery:

$$2x + 6y \leq 72. \tag{3}$$

A further restriction is given by the fact that the numbers of chairs and sofas must be nonnegative:

$$x \geq 0, \qquad y \geq 0. \tag{4}$$

Now that we have written down the restrictions constraining x and y, let us express the profit (which is to be maximized) in terms of x and y. The profit comes from two sources—chairs and sofas. Therefore,

$$\begin{aligned}[\text{profit}] &= [\text{profit from chairs}] + [\text{profit from sofas}] \\ &= [\text{profit per chair}] \cdot [\text{number of chairs}] \\ &\quad + [\text{profit per sofa}] \cdot [\text{number of sofas}] \\ &= 80x + 70y.\end{aligned} \tag{5}$$

Combining (1) to (5), we arrive at the following:

Furniture Manufacturing Problem—Mathematical Formulation Find numbers x and y for which $80x + 70y$ is as large as possible, and for which all the following inequalities hold simultaneously:

$$\begin{cases} 6x + 3y \leq 96 \\ x + y \leq 18 \\ 2x + 6y \leq 72 \\ x \geq 0, \quad y \geq 0. \end{cases} \tag{6}$$

We may describe this mathematical problem in the following general way. We are required to maximize an expression in a certain number of variables, where the variables are subject to restrictions in the form of one or more inequalities. Problems of this sort are called *mathematical programming problems.* Actually, general mathematical programming problems can be quite involved, and their solutions may require very sophisticated mathematical ideas. However, this is not the case with the furniture manufacturing problem. What makes it a rather simple mathematical programming problem is that both the expression to be maximized and the inequalities are linear. For this reason the furniture manufacturing problem is called a *linear programming problem.* The theory of linear programming is a fairly recent advance in mathematics. It was developed over the last 50 years to deal with the increasingly more complicated problems of our technological society. The 1975 Nobel Prize in economics was awarded to Kantorovich and Koopmans for their pioneering work in the field of linear programming.

We will solve the furniture manufacturing problem in Section 3.2, where we develop a general technique for handling similar linear programming problems. At this point it is worthwhile to attempt to gain some insights into the problem and possible methods for attacking it.

It seems clear that a factory will operate most efficiently when its labor is fully utilized. Let us therefore take the operations one at a time and determine the conditions on x and y that fully utilize the three kinds of labor. The restriction on carpentry asserts that

$$6x + 3y \leq 96.$$

If x and y were chosen so that $6x+3y$ is actually *less* than 96, we would leave the carpenters idle some of the time, a waste of labor. Thus it would seem reasonable to choose x and y to satisfy

$$6x + 3y = 96.$$

Similarly, to utilize all the finishers' time, x and y must satisfy

$$x + y = 18,$$

and to utilize all the upholsterers' time, we must have

$$2x + 6y = 72.$$

Thus, if no labor is to be wasted, then x and y must satisfy the system of equations

$$\begin{cases} 6x + 3y = 96 \\ x + \ \ y = 18 \\ 2x + 6y = 72. \end{cases} \tag{7}$$

Let us now graph the three equations of (7), which represent the conditions for full utilization of all forms of labor. (See the chart below and Fig. 1.)

Equation	*Standard form*	*x-Intercept*	*y-Intercept*
$6x + 3y = 96$	$y = -2x + 32$	$(16, 0)$	$(0, 32)$
$x + y = 18$	$y = -x + 18$	$(18, 0)$	$(0, 18)$
$2x + 6y = 72$	$y = -\frac{1}{3}x + 12$	$(36, 0)$	$(0, 12)$

What does Fig. 1 say about the furniture manufacturing problem? Each particular pair of numbers (x, y) is called a *production schedule.* Each of the lines in Fig. 1 gives the production schedules that fully utilize one of the types of labor. Notice that the three lines do not have a common intersection point. This means that there is *no* production schedule that *simultaneously* makes full use of all three types of labor. In any production schedule at least some of the labor-hours must be wasted. This is not a solution to the furniture manufacturing problem, but it is a valuable insight. It says that in the inequalities of (6) not all of the corresponding equations can hold. This suggests that we take a closer look at the system of inequalities.

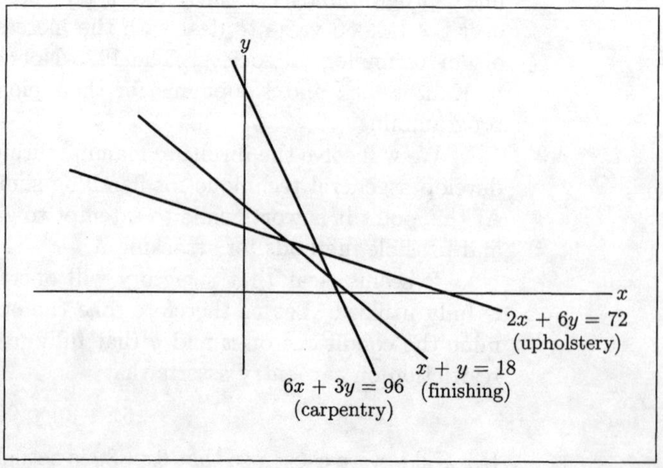

FIGURE 1

The standard forms of the inequalities (6) are

$$\begin{cases} y \leq -2x + 32 \\ y \leq \quad -x + 18 \\ y \leq -\frac{1}{3}x + 12 \\ x \geq 0, \quad y \geq 0. \end{cases}$$

By using the techniques of Section 1.2, we arrive at a feasible set for the system of inequalities above, as shown in Fig. 2.

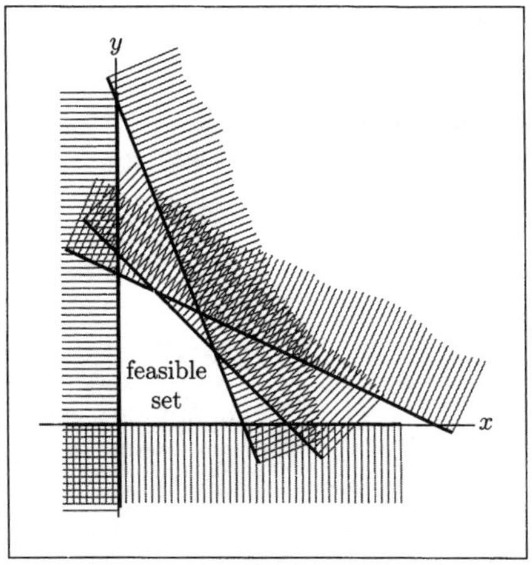

FIGURE 2

The feasible set for the furniture manufacturing problem is a bounded, five-sided region. The points on and inside the boundary of this feasible set give the production schedules that satisfy all the restrictions. In Section 3.2 we show how to pick out the particular point of the feasible set that corresponds to a maximum profit.

Practice Problems 1

1. Determine whether the following points are in the feasible set of the furniture manufacturing problem: (a) (10, 9); (b) (14, 4).

2. A physical fitness enthusiast decides to devote her exercise time to a combination of jogging and cycling. She wants to earn aerobic points (a measure of the benefit of the exercise to strengthening the heart and lungs) and also to achieve relaxation and enjoyment. She jogs at 6 miles per hour and cycles at 18 miles per hour. An hour of jogging earns 12 aerobic points and an hour of cycling earns 9 aerobic points. Each week she would like to earn at least 36 aerobic points, cover at least 54 miles, and cycle at least as much as she jogs.

   (a) Fill in the chart below.

	One hour of jogging	*One hour of cycling*	*Requirement*
Miles covered			
Aerobic points			
Time required			

(b) Let x be the number of hours of jogging and y the number of hours of cycling each week. Referring to the chart, give the inequalities that x and y must satisfy due to miles covered and aerobic points.

(c) Give the inequalities that x and y must satisfy due to her preference for cycling and also due to the fact that x and y cannot be negative.

(d) Express the time required as a linear function of x and y.

(e) Graph the feasible set for the system of linear inequalities.

EXERCISES 1

In Exercises 1–4, determine whether the given point is in the feasible set of the furniture manufacturing problem. (The inequalities are given below.)

$$\begin{cases} 6x + 3y \leq 96 \\ x + \ \ y \leq 18 \\ 2x + 6y \leq 72 \\ x \geq 0, \quad y \geq 0. \end{cases}$$

1. $(8, 7)$

2. $(14, 3)$

3. $(9, 10)$

4. $(16, 0)$

5. **(*Shipping*)** A truck traveling from New York to Baltimore is to be loaded with two types of cargo. Each crate of cargo A is 4 cubic feet in volume, weighs 100 pounds, and earns $13 for the driver. Each crate of cargo B is 3 cubic feet in volume, weighs 200 pounds, and earns $9 for the driver. The truck can carry no more than 300 cubic feet of crates and no more than 10,000 pounds. Also, the number of crates of cargo B must be less than or equal to twice the number of crates of cargo A.

(a) Fill in the following chart.

	A	B	*Truck capacity*
Volume			
Weight			
Earnings			

(b) Let x be the number of crates of cargo A and y the number of crates of cargo B. Referring to the chart, give the two inequalities that x and y must satisfy because of the truck's capacity for volume and weight.

(c) Give the inequalities that x and y must satisfy because of the last sentence of the problem and also because x and y cannot be negative.

(d) Express the earnings from carrying x crates of cargo A and y crates of cargo B.

(e) Graph the feasible set for the shipping problem.

6. **(*Mining*)** A coal company owns mines in two different locations. Each day mine 1 produces 4 tons of anthracite (hard coal), 4 tons of ordinary coal, and 7 tons of bituminous (soft) coal. Each day mine 2 produces 10 tons of anthracite, 5 tons of ordinary coal, and 5 tons of bituminous coal. It costs the company $150 per day to operate mine 1 and $200 per day to operate mine 2. An order is received for 80 tons of anthracite, 60 tons of ordinary coal, and 75 tons of bituminous coal.

(a) Fill in the chart below.

	Mine 1	*Mine 2*	*Ordered*
Anthracite			
Ordinary			
Bituminous			
Daily cost			

(b) Let x be the number of days mine 1 should be operated and y the number of days mine 2 should be operated. Refer to the chart and give three inequalities that x and y must satisfy to fill the order.

(c) Give other requirements that x and y must satisfy.

(d) Find the cost of operating mine 1 for x days and mine 2 for y days.

(e) Graph the feasible set for the mining problem.

7. **(*Exam Strategy*)** A student is taking an exam consisting of 10 essay questions and 50 short-answer questions. He has 90 minutes to take the exam and knows he cannot possibly answer every question. The essay questions are worth 20 points each and the short-answer questions are worth 5 points each. An essay question takes 10 minutes to answer and a short-answer question takes 2 minutes. The student must do at least 3 essay questions and at least 10 short-answer questions.

(a) Fill in the following chart.

	Essay questions	Short-answer questions	Available
Time to answer			
Quantity			
Required			
Worth			

(b) Let x be the number of essay questions to be answered and y the number of short-answer questions to be answered. Refer to the chart and give the inequality that x and y must satisfy due to the amount of time available.

(c) Give the inequalities that x and y must satisfy because of the numbers of each type of question and also because of the minimum number of each type of question that must be answered.

(d) Give an expression for the score obtained from answering x essay questions and y short-answer questions.

(e) Graph the feasible set for the exam strategy problem.

8. (***Political Campaign—Resource Allocation***) A local politician has budgeted at most \$80,000 for her media campaign. She plans to distribute these funds between TV ads and radio ads. Each 1-minute TV ad is expected to be seen by 20,000 viewers, and each 1-minute radio ad is expected to be heard by 4000 listeners. Each minute of TV time costs \$8000 and each minute of radio time costs \$2000. She has been advised to use at most 90% of her media campaign budget on television ads.

(a) Fill in the chart below. (*Note*: Fill in only the first entry of the last column.)

	One-minute TV ads	One-minute radio ads	Money available
Cost			
Audience reached			

(b) Let x be the number of minutes of TV ads and let y be the number of minutes of radio ads. Refer to the chart and give an inequality that x and y must satisfy due to the amount of money available.

(c) Give the inequality that x must satisfy due to the limitation on the amount of money to be spent on TV ads. Also, give the inequalities that x and y must satisfy because x and y cannot be negative.

(d) Give an expression for the audience reached by x minutes of TV ads and y minutes of radio ads.

(e) Graph the feasible set for the political campaign problem.

9. (***Nutrition—Dairy Cows***) A dairy farmer concludes that his small herd of cows will need at least 4550 pounds of protein in their winter feed, at least 26,880 pounds of total digestible nutrients (TDN), and at least 43,200 international units (IUs) of vitamin A. Each pound of alfalfa hay provides .13 pound of protein, .48 pound of TDN, and 2.16 IUs of vitamin A. Each pound of ground ears of corn supplies .065 pound of protein, .96 pound of TDN, and no vitamin A. Alfalfa hay costs \$1 per 100-pound sack. Ground ear corn costs \$1.60 per 100-pound sack.

(a) Fill in the chart below.

	Alfalfa	Corn	Requirements
Protein			
TDN			
Vitamin A			
Cost/lb			

(b) Let x be the number of pounds of alfalfa hay and y be the number of pounds of ground ears of corn to be bought. Give the inequalities that x and y must satisfy.

(c) Graph the feasible set for the system of linear inequalities.

(d) Express the cost of buying x pounds of alfalfa hay and y pounds of ground ears of corn.

Solutions to Practice Problems 1

1. A point is in the feasible set of a system of inequalities if it satisfies every inequality. Either the original form or the standard form of the inequalities may be used. The original form of the inequalities of the furniture manufacturing problem is

$$\begin{cases} 6x + 3y \leq 96 \\ x + y \leq 18 \\ 2x + 6y \leq 72 \\ x \geq 0, \quad y \geq 0. \end{cases}$$

(a) (10, 9)

$$\begin{cases} 6(10) + 3(9) \le 96 \\ 10 + 9 \le 18 \\ 2(10) + 6(9) \le 72 \\ 10 \ge 0, \quad 9 \ge 0; \end{cases} \qquad \begin{cases} 87 \le 96 & \text{true} \\ 19 \le 18 & \text{false} \\ 74 \le 72 & \text{false} \\ 10 \ge 0, \quad 9 \ge 0. & \text{true} \end{cases}$$

(b) (14, 4)

$$\begin{cases} 6(14) + 3(4) \le 96 \\ 14 + 4 \le 18 \\ 2(14) + 6(4) \le 72 \\ 14 \ge 0, \quad 4 \ge 0; \end{cases} \qquad \begin{cases} 96 \le 96 & \text{true} \\ 18 \le 18 & \text{true} \\ 52 \le 72 & \text{true} \\ 14 \ge 0, \quad 4 \ge 0. & \text{true} \end{cases}$$

Therefore, (14, 4) is in the feasible set and (10, 9) is not.

2. (a)

	One hour of jogging	*One hour of cycling*	*Requirement*
Miles covered	6	18	54
Aerobic points	12	9	36
Time required	1	1	

(b) Miles covered: $6x + 18y \ge 54$.
Aerobic points: $12x + 9y \ge 36$.

(c) $y \ge x$, $x \ge 0$. It is not necessary to list $y \ge 0$ since this is automatically assured if the other two inequalities hold.

(d) $x+y$. (An objective of the exercise program might be to minimize $x+y$.)

(e)

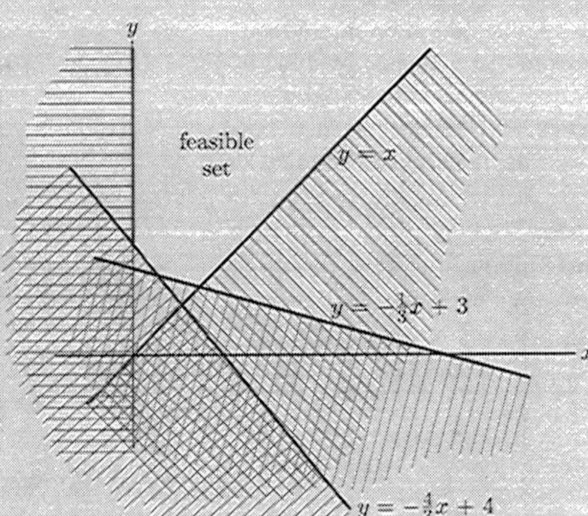

3.2 Linear Programming I

We have shown that the feasible set for the furniture manufacturing problem—that is, the set of points corresponding to production schedules satisfying all five restriction inequalities—consists of the points in and on the interior and boundary of the five-sided region drawn in Fig. 1. For reference, we have labeled each line segment with the equation of the line to which it belongs. The line segments intersect in five points, each of which is a corner of the feasible set. Such a corner is called a *vertex*. Somehow, we must pick out of the feasible set

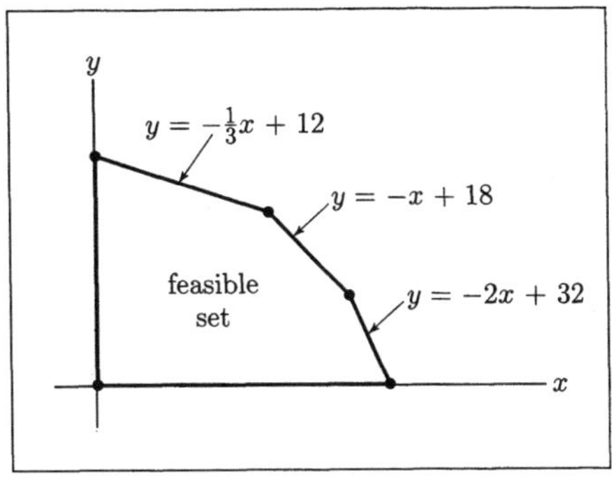

FIGURE 1

an *optimal point*—that is, a point corresponding to a production schedule which yields a maximum profit. To assist us in this task, we have the following result:[1]

> ***Fundamental Theorem of Linear Programming*** The maximum (or minimum) value of the objective function is achieved at one of the vertices of the feasible set.

This result does not completely solve the furniture manufacturing problem for us, but it comes close. It tells us that an optimal production schedule (a, b) corresponds to one of the five points labeled A–E in Fig. 2. So to complete the solution of the furniture manufacturing problem, it suffices to find the coordinates of the five points, evaluate the profit at each, and then choose the point corresponding to the maximum profit.

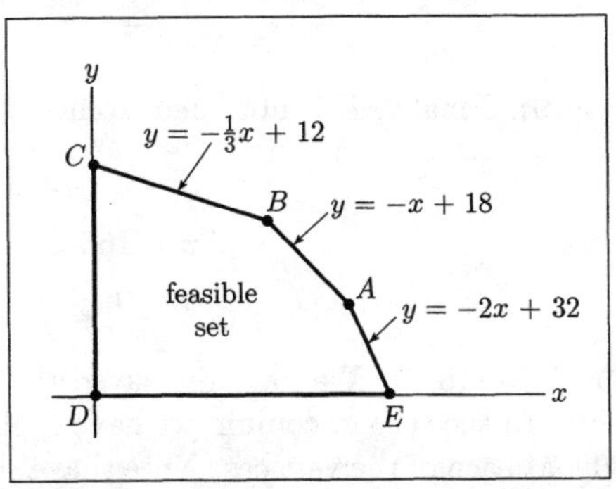

FIGURE 2

Solution of the Furniture Manufacturing Problem Let us begin by determining the coordinates of the points A–E in Fig. 2. Remembering that the x-axis has the equation $y = 0$ and the y-axis the equation $x = 0$, we see from Fig. 2 that

[1]For a proof, see Section 3.3.

the coordinates of A–E can be found as intersections of the following lines:

$$A: \begin{cases} y = -x + 18 \\ y = -2x + 32 \end{cases}$$

$$B: \begin{cases} y = -x + 18 \\ y = -\frac{1}{3}x + 12 \end{cases}$$

$$C: \begin{cases} y = -\frac{1}{3}x + 12 \\ x = 0 \end{cases}$$

$$D: \begin{cases} y = 0 \\ x = 0 \end{cases}$$

$$E: \begin{cases} y = 0 \\ y = -2x + 32. \end{cases}$$

The point D is clearly $(0, 0)$, and C is clearly the point $(0, 12)$. We obtain A from

$$\begin{aligned} -x + 18 &= -2x + 32 \\ x &= 14 \\ y &= -14 + 18 = 4. \end{aligned}$$

Hence $A = (14, 4)$. Similarly, we obtain B from

$$\begin{aligned} -x + 18 &= -\tfrac{1}{3}x + 12 \\ -\tfrac{2}{3}x &= -6 \\ x &= 9 \\ y &= -9 + 18 = 9, \end{aligned}$$

so $B = (9, 9)$. Finally, E is obtained from

$$\begin{aligned} 0 &= -2x + 32 \\ x &= 16 \\ y &= 0, \end{aligned}$$

and thus $E = (16, 0)$. We have displayed the vertices in Fig. 3 and listed them in Table 1. In the second column we have evaluated the profit, which is given by $80x + 70y$, at each of the vertices. Note that the largest profit occurs at the vertex $(14, 4)$, so the solution of the linear programming problem is $x = 14$, $y = 4$. In other words, the factory should produce 14 chairs and 4 sofas each day in order to achieve maximum profit, and the maximum profit is \$1400 per day.

The furniture manufacturing problem is one particular example of a linear programming problem. Generally, such problems involve finding the values of x and y that maximize (or minimize) a particular linear expression in x and y and where x and y are chosen so as to satisfy one or more restrictions in the form of linear inequalities. The expression that is to be maximized (or minimized) is called the *objective function.* On the basis of our experience with the furniture manufacturing problem, we can summarize the steps to be followed in approaching *any* linear programming problem.

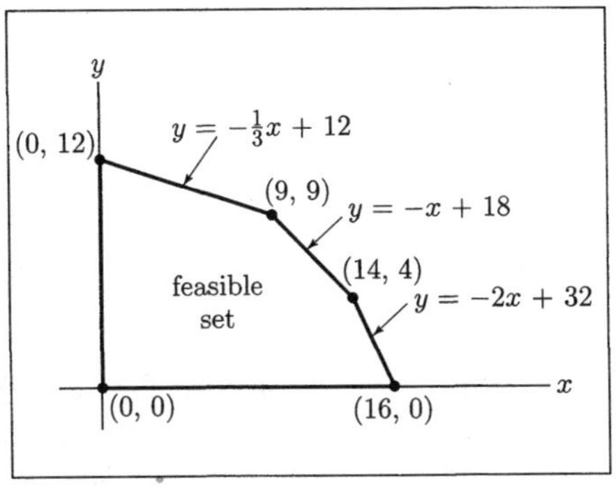

FIGURE 3

TABLE 1

Vertex	*Profit* $= 80x + 70y$
$(14, 4)$	$80(14) + 70(4) = 1400$
$(9, 9)$	$80(9) + 70(9) = 1350$
$(0, 12)$	$80(0) + 70(12) = 840$
$(0, 0)$	$80(0) + 70(0) = 0$
$(16, 0)$	$80(16) + 70(0) = 1280$

Step 1 Translate the problem into mathematical language.

A. Organize the data.
B. Identify the unknown quantities and define corresponding variables.
C. Translate the restrictions into linear inequalities.
D. Form the objective function.

Step 2 Graph the feasible set.

A. Put the inequalities in standard form.
B. Graph the straight line corresponding to each inequality.
C. Determine the side of the line belonging to the graph of each inequality. Cross out the other side. The remaining region is the feasible set.

Step 3 Determine the vertices of the feasible set.

Step 4 Evaluate the objective function at each vertex. Determine the optimal point.

Linear programming can be applied to many problems. The Army Corps of Engineers has used linear programming to plan the location of a series of dams so as to maximize the resulting hydroelectric power production. The restrictions were to provide adequate flood control and irrigation. Public transit companies have used linear programming to plan routes and schedule buses in order to maximize services. The restrictions in this case arose from the limitations on labor, equipment, and funding. The petroleum industry uses linear programming in the refining and blending of gasoline. Profit is maximized subject to restrictions on availability of raw materials, refining capacity, and product specifications. Some large advertising firms have used linear programming in media selection. The problem consists of determining how much to spend in each medium in order to maximize the number of consumers reached. The restrictions come from limitations on the budget and the relative costs of different media. Linear programming has been also used by psychologists to design an optimal battery of tests. The problem is to maximize the correlation between test scores and the characteristic that is to be predicted. The restrictions are imposed by the length and cost of the testing.

Linear programming is also used by dietitians in planning meals for large numbers of people. The object is to minimize the cost of the diet, and the restrictions reflect the minimum daily requirements of the various nutrients considered in the diet. The next example is representative of this type of problem. Whereas in actual practice many nutritional factors are considered, we shall simplify the problem by considering only three: protein, calories, and riboflavin.

EXAMPLE 1 (*Nutrition—People*) Suppose a person decides to make rice and soybeans part of his staple diet. The object is to design a lowest-cost diet that provides certain minimum levels of protein, calories, and vitamin B_2 (riboflavin). Suppose that one cup of uncooked rice costs 21 cents and contains 15 grams of protein, 810 calories, and $\frac{1}{9}$ milligram of riboflavin. On the other hand, one cup of uncooked soybeans costs 14 cents and contains 22.5 grams of protein, 270 calories, and $\frac{1}{3}$ milligram of riboflavin. Suppose that the minimum daily requirements are 90 grams of protein, 1620 calories, and 1 milligram of riboflavin. Design the lowest-cost diet meeting these specifications.

Solution We solve the problem by following steps 1–4. The first step is to translate the problem into mathematical language, and the first part of this step is to organize the data, preferably into a chart (Table 2).

TABLE 2

	Rice	*Soybeans*	*Required level per day*
Protein (grams/cup)	15	22.5	90
Calories (per cup)	810	270	1620
Riboflavin (milligrams/cup)	$\frac{1}{9}$	$\frac{1}{3}$	1
Cost (cents/cup)	21	14	

Now that we have organized the data, we ask for the unknowns. We wish to know how many cups each of rice and soybeans should comprise the diet, so we identify appropriate variables:

$$x = \text{number of cups of rice per day}$$
$$y = \text{number of cups of soybeans per day.}$$

Next, we obtain the restrictions on the variables. There is one restriction corresponding to each nutrient. That is, there is one restriction for each row of the chart. If x cups of rice and y cups of soybeans are consumed, the amount of protein is $15x + 22.5y$ grams. Thus, from the first row of the chart, $15x + 22.5y \geq 90$, a restriction expressing the fact that there must be at least 90 grams of protein per day. Similarly, the restrictions for calories and riboflavin lead to the inequalities $810x + 270y \geq 1620$ and $\frac{1}{9}x + \frac{1}{3}y \geq 1$, respectively. As in the furniture manufacturing problem, x and y cannot be negative, so there are two further restrictions: $x \geq 0$, $y \geq 0$. In all, there are five restrictions:

$$\begin{cases} 15x + 22.5y \geq 90 \\ 810x + 270y \geq 1620 \\ \frac{1}{9}x + \frac{1}{3}y \geq 1 \\ x \geq 0, \quad y \geq 0. \end{cases} \tag{1}$$

Now that we have the restrictions, we form the objective function, which tells us what we are out to maximize or minimize. Since we wish to minimize cost, we express cost in terms of x and y. Now x cups of rice cost $21x$ cents and y cups of soybeans cost $14y$ cents, so the objective function is given by

$$[\text{cost}] = 21x + 14y. \tag{2}$$

The problem can finally be stated in mathematical form: Minimize the objective function (2) subject to the restrictions (1). This completes the first step of the solution process.

The second step requires that we graph each of the inequalities (1). In Table 3 we have summarized all the steps necessary to obtain the information from which to draw the graphs.

TABLE 3

			Intercepts		
Inequality	*Standard form*	*Line*	x	y	*Graph*
$15x + 22.5y \geq 90$	$y \geq -\frac{2}{3}x + 4$	$y = -\frac{2}{3}x + 4$	$(6,0)$	$(0,4)$	above
$810x + 270y \geq 1620$	$y \geq -3x + 6$	$y = -3x + 6$	$(2,0)$	$(0,6)$	above
$\frac{1}{9}x + \frac{1}{3}y \geq 1$	$y \geq -\frac{1}{3}x + 3$	$y = -\frac{1}{3}x + 3$	$(9,0)$	$(0,3)$	above
$x \geq 0$	$x \geq 0$	$x = 0$	$(0,0)$	—	right
$y \geq 0$	$y \geq 0$	$y = 0$	—	$(0,0)$	above

We have sketched the graphs in Fig. 4. From Fig. 4(b) we see that the feasible set is an unbounded, five-sided region. There are four vertices, two of which are known from Table 3, since they are intercepts of boundary lines. Label the remaining two vertices A and B (Fig. 5).

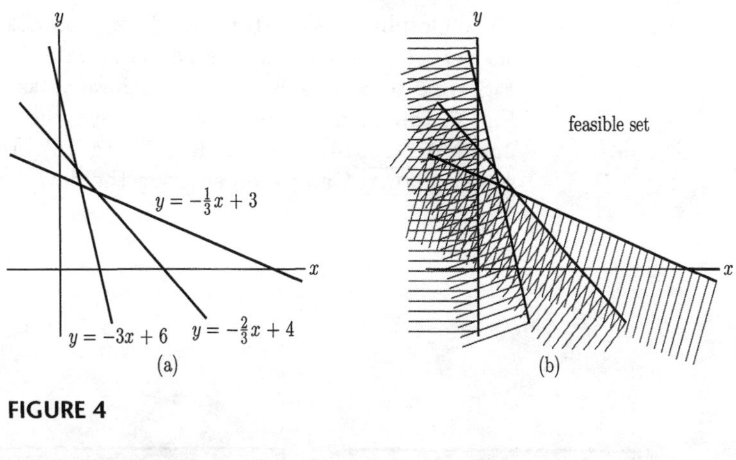

FIGURE 4

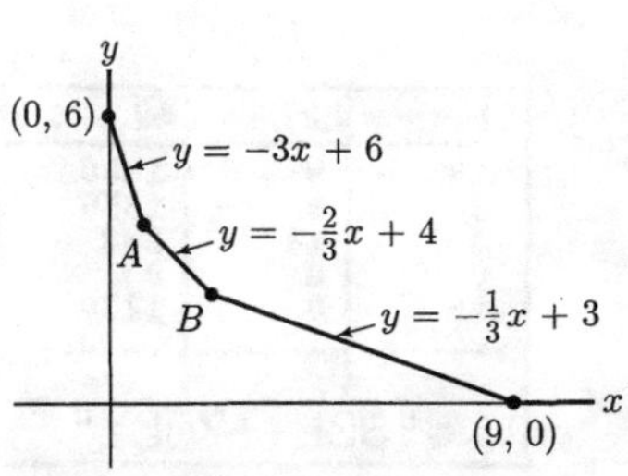

FIGURE 5

The third step of the solution process consists of determining the coordinates of A and B. From Fig. 5 these coordinates can be found by solving the following systems of equations:

$$A: \begin{cases} y = -3x + 6 \\ y = -\frac{2}{3}x + 4 \end{cases} \qquad B: \begin{cases} y = -\frac{2}{3}x + 4 \\ y = -\frac{1}{3}x + 3. \end{cases}$$

To solve the first system, equate the two expressions for y:

$$-\tfrac{2}{3}x + 4 = -3x + 6$$
$$3x - \tfrac{2}{3}x = 6 - 4$$
$$\tfrac{7}{3}x = 2$$
$$x = \tfrac{6}{7}$$
$$y = -3x + 6 = -3\left(\tfrac{6}{7}\right) + 6 = \tfrac{24}{7}$$
$$A = \left(\tfrac{6}{7}, \tfrac{24}{7}\right).$$

Similarly, we find B:

$$-\tfrac{2}{3}x + 4 = -\tfrac{1}{3}x + 3$$
$$-\tfrac{1}{3}x = -1$$
$$x = 3$$
$$y = -\tfrac{2}{3}(3) + 4 = 2$$
$$B = (3, 2).$$

TABLE 4

Vertex	*Cost* $= 21x + 14y$
$(0, 6)$	$21 \cdot 0 + 14 \cdot 6 = 84$
$\left(\frac{6}{7}, \frac{24}{7}\right)$	$21 \cdot \frac{6}{7} + 14 \cdot \frac{24}{7} = 66$
$(3, 2)$	$21 \cdot 3 + 14 \cdot 2 = 91$
$(9, 0)$	$21 \cdot 9 + 14 \cdot 0 = 189$

The fourth step consists of evaluating the objective function, in this case $21x+14y$, at each vertex. From Table 4 we see that the minimum cost is achieved at the vertex $\left(\frac{6}{7}, \frac{24}{7}\right)$. So the optimal diet—that is, the one that gives nutrients at the desired levels but at minimum cost—is the one that has $\frac{6}{7}$ cup of rice per day and $\frac{24}{7}$ cups of soybeans per day. ▮

Note We have assumed that all linear programming problems have solutions. Although every linear programming problem presented in this text has a solution, there are problems that have no optimal feasible solution. This can happen in two ways. First, there might be no points in the feasible set. Second, feasible solutions to the system of inequalities might exist, but the objective function might not have a maximum (or minimum) value within the feasible set. See Exercises 36 and 37.

GC Graphing calculators can draw (and often shade) a feasible set, determine the vertices of the feasible set, and evaluate the objective function at the vertices. Figures 6 to 8 show how some of these tasks can be carried out on a TI-83. In Fig. 6, the shading was produced with the three commands `Shade(0,Y₁,0,9)`, `Shade(0,Y₂,9,14)`, and `Shade(0,Y₃,14,16)`. In Fig. 8, the table was created by pressing STAT 1 and then entering the data for L1 and L2.

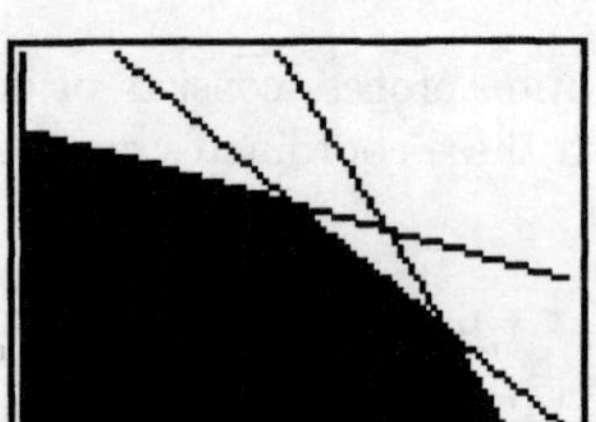

FIGURE 6

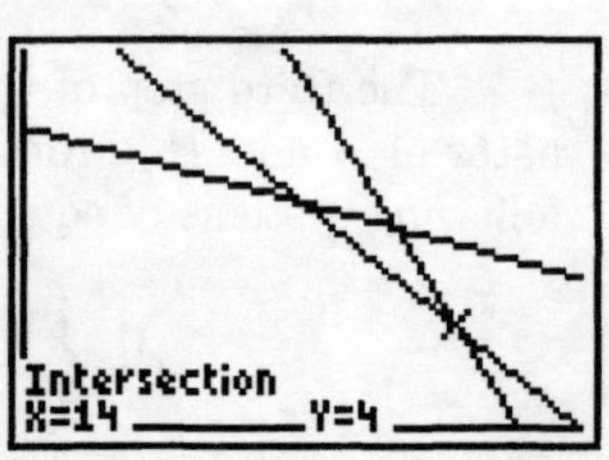

FIGURE 7

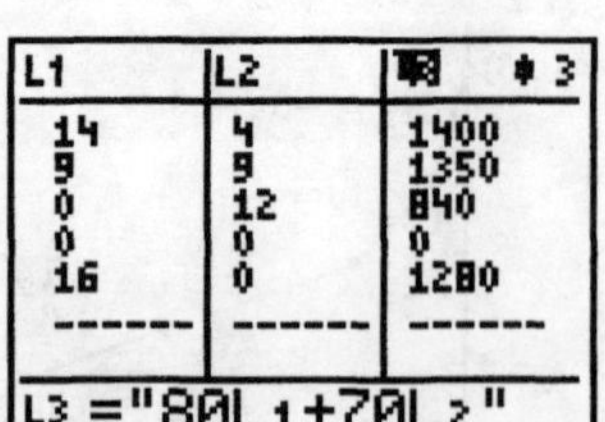

FIGURE 8

Practice Problems 2

1. The feasible set for the nutrition problem is shown in Fig. 9. The cost is $21x + 14y$. *Without* using the fundamental theorem of linear programming, explain why the cost could not possibly be minimized at the point $(4, 4)$.

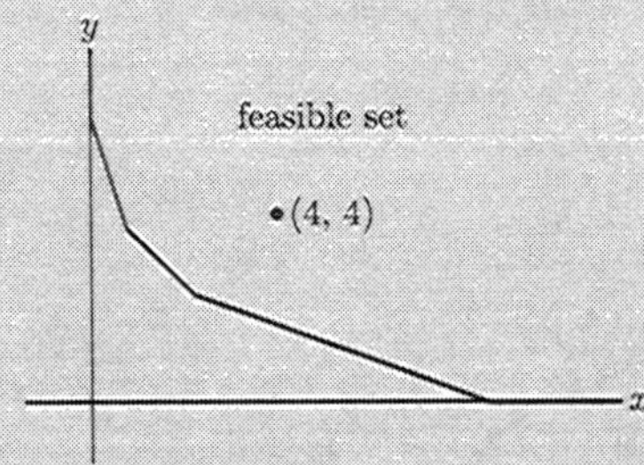

FIGURE 9

2. Rework the nutrition problem assuming that the cost of rice is changed to 7 cents per cup.

EXERCISES 2

For each of the feasible sets in Exercises 1–4, determine x and y so that the objective function $4x + 3y$ is maximized.

1.

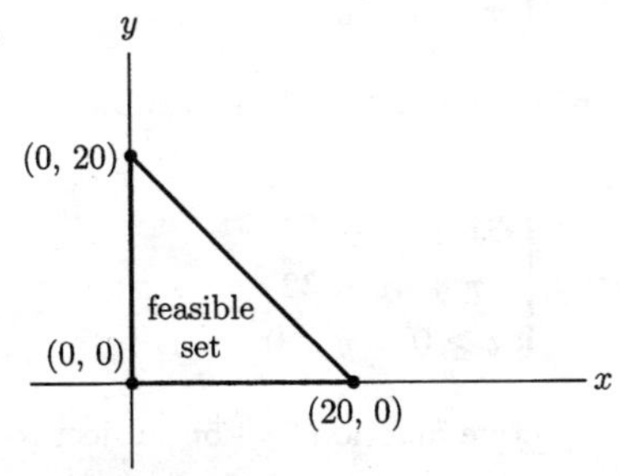

2.

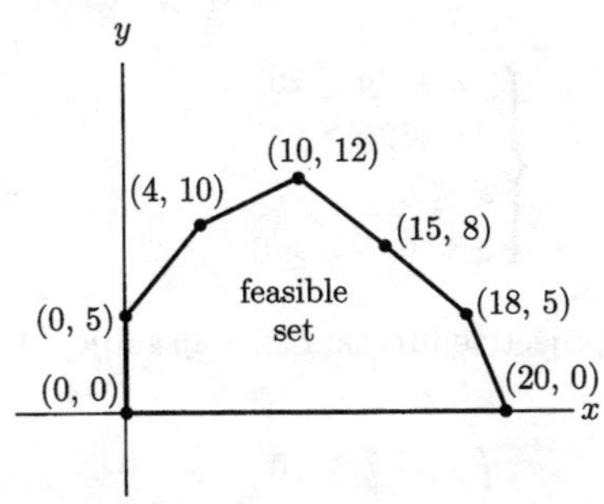

3.

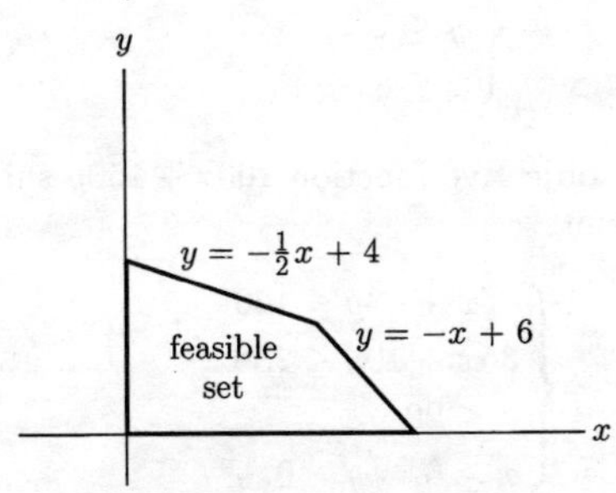

4.

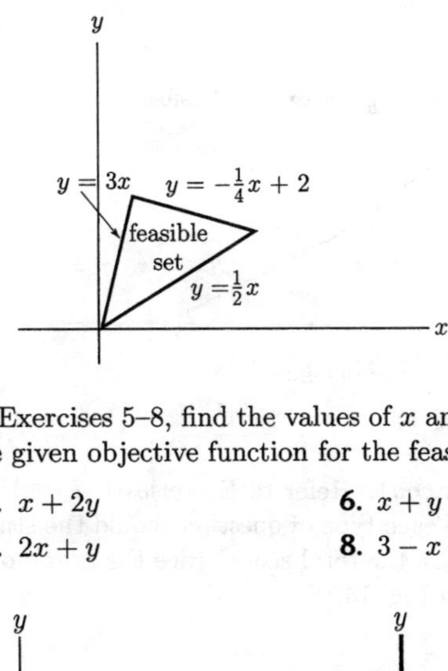

In Exercises 5–8, find the values of x and y that maximize the given objective function for the feasible set in Fig. 10.

5. $x + 2y$ **6.** $x + y$

7. $2x + y$ **8.** $3 - x - y$

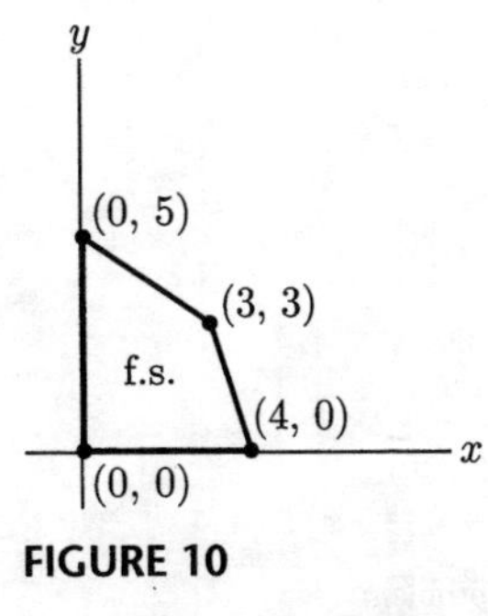

FIGURE 10

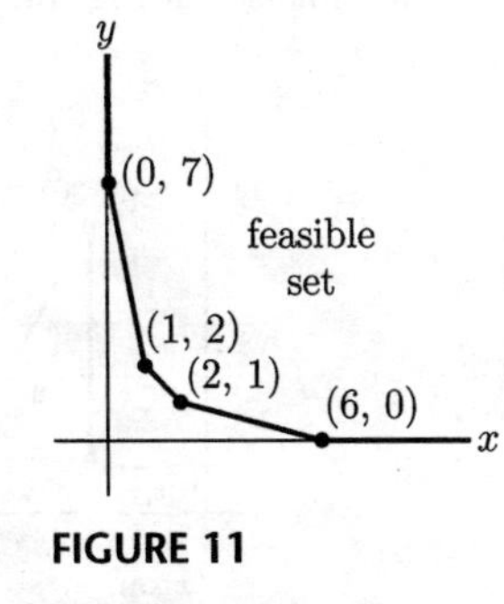

FIGURE 11

In Exercises 9–12, find the values of x and y that minimize the given objective function for the feasible set in Fig. 11.

9. $8x + y$ **10.** $3x + 2y$

11. $2x + 3y$ **12.** $x + 8y$

13. (***Shipping***) Refer to Exercises 1, Problem 5. How many crates of each cargo should be shipped in order to satisfy the shipping requirements and yield the greatest earnings? (See the graph of the feasible set in Fig. 12.)

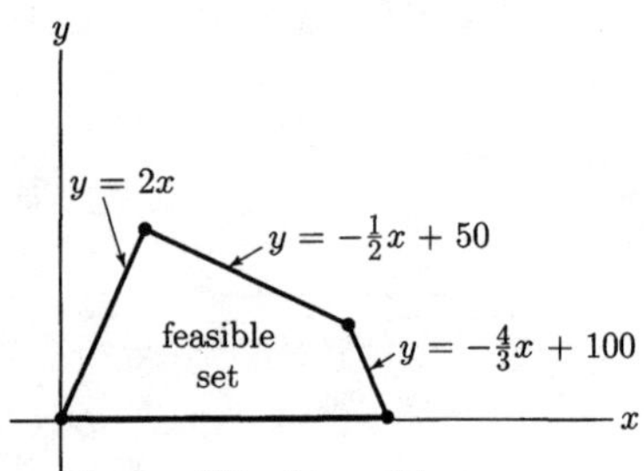

Shipping problem

FIGURE 12

14. (***Mining***) Refer to Exercises 1, Problem 6. Find the number of days that each mine should be operated in order to fill the order at the least cost. (See the graph of the feasible set in Fig. 13.)

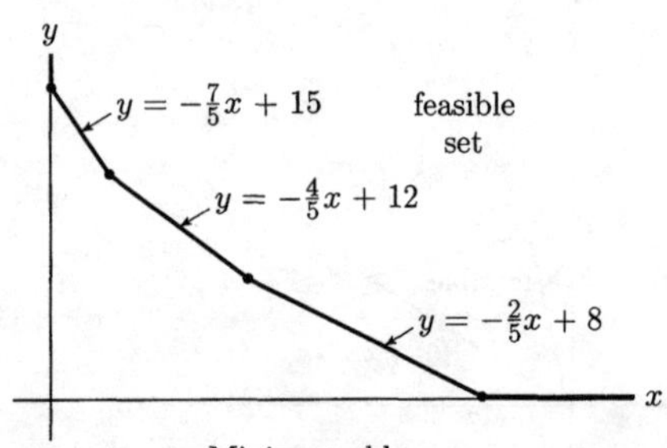

Mining problem

FIGURE 13

15. (***Exam Strategy***) Refer to Exercises 1, Problem 7. How many of each type of question should the student do to maximize the total score? (See the graph of the feasible set in Fig. 14.)

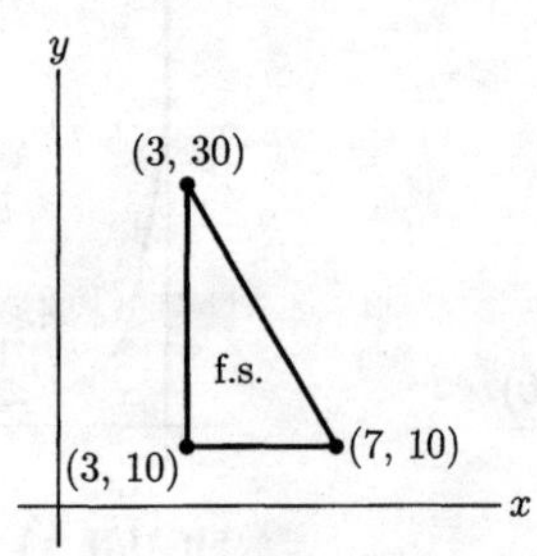

Exam strategy problem

FIGURE 14

16. (***Political Campaign—Resource Allocation***) Refer to Exercises 1, Problem 8. How should the media funds be allocated so as to maximize the total audience? (See the graph of the feasible set in Fig. 15.)

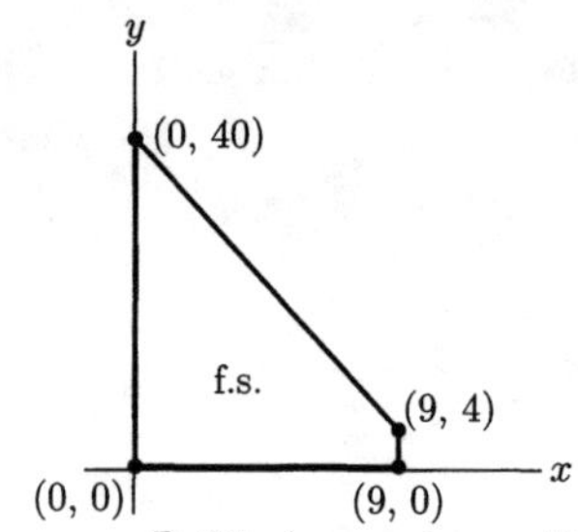

Political campaign problem

FIGURE 15

In Exercises 17 and 18, rework the furniture manufacturing problem, where everything is the same except that the profit per chair is changed to the given value. (See Table 1 for vertices.)

17. \$150

18. \$60

19. Minimize the objective function $3x + 4y$ subject to the constraints

$$\begin{cases} 2x + y \geq 10 \\ x + 2y \geq 14 \\ x \geq 0, \quad y \geq 0. \end{cases}$$

20. Maximize the objective function $7x + 4y$ subject to the constraints

$$\begin{cases} 3x + 2y \leq 36 \\ x + 4y \leq 32 \\ x \geq 0, \quad y \geq 0. \end{cases}$$

21. Maximize the objective function $2x + 5y$ subject to the constraints

$$\begin{cases} x + 2y \leq 20 \\ 3x + 2y \geq 24 \\ x \leq 6 \\ x \geq 0, \quad y \geq 0. \end{cases}$$

22. Minimize the objective function $2x + 3y$ subject to the constraints

$$\begin{cases} x + y \geq 10 \\ y \leq x \\ x \leq 8. \end{cases}$$

23. Maximize the objective function $100x + 150y$ subject to the constraints

$$\begin{cases} x + 3y \leq 120 \\ 35x + 10y \leq 780 \\ x \leq 20 \\ x \geq 0, \quad y \geq 0. \end{cases}$$

24. Minimize the objective function $\frac{1}{2}x + \frac{3}{4}y$ subject to the constraints

$$\begin{cases} 2x + 2y \geq 8 \\ 3x + 5y \geq 16 \\ x \geq 0, \quad y \geq 0. \end{cases}$$

25. Minimize the objective function $7x + 4y$ subject to the constraints

$$\begin{cases} y \geq -2x + 11 \\ y \leq -x + 10 \\ y \leq -\frac{1}{3}x + 6 \\ y \geq -\frac{1}{4}x + 4. \end{cases}$$

26. Maximize the objective function $x + 2y$ subject to the constraints

$$\begin{cases} y \leq -x + 100 \\ y \geq \frac{1}{3}x + 20 \\ y \leq x. \end{cases}$$

27. ***(Manufacturing—Resource Allocation)*** Infotron Inc. makes electronic hockey and soccer games. Each hockey game requires 2 labor-hours of assembly and 2 labor-hours of testing. Each soccer game requires 3 labor-hours of assembly and 1 labor-hour of testing. Each day there are 42 labor-hours available for assembly and 26 labor-hours available for testing. How many of each game should Infotron produce each day to maximize its total daily output?

28. ***(Manufacturing—Production Planning)*** An electronics company has factories in Cleveland and Toledo that manufacture three-head and four-head VCRs. Each day the Cleveland factory produces 500 three-head VCRs and 300 four-head VCRs at a cost of $18,000. Each day the Toledo factory produces 300 of each type of VCR at a cost of $15,000. An order is received for 25,000 three-head VCRs and 21,000 four-head VCRs. For how many days should each factory operate to fill the order at the least cost?

29. ***(Construction—Resource Allocation)*** A contractor builds two types of homes. The first type requires one lot, $12,000 capital, and 150 labor-days to build and is sold for a profit of $2400. The second type of home requires one lot, $32,000 capital, and 200 labor-days to build and is sold for a profit of $3400. The contractor owns 150 lots and has available for the job $2,880,000 capital and 24,000 labor-days. How many homes of each type should she build to realize the greatest profit?

30. ***(Nutrition—People)*** A nutritionist, working for NASA, must meet certain nutritional requirements and yet keep the weight of the food at a minimum. He is considering a combination of two foods, which are packaged in tubes. Each tube of food A contains 4 units of protein, 2 units of carbohydrate, and 2 units of fat and weighs 3 pounds. Each tube of food B contains 3 units of protein, 6 units of carbohydrate, and 1 unit of fat and weighs 2 pounds. The requirement calls for 42 units of protein, 30 units of carbohydrate, and 18 units of fat. How many tubes of each food should be supplied to the astronauts?

31. ***(Packaging—Product Mix)*** The Beautiful Day Fruit Juice Company makes two varieties of fruit drink. Each can of Fruit Delight contains 10 ounces of pineapple juice, 3 ounces of orange juice, and 1 ounce of apricot juice and makes a profit of 20 cents. Each can of Heavenly Punch contains 10 ounces of pineapple juice, 2 ounces of orange juice, and 2 ounces of apricot juice and makes a profit of 30 cents. Each week the company has available 9000 ounces of pineapple juice, 2400 ounces of orange juice, and 1400 ounces of apricot juice. How many cans of Fruit Delight and of Heavenly Punch should be produced each week to maximize profits?

32. ***(Manufacturing—Resource Allocation)*** The Bluejay Lacrosse Stick Company makes two kinds of lacrosse sticks. Type A sticks require 2 labor-hours for cutting, 1 labor-hour for stringing, and 2 labor-hours for finishing and are sold for a profit of $8. Type B sticks require 1 labor-hour for cutting, 3 labor-hours for stringing, and 2 labor-hours for finishing and are sold for a profit of $10. Each day the company has available 120 labor-hours for cutting, 150 labor-hours for stringing, and 140 labor-hours for finishing. How many lacrosse sticks of each kind should be manufactured each day to maximize profits?

33. ***(Agriculture—Crop Planning)*** A farmer has 100 acres on which to plant oats or corn. Each acre of oats requires $18 capital and 2 hours of labor. Each acre of corn requires $36 capital and 6 hours of labor. Labor costs are $8 per hour. The farmer has $2100 available for capital and $2400 available for labor. If the revenue is $55 from each acre of oats and $125 from each acre of corn, what planting combination will produce the greatest total profit? (Profit here is revenue plus leftover capital and labor cash reserve.) What is the maximum profit?

34. ***(Agriculture—Crop Planning)*** Suppose that the farmer of Exercise 33 can allocate the $4500 available for capital and labor however he or she wants.

(a) Without solving the linear programming problem, explain why the optimal profit cannot be less than what was found in Exercise 33.

(b) Find the optimal solution in the new situation. Does it provide more profit than in Exercise 33?

35. ***(Manufacturing—Resource Allocation)*** A company makes two items, I_1 and I_2, from three raw materials, M_1, M_2, and M_3. Item I_1 uses 3 ounces of M_1, 2 ounces of M_2, and 2 ounces of M_3. Item I_2 uses

4 ounces of M_1, 1 ounce of M_2, and 3 ounces of M_3. The profit on item I_1 is \$8 and on item I_2 is \$6. The company has a daily supply of 40 ounces of M_1, 20 ounces of M_2, and 60 ounces of M_3.

(a) How many of items I_1 and I_2 should be made each day to maximize profit?

(b) What is the maximum profit?

(c) How many ounces of each raw material are used?

(d) If the profit on item I_1 increases to \$13, how many of items I_1 and I_2 should be made each day to maximize profit?

36. Consider the following linear programming problem: Maximize $M = 10x + 6y$ subject to the constraints

$$\begin{cases} x + y \geq 6 \\ x \geq 0, \quad y \geq 0. \end{cases}$$

(a) Sketch the feasible set.

(b) Determine three points in the feasible set and calculate M at each of them.

(c) Show that the objective function attains no maximum value for points in the feasible set.

37. Consider the following linear programming problem: Minimize $M = 10x + 6y$ subject to the constraints

$$\begin{cases} x + \ \ y \geq 6 \\ 4x + 3y \leq 4 \\ x \geq 0, \quad y \geq 0. \end{cases}$$

(a) Sketch the feasible set for the linear programming problem.

(b) Determine a point of the feasible set.

Solutions to Practice Problems 2

1. The point P has a smaller value of x and a smaller value of y than $(4, 4)$ and is still in the feasible set. It therefore corresponds to a lower cost than $(4, 4)$ and still meets the requirements. We conclude that no interior point of the feasible set could possibly be an optimal point. This geometric argument indicates that an optimal point might be one that juts out far—that is, a vertex (Fig. 16).

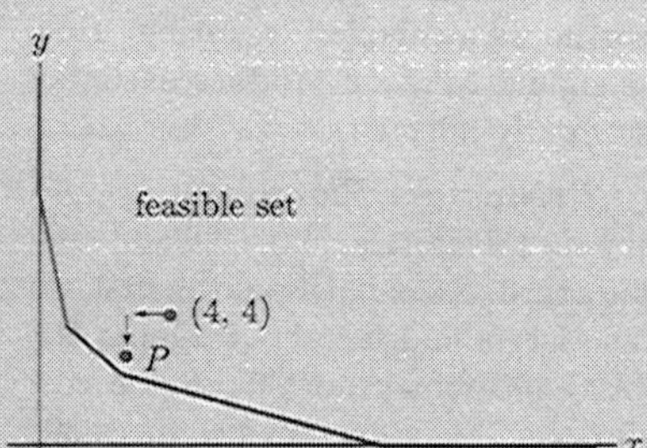

FIGURE 16

2. The system of linear inequalities, feasible set, and vertices will all be the same as before. Only the objective function changes. The new objective function is $7x + 14y$. The minimum cost occurs when using 3 cups of rice and 2 cups of soybeans.

Vertex	*Cost* $= 7x + 14y$
$(0, 6)$	84
$(\frac{6}{7}, \frac{24}{7})$	54
$(3, 2)$	49
$(9, 0)$	63

3.3 Linear Programming II

In this section we apply the technique of linear programming to the design of a portfolio for a retirement fund and to the transportation of goods from warehouses to retail outlets. The significant new feature of each of these problems is that, on the surface, they appear to involve more than two variables. However, they can be translated into mathematical language so that only two variables are required.

EXAMPLE 1 (*Investment Planning*) A pension fund has \$30 million to invest. The money is to be divided among Treasury notes, bonds, and stocks. The rules for administration of the fund require that at least \$3 million be invested in each type of investment, that at least half the money be invested in Treasury notes and bonds, and that the amount invested in bonds not exceed twice the amount invested in Treasury notes. The annual yields for the various investments are 7% for Treasury notes, 8% for bonds, and 9% for stocks. How should the money be allocated among the various investments to produce the largest return?

Solution First, let us agree that all numbers stand for millions. That is, we write 30 to stand for 30 million dollars. This will save us from writing many zeros. In examining the problem, we find that very little organization needs to be done. The rules for administration of the fund are written in a form from which inequalities can be read right off. Let us just summarize the remaining data in the first row of a chart (Table 1).

TABLE 1

	Treasury notes	*Bonds*	*Stocks*
Yield	.07	.08	.09
Variables	x	y	$30-(x+y)$

There appear to be three variables—the amounts to be invested in each of the three categories. However, since the three investments must total 30, we need only two variables. Let $x =$ the amount to be invested in Treasury notes and $y =$ the amount to be invested in bonds. Then the amount invested in stocks is $30-(x+y)$. We have displayed the variables in Table 1.

Now for the restrictions. Since at least 3 (million dollars) must be invested in each category, we have the three inequalities

$$x \geq 3$$
$$y \geq 3$$
$$30-(x+y) \geq 3.$$

Moreover, since at least half the money, or 15, must be invested in Treasury notes and bonds, we must have

$$x+y \geq 15.$$

Finally, since the amount invested in bonds must not exceed twice the amount invested in Treasury notes, we must have

$$y \leq 2x.$$

(In this example we do not need to state that $x \geq 0$, $y \geq 0$, since we have already required that they be greater than or equal to 3.) Thus there are five restriction inequalities:

$$\begin{cases} x \geq 3, \quad y \geq 3 \\ 30 - (x + y) \geq \ 3 \\ \qquad x + y \geq 15 \\ \qquad\quad y \leq 2x. \end{cases} \tag{1}$$

Next, we form the objective function, which in this case equals the total return on the investment. Since x dollars is invested at 7%, y dollars at 8%, and $30 - (x + y)$ dollars at 9%, the total return is

$$\begin{aligned} [\text{return}] &= .07x + .08y + .09[30 - (x + y)] \\ &= .07x + .08y + 2.7 - .09x - .09y \\ &= 2.7 - .02x - .01y. \end{aligned} \tag{2}$$

So the mathematical statement of the problem is: Maximize the objective function (2) subject to the restrictions (1).

The next step of the solution is to graph the inequalities (1). The necessary information is tabulated in Table 2.

TABLE 2

Inequality	*Standard form*	*Equation*	*Intercepts* x	*Intercepts* y	*Graph*
$x \geq 3$	$x \geq 3$	$x = 3$	$(3, 0)$	—	Right of line
$y \geq 3$	$y \geq 3$	$y = 3$	—	$(0, 3)$	Above line
$30 - (x + y) \geq 3$	$y \leq -x + 27$	$y = -x + 27$	$(27, 0)$	$(0, 27)$	Below line
$x + y \geq 15$	$y \geq -x + 15$	$y = -x + 15$	$(15, 0)$	$(0, 15)$	Above line
$y \leq 2x$	$y \leq 2x$	$y = 2x$	$(0, 0)$	$(0, 0)$	Below line

One point about the chart is worth noting: It contains enough data to graph each of the lines, with the exception of $y = 2x$. The reason is that the x- and y-intercepts of this line are the same, $(0, 0)$. So to graph $y = 2x$, we must find an additional point on the line. For example, if we set $x = 2$, then $y = 4$, so $(2, 4)$ is on the line. In Fig. 1(a) we have drawn the various lines, and in Fig. 1(b) we have crossed out the appropriate regions to produce the graph of the system. The feasible set, as well as the equations of the various lines that make up its boundary, are shown in Fig. 2. From Fig. 2 we find the pairs of equations that determine each of the vertices A–D. This is the third step of the solution procedure.

$$A: \begin{cases} y = 3 \\ y = -x + 15 \end{cases} \qquad B: \begin{cases} y = 3 \\ y = -x + 27 \end{cases}$$

$$C: \begin{cases} y = -x + 27 \\ y = 2x \end{cases} \qquad D: \begin{cases} y = 2x \\ y = -x + 15. \end{cases}$$

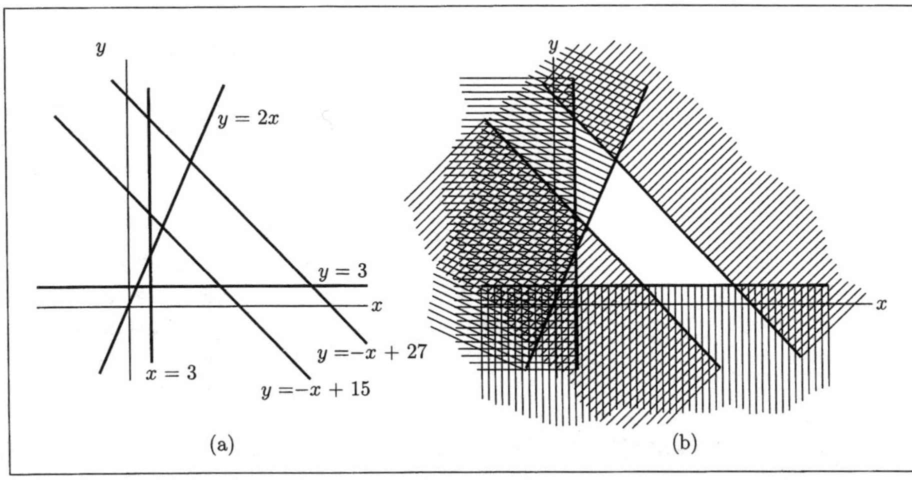

FIGURE 1

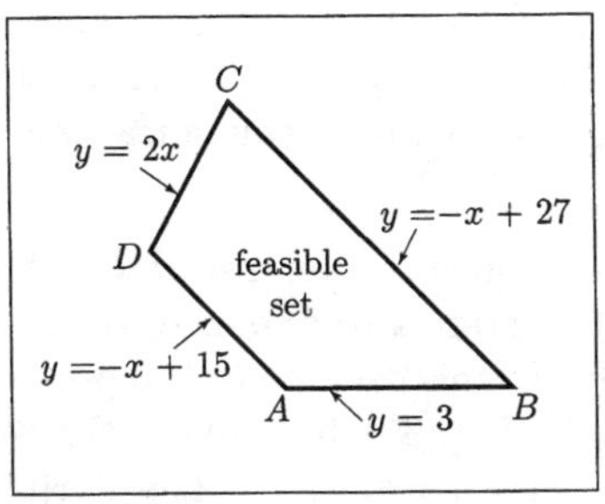

FIGURE 2

A and B are the easiest to determine. To find A, we must solve

$$3 = -x + 15$$
$$x = 12$$
$$y = 3$$
$$A = (12, 3).$$

Similarly, $B = (24, 3)$. To find C, we must solve

$$2x = -x + 27$$
$$3x = 27$$
$$x = 9$$
$$y = 2(9) = 18$$
$$C = (9, 18).$$

Similarly, $D = (5, 10)$.

Finally, we list the four vertices A, B, C, D and evaluate the objective function (2) at each one. The results are summarized in Table 3.

TABLE 3

Vertex	*Return* $= 2.7 - .02x - .01y$
$(5, 10)$	$2.7 - .02(5) - .01(10) = \$2.5$ million
$(9, 18)$	$2.7 - .02(9) - .01(18) = \$2.34$ million
$(24, 3)$	$2.7 - .02(24) - .01(3) = \$2.19$ million
$(12, 3)$	$2.7 - .02(12) - .01(3) = \$2.43$ million

It is clear that the largest return occurs when $x = 5$, $y = 10$. In other words, \$5 million should be invested in Treasury notes, \$10 million in bonds, and $30 - (x + y) = 30 - (5 + 10) = \$15$ million in stocks. ▮

Linear programming is of use not only in analyzing investments but in the fields of transportation and shipping. It is often used to plan routes, determine locations of warehouses, and develop efficient procedures for getting goods to people. Many linear programming problems of this variety can be formulated as *transportation problems.* A typical transportation problem involves determining the least-cost scheme for delivering a commodity stocked in a number of different warehouses to a number of different locations, say retail stores. Of course, in practical applications, it is necessary to consider problems involving perhaps dozens or even hundreds of warehouses, and possibly just as many delivery locations. For problems on such a grand scale, the methods developed so far are inadequate. For one thing, the number of variables required is usually more than two. We must wait until Chapter 4 for methods that apply to such problems. However, the next example gives an instance of a transportation problem that does not involve too many warehouses or too many delivery points. It gives the flavor of general transportation problems.

EXAMPLE 2 (*Transportation—Shipping*) Suppose that a Maryland TV dealer has stores in Annapolis and Rockville and warehouses in College Park and Baltimore. The cost of shipping a TV set from College Park to Annapolis is \$6; from College Park to Rockville, \$3; from Baltimore to Annapolis, \$9; and from Baltimore to Rockville, \$5. Suppose that the Annapolis store orders 25 TV sets and the Rockville store 30. Suppose further that the College Park warehouse has a stock of 45 sets and the Baltimore warehouse 40. What is the most economical way to supply the requested TV sets to the two stores?

Solution The first step in solving a linear programming problem is to translate it into mathematical language. And the first part of this step is to organize the information given, preferably in the form of the chart. In this case, since the problem is geographic, we draw a schematic diagram, as in Fig. 3, which shows the flow of goods between warehouses and retail stores. By each route, we have written the cost. Below each warehouse we have written down its stock and below each retail store the number of TV sets it ordered.

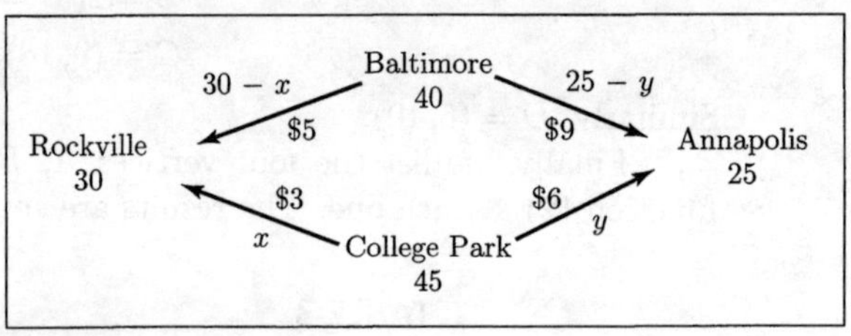

FIGURE 3

Next, let us determine the variables. It appears initially that four variables are required, namely the number of TV sets to be shipped over each route. However, a closer look shows that only two variables are required. For if x denotes the number of TV sets to be shipped from College Park to Rockville, then since Rockville ordered 30 sets, the number shipped from Baltimore to Rockville is $30 - x$. Similarly, if y denotes the number of sets shipped from College Park to Annapolis, then the number shipped from Baltimore to Annapolis is $25 - y$. We have written the appropriate shipment sizes beside the various routes in Fig. 3.

As the third part of the translation process, let us write down the restrictions on the variables. Basically, there are two kinds of restrictions: None of x, y, $30 - x$, $25 - y$ can be negative, and a warehouse cannot ship more TV sets than it has in stock. Referring to Fig. 3, we see that College Park ships $x + y$ sets, so that $x + y \leq 45$. Similarly, Baltimore ships $(30 - x) + (25 - y)$ sets, so that $(30 - x) + (25 - y) \leq 40$. Simplifying this inequality, we get

$$\begin{aligned} 55 - x - y &\leq 40 \\ -x - y &\leq -15 \\ x + y &\geq 15. \end{aligned}$$

The inequality $30 - x \geq 0$ can be simplified to $x \leq 30$, and the inequality $25 - y \geq 0$ can be written $y \leq 25$. So our restriction inequalities are these:

$$\begin{cases} x \geq 0, \quad y \geq 0 \\ x \leq 30, \quad y \leq 25 \\ x + y \geq 15 \\ x + y \leq 45. \end{cases} \tag{3}$$

The final step in the translation process is to form the objective function. In this problem we are attempting to minimize cost, so the objective function must express the cost in terms of x and y. Refer again to Fig. 3. There are x sets going from College Park to Rockville, and each costs \$3 to transport, so the cost of delivering these x sets is $3x$. Similarly, the costs of making the other deliveries are $6y$, $5(30 - x)$, and $9(25 - y)$, respectively. Thus the objective function is

$$\begin{aligned} [\text{cost}] &= 3x + 6y + 5(30 - x) + 9(25 - y) \\ &= 3x + 6y + 150 - 5x + 225 - 9y \\ &= 375 - 2x - 3y. \end{aligned} \tag{4}$$

So the mathematical problem we must solve is: Find x and y that minimize the objective function (4) and satisfy the restrictions (3).

To solve the mathematical problem, we must graph the system of inequalities in (3). Four of the inequalities have graphs determined by horizontal and vertical lines. The only inequalities involving any work are $x + y \geq 15$ and $x + y \leq 45$. And even these are very easy to graph. The result is the graph in Fig. 4.

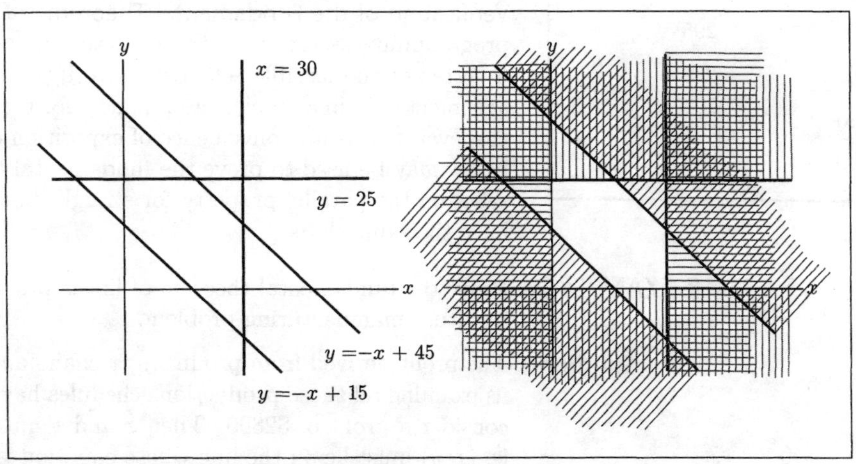

FIGURE 4

In Fig. 5 we have drawn the feasible set and have labeled each boundary line with its equation. The vertices A–F are now simple to determine. First, A and F are the intercepts of the line $y = -x + 15$. Therefore, $A = (15, 0)$ and $F = (0, 15)$. Since B is the x-intercept of the line $x = 30$, we have $B = (30, 0)$. Similarly, $E = (0, 25)$. Since C is on the line $x = 30$, its x-coordinate is 30.

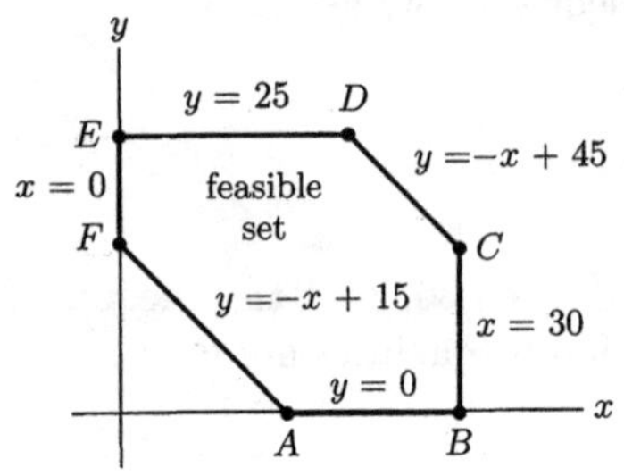

FIGURE 5

TABLE 4

Vertex	*Cost* = 375 − 2x − 3y
(0, 25)	300
(0, 15)	330
(15, 0)	345
(30, 0)	315
(30, 15)	270
(20, 25)	260

TABLE 5

Vertex	*Cost* = 325 − 2x − y
(0, 25)	300
(0, 15)	310
(15, 0)	295
(30, 0)	265
(30, 15)	250
(20, 25)	260

Its y-coordinate is $y = -30 + 45 = 15$, so $C = (30, 15)$. Similarly, since D has y-coordinate 25, its x-coordinate is given by $25 = -x + 45$ or $x = 20$. Thus $D = (20, 25)$.

We have listed in Table 4 the vertices A–F as well as the cost corresponding to each one. The minimum cost of \$260 occurs at the vertex $(20, 25)$. So $x = 20$, $y = 25$ yields the minimum of the objective function. In other words, 20 TV sets should be shipped from College Park to Rockville and 25 from College Park to Annapolis, $30 - x = 10$ from Baltimore to Rockville, and $25 - y = 0$ from Baltimore to Annapolis. This solves our problem. ∎

Remarks Concerning the Transportation Problem Note that the highest-cost route is the one from Baltimore to Annapolis. The solution we have obtained eliminates any shipments over this route. One might infer from this that one should always avoid the most expensive route. But this is not correct reasoning. To see why, reconsider Example 2, except change the cost of transporting a TV set from Baltimore to Annapolis from \$9 to \$7. The Baltimore–Annapolis route is still the most expensive. However, in this case the minimum cost is not obtained by eliminating the Baltimore–Annapolis route. For the revised problem, the linear inequalities stay the same. So the feasible set and the vertices remain the same. The only change is in the objective function, which now is given by

$$[\text{cost}] = 3x + 6y + 5(30 - x) + 7(25 - y) = 325 - 2x - y.$$

Therefore, the costs at the various vertices are given in Table 5. So the minimum cost of \$250 is achieved when $x = 30$, $y = 15$, $30 - x = 0$, and $25 - y = 10$. Note that 10 sets are being shipped from Baltimore to Annapolis, even though this is the most expensive route.

It is even possible for the cost function to be optimized simultaneously at two different vertices. For example, if the cost from Baltimore to Annapolis is \$8 and all other data are the same as in Example 2, then the optimal cost is \$260 and is achieved at both vertices $(30, 15)$ and $(20, 25)$.

Verification of the Fundamental Theorem The fundamental theorem of linear programming asserts that the objective function assumes its optimal value at a vertex of the feasible set. Let us verify this fact. For simplicity, we give the argument only in a special case, namely for the furniture manufacturing problem. However, this is for convenience of exposition only. The same argument as given below may be used to prove the fundamental theorem in general. Our argument relies on the parallel property for straight lines, which asserts that parallel lines have the same slope.

EXAMPLE 3 Prove the fundamental theorem of linear programming in the special case of the furniture manufacturing problem.

Solution The profit derived from producing x chairs and y sofas is $80x + 70y$ dollars. Let us examine all those production schedules having a given profit. As an example, consider a profit of \$2800. Then x and y must satisfy $80x + 70y = 2800$. That is, (x, y) must lie on the line whose equation is $80x + 70y = 2800$, or in standard form, $y = -\frac{8}{7}x + 40$. The slope of this line is $-\frac{8}{7}$ and its y-intercept is $(0, 40)$. We have drawn this line in Fig. 6(a), in which we have also drawn the feasible set for the furniture manufacturing problem. Note two fundamental facts: (1) Every production schedule on the line corresponds to a profit of \$2800. (2) The line lies above the feasible set. In particular, no production schedule on the line satisfies all the restrictions of the problem. The difficulty is that \$2800 is too high a profit for which to ask.

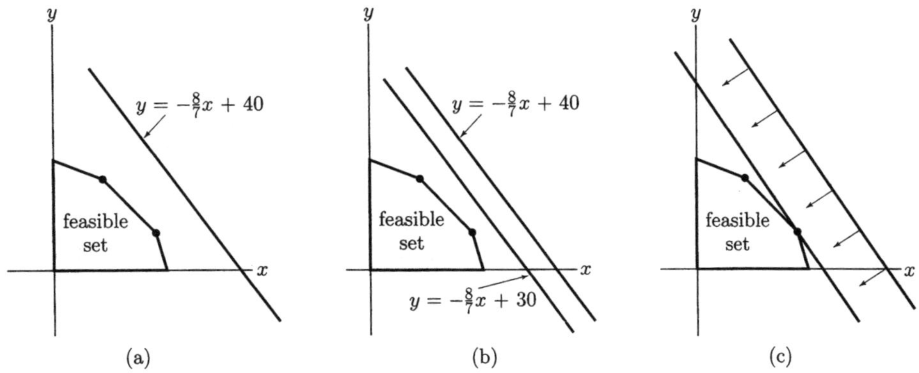

FIGURE 6

So now lower the profit, say, to \$2100. In this case the production schedule (x, y) lies on the line $80x + 70y = 2100$, or in standard form, $y = -\frac{8}{7}x + 30$. This line is drawn in Fig. 6(b). Note that since both lines have slope $-\frac{8}{7}$, they are parallel by the parallel property. Actually, if we look at the production schedules yielding any fixed profit p, then they will lie along a line of slope $-\frac{8}{7}$, which is parallel to the two lines already drawn. For if the production schedule (x, y) yields a profit p, then

$$80x + 70y = p \quad \text{or} \quad y = -\frac{8}{7}x + \frac{p}{70}.$$

In other words, (x, y) lies on a line of slope $-\frac{8}{7}$ and y-intercept $(0, p/70)$. In particular, all the "lines of constant profit" are parallel to one another. So let us go back to the line of \$2800 profit. It does not touch the feasible set. So now lower the profit and therefore translate the line downward parallel to itself. Next lower the profit until we first touch the feasible set. This line now touches the feasible set at a vertex [Fig. 6(c)]. And this vertex corresponds to the optimal production schedule, since any other point of the feasible set lies on a "line of constant profit" corresponding to an even-lower profit. This shows why the fundamental theorem of linear programming is true. ▮

If the objective function is parallel to one of the boundary lines of the feasible set, there might be infinitely many solutions—all points on that boundary line provide optimal values for the objective function. Two such points are vertices of the feasible set.

▮ EXAMPLE 4 Reconsider the TV shipping problem discussed in Example 2 with the cost of shipping from Baltimore to Annapolis now \$8 and all other data the same as in Example 2. Minimize the shipping costs.

Solution The cost function to be minimized is now

$$[\text{cost}] = 350 - 2x - 2y.$$

Note that for any fixed value of the cost, the slope of the objective function is -1. The feasible set is identical to that of Example 2 (Fig. 5). The slope of the boundary line $y = -x + 45$ is also -1. A check of the vertices of the feasible set (Table 6) shows that the minimal cost of \$260 is achieved at the two vertices $C = (30, 15)$ and $D = (20, 25)$. In fact, the cost of \$260 is achieved at every point on the boundary line joining these two vertices. Let's verify this in two cases. The points $(25, 20)$ and $(28, 17)$ both lie on the line $y = -x + 45$ and produce a cost of \$260. Figure 7 illustrates the result. ▮

TABLE 6

Vertex	*Cost* $= 350 - 2x - 2y$
$(0, 25)$	300
$(0, 15)$	320
$(15, 0)$	320
$(30, 0)$	290
$(30, 15)$	260
$(20, 25)$	260

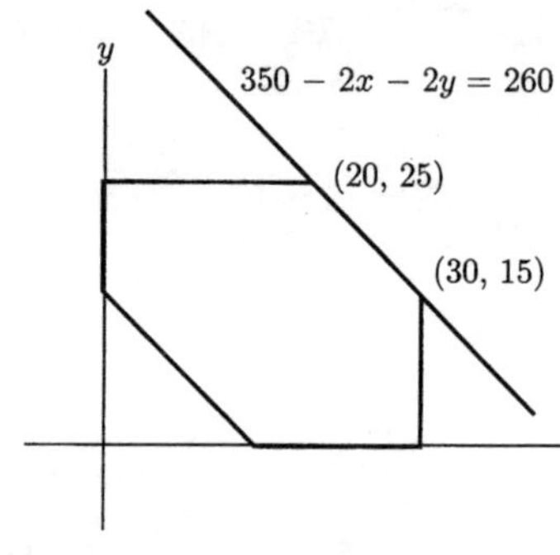

FIGURE 7

FIGURE 8

GC Graphing calculators can draw accurate feasible sets and accurate lines of constant profit (or cost). These precise drawings often permit us to select the optimal vertex by inspection. Figure 8 shows the screen for the shipping problem of Example 2. The thick line is the line of constant cost, \$253. The shaded region was produced with the commands `Shade(15 – X, 25, 0, 15)`, `Shade(0, 25, 15, 20)`, and `Shade(0, 45 – X, 20, 30)`. As the thick line is lowered, the point $(20, 25)$ is selected.

Practice Problems 3

Problems 1 and 2 refer to Example 1. Translate the statement into an inequality.

1. The amount to be invested in bonds is at most \$5 million more than the amount to be invested in Treasury notes.

2. No more than \$25 million should be invested in stocks and bonds.

3. Rework Example 1, assuming that the yield for Treasury notes goes up to 8%.

4. A linear programming problem has objective function $\lceil \text{cost} \rceil = 5x + 10y$, which is to be minimized. Figure 9 shows the feasible set and the straight line of all combinations of x and y for which $\lceil \text{cost} \rceil = \$20$.

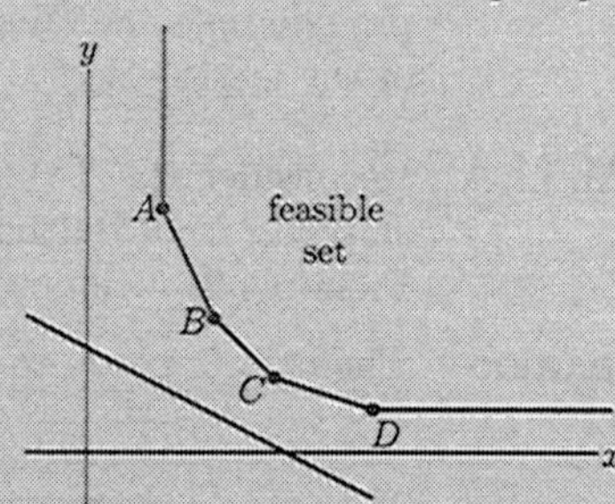

FIGURE 9

(a) Give the linear equation (in standard form) of the line of constant cost c.

(b) As c increases, does the line of constant cost c move up or down?

(c) By inspection, find the vertex of the feasible set that gives the optimal solution.

EXERCISES 3

1. Figure 10(a) shows the feasible set of the nutrition problem of Section 3.2 and the straight line of all combinations of rice and soybeans for which the cost is 42 cents.
   (a) The objective function is $21x+14y$. Give the linear equation (in standard form) of the line of constant cost c.
   (b) As c increases, does the line of constant cost move up or down?
   (c) By inspection, find the vertex of the feasible set that gives the optimal solution.

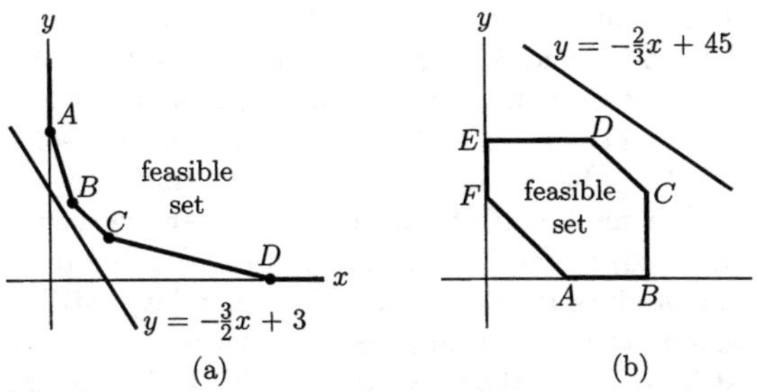

FIGURE 10

2. Figure 10(b) shows the feasible set of the transportation problem of Example 2 and the straight line of all combinations of shipments for which the transportation cost is \$240.
   (a) The objective function is $[\text{cost}] = 375 - 2x - 3y$. Give the linear equation (in standard form) of the line of constant cost c.
   (b) As c increases, does the line of constant cost move up or down?
   (c) By inspection, find the vertex of the feasible set that gives the optimal solution.

Consider the feasible set in Fig. 11(a). In Exercises 3–6, find an objective function of the form $ax + by$ which has its greatest value at the given point.

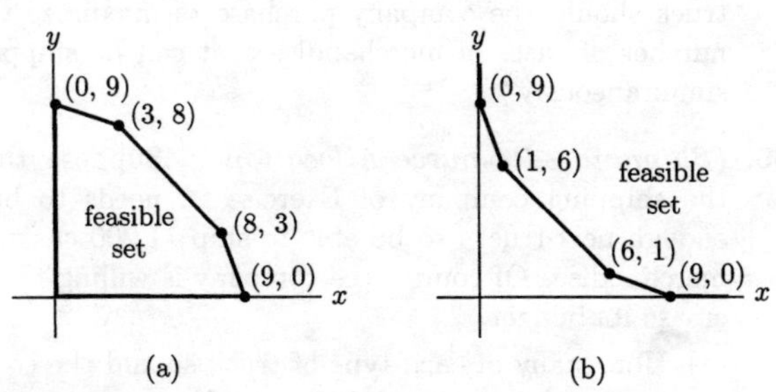

FIGURE 11

3. $(9,0)$ 4. $(3,8)$ 5. $(8,3)$ 6. $(0,9)$

Consider the feasible set in Fig. 11(b). In Exercises 7–10, find an objective function of the form $ax + by$ which has its least value at the given point.

7. $(9,0)$ 8. $(1,6)$ 9. $(6,1)$ 10. $(0,9)$

Consider the feasible set in Fig. 12(a), where three of the boundary lines are labeled with their slopes. In Exercises 11–14, find the point at which the given objective function has its greatest value.

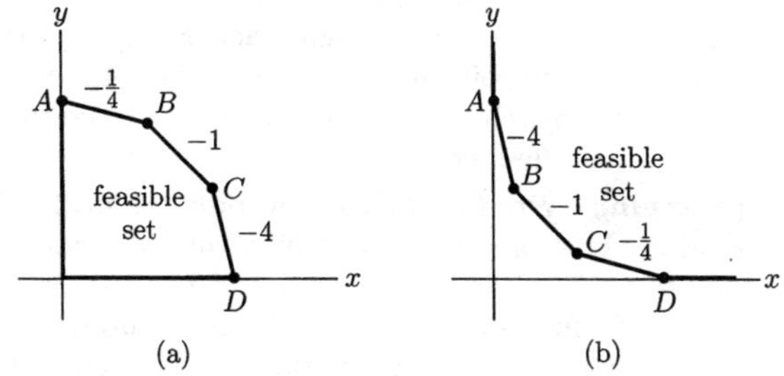

FIGURE 12

11. $3x + 2y$
12. $2x + 10y$
13. $10x + 2y$
14. $2x + 3y$

Consider the feasible set in Fig. 12(b), where three of the boundary lines are labeled with their slopes. In Exercises 15–18, find the point at which the given objective function has its least value.

15. $2x + 10y$
16. $10x + 2y$
17. $2x + 3y$
18. $3x + 2y$

19. Consider the feasible set in Fig. 13. For what values of k will the objective function $x + ky$ be maximized at the vertex $(3,4)$?

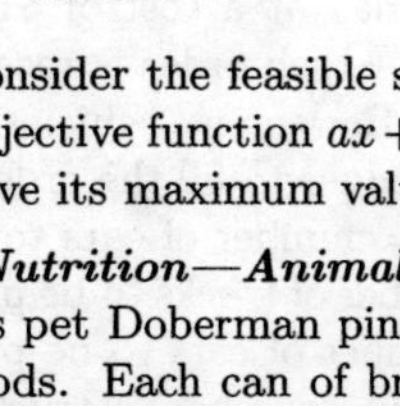

FIGURE 13

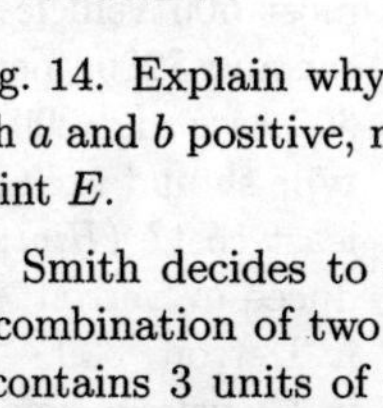

FIGURE 14

20. Consider the feasible set in Fig. 14. Explain why the objective function $ax+by$, with a and b positive, must have its maximum value at point E.

21. ***(Nutrition—Animal)*** Mr. Smith decides to feed his pet Doberman pinscher a combination of two dog foods. Each can of brand A contains 3 units of protein, 1 unit of carbohydrates, and 2 units of fat and costs 80 cents. Each can of brand B contains 1 unit

of protein, 1 unit of carbohydrates, and 6 units of fat and costs 50 cents. Mr. Smith feels that each day his dog should have at least 6 units of protein, 4 units of carbohydrates, and 12 units of fat. How many cans of each dog food should he give to his dog each day to provide the minimum requirements at the least cost?

22. (***Oil Production***) An oil company owns two refineries. Refinery I produces each day 100 barrels of high-grade oil, 200 barrels of medium-grade oil, and 300 barrels of low-grade oil and costs $10,000 to operate. Refinery II produces each day 200 barrels of high-grade, 100 barrels of medium-grade, and 200 barrels of low-grade oil and costs $9000 to operate. An order is received for 1000 barrels of high-grade oil, 1000 barrels of medium-grade oil, and 1800 barrels of low-grade oil. How many days should each refinery be operated to fill the order at the least cost?

23. (***Shipping—Product Mix***) A produce dealer in Florida ships oranges, grapefruits, and avocados to New York by truck. Each truckload consists of 100 crates, of which at least 20 crates must be oranges, at least 10 crates must be grapefruits, at least 30 crates must be avocados, and there must be at least as many crates of oranges as grapefruits. The profit per crate is $5 for oranges, $6 for grapefruits, and $4 for avocados. How many crates of each type should be shipped to maximize the profit? [*Hint*: Let x = number of crates of oranges, y = number of crates of grapefruit. Then $100 - x - y$ = number of crates of avocados.]

24. (***Investment Planning***) Mr. Jones has $9000 to invest in three types of stocks: low-risk, medium-risk, and high-risk. He invests according to three principles. The amount invested in low-risk stocks will be at most $1000 more than the amount invested in medium-risk stocks. At least $5000 will be invested in low- and medium-risk stocks. No more than $7000 will be invested in medium- and high-risk stocks. The expected yields are 6% for low-risk stocks, 7% for medium-risk stocks, and 8% for high-risk stocks. How much money should Mr. Jones invest in each type of stock to maximize his total expected yield?

25. (***Manufacturing—Production Planning***) An automobile manufacturer has assembly plants in Detroit and Cleveland, each of which can assemble cars and trucks. The Detroit plant can assemble at most 800 vehicles in one day at a cost of $1200 per car and $2100 per truck. The Cleveland plant can assemble at most 500 vehicles in one day at a cost of $1000 per car and $2000 per truck. A rush order is received for 600 cars and 300 trucks. How many vehicles of each type should each plant produce to fill the order at the least cost? (*Hint*: Let x = number of cars to be produced in Detroit, y = number of trucks to be produced in Detroit, $600 - x$ = number of cars to be produced in Cleveland, and $300 - y$ = number of trucks to be produced in Cleveland.)

26. (***Transportation—Shipping***) A foreign-car dealer with warehouses in New York and Baltimore receives orders from dealers in Philadelphia and Trenton. The dealer in Philadelphia needs 4 cars and the dealer in Trenton needs 7. The New York warehouse has 6 cars and the Baltimore warehouse has 8. The cost of shipping cars from Baltimore to Philadelphia is $120 per car, from Baltimore to Trenton $90 per car, from New York to Philadelphia $100 per car, and from New York to Trenton $70 per car. Find the number of cars to be shipped from each warehouse to each dealer to minimize the shipping cost.

27. (***Manufacturing—Production Planning***) An oil refinery produces gasoline, jet fuel, and diesel fuel. The profits per gallon from the sale of these fuels are $.15, $.12, and $.10, respectively. The refinery has a contract with an airline to deliver a minimum of 20,000 gallons per day of jet fuel and/or gasoline (or some of each). It has a contract with a trucking firm to deliver a minimum of 50,000 gallons per day of diesel fuel and/or gasoline (or some of each). The refinery can produce 100,000 gallons of fuel per day, distributed among the fuels in any fashion. It wishes to produce at least 5000 gallons per day of each fuel. How many gallons of each should be produced to maximize the profit?

28. (***Manufacturing—Production Planning***) Suppose that a price war reduces the profits of gasoline in Problem 9 to $.05 per gallon and that the profits on jet fuel and diesel fuel are unchanged. How many gallons of each fuel should now be produced to maximize the profit?

29. (***Shipping—Resource Allocation***) A shipping company is buying new trucks. The high-capacity trucks cost $50,000 and hold 320 cases of merchandise. The low-capacity trucks cost $30,000 and hold 200 cases of merchandise. The company has budgeted $1,080,000 for the new trucks and has a maximum of 30 people qualified to drive the trucks. Due to availability limitations, the company can purchase at most 15 high-capacity trucks. How many of each type of truck should the company purchase to maximize the number of cases of merchandise that can be shipped simultaneously?

30. (***Shipping—Resource Allocation***) Suppose that the shipping company of Exercise 29 needs to buy enough new trucks to be able to ship 11,200 cases of merchandise. Of course, the company is willing to increase its budget.
   (a) How many of each type of truck should the company purchase to minimize cost?
   (b) What if the company hires 23 additional qualified drivers?

Solutions to Practice Problems 3

1. Amount invested in bonds $= y$. Five million dollars more than the amount invested in Treasury notes is $x + 5$. Therefore, $y \leq x + 5$.

2. Amount invested in stocks $= 30 - (x + y)$. Amount invested in bonds $= y$. Therefore,

$$30 - (x + y) + y \leq 25$$
$$30 - x \leq 25$$
$$x \geq 5.$$

3. The feasible set stays the same but the return becomes

$$[\text{return}] = .08x + .08y + .09[30 - (x + y)]$$
$$= .08x + .08y + 2.7 - .09x - .09y$$
$$= 2.7 - .01x - .01y.$$

When the return is evaluated at each of the vertices of the feasible set, the greatest return is achieved at two vertices. Either of these vertices yields an optimal solution.

(x, y)	$2.7 - .01x - .01y$
$(5, 10)$	2.55
$(12, 3)$	2.55
$(24, 3)$	2.43
$(9, 18)$	2.43

4. (a) The values of x and y for which the cost is c dollars satisfy $5x + 10y = c$. The standard form of this linear equation is $y = -\frac{1}{2}x + c/10$.

   (b) The line $y = -\frac{1}{2}x + c/10$ has slope $-\frac{1}{2}$ and y-intercept $(0, c/10)$. As c increases, the slope stays the same, but the y-intercept moves up. Therefore, the line moves up.

   (c) The line of constant cost \$20 does not contain any points of the feasible set, so such a low cost cannot be achieved. Increase the cost until the line of constant cost just touches the feasible set. As c increases, the line moves up (keeping the same slope) and first touches the feasible set at vertex C. Therefore, taking x and y to be the coordinates of C yields the minimum cost.

Chapter 3 Review of Fundamental Concepts

1. What is the nature of a linear programming problem?
2. What is the role of the objective function in a linear programming problem?
3. State the fundamental theorem of linear programming.
4. Give the four-step procedure for solving a linear programming problem.

CHAPTER 3 SUPPLEMENTARY EXERCISES

1. (*Travel—Resource Allocation*) Terrapin Airlines wants to fly 1400 members of a ski club to Colorado. The airline owns two types of planes. Type A can carry only 50 passengers, requires 3 flight attendants, and costs \$14,000 for the trip. Type B can carry 300 passengers, requires 4 flight attendants, and costs \$90,000 for the trip. If the airline must use at least as many type A planes as type B and has available only 42 flight attendants, how many planes of each type should be used to minimize the cost for the trip?

2. (*Nutrition—People*) A nutritionist is designing a new breakfast cereal using wheat germ and enriched oat flour as the basic ingredients. Each ounce of wheat germ contains 2 milligrams of niacin, 3 milligrams of iron, and .5 milligram of thiamin and costs 3 cents. Each ounce of enriched oat flour contains 3 milligrams of niacin, 3 milligrams of iron, and .25 milligram of thiamin and costs 4 cents. The nutritionist wants the cereal to have at least 7 milligrams of niacin, 9 milligrams of iron, and 1 milligram of thiamin. How many ounces of wheat germ and how many ounces of enriched oat flour should be used in each serving to meet the nutritional requirements at the least cost?

3. (*Manufacturing—Resource Allocation*) An automobile manufacturer makes hardtops and sports cars. Each hardtop requires 8 labor-hours to assemble, 2 labor-hours to paint, and 2 labor-hours to upholster and is sold for a profit of \$90. Each sports car requires 18 labor-hours to assemble, 2 labor-hours to paint, and 1 labor-hour to upholster and is sold for a profit of \$100. During each day 360 labor-hours are available to assemble, 50 labor-hours to paint, and 40 labor-hours to upholster automobiles. How many hardtops and sports cars should be produced each day to maximize the profit?

4. (*Packaging—Product Mix*) A confectioner makes two raisin–nut mixtures. A box of mixture A contains 6 ounces of peanuts, 1 ounce of raisins, and 4 ounces of cashews and sells for 50 cents. A box of mixture B contains 12 ounces of peanuts, 3 ounces of raisins, and 2 ounces of cashews and sells for 90 cents. He has available 5400 ounces of peanuts, 1200 ounces of raisins, and 2400 ounces of cashews. How many boxes of each mixture should he make to maximize revenue?

5. (*Publishing—Product Mix*) A textbook publisher puts out 72 new books each year, which are classified as elementary, intermediate, and advanced. The company's policy for new books is to publish at least four advanced books, at least three times as many elementary books as intermediate books, and at least twice as many intermediate books as advanced books. On the average, the annual profits are \$8000 for each elementary book, \$7000 for each intermediate book, and \$1000 for each advanced book. How many new books of each type should be published to maximize the annual profit while conforming to company policy?

6. (*Shipping—Resource Allocation*) A computer company has two manufacturing plants, one in Rochester and one in Queens. Transporting a computer from Rochester to the retail outlet takes 15 hours and costs \$15, while transporting a computer from Queens to the retail outlet takes 20 hours and costs \$30. The profit on each computer manufactured in Rochester is \$40 and the profit on each computer manufactured in Queens is \$30. The Rochester plant has 80 computers available and the Queens plant has 120 computers available. If there are 2100 hours and \$3000 allotted for transporting the computers, how many computers should be sent to the retail outlet from each of the two plants to maximize the company's profits?

7. (*Transportation—Shipping*) An appliance company has two warehouses and two retail outlets. Warehouse A has 400 refrigerators and warehouse B has 300 refrigerators. Outlet I needs 200 refrigerators and outlet II needs 300 refrigerators. It costs \$36 to ship a refrigerator from warehouse A to outlet I and \$30 to ship a refrigerator from warehouse A to outlet II. It costs \$30 to ship a refrigerator from warehouse B to outlet I and \$25 to ship a refrigerator from warehouse B to outlet II. How should the company ship the refrigerators to minimize the cost?

Chapter 9
The Mathematics of Finance

10.1 Interest
10.2 Annuities
10.3 Amortization of Loans

This chapter presents several topics in the mathematics of finance, including compound and simple interest, annuities, and amortization. They are presented in the classical way, using methods that are centuries old but that lend themselves to modern computing techniques.

10.1 Interest

When you deposit money into a savings account, the bank pays you a fee for the use of your money. This fee is called *interest* and is determined by the amount deposited, the duration of the deposit, and the quoted interest rate. The amount deposited is called the *principal*, and the amount to which the principal grows (after the addition of interest) is called the *compound amount.*

The entries in a hypothetical bank passbook are shown in Table 1. Note the following facts about this passbook:

TABLE 1

Date	*Deposits*	*Withdrawals*	*Interest*	*Balance*
1/1/97	$100.00			$100.00
4/1/97			$1.00	101.00
7/1/97			1.01	102.01
10/1/97			1.02	103.03
1/1/98			1.03	104.06

1. The principal is $100.00. The compound amount after 1 year is $104.06.
2. Interest is being paid four times per year (or in financial language, *quarterly*).
3. Each quarter, the interest is 1% of the previous balance. That is, $1.00 is 1% of $100.00, $1.01 is 1% of $101.00, and so on. Since $4 \times 1\%$ is 4%, we say that the money is earning 4% *annual interest compounded quarterly.*

As in the passbook above, interest rates are usually stated as annual percentage rates, with the interest to be *compounded* (i.e., computed) a certain number of times per year. Some common frequencies for compounding are listed in Table 2. Of special importance is the *interest rate per period*, denoted i, which is calculated by dividing the annual percentage by the number of interest periods per year.

TABLE 2

Number of interest periods per year	*Length of each interest period*	*Interest compounded*
1	One year	Annually
2	Six months	Semiannually
4	Three months	Quarterly
12	One month	Monthly
52	One week	Weekly
365	One day	Daily

For example, in our passbook above, the annual percentage rate was 4%, the interest was compounded quarterly, and the interest per period was $4\%/4 = 1\%$.

> If interest is compounded m times per year and the annual interest rate is r, then the interest rate per period is
> $$i = \frac{r}{m}.$$

EXAMPLE 1 Determine the interest rate per period for each of the following interest rates.

(a) 5% interest compounded semiannually

(b) 6% interest compounded monthly

Solution (a) The annual percentage rate is 5% and the number of interest periods is 2. Therefore,

$$i = \frac{5\%}{2} = 2\tfrac{1}{2}\%.$$

Note: In decimal form, $i = .025$.

(b) The annual percentage rate is 6% and the number of interest periods is 12. Therefore,

$$i = \frac{6\%}{12} = \frac{1}{2}\%.$$

Note: In decimal form, $i = .005$. ▮

Consider a savings account in which the interest rate per period is i. Then the interest earned during a period is i times the previous balance. That is, at the end of an interest period, the new balance, B_{new}, is computed by adding this interest to the previous balance, $B_{previous}$. Therefore,

$$B_{new} = B_{previous} + i \cdot B_{previous}. \tag{1}$$

EXAMPLE 2 Compute the interest and the balance for the first two interest periods for a deposit of \$1000 at 4% interest compounded semiannually.

Solution Here $i = 2\%$, or .02. The interest for the first period is 2% of \$1000, or \$20. Let B_1 be the balance at the end of the first interest period and B_2 the balance after two interest periods. By formula (1),

$$B_1 = 1000 + .02(1000) = 1000 + 20 = \$1020.$$

Similarly, since 2% of \$1020 is \$20.40, the interest for the second period is \$20.40 and the balance is

$$B_2 = B_1 + .02B_1 = 1020 + .02(1020) = 1020 + 20.40 = \$1040.40.$$

Let us compute B_2 using another method:

$$\begin{aligned} B_2 &= B_1 + .02B_1 = 1 \cdot B_1 + .02B_1 = (1 + .02)B_1 \\ &= (1.02)B_1 = (1.02)1020 = \$1040.40. \end{aligned}$$ ▮

The alternative method for computing B_2 just presented can be generalized. Namely, we always have

$$B_{\text{new}} = B_{previous} + i \cdot B_{previous} = 1 \cdot B_{previous} + i \cdot B_{previous}.$$

Therefore,

$$B_{new} = (1 + i)B_{previous}.$$

This last result says that balances for successive time periods are computed by multiplying by $1 + i$.

The formula for the balance after any number of interest periods is now easily derived:

Principal	P
Balance after 1 interest period	$(1+i)P$
Balance after 2 interest periods	$(1+i)\cdot(1+i)P$ or $(1+i)^2P$
Balance after 3 interest periods	$(1+i)\cdot(1+i)^2P$ or $(1+i)^3P$
Balance after 4 interest periods	$(1+i)^4P$
$\vdots$	
Balance after n interest periods	$(1+i)^nP$.

Denote the compound amount by the letter F (suggestive of "future value"). Then, the compound amount after n interest periods is given by the formula

$$F=(1+i)^nP, \tag{2}$$

where i is the interest rate per period and P is the principal.

Values of $(1+i)^n$ for specific values of i and n are easily determined using either a calculator or a table. A brief table of useful values has been included as Table 2 of Appendix A. Financial calculators have a designated key to compute $(1+i)^n$. On an ordinary calculator, the following keystrokes will compute the value of $(1+.005)^{24}$.

[1] [+] [.005] [=] [x^y] [24] [=]

On the home screen of a graphing calculator such as the TI-82 or TI-83, use

[(] [1] [+] [.005] [)] [^] [24] [ENTER]

EXAMPLE 3 Apply formula (2) to the savings account passbook discussed at the beginning of this section and calculate the compound amount after 1 year and after 5 years.

Solution The principal P is \$100. Since the interest rate is 4% compounded quarterly, we have $i = 1\%$, or .01. One year consists of four interest periods, so $n = 4$. Therefore, the compound amount after 1 year is

$$F=(1+.01)^4\cdot 100=(1.01)^4\cdot 100=(1.04060401)\cdot 100 \qquad \text{(from Table 2)}$$

$$=104.060401=\$104.06 \qquad \text{(after rounding to the nearest cent).}$$

Five years consists of $n = 5\times 4 = 20$ interest periods. Therefore, the compound amount after 5 years is

$$F=(1.01)^{20}\cdot 100=(1.22019004)\cdot 100=\$122.02.$$

The Excel spreadsheet in Table 3 shows the effects of interest rates (compounded quarterly) on the compound amount. [For those familiar with Excel, the entries in cells B5 and C5 are `$B$1*(1+A5/4)^20` and `$B$1*(1+A5/4)^40`, respectively. The remaining entries in the B and C columns were created with the AutoFill option from the Edit/Fill/Series dialog box. The dollar sign in `$B$1` prevents references to this location from being changed by AutoFill.]

The next example is a variation of the previous one and introduces a new concept, present value.

EXAMPLE 4 How much money must be deposited now in order to have \$1000 after 5 years if interest is paid at a 4% annual rate compounded quarterly?

TABLE 3

	A	B	C
1	Principal	$100.00	
2			
3		Compound Amount	
4	Interest Rate	5 Years	10 Years
5	3.00%	$116.12	$134.83
6	3.50%	$119.03	$141.69
7	4.00%	$122.02	$148.89
8	4.50%	$125.08	$156.44
9	5.00%	$128.20	$164.36
10	5.50%	$131.41	$172.68
11	6.00%	$134.69	$181.40
12	6.50%	$138.04	$190.56
13	7.00%	$141.48	$200.16
14	7.50%	$144.99	$210.23
15	8.00%	$148.59	$220.80

Solution As in Example 3, we have $i = .01$ and $n = 20$. However, now we are given F and are asked to solve for P.

$$F = (1+i)^n P$$
$$1000 = (1.01)^{20} P$$
$$P = \frac{1000}{(1.01)^{20}}.$$

From Table 2 of Appendix A, $(1.01)^{20} = 1.22019004$. However, the cumbersome arithmetic can be avoided by using Table 3 of Appendix A, which tabulates values of $1 \div (1+i)^n$ for various values of i and n.

$$P = \frac{1}{(1.01)^{20}} \cdot 1000 = (.81954447) \cdot 1000 = \$819.54.$$

We say that \$819.54 is the present value of \$1000, 5 years from now, at 4% interest compounded quarterly. ▮

In general, the present value of F dollars at a given interest rate and given length of time is the amount of money P that must be deposited now in order for the compound amount to grow to F dollars in the given length of time. From formula (2), we see that the present value P may be computed from the following formula:

$$P = \frac{1}{(1+i)^n} \cdot F. \tag{3}$$

▮ EXAMPLE 5 Determine the present value of a \$10,000 payment to be received on January 1, 2004, if it is now May 1, 1995, and money can be invested at 6% interest compounded monthly.

Solution Here $F = 10{,}000$, $n = 104$ (the number of months between the two given dates), and $i = \frac{1}{2}\% = .005$. By formula (3),

$$P = \frac{1}{(1+i)^n} \cdot F = (.59529136)10{,}000 = \$5952.91.$$

Therefore, \$5952.91 invested on May 1, 1995, will grow to \$10,000 by January 1, 2004. ▮

The interest that we have been discussing so far is the most prevalent type of interest and is known as *compound interest.* There is another type of interest, called *simple interest*, which is used in some financial circumstances. Let us now discuss this type of interest.

Interest rates for simple interest are given as an annual percentage rate r. Interest is earned *only* on the principal P and the interest is rP for each year. Therefore, the interest earned in n years is nrP. So the amount A after n years is the original amount plus the interest earned. That is,

$$A = P + nrP = 1 \cdot P + nrP = (1 + nr)P. \tag{4}$$

EXAMPLE 6 Calculate the amount after 4 years if \$1000 is invested at 5% simple interest.

Solution Apply formula (4) with $P = \$1000$, $n = 4$, and $r = 5\%$ or .05.

$$A = (1 + nr)P = [1 + 4(.05)]1000 = (1.2)1000 = \$1200.$$

In Example 6, had the money been invested at 5% compound interest with annual compounding, then the compound amount would have been \$1215.51. Money invested at simple interest is earning interest only on the principal amount. However, with compound interest, after the first interest period, the interest is also earning interest. Thus, if we compare two savings accounts, each earning interest at the same stated annual rate, but with one earning simple interest and the other earning compound interest, the latter will grow at a faster rate.

Let us close this section with a summary of the key formulas we have developed so far.

Compound Interest

Compound amount: $F = (1 + i)^n P$

Present value: $P = \left[\dfrac{1}{(1 + i)^n}\right] F$

where i is the interest rate per period and n is the number of interest periods.

Simple Interest

Amount: $A = (1 + nr)P$

where r is the interest rate and n is the number of years.

GC Graphing calculators can easily display successive balances in a savings account. Consider the situation from Table 1 in which \$100 is deposited at 4% interest compounded quarterly. In Fig. 1 successive balances are displayed in the home screen. In Fig. 2 successive balances are graphed, and in Fig. 3 they are displayed in a table.

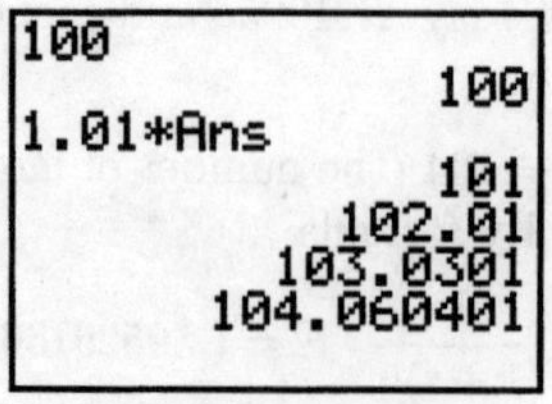

FIGURE 1

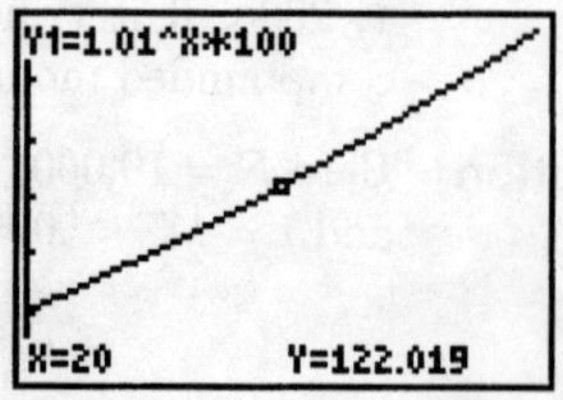

FIGURE 2

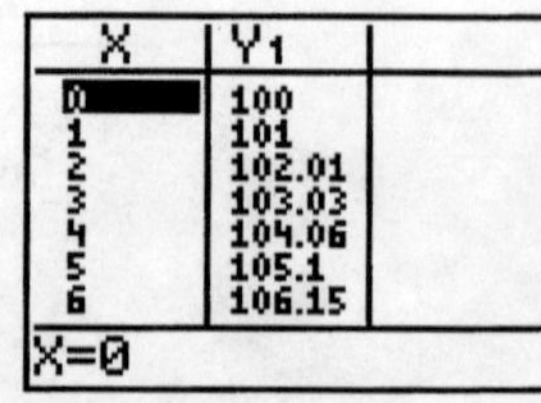

FIGURE 3

Displaying successive balances on the home screen Successive balances can be determined with the relation $B_{new} = (1+i)B_{previous}$. In Fig. 1, after the principal (`100`) is entered, the last displayed value (`100`) is assigned to **`Ans`**. The instruction `1.01*Ans` generates the next balance (`101`) and assigns it to **`Ans`**. Each subsequent press of the ENTER key generates another balance.

Graphs with the TI-82 or TI-83 Figure 4 shows the `Y=` editor, where $\texttt{Y}_1$ has been set equal to `1.01^X*100`; that is, $(1+i)^{\text{X}} \cdot P$. Here `X`, instead of n, is used to represent the number of interest periods. In Fig. 5, the function $\texttt{Y}_1$ is used in two ways to calculate future values (after 40 interest periods) in the home screen. (*Note*: To display the arrow, press STO ▶. To display $\texttt{Y}_1$ with a TI-83, press VARS ▶ **1 1**. With a TI-82 press 2nd [Y-VARS] **1 1**.) In Fig. 6, the WINDOW screen is set up to display the growth of the balance for 10 years, that is, 40 interest periods. `Xscl` was set to 0 since the x-axis is not being displayed. `Ymin` was set less than 100 to allow room for coordinates of points to be displayed at the bottom of the screen. Pressing GRAPH produces the screen of Fig. 2.

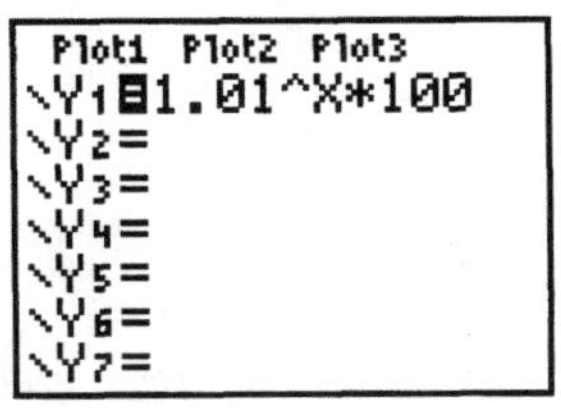

FIGURE 4

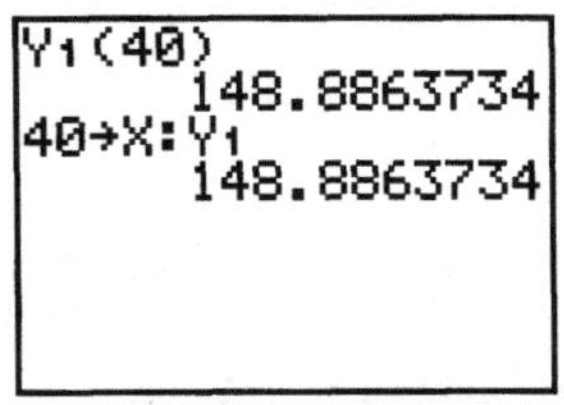

FIGURE 5

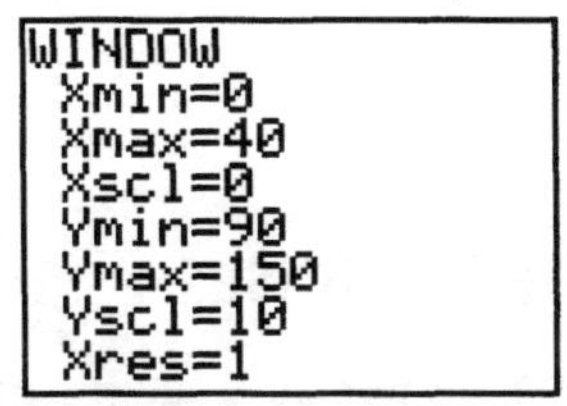

FIGURE 6

FIGURE 7

Tables with the TI-82 or TI-83 Pressing 2nd [TBLSET] brings up the TABLE SETUP screen of Fig. 7. The table determined by the settings will have a column for `X` and a column for the values of each function that is selected in the `Y=` editor. The values for `X` will begin with the setting of `TblStart` and increase by the setting of `ΔTbl`. For our purposes, the settings for `Indpnt` and `Depend` should always be **`Auto`**. After you press 2nd [TABLE] to display the table (see Fig. 3), you can use the down-arrow key to generate further balances.

EXAMPLE 7 Consider the situation from Table 1 in which \$100 is deposited at 4% interest compounded quarterly. Determine when the balance will exceed \$130.

Solution There are three ways to answer this question with a graphing calculator—with the home screen, with a graph, and with a table.

1. (Home screen) Generate successive balances (while keeping track of the number of times the ENTER key has been pressed) until a balance exceeding 130 is reached. See Fig. 8.

2. (Graph) Set $\texttt{Y}_2 = 130$ and find its intersection with the graph of $\texttt{Y}_1 =$ `1.01^X*100`. See Fig. 9.

3. (Table) Scroll down the table for $\texttt{Y}_1$ until the balance exceeds 130. See Fig. 10.

In each case we see that the balance will exceed \$130 after 27 interest periods. ▮

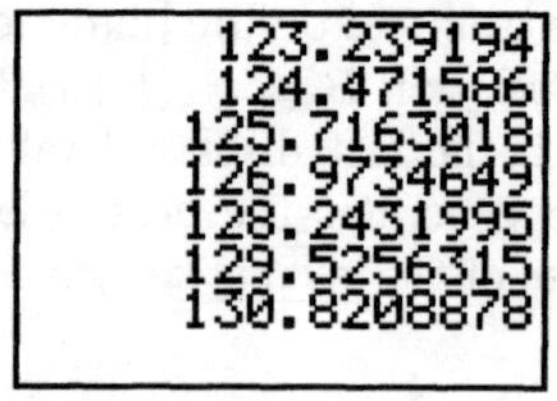

FIGURE 8

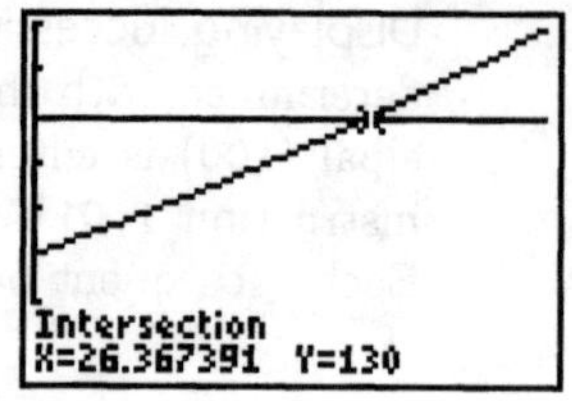

FIGURE 9

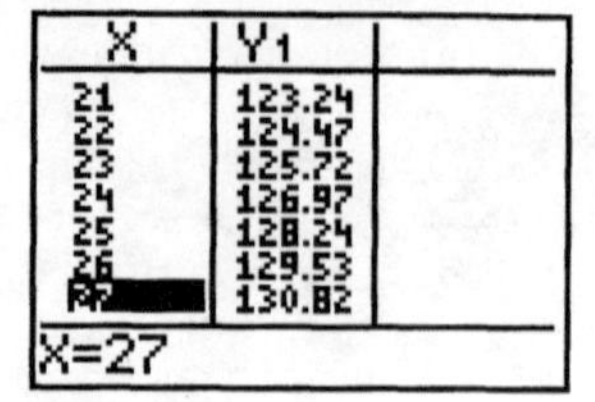

FIGURE 10

Practice Problems 1

1. (a) In Table 3 of Appendix A, look up the value of $1/(1+i)^n$ for $i = 6\%$ and $n = 10$.

   (b) Calculate the present value of \$1000 to be received 10 years in the future at 6% interest compounded annually.

2. Calculate the compound amount after 2 years of \$1 at 26% interest compounded weekly.

3. Calculate the future amount of \$2000 after 6 months if invested at 6% simple interest.

EXERCISES 1

1. Determine i and n for the following situations.
   (a) 12% interest compounded monthly for 2 years
   (b) 8% interest compounded quarterly for 5 years
   (c) 10% interest compounded semiannually for 20 years

2. Determine i and n for the following situations.
   (a) 6% interest compounded annually for 3 years
   (b) 6% interest compounded monthly from January 1, 1998, to July 1, 1999
   (c) 9% interest compounded quarterly from January 1, 1998, to October 1, 1998

3. Determine i, n, P, and F for the following situations.
   (a) \$500 invested at 6% interest compounded annually grows to \$631.24 in 4 years.
   (b) \$800 invested on January 1, 1998, at 6% interest compounded monthly grows to \$1455.52 by January 1, 2008.
   (c) \$6177.88 is deposited on January 1, 1998. The balance on July 1, 2007, is \$9000 and the interest is 4% interest compounded semiannually.

4. Determine i, n, P, and F for each of the following situations.
   (a) The amount of money that must be deposited now at 4.5% interest compounded weekly in order to have \$7500 in 1 year is \$7170.12.
   (b) \$3000 deposited at 6% interest compounded monthly will grow to \$18,067.73 in 30 years.
   (c) In 1626, Peter Minuit, the first director-general of New Netherlands province, purchased Manhattan Island for trinkets and cloth valued at about \$24. Had this money been invested at 8% interest compounded quarterly, it would have amounted to \$150,406,580,000,000 by 1998.

5. Calculate the compound amount of \$1000 after 2 years if deposited at 6% interest compounded monthly.

6. Calculate the present value of \$10,000 payable in 5 years at 4% interest compounded semiannually.

7. Calculate the present value of \$100,000 payable in 25 years at 6% interest compounded monthly.

8. Calculate the compound amount of \$1000 after 1 year if deposited at 7.3% interest compounded daily.

9. Six thousand dollars is deposited in a savings account at 6% interest compounded monthly. Find the balance after 3 years and the amount of interest earned during that time.

10. Two thousand dollars is deposited in a savings account at 4% interest compounded semiannually. Find the balance after 7 years and the amount of interest earned during that time.

11. (a) How much money would have to be deposited now at 6% interest compounded monthly to accumulate to \$4000 in 2 years?
   (b) How much of the \$4000 would be interest?
   (c) Prepare a table showing the growth of the account for the first 3 months.

12. (a) How much money would have to be deposited now at 8% interest compounded quarterly to accumulate to $10,000 in 15 years?

(b) How much of the $10,000 would be interest?

(c) Prepare a table showing the growth of the account for the first 3 quarters.

13. If you had invested $10,000 on January 1, 1994, at 4% interest compounded quarterly, how much would you have on April 1, 2000?

14. In order to have $10,000 on his 25th birthday, how much would a person just turned 21 have to invest if the money will earn 6% interest compounded monthly?

15. Mr. Smith wishes to purchase a $10,000 sailboat upon his retirement in 3 years. He has just won the state lottery and would like to set aside enough cash in a savings account paying 4% interest compounded quarterly to buy the boat upon retirement. How much should he deposit?

16. Ms. Jones has just invested $100,000 at 6% interest compounded annually. How much money will she have in 20 years?

17. Consider the following savings account passbook:

	Date		
	1/1/98	*2/1/98*	*3/1/98*
Deposit	$1000.00		
Withdrawal			
Interest		$5.00	$5.03
Balance	$1000.00	$1005.00	$1010.03

(a) What annual interest rate is this bank paying?

(b) Give the interest and balance on 4/1/98.

(c) Give the interest and balance on 1/1/2000.

18. Consider the following savings account passbook:

	Date		
	1/1/97	*7/1/97*	*1/1/98*
Deposit	$10,000.00		
Withdrawal			
Interest		$200.00	$204.00
Balance	$10,000.00	$10,200.00	$10,404.00

(a) Give the interest and balance on 7/1/98.

(b) Give the interest and balance on 1/1/2006.

19. Is it more profitable to receive $1000 now or $1700 in 9 years? Assume that money can earn 6% interest compounded annually.

20. Is it more profitable to receive $7000 now or $10,000 in 9 years? Assume that money can earn 4% interest compounded quarterly.

21. In 1995 the Dow Jones industrial average grew at an interest rate of about 26% compounded weekly. Is this interest rate better or worse than 30% compounded annually?

22. Would you rather earn 8% interest compounded annually or 7.3% interest compounded daily?

23. On January 1, 1995, a deposit was made into a savings account paying interest compounded quarterly. The balance on January 1, 1998, was $10,000 and the balance on April 1, 1998, was $10,100. How large was the deposit?

24. During the 1970s a deposit was made into a savings account paying 4% interest compounded quarterly. On January 1, 1998, the balance was $2020. What was the balance on October 1, 1997?

Exercises 25–36 concern simple interest.

25. Determine r, n, P, and A for each of the following situations.

(a) $500 invested at 4% simple interest grows to $510.00 in 6 months.

(b) In order to have $550 after 2 years at 5% simple interest, $500 must be deposited.

26. Determine r, n, P, and A for each of the following situations.

(a) At 6% simple interest, $1000 deposited on January 1, 1998, will be worth $1040 on September 1, 1998.

(b) At 4% simple interest, in order to have $3600 in 5 years, $3000 must be deposited now.

27. Calculate the amount after 3 years if $1000 is deposited at 5% simple interest.

28. Calculate the amount after 18 months if $2000 is deposited at 6% simple interest.

29. Find the present value of $3000 in 2 years at 10% simple interest.

30. Find the present value of $2000 in 4 years at 7% simple interest.

31. Determine the (simple) interest rate at which \$980 grows to \$1000 in 6 months.

32. How many years are required for \$1000 to grow to \$1210 at 7% simple interest?

33. Determine the amount of time required for money to double at 5% simple interest.

34. Derive the formula for the (simple) interest rate at which P dollars grows to A dollars in n years. That is, express r in terms of P, A, and n.

35. Derive the formula for the present value of A dollars in n years at simple interest rate r.

36. Derive the formula for the number of years required for P dollars to grow to A dollars at simple interest rate r.

37. Suppose a principal of \$100 is deposited for 5 years in a savings account at 6% interest compounded semiannually. Which of items (a)–(d) can be used to fill in the blank in the following statement? (*Note*: Before computing, use your intuition to guess the correct answer.)

If the ______________ is doubled, then the compound amount will double.

(a) principal

(b) interest rate

(c) number of interest periods per year

(d) number of years

38. Suppose a principal of \$100 is deposited for 5 years in a savings account at 6% simple interest. Which of items (a)–(d) can be used to fill in the blank in the following statement? (*Note*: Before computing, use your intuition to guess the correct answer.)

If the ______________ is doubled, then the amount will double.

(a) principal

(b) interest rate

(c) number of years

39. Compute the compound amount after 1 year for \$100 invested at 4% interest compounded quarterly. What simple interest rate will yield the same amount in 1 year?

40. Compute the compound amount after 1 year for \$100 invested at 6% interest compounded monthly. What simple interest rate will yield the same amount in 1 year?

Suppose that interest is compounded. Then the annual percentage rate is called the *nominal rate*. The simple interest rate that yields the same amount after 1 year is called the *effective rate*. For example, refer to Exercise 39. The nominal rate there is 4% and corresponds to an effective rate of 4.06%.

41. Calculate the effective rate for 4% interest compounded semiannually.

42. Calculate the effective rate for 26% interest compounded weekly.

43. Calculate the effective rate for 6% interest compounded monthly.

44. Calculate the effective rate for 8% interest compounded quarterly.

45. Suppose that an investment earns a nominal rate of r (expressed as a decimal) compounded twice a year. What is the effective rate?

46. Suppose that an investment of P dollars earns a nominal rate of r compounded n times per year. What is the compound amount after 1 year?

47. Suppose that an investment of P dollars earns a nominal rate of r compounded n times per year. What is the effective rate?

Exercises 48–51 require the use of a graphing calculator or a computer.

48. One thousand dollars is deposited into a savings account at 6.5% interest compounded annually. What are the balances after 1, 2, and 3 years? How many years are required for the balance to reach \$1655? After how many years will the balance exceed \$2000?

49. Ten thousand dollars is deposited into a savings account at 6% interest compounded monthly. What are the balances after 5, 10, and 15 months? How many months are required for the balance to reach \$11,614? After how many months will the balance exceed \$13,000?

50. Tom invests \$100,000 at 8% interest compounded annually. When will Tom be a millionaire?

51. How many years are required for \$100 to double if deposited at 6% interest compounded quarterly?

Solutions to Practice Problems 1

1. (a) Look up the entry in Table 3 in Appendix A in the row labeled "10" and the column labeled 6%. That entry is .55839478.

   (b) Here we are given the value in the future $F = \$1000$ and are asked to find the present value P. Interest compounded annually has just one interest period per year, so $n = 10 \cdot 1 = 10$ and $i = 6\%/1 = 6\%$. By formula (3),

$$\begin{aligned} P &= \frac{1}{(1+i)^n}F \\ &= \frac{1}{(1+i)^n} \cdot 1000 && \text{(with } i = 6\% \text{ and } n = 10) \\ &= (.55839478) \cdot 1000 && \text{[from part (a)]} \\ &= 558.39478 \\ &= \$558.39 && \text{(rounding to the nearest cent).} \end{aligned}$$

2. Here we are given the present value, $P = 1$, and are asked to find the value F at a future time. Interest compounded weekly has 52 interest periods each year, so $n = 2 \times 52 = 104$ and $i = 26\%/52 = \frac{1}{2}\%$. By formula (2), we have

$$\begin{aligned} F &= (1+i)^n P \\ &= (1+i)^n \cdot 1 && \text{(with } i = 1/2\% \text{ and } n = 104) \\ &= (1.67984969) \cdot 1 && \text{(using Table 2)} \\ &= \$1.68. \end{aligned}$$

   [*Note*: In general, whenever $P = \$1$, then $F = (1+i)^n \cdot 1 = (1+i)^n$. This explains why $(1+i)^n$ is often referred to as the *compound amount* of \$1.]

3. In simple interest problems, time should be expressed in terms of years. Since 6 months is $\frac{1}{2}$ of a year, formula (4) gives

$$A = (1 + nr)P = [1 + \tfrac{1}{2}(.06)]2000 = (1.03)2000 = \$2060.$$

10.2 Annuities

An annuity is a sequence of equal payments made at regular intervals of time. Here are two illustrations.

1. As the proud parent of a newborn daughter, you decide to save for her college education by depositing \$100 at the end of each month into a savings account paying 6% interest compounded monthly. Eighteen years from now, after you make the last of 216 payments, the account will contain \$38,735.32.

2. Having just won the state lottery, you decide not to work for the next 5 years. You want to deposit enough money into the bank so that you can withdraw \$5000 at the end of each month for 60 months. If the bank pays 6% interest compounded monthly, you must deposit \$258,627.80.

The payments in the foregoing financial transactions are called *rent*. The amount of a typical rent payment is denoted by the letter R. Thus, in the examples above, we have $R = \$100$ and $R = \$5000$, respectively.

In illustration 1, you make equal payments to a bank in order to generate a large sum of money in the future. This sum, namely \$38,735.32, is called the *future value of the annuity*.

In illustration 2, the bank will make equal payments to you in order to pay back the sum of money that you currently deposit. The value of the current deposit, namely \$258,627.80, is called the *present value* of the annuity.

We now derive formulas for future values and present values of annuities. Suppose that an annuity like that of illustration 1 consists of a sequence of n equal payments, each of R dollars. Suppose that the annuity payments are deposited into an account paying compound interest at the rate of i per interest period. We will further suppose that there is a single annuity payment per interest period and that the payment is made at the end of the interest period. Let us derive a formula for the future value of the annuity: that is, a formula for the balance of the account immediately after the last payment.

TABLE 1

Payment	*Amount*	*Number of interest periods on deposit*	*Compound amount*
1	R	$n-1$	$(1+i)^{n-1}R$
2	R	$n-2$	$(1+i)^{n-2}R$
$\vdots$	$\vdots$	$\vdots$	$\vdots$
$n-2$	R	2	$(1+i)^2R$
$n-1$	R	1	$(1+i)R$
n	R	0	R

Each payment accumulates interest for a different number of interest periods, so let us calculate the balance in the account as the sum of n compound amounts, one corresponding to each payment (Table 1). Denote the future value of the annuity by F. Then F is the sum of the numbers in the right-hand column:

$$F = R + (1+i)R + (1+i)^2R + (1+i)^3R + \cdots + (1+i)^{n-1}R.$$

A compact expression for F can be obtained by multiplying both sides of this equation by $(1 + i)$ and then subtracting the original equation from the new equation.

$$\begin{aligned}(1+i)F &= \phantom{R+{}} (1+i)R + (1+i)^2R + (1+i)^3R + \cdots + (1+i)^{n-1}R + (1+i)^nR \\ F &= R + (1+i)R + (1+i)^2R + (1+i)^3R + \cdots + (1+i)^{n-1}R \\ \hline iF &= (1+i)^nR - R\end{aligned}$$

The last equation can be written $iF = [(1+i)^n - 1] \cdot R$. Dividing both sides by i yields

$$F = \frac{(1+i)^n - 1}{i} \cdot R.$$

The expression $[(1+i)^n - 1]/i$ occurs often in financial analysis and is denoted by the special symbol $s_{\overline{n}|i}$ (read "s sub n angle i"). We may summarize our calculations above as follows.

Suppose that an annuity consists of n payments of R dollars each, deposited at the ends of consecutive interest periods into an account with interest compounded at a rate i per period. Then the future value F of the annuity is given by the formula

$$F = s_{\overline{n}|i}R. \quad (1)$$

Values of $s_{\overline{n}|i}$ may be computed using a calculator or looked up in an appropriate table. We have included a brief table as Table 4 of Appendix A. Financial calculators have a designated key to compute $s_{\overline{n}|i}$. On a typical ordinary calculator, the following keystrokes will compute the value of $s_{\overline{24}|.005}$.

[1.005] [x^y] [24] [=] [−] [1] [=] [÷] [.005] [=]

On the home screen of a graphing calculator such as the TI-82 or TI-83, use

[(] [(] [1.005] [^] [24] [)] [−] [1] [)] [÷] [.005] [ENTER]

EXAMPLE 1 Calculate the future value of an annuity of \$100 per month for 5 years at 6% interest compounded monthly.

Solution Here $R = 100$ and $i = \frac{1}{2}\%$. Since a payment is made at the end of each month for 5 years, there will be $5 \times 12 = 60$ payments. So $n = 60$. Therefore,

$$F = s_{\overline{n}|i}R = s_{\overline{60}|1/2\%} \cdot 100 = (69.77003051) \cdot 100 \qquad \text{(by Table 4)}$$
$$= \$6977.00.$$

We can easily derive a formula that illustrates exactly how the future value of an annuity changes from period to period. Each new balance can be computed from the previous balance with the formula

$$B_{new} = (1 + i)B_{previous} + R.$$

That is, the new balance equals the growth of the previous balance due to interest, plus the amount paid. So, for instance, if B_1 is the balance after the first payment is made, B_2 is the balance after the second payment is made, and so on, then

$$B_1 = R$$
$$B_2 = (1 + i)B_1 + R$$
$$B_3 = (1 + i)B_2 + R$$
$$\vdots$$

Successive unpaid balances are computed by multiplying by $(1 + i)$ and adding R.

If \$$R$ is deposited at the end of each interest period at interest rate i per period,

$$B_{new} = (1 + i)B_{previous} + R \quad (2)$$

can be used to calculate each new balance, B_{new}, from the previous balance, $B_{previous}$.

EXAMPLE 2 Consider the annuity of Example 1. Determine the future value after 61 months.

Solution From Example 1, $i = .005$, the balance after 60 months is \$6977.00, and $R = 100$. Therefore, by (2)

$$\begin{aligned}[\text{balance after 61 months}] &= (1 + .005)[\text{balance after 60 months}] + 100 \\ &= 1.005(6977) + 100 = 7111.89.\end{aligned}$$

Therefore, the annuity will have accumulated to \$7111.89 after 61 months. ▮

Formula (1) can also be used to determine the rent necessary to achieve a certain future value:

$$F = s_{\overline{n}|i} R$$

$$R = \frac{F}{s_{\overline{n}|i}} = \frac{1}{s_{\overline{n}|i}} F.$$

Thus we have established the following result.

> Suppose that an annuity of n payments has future value F and has interest compounded at the rate i per period. Then the rent R is given by
>
> $$R = \frac{1}{s_{\overline{n}|i}} \cdot F. \qquad (3)$$

Table 5 of Appendix A gives values of $1/s_{\overline{n}|i}$ for various values of n and i.

EXAMPLE 3 Ms. Adams would like to buy a \$30,000 airplane when she retires in 8 years. How much should she deposit at the end of each half-year into an account paying 4% interest compounded semiannually so that she will have enough money to purchase the airplane?

Solution Here $n = 16$, $i = 2\%$, and $F = 30{,}000$. Therefore,

$$\begin{aligned}R = \frac{1}{s_{\overline{n}|i}} \cdot F = \frac{1}{s_{\overline{16}|2\%}} \cdot 30{,}000 &= (.05365013) \cdot 30{,}000 \qquad \text{(by Table 5)} \\ &= \$1609.50.\end{aligned}$$

She should deposit \$1609.50 at the end of each half-year period. ▮

The *present value* of an annuity like that of illustration 2 is the amount of money it would take to finance the sequence of annuity payments. More specifically, the present value of an annuity is the amount you would need to deposit in order to provide the desired sequence of annuity payments and leave a balance of zero at the end of the term. Let us now find a formula for the present value of an annuity. There are two ways to proceed. On the one hand, the desired present value could be computed as the sum of the present values of the various annuity payments. This computation would make use of the formula for the sum of the first n terms of a geometric progression. However, there is a much "cleaner" derivation that proceeds indirectly to obtain a formula relating the present value P and the rent R.

As before, let us assume that our annuity consists of n payments made at the ends of interest periods, with interest compounded at a rate i per interest period.

***Situation 1*:** Suppose that the P dollars were just left in the account and that the annuity payments were not withdrawn. At the end of the n interest periods, there would be $(1 + i)^n P$ dollars in the account.

Situation 2: Suppose that the payments are withdrawn but are immediately redeposited into another account having the same rate of interest. At the end of the n interest periods, there would be $s_{\overline{n}|i}R$ dollars in the new account.

In both of these situations, P dollars is deposited and it, together with all the interest generated, is earning income at the same interest rate for the same amount of time. Therefore, the final amounts of money in the accounts should be the same. That is,

$$(1+i)^n P = s_{\overline{n}|i}R \qquad P = \frac{s_{\overline{n}|i}}{(1+i)^n} \cdot R.$$

Let us now substitute the value of $s_{\overline{n}|i}$, namely $[(1+i)^n - 1]/i$, into the formula above. If we denote by $a_{\overline{n}|i}$ the expression

$$a_{\overline{n}|i} = \frac{(1+i)^n - 1}{i(1+i)^n},$$

then the formula above may be written in the simple form

$$P = a_{\overline{n}|i}R \qquad \text{or} \qquad R = \frac{1}{a_{\overline{n}|i}}P.$$

Tables 6 and 7 of Appendix A give, respectively, the values of $a_{\overline{n}|i}$ and $1/a_{\overline{n}|i}$ for various values of n and i. Let us record the main result of our preceding discussion.

> The present value P and the rent R of an annuity of n payments with interest compounded at a rate i per interest period are related by the formulas
>
> $$P = a_{\overline{n}|i}R, \qquad R = \frac{1}{a_{\overline{n}|i}}P.$$

EXAMPLE 4 How much money must you deposit now at 6% interest compounded quarterly in order to be able to withdraw \$3000 at the end of each quarter year for 2 years?

Solution Here $R = 3000$, $i = 1.5\%$, and $n = 8$. We are asked to calculate the present value of the sequence of payments.

$$P = a_{\overline{n}|i}R = a_{\overline{8}|1.5\%} \cdot 3000 = (7.48592508) \cdot 3000 \qquad \text{(by Table 6)}$$
$$= \$22{,}457.78.$$

The Excel spreadsheet in Table 2 shows the effect of interest rates (compounded quarterly) on the compound amount where the duration of the annuity is two years. [For those familiar with Excel, the entry in cell B5 was set to `=PV(A5/4,8,-$B$1)`. The form of the present value function is `PV(`*`rate,nper, pmt`*`)`. The payment was specified as a negative number since money is being withdrawn from the account. The remainder of the B column was created with the AutoFill option from the Edit/Fill/Series dialog box. The dollar sign in `$B$1` prevents references to this location from being changed by AutoFill.]

EXAMPLE 5 If you deposit \$10,000 into a fund paying 6% interest compounded monthly, how much can you withdraw at the end of each month for 1 year?

Solution Here $P = 10{,}000$, $i = \frac{1}{2}\%$, and $n = 12$. We are asked to calculate the rent for the sequence of payments.

$$R = \frac{1}{a_{\overline{n}|i}}P = \frac{1}{a_{\overline{12}|1/2\%}} \cdot 10{,}000 = (.08606643) \cdot 10{,}000 \qquad \text{(by Table 7)}$$
$$= \$860.66.$$

TABLE 2

	A	B
1	Rent	$3,000.00
2		
3	Interest	Present
4	rate	value
5	3.00%	$23,209.84
6	3.50%	$23,081.91
7	4.00%	$22,955.03
8	4.50%	$22,829.19
9	5.00%	$22,704.37
10	5.50%	$22,580.57
11	6.00%	$22,457.78
12	6.50%	$22,335.97
13	7.00%	$22,215.16
14	7.50%	$22,095.32
15	8.00%	$21,976.44

Remark In this section we have considered only annuities with payments made at the end of each interest period. Such annuities are called *ordinary annuities.* Annuities that have payments at the beginning of the interest period are called *annuities due.* Annuities whose payment period is different from the interest period are called *general annuities.*

GC As a time-saving device, the formulas for $s_{\overline{n}|i}$, $\frac{1}{s_{\overline{n}|i}}$, $a_{\overline{n}|i}$, and $\frac{1}{a_{\overline{n}|i}}$ can be assigned to functions in the `Y=` editor. In Fig. 1, the formulas have been stored in $\texttt{Y}_4$ through $\texttt{Y}_7$ to correspond to the table numbers in Appendix A. Here `X`, instead of n, is used to represent the number of periods. (*Note*: To display $\texttt{Y}_4$ with a TI-83, press [VARS] [▶] **1 4**. With a TI-82, press [2nd] [Y-VARS] **1 4**.) These functions have been deselected since we want to use them, but not graph them. Figure 2 shows two ways the value of $s_{\overline{60}|1/10\%}$ can be displayed on the home screen.

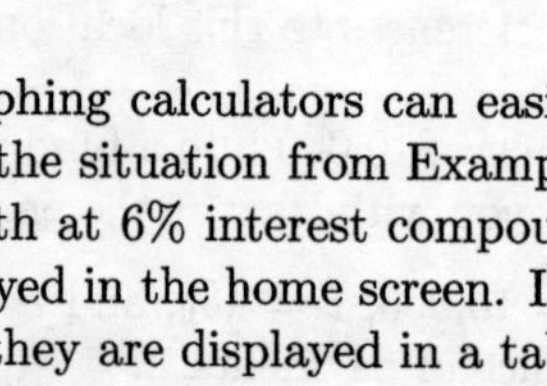

FIGURE 1

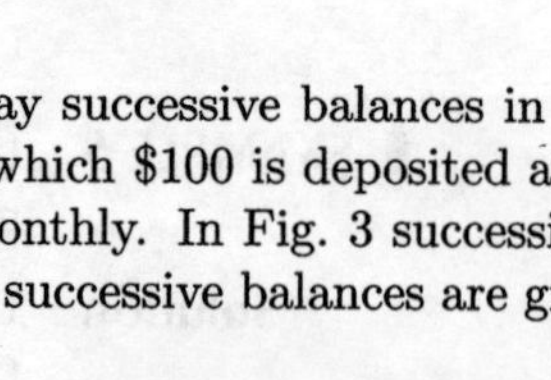

FIGURE 2

Graphing calculators can easily display successive balances in an annuity. Consider the situation from Example 1 in which $100 is deposited at the end of each month at 6% interest compounded monthly. In Fig. 3 successive balances are displayed in the home screen. In Fig. 4 successive balances are graphed, and in Fig. 5 they are displayed in a table.

Displaying successive balances on the home screen as in Fig. 3 Successive balances can be determined with the relation $B_{new} = (1+i)B_{previous} + R$. After the first payment (100) is entered, the last displayed value (100) is assigned to `Ans`.

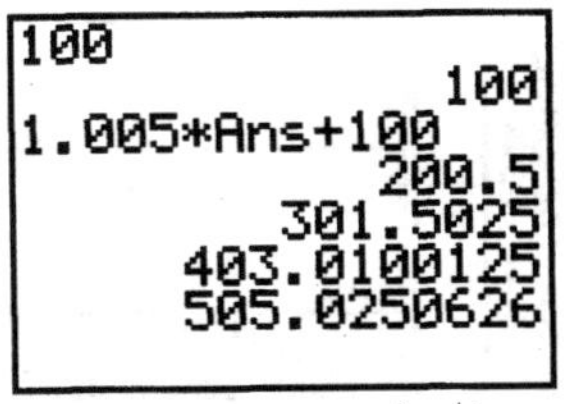

FIGURE 3

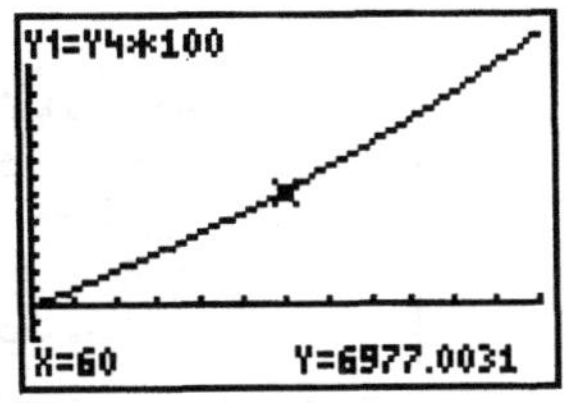

FIGURE 4

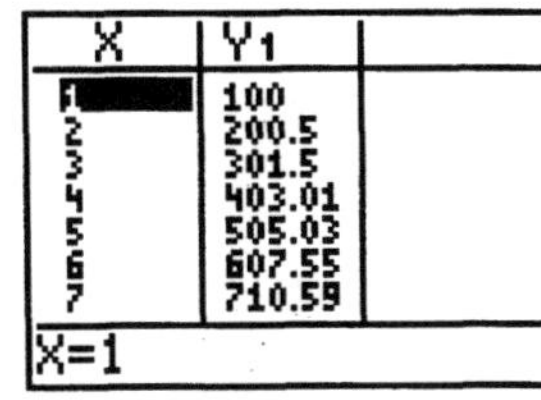

FIGURE 5

The instruction `1.005*Ans+100` generates the next balance (`200.5`) and assigns it to `Ans`. Each subsequent press of the [ENTER] key generates another balance.

Graphing successive balances with the TI-82 or TI-83 as in Fig.4 Figure 6 shows the `Y=` editor, where Y_1 has been set equal to `Y4*100`, that is, $s_{\overline{n}|i}\cdot R$. In Fig. 7 the value `.005` is assigned to `I` from the home screen. (*Note*: To display the arrow, press [STO ▶].) In Fig. 8 the WINDOW screen is set up to display the growth of the balance for 10 years, that is, 120 months. `Ymin` was set to a negative value to allow room for coordinates of points to be displayed at the bottom of the screen. Pressing [GRAPH] produces the screen of Fig. 4.

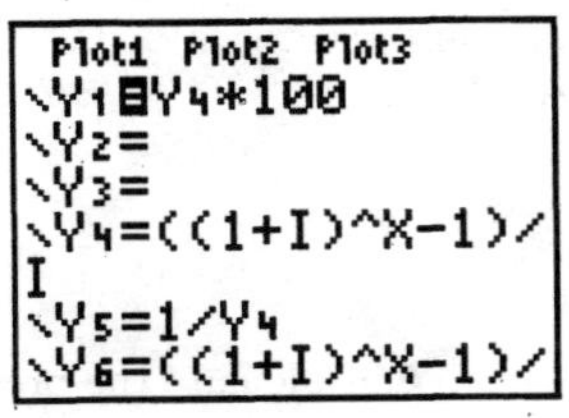

FIGURE 6

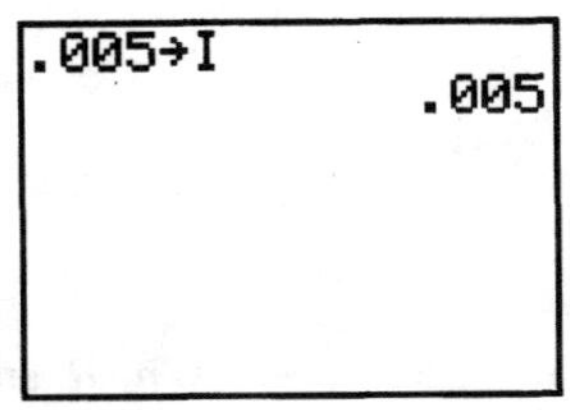

FIGURE 7

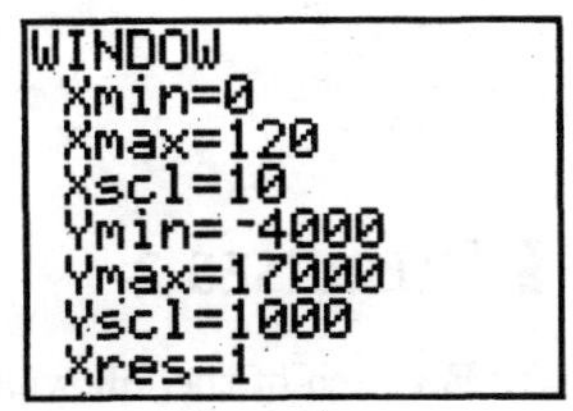

FIGURE 8

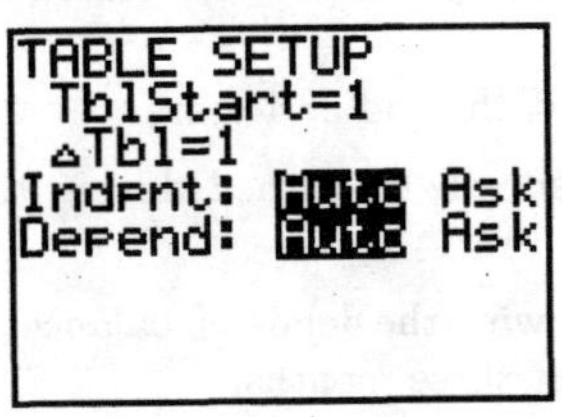

FIGURE 9

Obtaining a table of successive balances with the TI-82 or TI-83 as in Fig. 5 Pressing [2nd] [TBLSET] brings up the TABLE SETUP screen of Fig. 9. The table determined by the settings will have a column for `X` and a column for the values of each function that is selected in the `Y=` editor. The values for `X` will begin with the setting of `TblStart` (`TblMin` on TI-82) and increase by the setting of `ΔTbl`. For our purposes, the settings for `Indpnt` and `Depend` should always be `Auto`. After you press [2nd] [TABLE] to display the table (see Fig. 5), you can use the down-arrow key to generate further balances.

EXAMPLE 6 Consider the situation from Example 1 in which \$100 is deposited monthly at 6% interest compounded monthly. Determine when the balance will exceed \$1000.

Solution There are three ways to answer this question with a graphing calculator—with the home screen, with a graph, and with a table.

1. (Home screen) Generate successive balances (while keeping track of the number of times the [ENTER] key has been pressed) until a balance exceeding 10,000 is reached. See Fig. 10.
2. (Graph) Set $Y_2 = 10000$ and find its intersection with the graph of $Y_1 =$ Y_4`*100`. See Fig. 11.
3. (Table) Scroll down the table for Y_1 until the balance exceeds 10,000. See Fig. 12.

In each case we see that the balance will exceed \$10,000 after 82 months; that is, after 6 years and 10 months. ▮

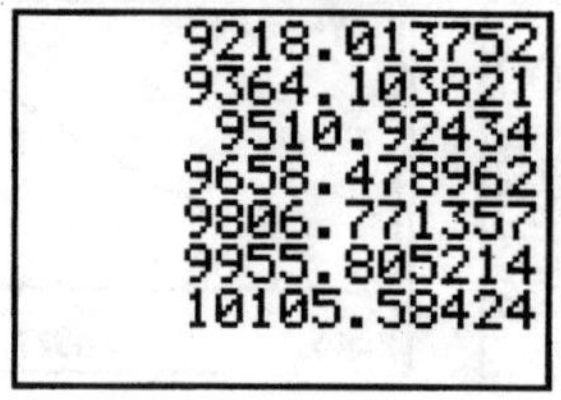

FIGURE 10

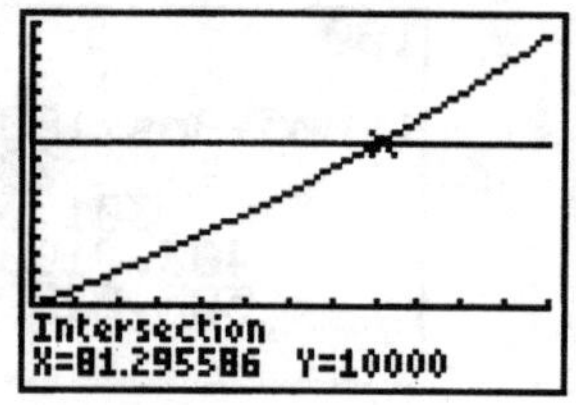

FIGURE 11

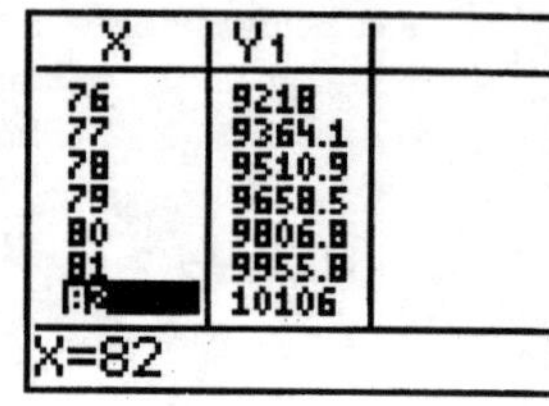

FIGURE 12

Practice Problems 2

Decide whether or not each of the following annuities is an ordinary annuity, that is, the type of annuities considered in this section. If so, identify n, i, and R and calculate the present value or the future value, whichever is appropriate.

1. You make a deposit at 9% interest compounded monthly into a fund that pays you \$1 at the end of each month for 5 years.
2. At the end of each week for 2 years you deposit \$10 into a savings account earning 6% interest compounded monthly.
3. At the end of each month for 2 years, you deposit \$10 into a savings account earning 6% interest compounded monthly.

EXERCISES 2

1. For each of the following annuities, specify i, n, R, and F.

(a) If at the end of each month, \$50 is deposited into a savings account paying 6% interest compounded monthly, the balance after 10 years will be \$8193.97.

(b) Mr. Smith is saving to buy a \$65,000 yacht in the year 2000. Since 1990, he has been depositing \$2675.19 at the end of each half-year into a fund paying 4% interest compounded semiannually.

2. For each of the following annuities, specify i, n, R, and P.

(a) A retiree deposits \$72,582 into a bank paying 6% interest compounded monthly and withdraws \$520 at the end of each month for 20 years.

(b) In order to receive \$700 at the end of each quarter-year from 1998 until 2003, Ms. Jones deposited \$11,446 into a savings account paying 8% interest compounded quarterly.

3. Calculate the future value of an annuity of \$100 per month for 5 years at 6% interest compounded monthly.

4. Calculate the rent of an annuity at 4% interest compounded semiannually if payments are made every half-year and the future value after 7 years is \$10,000.

5. Calculate the rent of an annuity at 8% interest compounded quarterly with payments made every quarter-year for 7 years and present value \$100,000.

6. Calculate the present value of an annuity of \$1000 per year for 10 years at 6% interest compounded annually.

7. A person deposits \$500 into a savings account at the end of every month for 4 years at 6% interest compounded monthly.

(a) Find the balance at the end of 4 years.

(b) How much interest will be earned during the 4 years?

(c) Prepare a table showing the deposits, balance, and interest for the first three months.

8. A person deposits \$800 into a savings account at the end of every quarter-year for 5 years at 6% interest compounded quarterly.

(a) Find the balance at the end of 5 years.

(b) How much interest will be earned during the 5 years?

(c) Prepare a table showing the deposits, balance, and interest for the first three quarters.

9. How much must Jim save each month in order to have \$12,000 to buy a new car in 3 years if the interest rate is 6% compounded monthly? How much of the \$12,000 does Jim actually deposit and how much of it is interest?

10. Valerie needs \$5000 3 years from now in order to pay off a loan. How much must she save each quarter for the next 3 years if interest rates are 8% compounded quarterly? How much of the \$5000 will be interest?

11. When Bridget takes a new job, she is offered a $2000 bonus now or the option of an extra $200 each month for the next year. If interest rates are 6% compounded monthly, which choice is better and by how much?

12. When Michael started college, his father gave him $5000 deposited in a savings account earning 6% interest compounded monthly. Michael withdraws $500 a month starting at the beginning of the second month. Prepare a schedule showing the monthly balance in the account for the first four months.

13. A city has a debt of $1,000,000 falling due in 15 years. How much money must it deposit at the end of each half-year into a savings fund at 4% interest compounded semiannually in order to raise this amount?

14. On January 1, 1997, Tom decided to save for exactly 1 year for a 10-speed bike by depositing $10 at the end of each month into a savings account paying 6% interest compounded monthly. How much did he accumulate?

15. During Jack's first year at college, his father had been sending him $100 per month for incidental expenses. For the sophomore year, his father decided instead to make a deposit into a savings account on August 1 and have his son withdraw $100 on the first of each month from September 1 to May 1. If the bank pays 6% interest compounded monthly, how much should Jack's father deposit?

16. Suppose that a magazine subscription costs $9 per year and that you receive a magazine at the end of each month. At an interest rate of 6% compounded monthly, how much are you actually paying for each issue?

17. Is it more profitable to receive $1000 at the end of each month for 10 years or to receive a lump sum of $163,000 at the end of 10 years? Assume that money can earn 6% interest compounded monthly.

18. Is it more profitable to receive a lump sum of $10,000 at the end of 3 years or to receive $750 at the end of each quarter-year for 3 years? Assume that money can earn 8% interest compounded quarterly.

19. Suppose that you deposited $1000 into a savings account on January 1, 1990, and deposited an additional $100 into the account at the end of each quarter-year. If the bank pays 8% interest compounded quarterly, how much will be in the account on January 1, 1999?

20. Suppose that you opened a savings account on January 1, 1995, and made a deposit of $100. In 1996 you began depositing $10 into the account at the end of each month. If the bank pays 6% interest compounded monthly, how much money will be in the account on January 1, 2000?

21. Ms. Jones deposited $100 at the end of each month for 10 years into a savings account earning 6% interest compounded monthly. However, she deposited an additional $1000 at the end of the seventh year. How much money was in the account at the end of the tenth year?

22. Redo Exercise 21 for the situation where Ms. Jones withdrew $1000 at the end of the seventh year instead of depositing it.

23. Suppose you deposit $1200 per year for 5 years into an annuity at 4% interest compounded semiannually. Assume that equal amounts are deposited at the end of each interest period. Which of items (a)–(d) can be used to fill in the blank in the following statement? (*Note*: Before computing, use your intuition to guess the correct answer.)

    If the ____________ doubles, then the amount accumulated will double.

    (a) total amount deposited each year

    (b) interest rate

    (c) number of interest periods per year

    (d) number of years

24. Suppose you deposit $100,000 into an annuity at 4% interest compounded semiannually and withdraw an equal amount at the end of each interest period so that the account is depleted after 10 years. Which of items (a)–(d) can be used to fill in the blank in the following statement? (*Note*: Before computing, use your intuition to guess the correct answer.)

    If the ____________ doubles, then the total amount withdrawn each year will double.

    (a) amount deposited

    (b) interest rate

    (c) number of interest periods per year

    (d) number of years

A *perpetuity* is an annuity whose payments are to continue forever. Exercises 25 and 26 concern such annuities.

25. A grateful alumnus decides to donate a permanent scholarship of $1200 per year. How much money should be deposited in the bank at 5% interest compounded annually in order to be able to supply the money for the scholarship at the end of each year?

26. Show that to establish a perpetuity paying R dollars at the end of each interest period, it requires a deposit of R/i dollars, where i is the interest rate per interest period.

A *deferred annuity* is an annuity whose term is to start at some future date. Exercises 27–30 concern such annuities.

27. On his tenth birthday, a boy inherits $10,000, which is to be used for his college education. The money is deposited into a trust fund that will pay him R dollars on his 18th, 19th, 20th, and 21st birthdays. Find R if the money earns 6% interest compounded annually.

28. Refer to Exercise 27. Find the size of the inheritance that would result in \$10,000 per year during the college years (ages 18–21, inclusive).

29. On December 1, 1990, a philanthropist set up a permanent trust fund to buy Christmas presents for needy children. The fund will provide \$6000 each year beginning on December 1, 2000. How much must have been set aside in 1990 if the money earns 6% interest compounded annually?

30. Show that the rent paid by a deferred annuity of n payments that are deferred by m interest periods is given by the formula

$$R = \frac{i(1+i)^{n+m}}{(1+i)^n - 1} \cdot P.$$

31. One dollar is deposited in a savings account with an interest rate of i per interest period. At the end of each interest period, the earned interest is withdrawn from the account and deposited into a second account earning the same rate of interest.

(a) How much money will be in the first account after n interest periods?

(b) How much money will be in the second account after n interest periods?

(c) Since both the original deposit and the interest are all on deposit and earning interest throughout the entire n interest periods, the amounts in parts (a) and (b) must add up to $(1+i)^n$. Use this fact to derive

$$s_{\overline{n}|i} = \frac{(1+i)^n - 1}{i}.$$

32. Show that $\dfrac{1}{a_{\overline{n}|i}} - \dfrac{1}{s_{\overline{n}|i}} = i.$

33. Show that $s_{\overline{n+1}|i} = (1+i)s_{\overline{n}|i} + 1.$

34. One hundred dollars is deposited in the bank at the end of each year for 20 years. During the first 5 years, the bank paid 6% interest compounded annually and after that paid 7% interest compounded annually. Show that the account balance after 20 years is

$$(1.07)^{15} \cdot 100 s_{\overline{5}|6\%} + 100 s_{\overline{15}|7\%}.$$

35. A municipal bond pays 8% interest compounded semiannually on its face value of \$5000. The interest is paid at the end of every half-year period. Fifteen years from now the face value of \$5000 will be returned. The current market interest rate is 12% compounded semiannually. How much should you pay for the bond?

36. A businessperson wishes to lend money for a second mortgage on some local real estate. Suppose that the mortgage pays \$500 per month for a 5-year period. Suppose that you can invest your money in certificates of deposit paying 6% interest compounded monthly. How much should you offer for the mortgage?

37. Suppose that a business note for \$50,000 carries an interest rate of 18% compounded monthly. Suppose that the business pays only interest for the first 5 years and then repays the loan amount plus interest in equal monthly installments for the next 5 years.

(a) Calculate the monthly payments during the second 5-year period.

(b) Assume that the current market interest rate for such loans is 12%. How much should you be willing to pay for such a note?

38. A lottery winner is to receive \$1000 a month for the next 5 years. How much is this sequence of payments worth today if interest rates are 6% compounded monthly? How is the difference between this amount and the \$60,000 paid out beneficial to the agency running the lottery?

Exercises 39–42 require the use of a graphing calculator or a computer.

39. A person deposits \$1000 at the end of each year into an annuity earning 5% interest compounded annually. What are the balances (that is, future values) after 1, 2, and 3 years? How many years are required for the balance to reach \$30,539? After how many years will the balance exceed \$50,000?

40. A person deposits \$700 at the end of each quarter-year into an annuity earning 8% interest compounded quarterly. What are the balances (that is, future values) after 1, 2, and 3 quarters? How many quarters are required for the balance to reach \$5204? After how many quarters will the balance exceed \$10,000?

41. Jane has taken a part-time job to save for a \$503 racing bike. If she puts \$15 each week into a savings account paying 5.2% interest compounded weekly, when will she be able to buy the bike?

42. Bob needs \$3228 to have some repairs done on his house. He decides to deposit \$100 at the end of each month into an annuity earning 6% interest compounded monthly. When will he be able to do the repairs?

Solutions to Practice Problems 2

1. An ordinary annuity with $n = 60$, $i = \frac{3}{4}\%$, and $R = 1$. You will make a deposit now, in the present, and then withdraw money each month. The amount of this deposit is the present value of the annuity.

$$P = a_{\overline{n}|\,i}R = a_{\overline{60}|\,.75\%} \cdot 1 = (48.17337352) \cdot 1 \quad \text{(by Table 6)}$$
$$= \$48.17.$$

[*Notes*: (1) When $R = 1$, we have $P = a_{\overline{n}|\,i}$. This explains why $a_{\overline{n}|\,i}$ is often called the *present value of $1*. (2) For this transaction, the future value of the annuity has no significance. At the end of 5 years, the fund will have a balance of 0.]

2. Not an ordinary annuity since the payment period (1 week) is different from the interest period (1 month).

3. An ordinary annuity with $n = 24$, $i = .5\%$, and $R = \$10$. There is no money in the account now, in the present. However, in 2 years, in the future, money will have accumulated. So for this annuity, only the future value has significance.

$$F = s_{\overline{n}|\,i}R = s_{\overline{24}|\,.5\%} \cdot 10 = (25.43195524) \cdot 10 \quad \text{(by Table 4)}$$
$$= \$254.32.$$

10.3 Amortization of Loans

In this section we analyze the mathematics of paying off loans. The loans we shall consider all will be repaid in a sequence of equal payments at regular time intervals, with the payment intervals coinciding with the interest periods. The process of paying off such a loan is called *amortization.* In order to obtain a feeling for the amortization process, let us consider a particular case, the amortization of a $563 loan to buy a color television set. Suppose that this loan charges interest at a 12% rate with interest compounded monthly and that the monthly payments are $116 for 5 months. The repayment process is summarized in Table 1. Note the following facts about the financial transactions.

1. Payments are made at the end of each month. The payments have been carefully calculated to pay off the debt, with interest, in the specified time interval.

TABLE 1

Payment number	*Amount*	*Interest*	*Applied to principal*	*Unpaid balance*
1	$116	$5.63	$110.37	$452.63
2	116	4.53	111.47	341.16
3	116	3.41	112.59	228.57
4	116	2.29	113.71	114.85
5	116	1.15	114.85	0.00

2. Since $i = 1\%$, the interest to be paid each month is 1% of the unpaid balance at the end of the previous month. That is, 5.63 is 1% of 563, 4.53 is 1% of 452.63, and so on.
3. Although we write just one check each month for \$116, we regard part of the check as being applied to payment of that month's interest. The remainder, namely 116 – [interest], is regarded as being applied to repayment of part of the principal amount.
4. The unpaid balance at the end of each month is the previous unpaid balance minus the portion of the payment applied to the principal. A loan can be paid off early by paying just the current unpaid balance.

The four factors that describe the amortization process above are:

the principal	\$563
the interest rate	12% compounded monthly
the term	5 months
the monthly payment	\$116

The important fact to recognize is that the sequence of payments in the amortization above constitutes an annuity, with the person taking out the loan paying the interest. Therefore, the mathematical tools developed in Section 10.2 suffice to analyze the amortization. In particular, we could determine the monthly payment or the principal once the other three factors have been specified.

EXAMPLE 1 Suppose that a loan has an interest rate of 12% compounded monthly and a term of 5 months.

(a) Given that the principal is \$563, calculate the monthly payment.

(b) Given that the monthly payment is \$116, calculate the principal.

(c) Given that the monthly payment is \$116, calculate the unpaid balance after 3 months.

Solution The sequence of payments constitutes an annuity with the monthly payments as rent and the principal as the present value. Also, $n = 5$ and $i = 1\%$.

(a) Since $R = (1/a_{\overline{n}|\,i})P$, we see that

$$R = \frac{1}{a_{\overline{5}|\,1\%}} \cdot 563 = (.20603980) \cdot 563 \qquad \text{(by Table 7)}$$
$$= \$116.$$

(b) Since $P = a_{\overline{n}|\,i}R$, we see that

$$P = a_{\overline{5}|\,1\%} \cdot 116 = (4.85343124) \cdot 116 = \$563.00.$$

(c) The unpaid balance is most easily calculated by regarding it as the amount necessary to retire the debt. Therefore, it must be sufficient to generate, with interest, the sequence of two remaining payments. That is, the unpaid balance is the present value of an annuity of two payments of \$116. So we see that

$$[\text{unpaid balance after 3 months}] = a_{\overline{2}|\,1\%} \cdot 116 = (1.97039506) \cdot 116$$
$$= \$228.57. \blacksquare$$

A mortgage is a long-term loan used to purchase real estate. The real estate is used as collateral to guarantee the loan.

EXAMPLE 2 On December 31, 1990, a house was purchased with the buyer taking out a 30-year, $112,475 mortgage at 9% interest, compounded monthly. The mortgage payments are made at the end of each month.

(a) Calculate the amount of the monthly payment.

(b) Calculate the unpaid balance of the loan on December 31, 2016, just after the 312th payment.

(c) How much interest will be paid during the month of January 2017?

(d) How much of the principal will be paid off during the year 2016?

(e) How much interest will be paid during the year 2016?

Solution (a) Denote the monthly payment by R. Then $112,475 is the present value of an annuity of $n = 360$ payments, with $i = \frac{3}{4}\%$. Therefore,

$$\begin{aligned} R &= \frac{1}{a_{\overline{n}|\,i}} P = \frac{1}{a_{\overline{360}|\,.75\%}} \cdot 112{,}475 \\ &= (.00804623) \cdot 112{,}475 \qquad \text{(by Table 7)} \\ &= \$905.00. \end{aligned}$$

(b) The remaining payments constitute an annuity of 48 payments. Therefore, the unpaid balance is the present value of that annuity.

$$\begin{aligned} [\text{unpaid balance}] &= a_{\overline{n}|\,i} R = a_{\overline{48}|\,.75\%} \cdot 905 \\ &= (40.18478189) \cdot 905 \qquad \text{(by Table 6)} \\ &= \$36{,}367.23. \end{aligned}$$

(c) The interest paid during 1 month is i, the interest rate per month, times the unpaid balance at the end of the preceding month. Therefore,

$$\begin{aligned} [\text{interest for January 2017}] &= \tfrac{3}{4}\% \text{ of } \$36{,}367.23 \\ &= .0075 \cdot 36{,}367.23 \\ &= \$272.75. \end{aligned}$$

(d) Since the portions of the monthly payments applied to repay the principal have the effect of reducing the unpaid balance, this question may be answered by calculating how much the unpaid balance will be reduced during 2016. Reasoning as in part (b), we determine that the unpaid balance on December 31, 2015 (just after the 300th payment), is equal to $43,596.90. Therefore,

$$\begin{aligned} &[\text{amount of principal repaid in 2016}] \\ &\quad = [\text{unpaid balance Dec. 31, 2015}] - [\text{unpaid balance Dec. 31, 2016}] \\ &\quad = \$43{,}596.90 - \$36{,}367.23 = \$7229.67. \end{aligned}$$

(e) During the year 2016, the total amount paid is $12 \times 905 = \$10{,}860$. But by part (d), $7229.67 is applied to repayment of principal, the remainder being applied to interest.

$$\begin{aligned} [\text{interest in 2016}] &= [\text{total amount paid}] - [\text{principal repaid in 2016}] \\ &= \$10{,}860 - \$7229.67 = \$3630.33. \end{aligned}$$ ∎

In the early years of a mortgage, most of each monthly payment is applied to interest. For the mortgage above, the interest portion will exceed the principal portion until the twenty-sixth year.

Each new (unpaid) balance can be computed from the previous balance with the formula

$$B_{new} = (1 + i)B_{previous} - R.$$

That is, the new balance equals the growth of the previous balance due to interest, minus the amount paid. Successive balances are computed by multiplying by $(1 + i)$ and subtracting R.

> If $\$P$ is borrowed at interest rate i per period and $\$R$ is paid back at the end of each interest period, then the formula
>
> $$B_{new} = (1 + i)B_{previous} - R \qquad (1)$$
>
> can be used to calculate each new balance, B_{new}, from the previous balance, $B_{previous}$.

The Excel spreadsheet in Table 2 calculates the payment per period and shows the successive balances for the annuity of Example 2. [For those familiar with Excel, the entry in cell B4 was set to `=PMT(B1,B2,-B3)`. The form of the payment function is `PMT(`***`rate,nper,pv`***`)`. The present value, that is, the principal, was specified as a negative number since the money is owed to the bank. B10 was set to B3 and B11 was set to `(1+$B$1)*B10-$B$4`. The remainder of the B column was created with the AutoFill option from the Edit/Fill/Series dialog box. C11 was set to `$B$1*B10` and D11 was set to `$B$4-C11`. The remainder of the C and D columns was created with the AutoFill option.]

EXAMPLE 3 Refer to Example 2. Compute the unpaid balance of the loan on January 31, 2017, just after the 313th payment.

TABLE 2 Amortization Table

	A	B	C	D
1	Interest rate	0.75%		
2	Number of periods	360		
3	Principal	$112,475.00		
4	Payment per period	$905.00		
5				
6		Amortization Table		
7				
8		Unpaid		Applied to
9	Payment number	balance	Interest	principal
10	0	$112,475.00		
11	1	$112,413.56	$843.56	$61.44
12	2	$112,351.66	$843.10	$61.90
13	3	$112,289.30	$842.64	$62.36
14	4	$112,226.47	$842.17	$62.83
15	5	$112,163.17	$841.70	$63.30
16	6	$112,099.39	$841.22	$63.78
17	7	$112,035.14	$840.75	$64.25
18	8	$111,970.40	$840.26	$64.74
19	9	$111,905.18	$839.78	$65.22
20	10	$111,839.47	$839.29	$65.71
21	11	$111,773.27	$838.80	$66.20
22	12	$111,706.57	$838.30	$66.70

	A	B	C	D
359	349	$9,521.20	$77.61	$827.38
360	350	$8,687.61	$71.41	$833.59
361	351	$7,847.77	$65.16	$839.84
362	352	$7,001.63	$58.86	$846.14
363	353	$6,149.14	$52.51	$852.49
364	354	$5,290.26	$46.12	$858.88
365	355	$4,424.94	$39.68	$865.32
366	356	$3,553.13	$33.19	$871.81
367	357	$2,674.78	$26.65	$878.35
368	358	$1,789.84	$20.06	$884.94
369	359	$898.26	$13.42	$891.58
370	360	$0.00	$6.74	$898.26
371		Unpaid		Applied to
372	Payment number	balance	Interest	principal

Solution By formula (1),

$$B_{\text{Jan}} = (1+i)B_{\text{Dec}} - R = (1.0075) \cdot 36{,}367.23 - 905$$
$$= \$35{,}734.98. \quad \blacksquare$$

Sometimes amortized loans stipulate a *balloon payment* at the end of the term. For instance, you might pay \$200 at the end of each quarter-year for 3 years and an additional \$1000 at the end of the third year. The \$1000 is a balloon payment.

EXAMPLE 4 How much money can you borrow at 8% interest compounded quarterly if you agree to pay \$200 at the end of each quarter-year for 3 years and in addition a balloon payment of \$1000 at the end of the third year?

Solution Here you are borrowing in the present and repaying in the future. The amount of the loan will be the present value of *all* the future payments. The future payments consist of an annuity and a lump-sum payment. Let us calculate the present values of each of these separately. Now, $i = 2\%$ and $n = 12$.

$$[\text{present value of annuity}] = a_{\overline{n}|\,i}R = a_{\overline{12}|\,2\%} \cdot 200$$
$$= (10.57534122) \cdot 200 \quad \text{(by Table 6)}$$
$$= \$2115.07.$$

$$[\text{present value of balloon payment}] = \frac{1}{(1+i)^n} \cdot F$$
$$= \frac{1}{(1+i)^n} \cdot 1000 \quad (\text{for } i = 2\% \text{ and } n = 12)$$
$$= .78849318 \cdot (1000) = \$788.49.$$

Therefore, the amount you can borrow is

$$\$2115.07 + \$788.49 = \$2903.56. \quad \blacksquare$$

GC As in Section 10.2, we will assume that the formulas for $s_{\overline{n}|\,i}$, $\frac{1}{s_{\overline{n}|\,i}}$, $a_{\overline{n}|\,i}$, and $\frac{1}{a_{\overline{n}|\,i}}$ have been assigned to the functions Y_4 through Y_7.

Graphing calculators can easily display successive balances for a loan. Consider the situation from Example 2 in which \$112,475 is borrowed at 9% interest compounded monthly and repaid with 360 monthly payments of \$905. In Fig. 1 successive balances are displayed in the home screen. In Fig. 2 successive balances are graphed, and in Fig. 3 they are displayed as part of an amortization table. (The Y_2 column gives the interest portion of the monthly payment.)

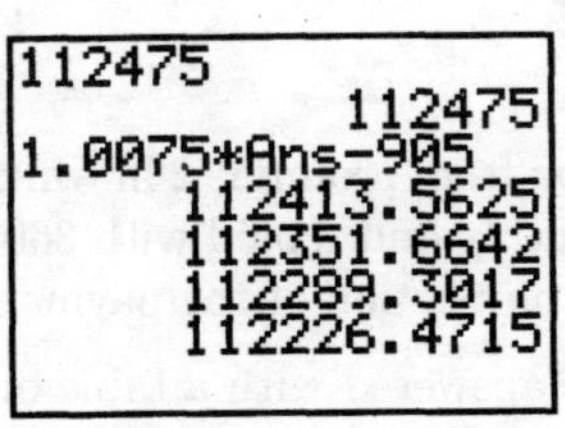

FIGURE 1

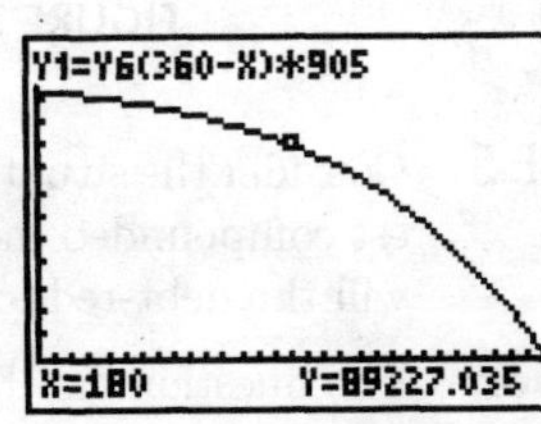

FIGURE 2

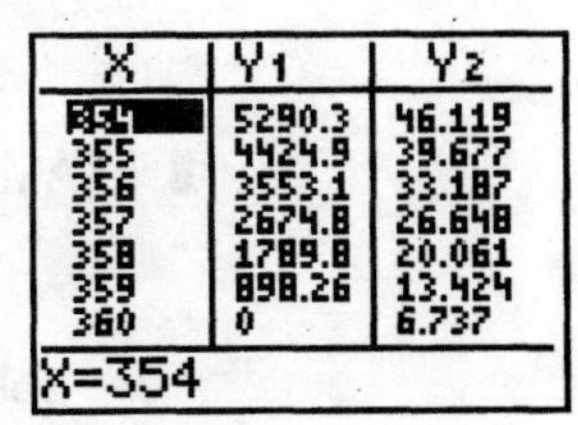

FIGURE 3

Displaying successive balances on the home screen as in Fig. 1 Successive balances can be determined with the relation $B_{new} = (1+i)B_{previous} - R$. After the principal (`112475`) is entered, the last displayed value (`112475`) is assigned to `Ans`. The instruction `1.0075*Ans-905` generates the next balance (`112413.5625`) and assigns it to `Ans`. Each subsequent press of the ENTER key generates another balance.

Graphing successive balances with the TI-82 or TI-83 as in Fig. 2 Figure 4 shows the `Y=` editor where Y_1 has been set equal to Y_6`(360-X)*905`, that is, $a_{\overline{360-n}|i} \cdot R$. In Fig. 5 the value .0075 is assigned to `I` from the home screen. In Fig. 6 the WINDOW screen is set up to display the decline of the balance for 30 years, that is, 360 months. `Ymin` was set to a negative value to allow room for coordinates of points to be displayed at the bottom of the screen. Pressing GRAPH produces the screen of Fig. 2.

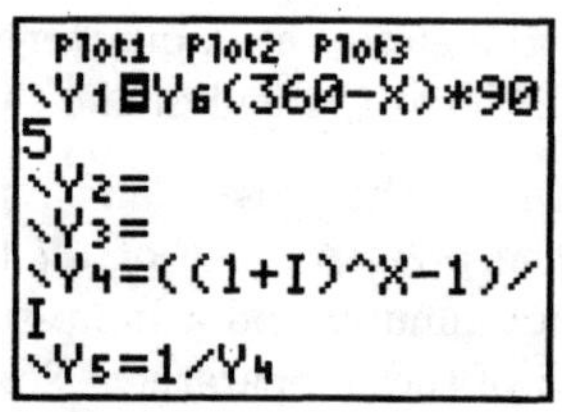

FIGURE 4

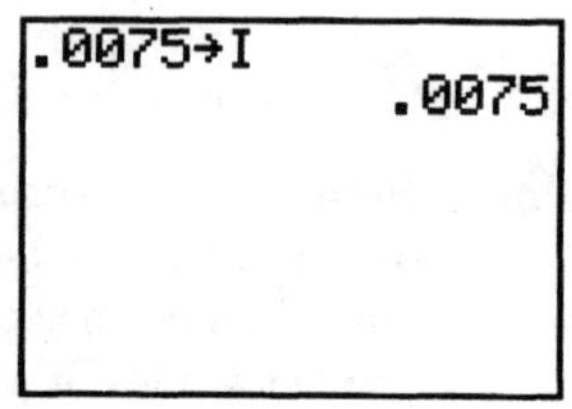

FIGURE 5

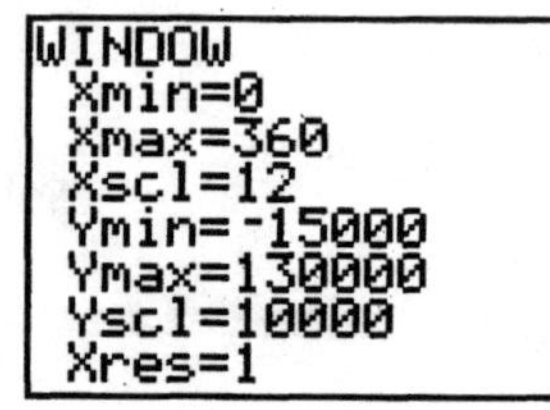

FIGURE 6

Obtaining a table of successive balances with the TI-82 or TI-83 as in Fig. 3 In the `Y=` editor, set Y_2`=I*`Y_1`(X-1)`, that is, the interest rate times the amount of the previous balance. Also, set Y_3`=905-`Y_2. Then the columns for Y_2 and Y_3 will give the interest and debt-reduction portions of the monthly payment. In Fig. 7 we have set `TblStart=354` to view the final 7 months of the mortgage. Figure 3 shows the resulting table. If you press ▶ three times, you also will see a column for Y_3. See Fig. 8. (*Note*: The values shown in Fig. 3 agree with the values in the Excel spreadsheet of Table 2. However, early values for the balance in the calculator table will differ slightly from those generated by the Excel spreadsheet and those shown in Fig. 1. This is due both to the effects of roundoff and the different methods used to generate the numbers. The values for interest and debt-reduction will be the same throughout.)

FIGURE 7

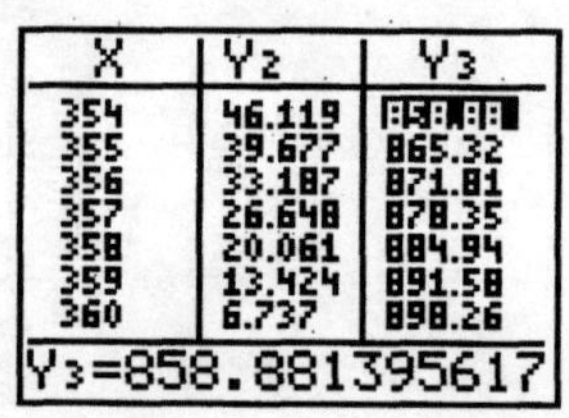

FIGURE 8

EXAMPLE 5 Consider the situation from Example 2 in which \$112,475 is borrowed at 9% interest compounded monthly and repaid with 360 monthly payments of \$905. When will the debt-reduction portion of the payment surpass the interest portion?

Solution The question can be answered with a table or a graph.

1. (Table) Consider the table of Fig. 3. Scroll up until an interest greater than half the monthly payment, that is, 452.50 is reached. Then press ▶ three times to obtain Fig. 9.

2. (Graph) Let Y_2 and Y_3 be the interest and debt-reduction functions as in Part 1. Set the window as in Fig. 10, graph the two functions, and find their point of intersection. See Fig. 11.

In each case we see that debt reduction will exceed interest after 269 months; that is, after 22 years and 5 months. ❚

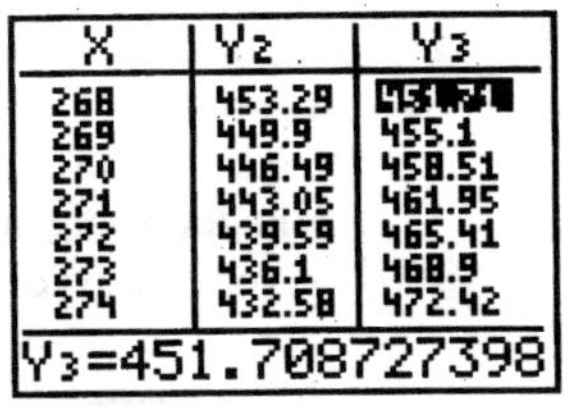

FIGURE 9

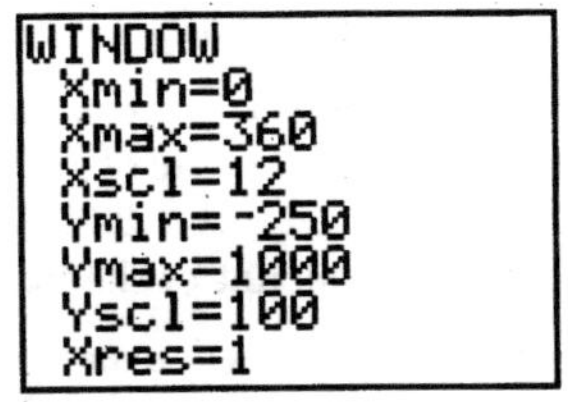

FIGURE 10

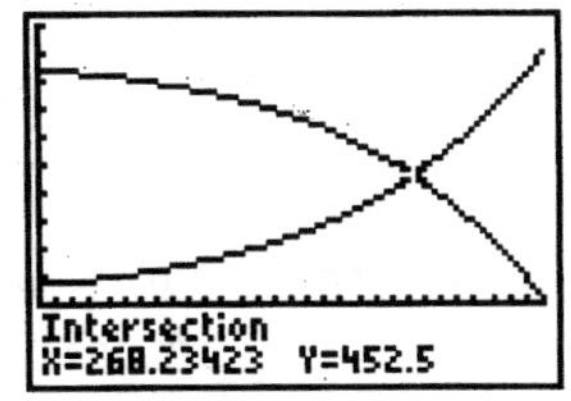

FIGURE 11

Practice Problems 3

1. The word "amortization" comes from the French "a mort" meaning "at the point of death." Justify the word.

2. Explain why only present values and not future values arise in amortization problems.

❚ EXERCISES 3

1. A loan of $10,000 is to be repaid with monthly payments for 5 years at 6% interest compounded monthly. Calculate the monthly payment.

2. Find the monthly payment on a $100,000, 25-year mortgage at 12% interest compounded monthly.

3. How much money can you borrow at 12% interest compounded semiannually if the loan is to be repaid at half-year intervals for 10 years and you can afford to pay $1000 per half-year?

4. You buy a car with a down payment of $500 and $100 per month for 3 years. If the interest rate is 9% compounded monthly, how much did the car cost?

5. Consider a $58,331, 30-year mortgage at interest rate 12% compounded monthly with a $600 monthly payment.
   (a) How much interest is paid the first month?
   (b) How much of the first month's payment is applied to paying off the principal?
   (c) What is the unpaid balance after 1 month?
   (d) What is the unpaid balance at the end of 25 years?
   (e) How much of the principal is repaid during the 26th year?
   (f) How much interest is paid during the 301st month?

6. Consider a $21,281.27 loan for 7 years at 8% interest compounded quarterly and a payment of $1000 per quarter-year.
   (a) Compute the unpaid balance after 5 years.
   (b) How much interest is paid during the fifth year?
   (c) How much principal is repaid in the first payment?
   (d) What is the total amount of interest paid on the loan?

7. Susie takes out a car loan for $8000 for a term of 3 years at 12% interest compounded monthly.
   (a) Find her monthly payments.
   (b) Find the total amount she pays for the car.
   (c) Find the total amount of interest she pays.
   (d) Find the amount she still owes after 1 year.
   (e) Find the amount she still owes after 2 years.
   (f) Find the total interest she pays in year 2.
   (g) Prepare an amortization schedule for the first 4 months.

8. James buys a house for $90,000. He puts $10,000 down and then finances the rest at 9% interest compounded monthly for 25 years.
   (a) Find his monthly payments.
   (b) Find the total amount he pays for the house.
   (c) Find the total amount of interest he pays.
   (d) Find the amount he still owes after 23 years.
   (e) Find the amount he still owes after 24 years.
   (f) Find the total amount of interest he pays in year 24.
   (g) Prepare the amortization schedule for the first 3 months.

9. A mortgage at 9% interest compounded monthly with a monthly payment of $1125 has an unpaid balance of $10,000 after 350 months. Find the unpaid balance after 351 months.

10. A loan with a quarterly payment of \$1500 has an unpaid balance of \$10,000 after 30 quarters and an unpaid balance of \$9000 after 31 quarters. If interest is compounded quarterly, find the interest rate.

11. A loan is to be amortized over an 8-year term at 12% interest compounded semiannually, payments of \$1000 every 6 months, and a balloon payment of \$10,000 at the end of the term. Calculate the amount of the loan.

12. A loan of \$105,504.50 is to be amortized over a 5-year term at 12% interest compounded monthly with monthly payments and a \$10,000 balloon payment at the end of the term. Calculate the monthly payment.

13. Write out a complete amortization schedule (as in Table 1 at the beginning of this section) for the amortization of a \$1000 loan with monthly payments at 12% interest compounded monthly for 4 months.

14. Write out a complete schedule (as in Table 1) for the amortization of a \$10,000 loan with payments every 6 months at 12% interest compounded semiannually for 1 year.

15. You purchase a \$120,000 house, pay \$20,000 down, and take out a 30-year mortgage with monthly payments, at an interest rate of 9% compounded monthly. How much money will you be paying each month?

16. In 1970 you purchased a house and took out a 25-year, \$50,000 mortgage at 6% interest compounded monthly. In 1980 you sold the house for \$150,000. How much money did you have left after you paid the bank the unpaid balance on the mortgage?

17. You are considering the purchase of a condominium to use as a rental property. You estimate that you can rent the condominium for \$1200 per month and that taxes, insurance, and maintenance costs will run about \$200 per month. If interest rates are 9% compounded monthly, how large a 25-year mortgage can you assume and still have the rental income cover the monthly expenses?

18. Consider formula (1). Derive an analogous formula for the balance of an annuity for which regular payments are deposited into a savings account and accrue with interest.

19. A car is purchased for \$6287.10 with \$2000 down and a loan to be repaid at \$100 a month for 3 years followed by a balloon payment. If the interest rate is 6% compounded monthly, how large will the balloon payment be?

20. A real estate speculator purchases a tract of land for \$1 million and assumes a 25-year mortgage at 12% interest compounded monthly.
    (a) What is his monthly payment?
    (b) Suppose that at the end of 5 years the mortgage is changed to a 10-year term for the remaining balance. What is the new monthly payment?
    (c) Suppose that after 5 more years, the mortgage is required to be repaid in full. How much will then be due?

21. Suppose you borrow \$10,000 at 4% interest compounded semiannually and pay off the loan in 7 years. Which of items (a)–(d) can be used to fill in the blank in the following statement? (*Note*: Before computing, use your intuition to guess the correct answer.)

    If the ________ is doubled, then the total amount paid will double.
    (a) amount of the loan
    (b) interest rate
    (c) number of interest periods per year
    (d) duration of the loan

22. Suppose you borrow money at 4% interest compounded semiannually and pay off the loan in payments of \$1000 per interest period for 5 years. Which of items (a)–(d) can be used to fill in the blank in the following statement? (*Note*: Before computing, use your intuition to guess the correct answer.)

    If the ________ is doubled, then the amount that can be borrowed will double.
    (a) amount paid per interest period
    (b) interest rate
    (c) number of interest periods per year
    (d) duration of the loan

23. Suppose that you make annual payments of \$5000 for 20 years into an annuity paying 6% interest compounded annually.
    (a) What is the value of the annuity at the end of the 20th year?
    (b) Suppose that you elect to have your annuity repaid to you over a 10-year period in annual installments. What is the annual payment you will receive?
    (c) Suppose that after you receive payments for 5 years you elect to have the remainder of your annuity paid to you in a lump sum. How much will you receive?

A *sinking fund* is a pool of money accumulated by a corporation or government to repay a specific debt at some future date.

24. A corporation wishes to deposit money into a sinking fund at the end of each half-year in order to repay \$50 million in bonds in 10 years. It can expect to receive a 12% (compounded semiannually) return on its deposits to the sinking fund. How much should the deposits be?

25. The Federal National Mortgage Association ("Fannie Mae") puts \$30 million at the end of each month into a sinking fund paying 12% interest compounded monthly. The sinking fund is to be used to repay debentures that mature 15 years from the creation of the fund. How large is the face amount of the debentures assuming the sinking fund will exactly pay them off?

26. A corporation borrows \$5 million to erect a new headquarters. The financing is arranged using industrial development bonds, to be repaid in 15 years. How much should it deposit into a sinking fund at the end of each quarter if the sinking fund earns 8% interest compounded quarterly?

27. A corporation sets up a sinking fund to replace some aging machinery. It deposits \$100,000 into the fund at the end of each month for 10 years. The sinking fund earns 12% interest compounded monthly. The equipment originally cost \$6 million. However, the cost of the equipment is rising 6% each year. Will the sinking fund be adequate to replace the equipment? If not, how much additional money is needed?

28. A corporation sets up a sinking fund to replace an aging warehouse. The cost of the warehouse today would be \$8 million. However, the corporation plans to replace the warehouse in 5 years. It estimates that the cost of the warehouse will increase by 10% annually. The sinking fund will earn 12% interest compounded monthly. What should be the monthly payments to the sinking fund?

Exercises 29–34 require the use of a graphing calculator or a computer.

29. Bill buys a computer for \$2188.91 and pays off the loan (at 9% interest compounded monthly) by paying \$100 at the end of each month. Determine the balance after 1, 2, and 3 months. After how many months will the loan be paid off?

30. Alice borrows \$20,000 to buy some medical equipment and pays off the loan (at 7% interest compounded annually) by paying \$4195.92 at the end of each year. Determine the balance after 1, 3, and 5 years. After how many years will the loan be paid off?

31. A loan of \$10,000 at 9% interest compounded monthly is repaid in 80 months with monthly payments of \$166.68. How much of the loan will have been paid off after 40 months? When will more than half of the loan have been repaid?

32. A loan of \$4000 at 6% interest compounded monthly is repaid in 8 years with monthly payments of \$52.57. After how many months will the loan be one-quarter paid off? One-half? Three-quarters?

33. A 25-year mortgage of \$124,188.57 at 8.5% interest compounded monthly has monthly payments of \$1000. After how many months will at least 75% of the monthly payment go toward debt reduction?

34. A 30-year mortgage of \$118,135.40 at 8.4% interest compounded monthly has monthly payments of \$900. After how many months will the amount applied to debt reduction be more than twice the amount applied to interest?

Solutions to Practice Problems 3

1. A portion of each payment is applied to reducing the debt, and by the end of the term the debt is totally annihilated.

2. The debt is formed when the creditor gives you a lump sum of money now, in the present. The lump sum of money is gradually repaid by you, with interest, thereby generating the annuity. At the end of the term, in the future, the loan is totally paid off, so there is no more debt. That is, the future value is always zero!

Remark You are actually functioning like a savings bank, since you are paying the interest. Think of the creditor as depositing the lump sum with you and then making regular withdrawals until the balance is 0.

Chapter 10 Review of Fundamental Concepts

1. What is meant by *principal*?
2. What is the difference between compound interest and simple interest?
3. What is meant by the *compound amount*?
4. How is interest per period determined from annual interest?

5. Explain how compound interest works. Give its formula.
6. Explain how simple interest works. Give its formula.
7. What is meant by the *present value* of a sum of money to be received in the future?
8. What is an annuity?
9. What is meant by the *future value* of an annuity? *present value*? *rent*?
10. Describe the two types of annuities discussed in this chapter. In each case, identify the present and future values.
11. Give the formula for computing a new balance from a previous balance for an annuity in which money is accumulated.
12. Define $s_{\overline{n}|i}$ and $a_{\overline{n}|i}$.
13. State the four formulas involving $s_{\overline{n}|i}$, $a_{\overline{n}|i}$, and their inverses.
14. What are the components of an amortization table of a loan, and how are they calculated?
15. Give the formula for computing a new balance from a previous balance for a loan.
16. What is a balloon payment?

CHAPTER 10 SUPPLEMENTARY EXERCISES

1. ***(From a GMAT exam)*** If \$1 were invested at 8 percent interest compounded annually, the total value of the investment, in dollars, at the end of 6 years would be
   (a) $(1.8)^6$ (b) $(1.08)^6$ (c) $6(1.08)$
   (d) $1 + (0.08)^6$ (e) $1 + 6(0.08)$
2. Mr. West wishes to purchase a condominium for \$80,000 cash upon his retirement 10 years from now. How much should he deposit at the end of each month into an annuity paying 6% interest compounded monthly in order to accumulate the required savings?
3. The income of a typical family in a certain city is currently \$19,200 per year. Family finance experts recommend that mortgage payments not exceed 25% of a family's income. Assuming a current mortgage interest rate of 9% compounded monthly for a 30-year mortgage with monthly payments, how large a mortgage can the typical family in that city afford?
4. Calculate the compound amount of \$50 after a year if deposited at 7.3% compounded daily.
5. Which is a better investment: 10% compounded annually or 9% compounded daily?
6. Ms. Smith deposits \$200 per month into a bond fund yielding 6% interest compounded monthly. How much are her holdings worth after 5 years?
7. A real estate investor takes out a \$200,000 mortgage subject to the following terms: For the first 5 years, the payments will be the same as the monthly payments on a 15-year mortgage at 12% interest compounded monthly. The balance will then be payable in full.
   (a) What are the monthly payments for the first 5 years?
   (b) What balance will be owed after 5 years?
8. College expenses at a private college currently average \$24,000 per year. It is estimated that these expenses are increasing at the rate of $\frac{1}{2}$% per month. What is the estimated cost of a year of college 10 years from now?
9. What is the present value of \$50,000 in 10 years at 6% interest compounded monthly?
10. An investment will pay \$10,000 in 2 years and then \$5,000 in 3 years. If the current market interest rate is 6% compounded monthly, what should a rational person be willing to pay for the investment?
11. A woman purchases a car for \$12,000. She pays \$3,000 as a down payment and finances the remaining amount at 6% interest compounded monthly for 4 years. What is her monthly car payment?
12. A businessman buys a \$100,000 piece of manufacturing equipment on the following terms: Interest will be charged at a rate of 4% compounded semiannu-

ally, but no payments will be made until 2 years after purchase. Beginning at that time equal semiannual payments will be made for 5 years. Determine the semiannual payment.

13. A retired person has set aside a fund of $105,003.50 for his retirement. This fund is in a bank account paying 6% interest compounded monthly. How much can he draw out of the account at the end of each month so that there is a balance of $30,000 at the end of 15 years? [*Hint*: First compute the present value of the $30,000.]

14. A business loan of $509,289.22 is to be paid off in monthly payments for 10 years with a $100,000 balloon payment at the end of the 10th year. The interest rate on the loan is 12% compounded monthly. Calculate the monthly payment.

15. Ms. Jones saves $100 per month for 30 years at 6% interest compounded monthly. How much are her accumulated savings worth?

16. An apartment building is currently generating an income of $2000 per month. Its owners are considering a 10-year second mortgage at 12% interest compounded monthly in order to pay for repairs. How large a second mortgage can the income of the apartment house support?

17. Investment A generates $1000 at the end of each year for 10 years. Investment B generates $5000 at the end of the fifth year and $5000 at the end of the tenth year. Assume a market rate of interest of 6% compounded annually. Which is the better investment?

18. A 5-year bond has a face value of $1000 and is currently selling for $800. The bond pays $5 interest at the end of each month and, in addition, will repay the $1000 face value at the end of the fifth year. The market rate of interest is currently 9% compounded monthly. Is the bond a bargain? Why or why not?

19. Calculate the effective rate for 10% interest compounded semiannually.

20. Calculate the effective rate for 18% interest compounded monthly.

21. A person makes an initial deposit of $10,000 into a savings account and then deposits $1000 at the end of each quarter-year for 15 years. If the interest rate is 8% compounded quarterly, how much money will be in the account after 15 years?

22. A $10,000 car loan at 6% interest compounded monthly is to be repaid with 36 equal monthly payments. Write out an amortization schedule for the first 6 months of the loan.

23. A person pays $200 at the end of each month for 10 years into a fund paying 1% interest per month compounded monthly. At the end of the 10th year, the payments cease, but the balance continues to earn interest. What is the value of the balance at the end of the 20th year?

24. A savings fund currently contains $300,000. It is decided to pay out this amount with 6% interest compounded monthly over a 5-year period. What are the monthly payments?

25. What is the monthly payment on a $150,000, 30-year mortgage at 9% interest?

APPENDIX A

TABLES

Table 1	Areas under the standard normal curve
Table 2	$(1 + i)^n$ Compound amount of \$1 invested for n interest periods at interest rate i per period
Table 3	$1/(1 + i)^n$ Present value of \$1. Principal that will accumulate to \$1 in n interest periods at a compound rate of i per period
Table 4	$s_{\overline{n}\rceil i}$ Future value of an ordinary annuity of n \$1 payments each, immediately after the last payment at compound interest rate of i per period
Table 5	$1/s_{\overline{n}\rceil i}$ Rent per period for an ordinary annuity of n payments, with compound interest rate i per period and future value \$1
Table 6	$a_{\overline{n}\rceil i}$ Present value of an ordinary annuity of n payments of \$1 one period before the first payment, with interest compounded at i per period

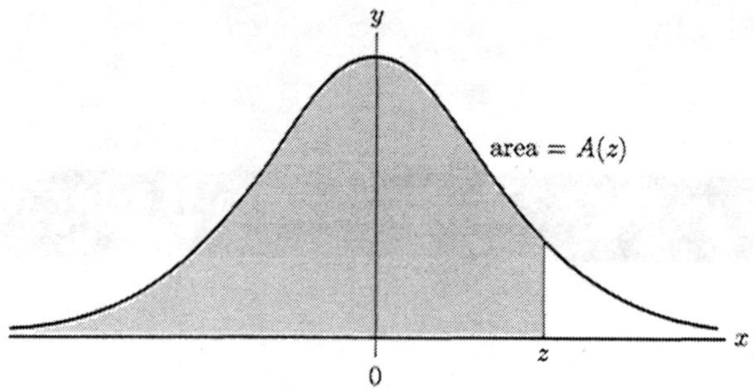

TABLE 1 Areas under the standard normal curve

z	A(z)	z	A(z)	z	A(z)	z	A(z)	z	A(z)
−3.50	.0002	−2.00	.0228	−.50	.3085	1.00	.8413	2.50	.9938
−3.45	.0003	−1.95	.0256	−.45	.3264	1.05	.8531	2.55	.9946
−3.40	.0003	−1.90	.0287	−.40	.3446	1.10	.8643	2.60	.9953
−3.35	.0004	−1.85	.0322	−.35	.3632	1.15	.8749	2.65	.9960
−3.30	.0005	−1.80	.0359	−.30	.3821	1.20	.8849	2.70	.9965
−3.25	.0006	−1.75	.0401	−.25	.4013	1.25	.8944	2.75	.9970
−3.20	.0007	−1.70	.0446	−.20	.4207	1.30	.9032	2.80	.9974
−3.15	.0008	−1.65	.0495	−.15	.4404	1.35	.9115	2.85	.9978
−3.10	.0010	−1.60	.0548	−.10	.4602	1.40	.9192	2.90	.9981
−3.05	.0011	−1.55	.0606	−.05	.4801	1.45	.9265	2.95	.9984
−3.00	.0013	−1.50	.0668	.00	.5000	1.50	.9332	3.00	.9987
−2.95	.0016	−1.45	.0735	.05	.5199	1.55	.9394	3.05	.9989
−2.90	.0019	−1.40	.0808	.10	.5398	1.60	.9452	3.10	.9990
−2.85	.0022	−1.35	.0885	.15	.5596	1.65	.9505	3.15	.9992
−2.80	.0026	−1.30	.0968	.20	.5793	1.70	.9554	3.20	.9993
−2.75	.0030	−1.25	.1056	.25	.5987	1.75	.9599	3.25	.9994
−2.70	.0035	−1.20	.1151	.30	.6179	1.80	.9641	3.30	.9995
−2.65	.0040	−1.15	.1251	.35	.6368	1.85	.9678	3.35	.9996
−2.60	.0047	−1.10	.1357	.40	.6554	1.90	.9713	3.40	.9997
−2.55	.0054	−1.05	.1469	.45	.6736	1.95	.9744	3.45	.9997
−2.50	.0062	−1.00	.1587	.50	.6915	2.00	.9772	3.50	.9998
−2.45	.0071	−.95	.1711	.55	.7088	2.05	.9798		
−2.40	.0082	−.90	.1841	.60	.7257	2.10	.9821		
−2.35	.0094	−.85	.1977	.65	.7422	2.15	.9842		
−2.30	.0107	−.80	.2119	.70	.7580	2.20	.9861		
−2.25	.0122	−.75	.2266	.75	.7734	2.25	.9878		
−2.20	.0139	−.70	.2420	.80	.7881	2.30	.9893		
−2.15	.0158	−.65	.2578	.85	.8023	2.35	.9906		
−2.10	.0179	−.60	.2743	.90	.8159	2.40	.9918		
−2.05	.0202	−.55	.2912	.95	.8289	2.45	.9929		

TABLE 2 $(1 + i)^n$ Compounded amount of $1 invested for n interest periods at interest rate i per period

	i				
n	*1/50%*	*1/2%*	*1%*	*2%*	*6%*
1	1.000200000	1.00500000	1.01000000	1.02000000	1.06000000
2	1.000400040	1.01002500	1.02010000	1.04040000	1.12360000
3	1.000600120	1.01507513	1.03030100	1.06120800	1.19101600
4	1.000800240	1.02015050	1.04060401	1.08243216	1.26247696
5	1.001000400	1.02525125	1.05101005	1.10408080	1.33822558
6	1.001200600	1.03037751	1.06152015	1.12616242	1.41851911
7	1.001400840	1.03552940	1.07213535	1.14868567	1.50363026
8	1.001601120	1.04070704	1.08285671	1.17165938	1.59384807
9	1.001601447	1.04591058	1.09368527	1.19509257	1.68947896
10	1.002001806	1.05114013	1.10462213	1.21899442	1.79084770
11	1.002202201	1.05639583	1.11566835	1.24337431	1.89829856
12	1.002402642	1.06167781	1.12682503	1.26824179	2.01219647
13	1.002603122	1.06698620	1.13809328	1.29360663	2.13292826
14	1.002803643	1.07232113	1.14947421	1.31947876	2.26090396
15	1.003004204	1.07768274	1.16096896	1.34586834	2.39655819
16	1.003204804	1.08307115	1.17257864	1.37278571	2.54035168
17	1.003405445	1.08848651	1.18430443	1.40024142	2.69277279
18	1.003505100	1.09392894	1.19614748	1.42824625	2.85433915
19	1.003806848	1.09939858	1.20810895	1.45681117	3.02559950
20	1.004007609	1.10489558	1.22019004	1.48594740	3.20713547
21	1.004208411	1.11042006	1.23239194	1.51566634	3.39956360
22	1.004409252	1.11597216	1.24471586	1.54597967	3.60353742
23	1.004610134	1.12155202	1.25716302	1.57689926	3.81974966
24	1.004811056	1.12715978	1.26973465	1.60843725	4.04893464
25	1.005012018	1.13279558	1.28243200	1.64060599	4.29187072
26	1.005213022	1.13845955	1.29525631	1.67341811	4.54938296
27	1.005414063	1.14415185	1.30820888	1.70688648	4.82234594
28	1.005615146	1.14987261	1.32129097	1.74102421	5.11168670
29	1.005816269	1.15562197	1.33450388	1.77584469	5.41838790
30	1.006017433	1.16140008	1.34784892	1.81136158	5.74349117
36	1.007225257	1.19668052	1.43076878	2.03988734	8.14725200
48	1.009645259	1.27048916	1.61222608	2.58707039	16.39387173
52	1.010453217	1.29609015	1.67768892	2.80032819	20.69688534
60	1.012071075	1.34885015	1.81669670	3.28103079	32.98769085
104	1.021015784	1.67984969	2.81464012	7.84183795	428.36106292
120	1.024287860	1.81939673	3.30038689	10.76516303	1,088.18774784
180	1.036652115	2.45409356	5.99580198	35.32083136	35,896.80101597
240	1.049165620	3.31020448	10.89255365	115.88873515	
300	1.061830136	4.46496981	19.78846626	380.23450806	
360	1.074647608	6.02257521	35.94964133	1,247.56112775	
365	1.075722685	6.17465278	37.78343433	1,377.40829197	

TABLE 3 $1/(1+i)^n$ Present value of \$1. Principal that will accumulate to \$1 in n interest periods at a compound rate of i per period

	i				
n	*1/50%*	*1/2%*	*1%*	*2%*	*6%*
1	.99980004	.99502488	.99009901	.98039216	.94339623
2	.99960012	.99007450	.98029605	.96116878	.88999644
3	.99940024	.98514876	.97059015	.94232233	.83961928
4	.99920040	.98024752	.96098034	.92384543	.79209366
5	.99900060	.97537067	.95146569	.90573081	.74725817
6	.99880084	.97051808	.94204524	.88797138	.70496054
7	.99860112	.96568963	.93271805	.87056018	.66505711
8	.99840144	.96088520	.92348322	.85349037	.62741237
9	.99820180	.95610468	.91433982	.83675527	.59189846
10	.99800220	.95134794	.90528695	.82034830	.55839478
11	.99780264	.94661487	.89632372	.80426304	.52678753
12	.99760312	.94190534	.88744923	.78849318	.49696936
13	.99740364	.93721924	.87866260	.77303253	.46883902
14	.99720420	.93255646	.86996297	.75787502	.44230096
15	.99700479	.92791688	.86134947	.74301473	.41726506
16	.99680543	.92330037	.85282126	.72844581	.39364628
17	.99660611	.91870684	.84437749	.71416256	.37136442
18	.99640683	.91413616	.83601731	.70015937	.35034379
19	.99620759	.90958822	.82773992	.68643076	.33051301
20	.99600839	.90506290	.81954447	.67297133	.31180473
21	.99580923	.90056010	.81143017	.65977582	.29415540
22	.99561010	.89607971	.80339621	.64683904	.27750510
23	.99541102	.89162160	.79544179	.63415592	.26179726
24	.99521198	.88718567	.78756613	.62172149	.24697855
25	.99501298	.88277181	.77976844	.60953087	.23299863
26	.99481401	.87837991	.77204796	.59757928	.21981003
27	.99461509	.87400986	.76440392	.58586204	.20736795
28	.99441621	.86966155	.75683557	.57437455	.19563014
29	.99421736	.86533488	.74934215	.56311231	.18455674
30	.99401856	.86102973	.74192292	.55207089	.17411013
36	.99282657	.83564492	.69892495	.49022315	.12274077
48	.99044688	.78709841	.62026041	.38653761	.06099840
52	.98965492	.77155127	.59605806	.35710100	.04831645
60	.98807290	.74137220	.55044962	.30478227	.03031434
104	.97941686	.59529136	.35528521	.12752113	.00233448
120	.97628805	.54963273	.30299478	.09289223	.00091896
180	.96464377	.40748243	.16678336	.02831190	.00002786
240	.95313836	.30209614	.09180584	.00862897	.00000084
300	.94177018	.22396568	.05053449	.00262996	.00000003
360	.93053759	.16604193	.02781669	.00080156	.00000000
365	.92960762	.16195243	.02646663	.00072600	.00000000

TABLE 4 $s_{\overline{n}|i}$ Future value of an ordinary annuity of n \$1 payments each, immediately after the last payment at compound interest rate of i per period

	i					
n	*1/2%*	*3/4%*	*1%*	*1.5%*	*2%*	*6%*
1	1.00000000	1.00000000	1.00000000	1.00000000	1.00000000	1.00000000
2	2.00500000	2.00750000	2.01000000	2.01500000	2.02000000	2.06000000
3	3.01502500	3.02255625	3.03010000	3.04522500	3.06040000	3.18360000
4	4.03010013	4.04522542	4.06040100	4.09090337	4.12160800	4.37461600
5	5.05025063	5.07556461	5.10100501	5.15226693	5.20404016	5.63709296
6	6.07550188	6.11363135	6.15201506	6.22955093	6.30812096	6.97531854
7	7.10587939	7.15948358	7.21353521	7.32299419	7.43428338	8.39383765
8	8.14140879	8.21317971	8.28567056	8.43283911	8.58296905	9.89746791
9	9.18211583	9.27477856	9.36852727	9.55933169	9.75462843	11.49131598
10	10.22802641	10.34433940	10.46221254	10.70272167	10.94972100	13.18079494
11	11.27916654	11.42192194	11.56683467	11.86326249	12.16871542	14.97164264
12	12.33556237	12.50758636	12.68250301	13.04121143	13.41208973	16.86994120
13	13.39724018	13.60139325	13.80932804	14.23682960	14.68033152	18.88213767
14	14.46422639	14.70340370	14.94742132	15.45038205	15.97393815	21.01506593
15	15.53654752	15.81367923	16.09689554	16.68213778	17.29341692	23.27596988
16	16.61423026	16.93228183	17.25786449	17.93236984	18.63928525	25.67252808
17	17.69730141	18.05927394	18.43044314	19.20135539	20.01207096	28.21287976
18	18.78578791	19.19471849	19.61474757	20.48937572	21.41231238	30.90565255
19	19.87971685	20.33867888	20.81089504	21.79671636	22.84055863	33.75999170
20	20.97911544	21.49121897	22.01900399	23.12366710	24.29736980	36.78559120
21	22.08401101	22.65240312	23.23919403	24.47052211	25.78331719	39.99272668
22	23.19443107	23.82229614	24.47158598	25.83757994	27.29898354	43.39229028
23	24.31040322	25.00096336	25.71630183	27.22514364	28.84496321	46.99582769
24	25.43195524	26.18847059	26.97346485	28.63352080	30.42186247	50.81557735
25	26.55911502	27.38488412	28.24319950	30.06302361	32.03029972	54.86451200
26	27.69191059	28.59027075	29.52563150	31.51396896	33.67090572	59.15638272
27	28.83037015	29.80469778	30.82088781	32.98667850	35.34432383	63.70576568
28	29.97452200	31.02823301	32.12909669	34.48147867	37.05121031	68.52811162
29	31.12439461	32.26094476	33.45038766	35.99870085	38.79223451	73.63979832
30	32.28001658	33.50290184	34.78489153	37.53868137	40.56807921	79.05818622
36	39.33610496	41.15271612	43.07687836	47.27596921	51.99436719	119.12086666
48	54.09783222	57.52071111	61.22260777	69.56521929	79.35351927	256.56452882
52	59.21803075	63.31106835	67.76889215	77.92489152	90.01640927	328.28142239
60	69.77003051	75.42413693	81.66966986	96.21465171	114.05153942	533.12818089
104	135.96993732	156.68432202	181.46401172	246.93411381	342.09189731	7,122.68438195
120	163.87934681	193.51427708	230.03868946	331.28819149	488.25815171	18,119.79579725
180	290.81871245	378.40576900	499.58019754	905.62451261	1,716.04156785	
240	462.04089516	667.88686993	989.25536539	2,308.85437027	5,744.43675765	
300	692.99396243	1,121.12193732	1,878.84662619	5,737.25330834	18,961.72540308	
360	1,004.51504245	1,830.74348307	3,494.96413277	14,113.58539279	62,328.05638744	
365	1,034.93055669	1,905.50947396	3,678.34343329	15,209.49204803	68,820.41459830	

TABLE 5 $1/s_{\overline{n}|i}$ Rent per period for an ordinary annuity of n payments, with compound interest rate i per period, and future value $1

	i					
n	1/2%	3/4%	1%	1.5%	2%	6%
1	1.00000000	1.00000000	1.00000000	1.00000000	1.00000000	1.00000000
2	.49875312	.49813200	.49751244	.49627792	.49504950	.48543689
3	.33167221	.33084579	.33002211	.32838296	.32675467	.31410981
4	.24813279	.24720501	.24628109	.24444479	.24262375	.22859149
5	.19800997	.19702242	.19603980	.19408932	.19215839	.17739640
6	.16459546	.16356891	.16254837	.16052521	.15852581	.14336263
7	.14072854	.13967488	.13862828	.13655616	.13451196	.11913502
8	.12282886	.12175552	.12069029	.11858402	.11650980	.10103594
9	.10890736	.10781929	.10674036	.10460982	.10251544	.08702224
10	.09777057	.09667123	.09558208	.09343418	.09132653	.07586796
11	.08865903	.08755094	.08645408	.08429384	.08217794	.06679294
12	.08106643	.07995148	.07884879	.07667999	.07455960	.05927703
13	.07464224	.07352188	.07241482	.07024036	.06811835	.05296011
14	.06913609	.06801146	.06690117	.06472332	.06260197	.04758491
15	.06436436	.06323639	.06212378	.05994436	.05782547	.04296276
16	.06018937	.05905879	.05794460	.05576508	.05365013	.03895214
17	.05650579	.05537321	.05425806	.05207966	.04996984	.03544480
18	.05323173	.05209766	.05098205	.04880578	.04670210	.03235654
19	.05030253	.04916740	.04805175	.04587847	.04378177	.02962086
20	.04766645	.04653063	.04541531	.04324574	.04115672	.02718456
21	.04528163	.04414543	.04303075	.04086550	.03878477	.02500455
22	.04311380	.04197748	.04086372	.03870332	.03663140	.02304557
23	.04113465	.03999846	.03888584	.03673075	.03466810	.02127848
24	.03932061	.03818474	.03707347	.03492410	.03287110	.01967900
25	.03765186	.03651650	.03540675	.03326345	.03122044	.01822672
26	.03611163	.03497693	.03386888	.03173196	.02969923	.01690435
27	.03468565	.03355176	.03244553	.03031527	.02829309	.01569717
28	.03336167	.03222871	.03112444	.02900108	.02698967	.01459255
29	.03212914	.03099723	.02989502	.02777878	.02577836	.01357961
30	.03097892	.02984816	.02874811	.02663919	.02464992	.01264891
36	.02542194	.02429973	.02321431	.02115240	.01923285	.00839483
48	.01848503	.01738504	.01633384	.01437500	.01260184	.00389765
52	.01688675	.01579503	.01475603	.01283287	.01110909	.00304617
60	.01433280	.01325836	.01224445	.01039343	.00876797	.00187572
104	.00735457	.00638226	.00551073	.00404966	.00292319	.00014040
120	00610205	.00516758	.00434709	.00301852	.00204810	.00005519
180	.00343857	.00264267	.00200168	.00110421	.00058274	.00000167
240	.00216431	.00149726	.00101086	.00043312	.00017408	.00000005
300	.00144301	.00089196	.00053224	.00017430	.00005274	.00000000
360	.00099551	.00054623	.00028613	.00007085	.00001604	.00000000
365	.00096625	.00052479	.00027186	.00006575	.00001453	.00000000

TABLE 6 $a_{\overline{n}|i}$ Present value of an ordinary annuity of n payments of $1 one period before the first payment, with interest compounded at i per period

	i					
n	*1/2%*	*3/4%*	*1%*	*1.5%*	*2%*	*6%*
1	.99502488	.99255583	.99009901	.98522167	.98039216	.94339623
2	1.98509938	1.97772291	1.97039506	1.95588342	1.94156094	1.83339267
3	2.97024814	2.95555624	2.94098521	2.91220042	2.88388327	2.67301195
4	3.95049566	3.92611041	3.90196555	3.85438465	3.80772870	3.46510561
5	4.92586633	4.88943961	4.85343124	4.78264497	4.71345951	4.21236379
6	5.89638441	5.84559763	5.79547647	5.69718717	5.60143089	4.91732433
7	6.86207404	6.79463785	6.72819453	6.59821396	6.47199107	5.58238144
8	7.82295924	7.73661325	7.65167775	7.48592508	7.32548144	6.20979381
9	8.77906392	8.67157642	8.56601758	8.36051732	8.16223671	6.80169227
10	9.73041186	9.59957958	9.47130453	9.22218455	8.98258501	7.36008705
11	10.67702673	10.52067452	10.36762825	10.07111779	9.78684805	7.88687458
12	11.61893207	11.43491267	11.25507747	10.90750521	10.57534122	8.38384394
13	12.55615131	12.34234508	12.13374007	11.73153222	11.34837375	8.85268296
14	13.48870777	13.24302242	13.00370304	12.54338150	12.10624877	9.29498393
15	14.41662465	14.13699495	13.86505252	13.34323301	12.84926350	9.71224899
16	15.33992502	15.02431261	14.71787378	14.13126405	13.57770931	10.10589527
17	16.25863186	15.90502492	15.56225127	14.90764931	14.29187188	10.47725969
18	17.17276802	16.77918107	16.39826858	15.67256089	14.99203125	10.82760348
19	18.08235624	17.64682984	17.22600850	16.42616837	15.67846201	11.15811649
20	18.98741915	18.50801969	18.04555297	17.16863879	16.35143334	11.46992122
21	19.88797925	19.36279870	18.85698313	17.90013673	17.01120916	11.76407662
22	20.78405896	20.21121459	19.66037934	18.62082437	17.65804820	12.04158172
23	21.67568055	21.05331473	20.45582113	19.33086145	18.29220412	12.30337898
24	22.56286622	21.88914614	21.24338726	20.03040537	18.91392560	12.55035753
25	23.44563803	22.71875547	22.02315570	20.71961120	19.52345647	12.78335616
26	24.32401794	23.54218905	22.79520366	21.39863172	20.12103576	13.00316619
27	25.19802780	24.35949286	23.55960759	22.06761746	20.70689780	13.21053414
28	26.06768936	25.17071251	24.31644316	22.72671671	21.28127236	13.40616428
29	26.93302423	25.97589331	25.06578530	23.37607558	21.84438466	13.59072102
30	27.79405397	26.77508021	25.80770822	24.01583801	22.39645555	13.76483115
36	32.87101624	31.44680525	30.10750504	27.66068431	25.48884248	14.62098713
48	42.58031778	40.18478189	37.97395949	34.04255365	30.67311957	15.65002661
52	45.68974664	42.92761812	40.39419423	35.92874185	32.14494992	15.86139252
60	51.72556075	48.17337352	44.95503841	39.38026889	34.76088668	16.16142771
104	80.94172854	72.03438325	64.47147918	52.49436634	43.62394373	16.62775868
120	90.07345333	78.94169267	69.70052203	55.49845411	45.35538850	16.65135068
180	118.50351467	98.59340884	83.32166399	62.09556231	48.58440478	16.66620237
240	139.58077168	111.14495403	90.81941635	64.79573209	49.56855168	16.66665259
300	155.20686401	119.16162216	94.94655125	65.90090069	49.86850220	16.66666624
360	166.79161439	124.28186568	97.21833108	66.35324174	49.95992180	16.66666665
365	167.60951473	124.61379021	97.35333747	66.37572674	49.96369994	16.66666666

Answers to Exercises

CHAPTER 6

Exercises 1.1

1., 3., 5.

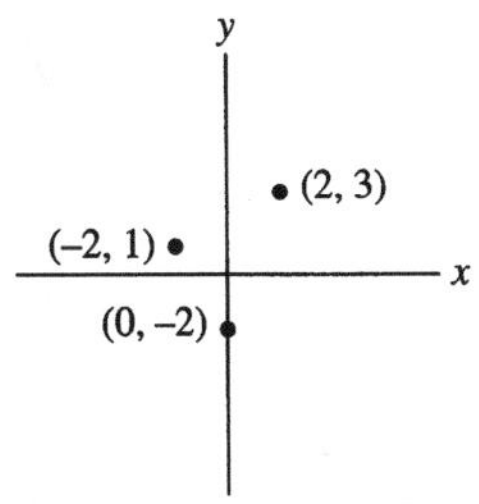

7.

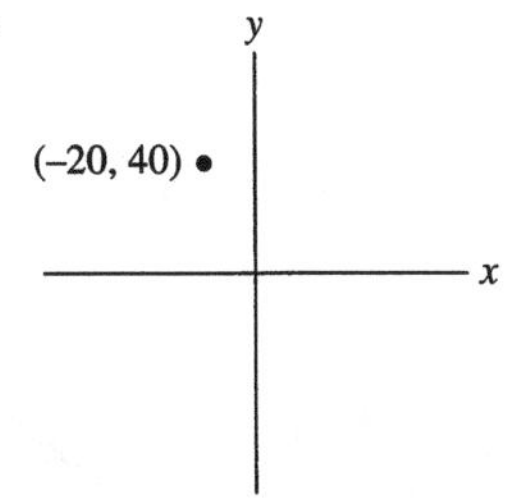

9. (e) **11.** $m = 5, b = 8$ **13.** $m = 0, b = 3$ **15.** $y = -2x + 3$ **17.** $x = \frac{5}{3}$

19. (2, 0), (0, 8) **21.** (7, 0), none

23.

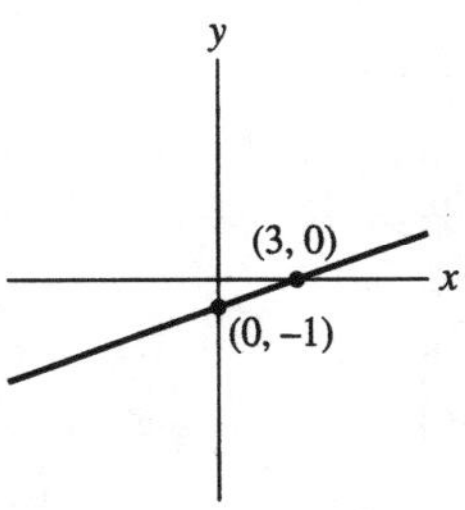

25.

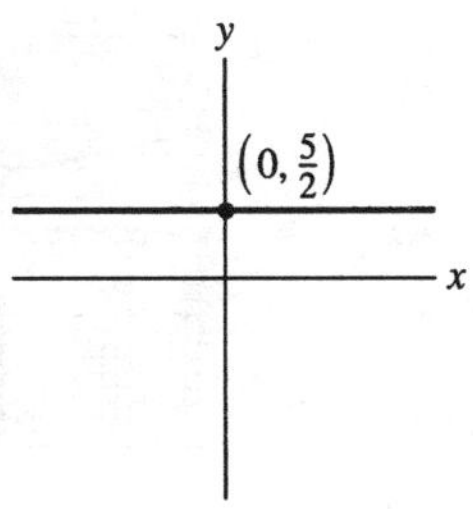

27.

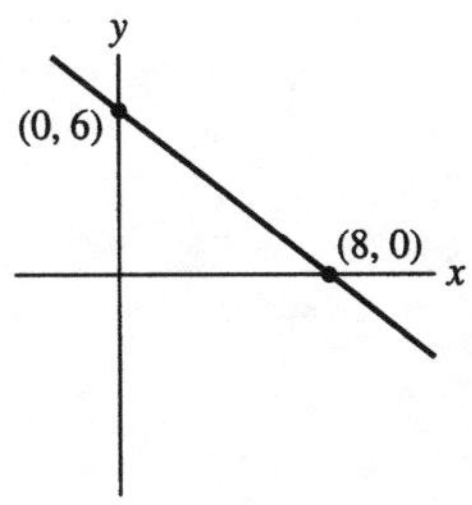

29. (a), (b), (c), (e) **31.** (a) L_3 (b) L_1 (c) L_2

33. (a) 4 minutes, 40 seconds (b) 72° water was placed in the kettle. (c) No

35. (a)

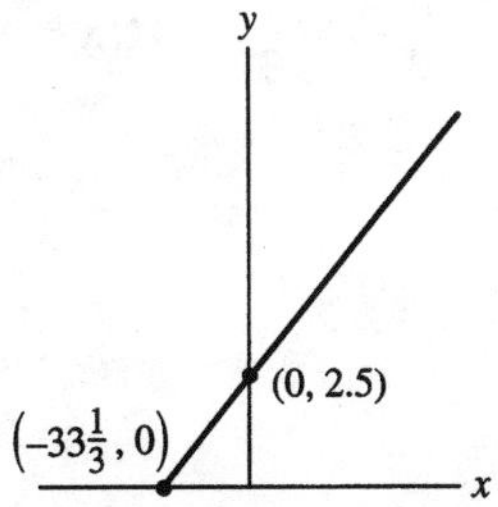

(b) In 1960, 2.5 trillion cigarettes were sold. (c) 1980 (d) 7 trillion

37. Up **39.** $y = 0$ **41.** No

Exercises 1.2

1. False **3.** True **5.** $x \geq 4$ **7.** $x \geq 3$ **9.** $y \leq -2x + 5$ **11.** $y \geq 15x - 18$ **13.** $x \geq -\frac{3}{4}$ **15.** Yes

17. No **19.** Yes **21.** Yes

23.

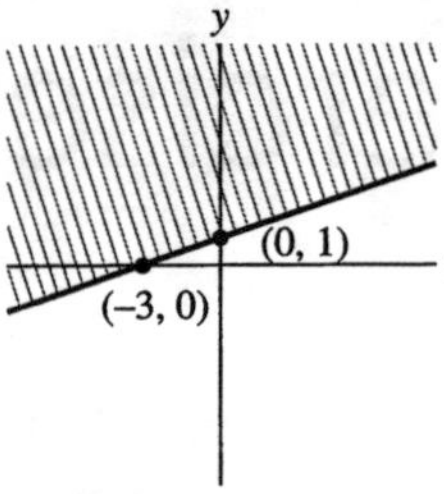

25.

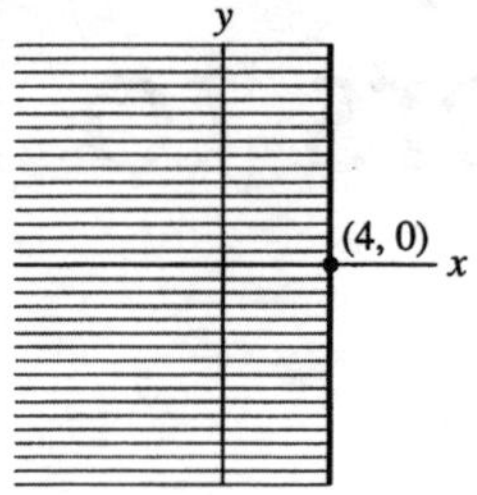

27.

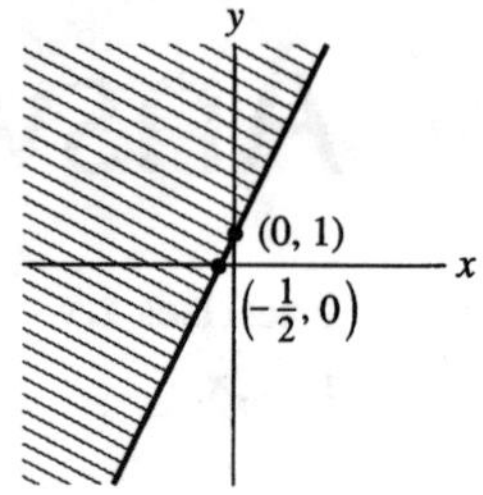

29.

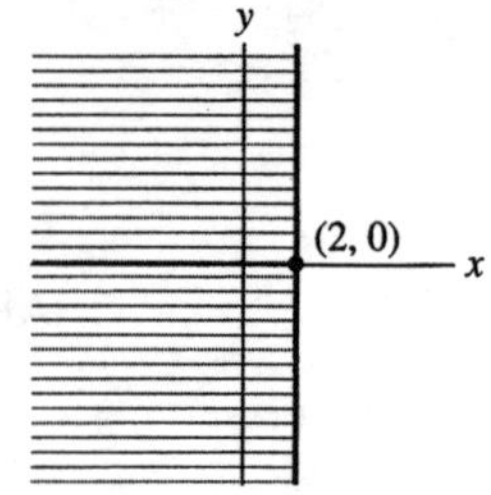

31.

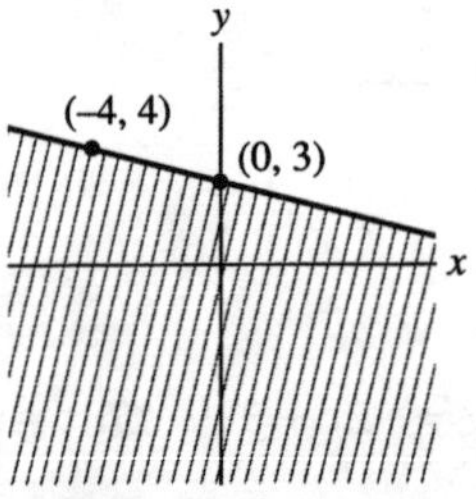

33.

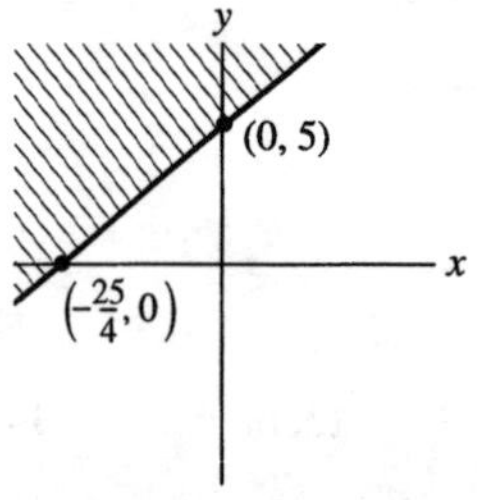

35.

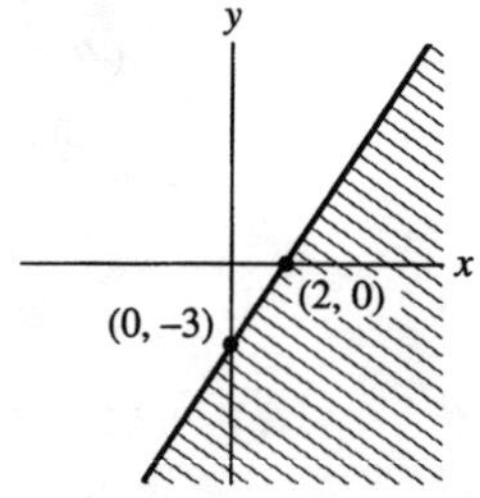

37.

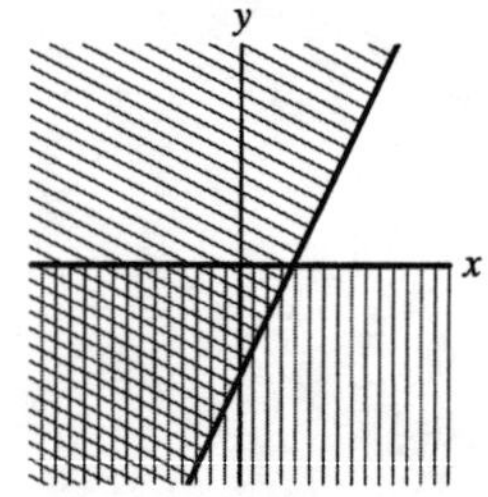

39.

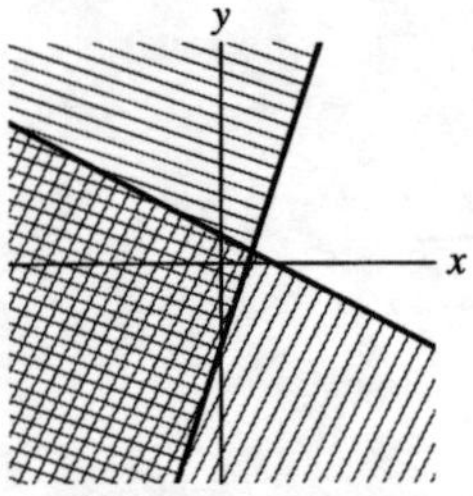

41.

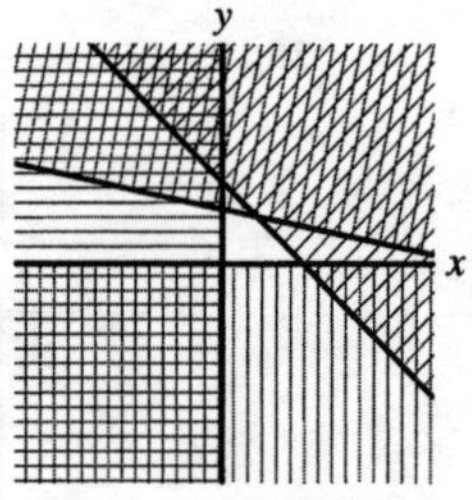

43. Yes **45.** No **47.** Below **49.** Above **51.** $\begin{cases} y \geq 2x - 1 \\ y \leq 2x \end{cases}$ **53.** (e) **55.** (a) (6, 2.5) (b) Above

57.

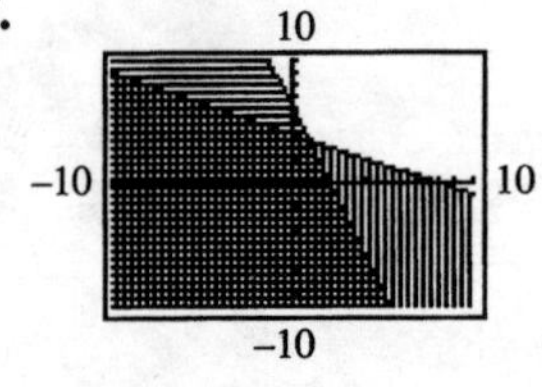

Exercises 1.3

1. (2, 3) **3.** (2, 1) **5.** (12, 3) **7.** Yes **9.** $x = \frac{10}{3}, y = \frac{1}{3}$ **11.** $x = -\frac{7}{9}, y = -\frac{22}{9}$ **13.** $A = (3, 4), B = (6, 2)$

15. $A = (0, 0), B = (2, 4), C = \left(5, \frac{11}{2}\right), D = (5, 0)$

17.

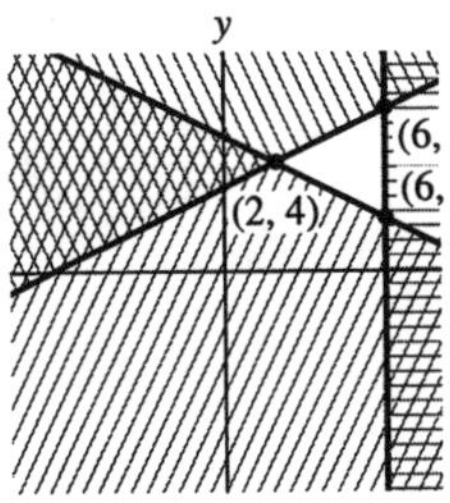

19.

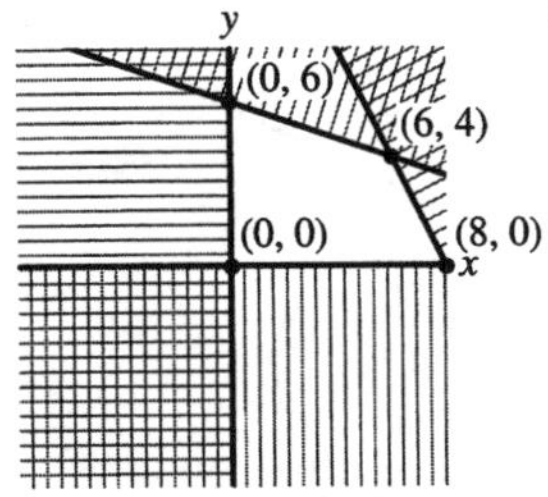

21.

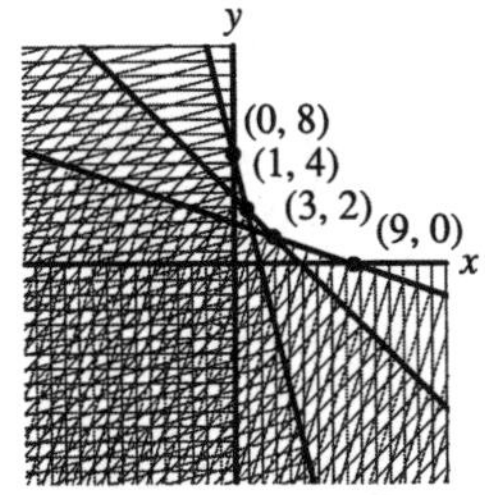

23. (a) 19,500 units (b) 5 cents **25.** (a) 3 dollars (b) 29,500 units **27.** Working: 32; Supervising: 8 **29.** (d)
31. (3.73, 2.23) **33.** (2.68, 1.92) **35.** (a)

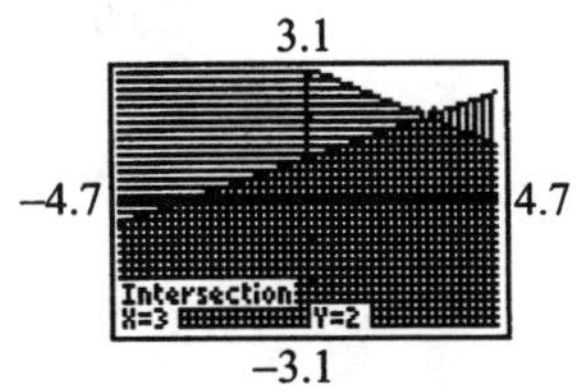

(b) (3, 2) (c)

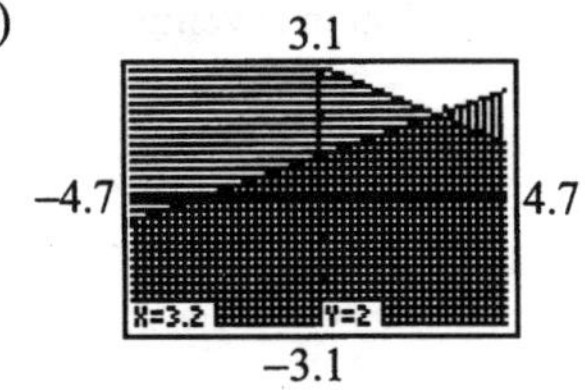

(d) No

Exercises 1.4

1. $\frac{2}{3}$ **3.** 5 **5.**

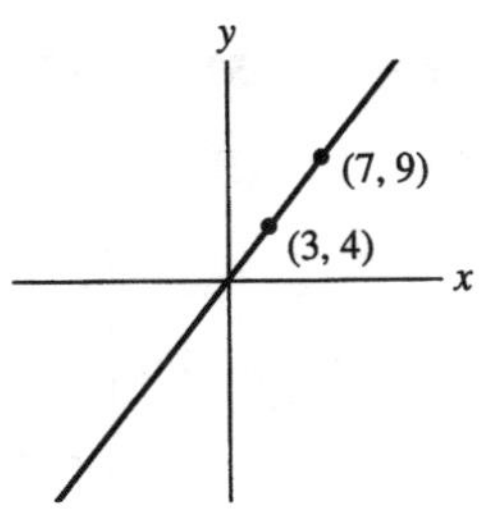

$\frac{5}{4}$ **7.**

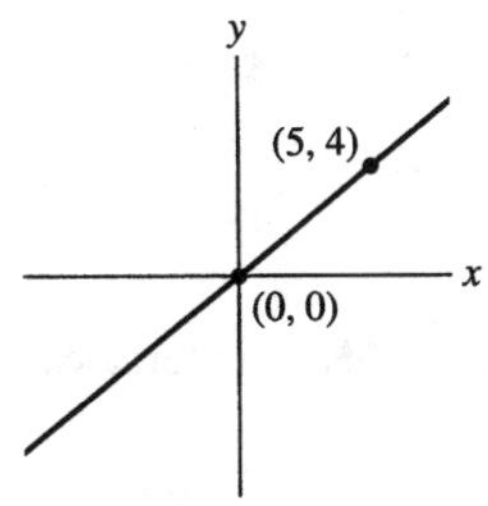

$\frac{4}{5}$ **9.** Undefined

11.

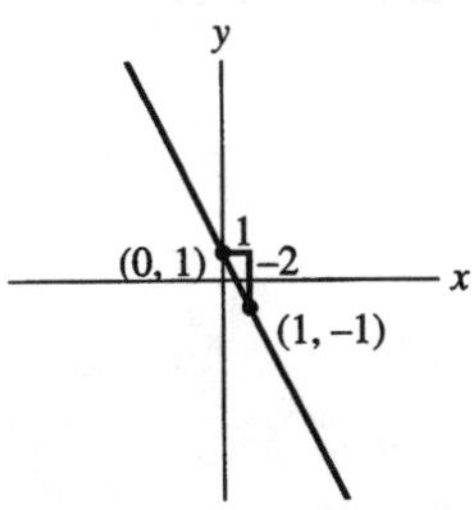

13.

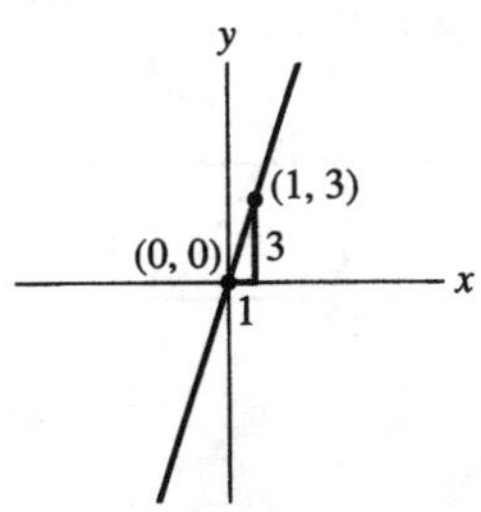

15. $y = -2x + 7$ **17.** $y = -2x + 4$

19. $y = \frac{1}{4}x + \frac{3}{2}$ **21.** $y = -x$ **23.** $y = 3$ **25.** (0, 3) **27.** Each unit sold yields a commission of \$5. In addition, she receives \$60 per week base pay. **29.** (a) (0, 1200); at \$1200 no one will buy the item. (b) (400, 0); even if the item is given away, only 400 will be taken. (c) −3; to sell an additional item, the price must be reduced by \$3. (d) \$150 (e) 300 items (f)

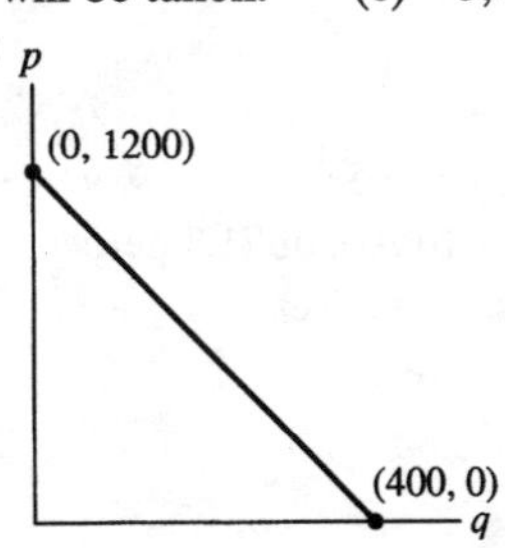

31. (a) $y = 90x + 5000$ (b) \$5000 (c) \$90 (d)

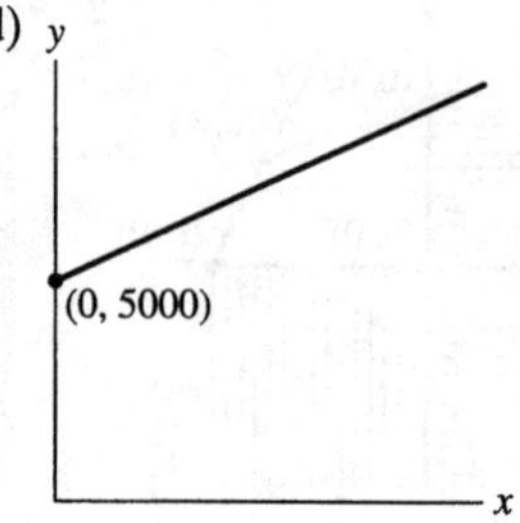

33. (a) \$30,000 (b) 60 coats (c) (0, 0); if no coats are sold, there is no revenue. (d) 100; each additional coat yields an additional \$100 in revenue. **35.**

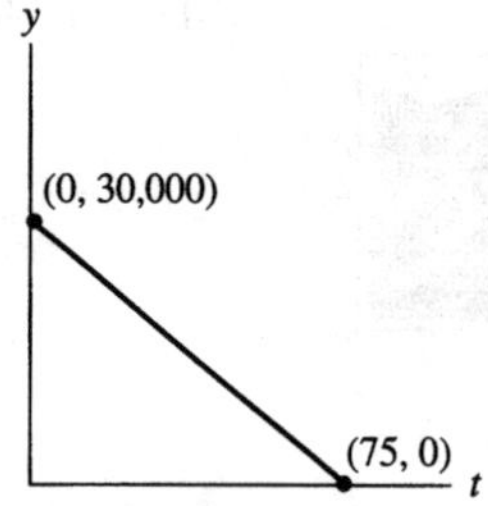

37. 12,000 gallons

39. (75, 0); the tank is empty after 75 days. **41.** \$50,000 **43.** $\left(15\frac{1}{3}, 0\right)$; the cash reserves will be depleted after $15\frac{1}{3}$ days.

45. July 11 **47.** 3 **49.** 1400 hours **51.** \$260 **53.** $y = 3x - 1$ **55.** $y = x + 1$ **57.** $y = -7x + 35$ **59.** $y = 4$

61. $y = \frac{1}{2}x$ **63.** $y = -2x$ **65.** 5; 1; −1 **67.** $-\frac{5}{4}$; $-\frac{3}{2}$; $-\frac{3}{4}$ **69.** (a) (C) (b) (B) (c) (D) (d) (A)

71. $F = \frac{9}{5}C + 32$ **73.** $y = \frac{1}{15}x + 6$; 3 million **75.** Counterclockwise **77.** $y \geq 4x + 3$ **79.** $\begin{cases} y \leq -\frac{1}{2}x + 4 \\ y \leq -x + 5 \\ y \geq x - 3 \\ x \geq 0, y \geq 0 \end{cases}$

81. $k = 9$ **83.** $a = -.05$ **87.** (c) **89.** (c) **91.**

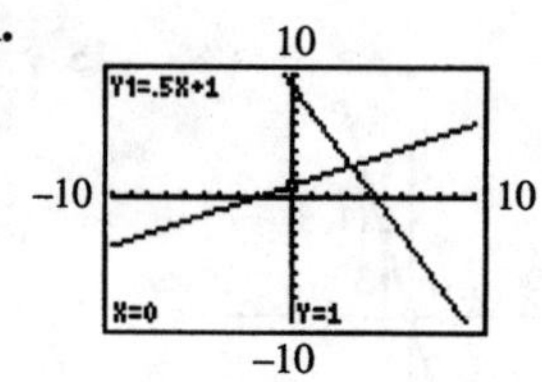

Do not appear perpendicular

93.

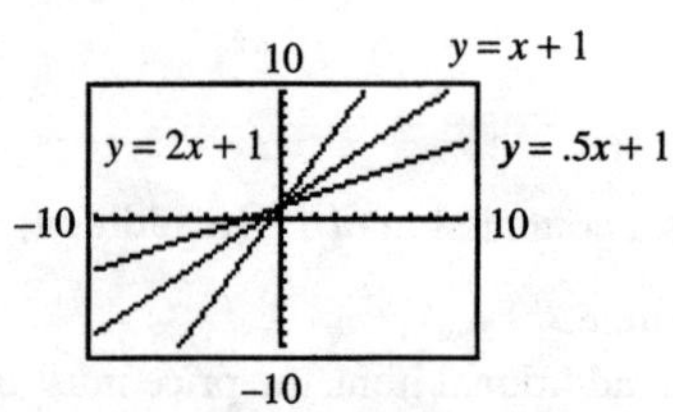

95. 1.5 units

Exercises 1.5

1. 4 **3.** 6.70 **5.** $m = -1.4$; $b = 8.5$ **7.** $y = 4.5x - 3$ **9.** $y = -2x + 11.5$ **11.** (a) $y = .338x + 21.6$ (b) About 393 **13.** (a) $y = .411x + 11.8$ (b) About 21.3 percent (c) 2007 **15.** (a) $y = 0.608x + 35.4$ (b) About 80.4 years (c) About 69.6 years **17.** Yes

Supplementary Exercises

1. $x = 0$ **2.**

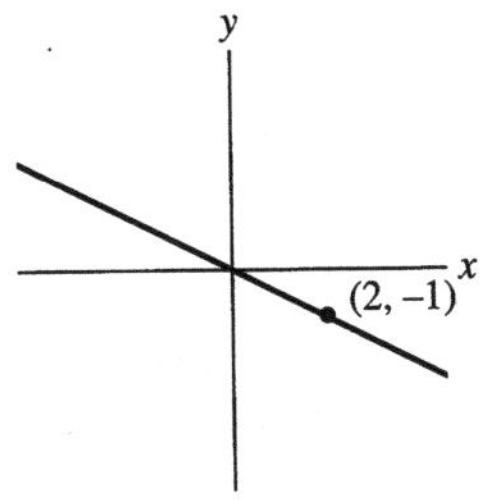

3. $\left(2, -\frac{4}{5}\right)$ **4.** $\frac{3}{4}$ **5.** $y = -\frac{1}{2}x + 5$ **6.**

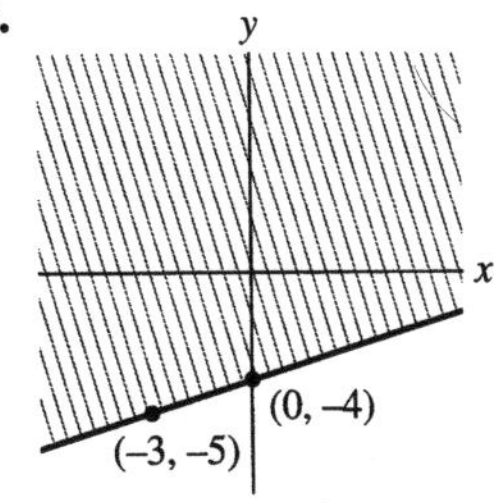

7. Yes **8.** (3, 5) **9.** $y = \frac{1}{5}x + 13$ **10.** 10 **11.** (5, 0) **12.**

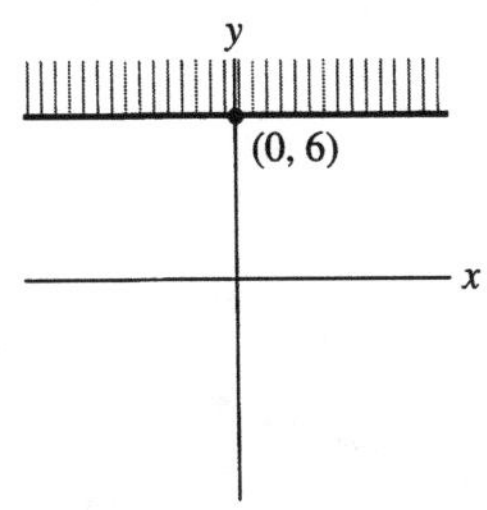

13. (7, 10)

14.

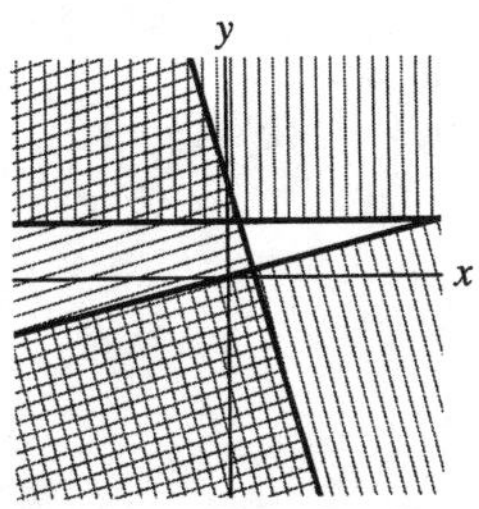

15. (0, 7) **16.** The rate is \$35 per hour plus a flat fee of \$20. **17.** No

18. $y = \frac{2}{3}x - 2$ **19.** (e) **20.** $y \le \frac{2}{3}x + \frac{3}{2}$ **21.** $y \ge 2.4x - 5.8$ **22.** $x = -.3, y = .4$ **23.** $y = -\frac{2}{5}x + \frac{7}{5}$

24. $x \ge \frac{6}{5}$ **25.** $m = -2$; y-intercept: (0, 8); x-intercept: (4, 0);

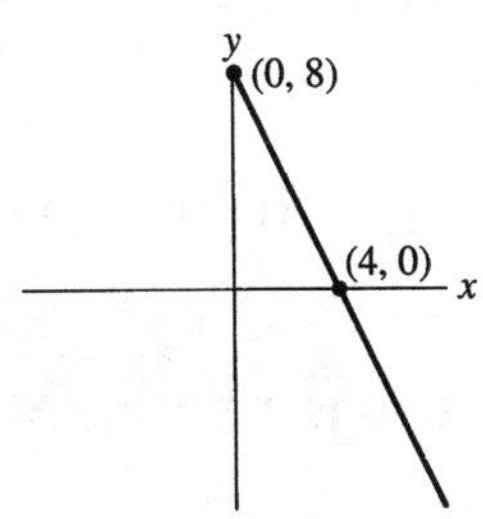

26. No

28. (a) (C) (b) (A) (c) (B) (d) (D) **29.** (a) L_3 (b) L_1 (c) L_2 **30.** $\begin{cases} y \le -\frac{7}{8}x + 5 \\ y \ge \frac{8}{7}x - \frac{43}{14} \\ x \ge 0, y \ge 0 \end{cases}$; $\left(\frac{43}{16}, 0\right)$

31. Supply curve: $q = 150p - 100$; demand curve: $q = -75p + 500$; $\left(\frac{8}{3}, 300\right)$ **32.** (0, 0), (10, 0), (9, 5), (6, 7), (0, 4)

33. $y = x + \frac{5}{2}$ **34.** (a) $y = 10x - 6000$

(b) x-intercept: $(600, 0)$, y-intercept: $(0, -6000)$ (c)

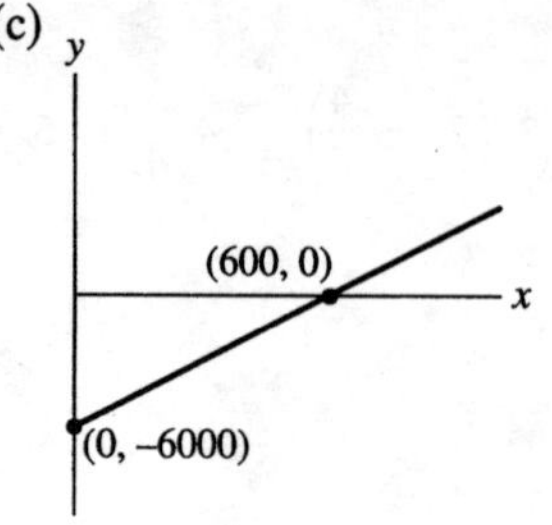

35. (a) A: $y = .1x + 50$; B: $y = .2x + 40$ (b) B (c) A (d) 100 miles **36.** (a) $y = -500x + 5000$
(b) \$3000 (c) 1992 **37.**

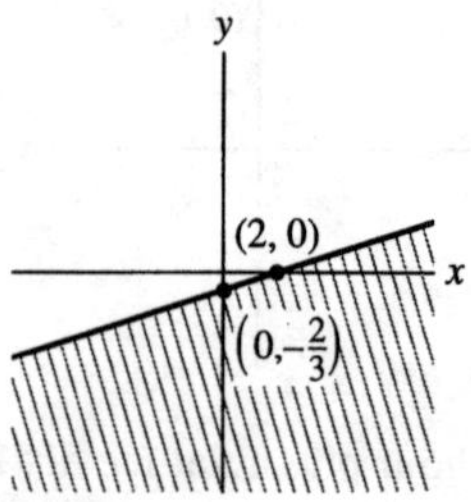

38. \$5000 **39.** $\begin{cases} y \le \frac{5}{4}x + 5 \\ y \le -\frac{2}{5}x + 2 \\ y \ge \frac{3}{5}x - 3 \\ y \ge -\frac{5}{2}x - 5 \end{cases}$ **40.** $\begin{cases} y \le -\frac{2}{3}x + 2 \\ y \ge -3 \\ x \ge -2 \\ x \le 4 \end{cases}$

41. (a) $y = 5.310x + 193.829$ (b) About 273.5 thousand (c) 2000 **42.** (a) $y = .152x - 3.063$ (b) About 21
(c) About 165 grams

CHAPTER 7

Exercises 2.1

1. $\xrightarrow{2[1]} \begin{cases} x - 6y = 4 \\ 5x + 4y = 1 \end{cases}$ **3.** $\xrightarrow{[2]+5[1]} \begin{cases} x + 2y = 3 \\ 14y = 16 \end{cases}$ **5.** $\xrightarrow{[3]+(-4)[1]} \begin{cases} x - 2y + z = 0 \\ y - 2z = 4 \\ 9y - z = 5 \end{cases}$

7. $\xrightarrow{[1]+\frac{1}{2}[2]} \left[\begin{array}{cc|c} 1 & 0 & 5 \\ 0 & 1 & 4 \end{array}\right]$ **9.** $[2] + 2[1]$ **11.** $[1] + (-2)[2]$ **13.** Interchange rows 1 and 2 or rows 1 and 3

15. $[1] + (-3)[3]$ **17.** $x = -1, y = 1$ **19.** $x = -\frac{8}{7}, y = -\frac{9}{7}, z = -\frac{3}{7}$ **21.** $x = -1, y = 1$ **23.** $x = 1, y = 2, z = -1$

25. $x = -2.5, y = 15$ **27.** $x = 1, y = -6, z = 2$ **29.** $x = -1, y = -2, z = 5$ **31.** (b) **33.** 137 adults, 463 children

35. $x = \$25{,}000, y = \$50{,}000, z = \$25{,}000$ **41.** $x = 3, y = -3, z = 2, w = -4$

Exercises 2.2

1. $\begin{bmatrix} 1 & -2 & 3 \\ 0 & 13 & -8 \end{bmatrix}$ **3.** $\begin{bmatrix} 9 & -1 & 0 & -7 \\ -\frac{1}{2} & \frac{1}{2} & 1 & 3 \\ 5 & -1 & 0 & -3 \end{bmatrix}$ **5.** $\begin{bmatrix} 1 & \frac{3}{2} \\ 0 & -9 \\ 0 & \frac{7}{2} \end{bmatrix}$ **7.** $\begin{bmatrix} 4 & 3 & 0 \\ 1 & 1 & 0 \\ \frac{1}{6} & \frac{1}{2} & 1 \end{bmatrix}$ **9.** y = any value, $x = 3 + 2y$

11. $x = 1, y = 2$ **13.** No solution **15.** z = any value, $x = -6 - z, y = 5$ **17.** No solution

19. z = any value, w = any value, $x = 2z + w, y = 5 - 3w$ **21.** $x = 5, y = 7$
23. Possible answers: $z = 0, x = -13, y = 9$; $z = 1, x = -8, y = 6$; $z = 2, x = -3, y = 3$
25. Possible answers: $y = 0, x = 23, z = 5$; $y = 1, x = 16, z = 5$; $y = 2, x = 9, z = 5$
27. x = food 1, y = food 2, z = food 3: z = any amount, $x = 300 - z, y = 100 - z$ $(0 \le z \le 100)$
29. 6 floral squares, the other 90 any mix of solid green and solid blue. **31.** $x = \pm 1, y = \pm 2, z = \pm 3$
33. No solution if $k \neq -12$. Infinitely many if $k = -12$. **35.** None **37.** (5, 6)
39. z = any value, $x = 11 - 21z, y = -6 + 12z$ **41.** No solution

Exercises 2.3

1. 2×3 **3.** 1×3, row matrix **5.** 1×1, square matrix **7.** $\begin{bmatrix} 9 & 3 \\ 7 & -1 \end{bmatrix}$ **9.** $\begin{bmatrix} 1 & 3 \\ 1 & 2 \\ 4 & -2 \end{bmatrix}$ **11.** [11] **13.** [10]

15. Yes; 3×5 **17.** No **19.** Yes; 3×1 **21.** $\begin{bmatrix} 6 & 17 \\ 6 & 10 \end{bmatrix}$ **23.** $\begin{bmatrix} 21 \\ -4 \\ 8 \end{bmatrix}$ **25.** $\begin{bmatrix} 5 & 6 \\ 7 & 8 \end{bmatrix}$ **27.** $\begin{bmatrix} .48 & .39 \\ .52 & .61 \end{bmatrix}$

29. $\begin{bmatrix} 25 & 17 & 2 \\ 3 & -1 & 2 \\ 1 & 1 & 4 \end{bmatrix}$ **31.** $\begin{cases} 2x + 3y = 6 \\ 4x + 5y = 7 \end{cases}$ **33.** $\begin{cases} x + 2y + 3z = 10 \\ 4x + 5y + 6z = 11 \\ 7x + 8y + 9z = 12 \end{cases}$ **35.** $\begin{bmatrix} 3 & 2 \\ 7 & -1 \end{bmatrix}\begin{bmatrix} x \\ y \end{bmatrix} = \begin{bmatrix} -1 \\ 2 \end{bmatrix}$

37. $\begin{bmatrix} 1 & -2 & 3 \\ 0 & 1 & 1 \\ 0 & 0 & 1 \end{bmatrix}\begin{bmatrix} x \\ y \\ z \end{bmatrix} = \begin{bmatrix} 5 \\ 6 \\ 2 \end{bmatrix}$ **43. a.** $\begin{bmatrix} 340 \\ 265 \end{bmatrix}$ **b.** Mike's clothes cost \$340; Don's clothes cost \$265.

45. a. I: 2.75, II: 2, III: 1.3 **b.** A: 74, B: 112, C: 128, D: 64, F: 22 **47.** 10,100 voting Deomcratic, 7900 voting Republican
49. Carpenters: \$1000, bricklayers: \$1050, plumbers: \$600 **51. a.** [162 150 143], number of units of each nutrient consumed at breakfast **b.** [186 200 239], number of units of each nutrient consumed at lunch
c. [288 300 344], number of units of each nutrient consumed at dinner **d.** [5 8], total number of ounces of each food that Mikey eats during a day **e.** [636 650 726], number of units of each nutrient consumed per day

53. a. $\begin{bmatrix} 720 \\ 646 \end{bmatrix}$ **b.** \$720 **55.** $x = 3, y = 4, z = 5$ **57.** $\begin{bmatrix} 3 & -2 & 1 \\ -5 & 6 & 7 \end{bmatrix}$

59. $\begin{bmatrix} 27.9 & 130.6 & -69.88 \\ 106.75 & -149.44 & 26.1 \\ -47.5 & 336.2 & -18.7 \end{bmatrix}$ **61.** $\begin{bmatrix} -69.14 & 147.9 & -43.26 \\ 158.05 & -3.69 & 33.46 \\ -176.1 & 259.5 & 59.3 \end{bmatrix}$ **63.** 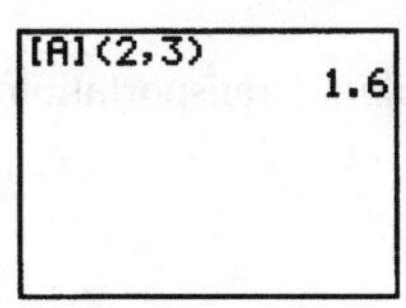

65. $\begin{bmatrix} 160.16 & -26.7 & 4 \\ 2.7 & 150.85 & -53 \\ 187.4 & -35.5 & 48.6 \end{bmatrix}$ **67.** They match.

Exercises 2.4

1. $x = 2, y = 0$ **3.** $\begin{bmatrix} 1 & -2 \\ -3 & 7 \end{bmatrix}$ **5.** $\begin{bmatrix} 1 & -1 \\ -\frac{5}{2} & 3 \end{bmatrix}$ **7.** $\begin{bmatrix} 1.6 & -.4 \\ -.6 & 1.4 \end{bmatrix}$ **9.** $\begin{bmatrix} 1 \\ 3 \end{bmatrix}$ **11.** $x = 4, y = -\frac{1}{2}$

13. $x = 32, y = -6$ **15. a.** $\begin{bmatrix} .8 & .3 \\ .2 & .7 \end{bmatrix}\begin{bmatrix} x \\ y \end{bmatrix} = \begin{bmatrix} m \\ s \end{bmatrix}$ **b.** $\begin{bmatrix} x \\ y \end{bmatrix} = \begin{bmatrix} 1.4 & -.6 \\ -.4 & 1.6 \end{bmatrix}\begin{bmatrix} m \\ s \end{bmatrix}$ **c.** 110,000 married; 40,000 single

d. 130,000 married; 20,000 single **17. a.** $\begin{bmatrix} .7 & .1 \\ .3 & .9 \end{bmatrix}\begin{bmatrix} x \\ y \end{bmatrix} = \begin{bmatrix} u \\ v \end{bmatrix}$ **b.** $\begin{bmatrix} x \\ y \end{bmatrix} = \begin{bmatrix} \frac{3}{2} & -\frac{1}{6} \\ -\frac{1}{2} & \frac{7}{6} \end{bmatrix}\begin{bmatrix} u \\ v \end{bmatrix}$ **c.** 8500; 4500

19. $x = 9, y = -2, z = -2$ **21.** $x = 1, y = 5, z = -4, w = 9$ **25. a.** $\begin{bmatrix} 1 & 2 \\ .9 & 0 \end{bmatrix}\begin{bmatrix} x \\ y \end{bmatrix} = \begin{bmatrix} a \\ b \end{bmatrix}$

b. After 1 year: 1,170,000 in group I, 405,000 in group II. After 2 years: 1,980,000 in group I, 1,053,000 in group II.

c. 700,000 in group I, 55,000 in group II. **27.** $\begin{bmatrix} -\frac{10}{73} & \frac{75}{292} \\ \frac{25}{73} & -\frac{5}{292} \end{bmatrix}$ **29.** $\begin{bmatrix} \frac{1020}{8887} & \frac{2910}{8887} & -\frac{500}{8887} \\ \frac{3050}{8887} & \frac{860}{8887} & \frac{1990}{8887} \\ \frac{125}{8887} & \frac{618}{8887} & \frac{810}{8887} \end{bmatrix}$ **31.** $x = -\frac{4}{5}, y = \frac{28}{5}, z = 5$

33. $x = 0, y = 2, z = 0, w = 2$ **35.** Displays ERR:SINGULAR MAT.

Exercises 2.5

1. $\begin{bmatrix} -2 & 3 \\ 5 & -7 \end{bmatrix}$ **3.** $\begin{bmatrix} \frac{1}{19} & \frac{3}{19} \\ \frac{3}{76} & -\frac{5}{38} \end{bmatrix}$ **5.** No inverse **7.** $\begin{bmatrix} -1 & 2 & -4 \\ 1 & -1 & 3 \\ 0 & 0 & 1 \end{bmatrix}$ **9.** No inverse **11.** $\begin{bmatrix} -5 & 6 & 0 & 0 \\ 1 & -1 & 0 & 0 \\ 0 & 0 & -\frac{1}{46} & \frac{1}{46} \\ 0 & 0 & \frac{25}{46} & -\frac{1}{23} \end{bmatrix}$

13. $x = 2, y = -3, z = 2$ **15.** $x = 4, y = -4, z = 3, w = -1$ **17.** Either no solutions or infinitely many solutions.

19. $\begin{bmatrix} -3 & 5 \\ 10 & -16 \end{bmatrix}$ **21.** $\begin{bmatrix} 2 & -1 \\ -7 & 4 \end{bmatrix}$ **23.** $\begin{bmatrix} -2.4 & 2.2 & 1.8 \\ .8 & -.4 & -.6 \\ .6 & -.8 & -.2 \end{bmatrix}$

Exercises 2.6

1. Coal: \$8.84 billion, steel: \$3.725 billion, electricity: \$9.895 billion
3. Computers: \$354 million, semiconductors: \$172 million **7.** Plastics: \$955,000, industrial equipment: \$590,000

9. Manufacturing: \$398 million, transportation: \$313 million, agricultural:\$452 million **11.** $\begin{bmatrix} 10.25 \\ 13.82 \\ 8.65 \end{bmatrix}$

Supplementary Exercises

1. $\begin{bmatrix} 1 & -2 & \frac{1}{3} \\ 0 & 8 & \frac{16}{3} \end{bmatrix}$ **2.** $\xrightarrow{[3]+(-6)[2]} \begin{bmatrix} 1 & 0 & 1 \\ 2 & 1 & 0 \\ -12 & 0 & 7 \end{bmatrix}$ **3.** $x = 4, y = 5$ **4.** $x = 50, y = 2, z = -12$

5. $x = -1,\ y = \frac{2}{3},\ z = \frac{1}{3}$ **6.** No solution **7.** $z =$ any value, $x = 1 - 3z, y = 4z, w = 5$ **8.** $x = 7, y = 3$ **9.** $\begin{bmatrix} 5 \\ 3 \\ 7 \end{bmatrix}$

10. $\begin{bmatrix} 6 & 17 \\ 12 & 26 \end{bmatrix}$ **11.** $x = -2, y = 3$ **12.** (a) $x = 13, y = 23, z = 19$ (b) $x = -4, y = 13, z = 14$ **13.** $\begin{bmatrix} -1 & 3 \\ \frac{1}{2} & -1 \end{bmatrix}$

14. $\begin{bmatrix} 5 & -1 & -1 \\ -3 & 1 & 0 \\ -1 & 0 & 1 \end{bmatrix}$ **15.** Corn: 500 acres, wheat: 0 acres, soybeans: 500 acres **16.** (a) A: 9400, 8980; B:7300, 7510 (b) A: 10,857, 12,082; B: 6571, 5959 **17.** (d) **18.** (c) **19.** Industry I: 20; industry II: 20

CHAPTER 8

Exercises 3.1

1. Yes **3.** No **5.** (a)

	A	*B*	*Truck capacity*
Volume	4 cubic feet	3 cubic feet	300 cubic feet
Weight	100 pounds	200 pounds	10,000 pounds
Earnings	$13	$9	

(b) $4x + 3y \le 300$; $100x + 20y \le 10{,}000$ (c) $y \le 2x, x \ge 0, y \ge 0$ (d) $13x + 9y$ (e)

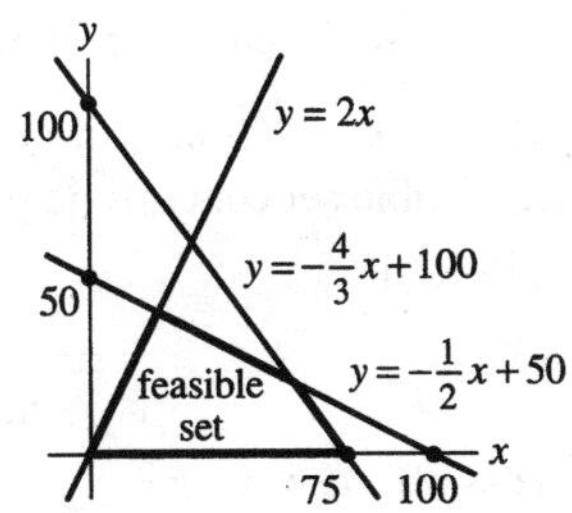

7. (a)

	Essay questions	*Short-answer questions*	*Available*
Time to answer	10 minutes	2 minutes	90 minutes
Quantity	10	50	
Required	3	10	
Worth	20 points	5 points	

(b) $10x + 2y \le 90$
(c) $x \ge 3, x \le 10, y \ge 10, y \le 50$

(d) $20x + 5y$ (e)

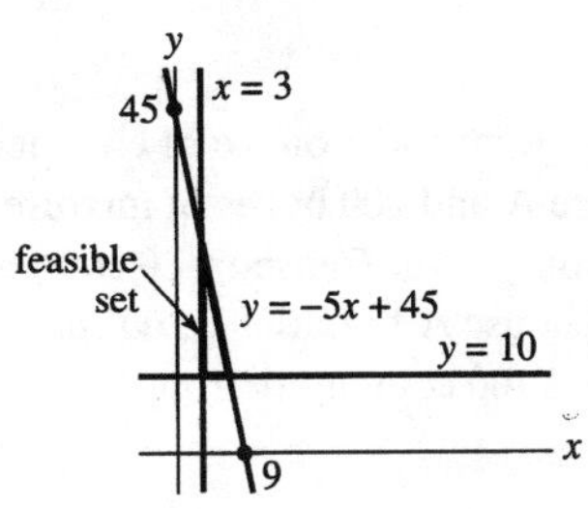

9. (a)

	Alfalfa	*Corn*	*Requirements*
Protein	.13 pound	.065 pound	4550 pounds
TDN	.48 pound	.96 pound	26,880 pounds
Vitamin A	2.16 IUs	0	43,200 IUs
Cost/lb	$0.01	$0.016	

(b) $.13x + .065y \geq 4550$; $.48x + .96y \geq 26{,}880$; $2.16x \geq 43{,}200$; $y \geq 0$ (c)

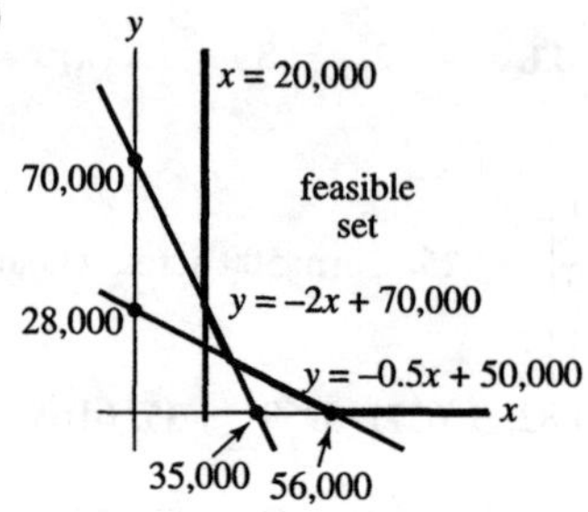

(d) $.01x + .016y$

Exercises 3.2

1. (20, 0) **3.** (6, 0) **5.** (0, 5) **7.** (3, 3) **9.** (0, 7) **11.** (2, 1) **13.** Ship 75 crates of cargo A and no crates of cargo B. **15.** Answer 3 essay questions and 30 short-answer questions. **17.** Make 16 chairs and no sofas. **19.** The minimum value is 30 and occurs at (2, 6). **21.** The maximum value is 49 and occurs at (2, 9). **23.** The maximum value is 6600 and occurs at (12, 36). **25.** The minimum value is 40 and occurs at (4, 3). **27.** Produce 9 hockey games and 8 soccer games each day. **29.** Make 80 homes of the first type and 60 homes of the second type. **31.** Make 400 cans of Fruit Delight and 500 cans of Heavenly Punch.

33. The farmer should plant $83\frac{1}{3}$ acres of oats and $16\frac{2}{3}$ acres of corn to make a profit of \$6933.33.

35. (a) Make 8 of item I_1 and 4 of item I_2. (b) \$88 (c) 40 ounces of M_1, 20 ounces of M_2, and 28 ounces of M_3 are used. (d) Make 10 of item I_1 and 0 of item I_2. **37.** The feasible set contains no points.

Exercises 3.3

1. (a) $y = -\frac{3}{2}x + \frac{c}{14}$ (b) Up (c) B **3.** Possible answer: $5x + y$ **5.** Possible answer: $2x + y$

7. Possible answer: $x + 5y$ **9.** Possible answer: $2x + 3y$ **11.** C **13.** D **15.** D **17.** C **19.** $\frac{1}{4} \leq k \leq 3$

21. Feed 1 can of brand A and 3 cans of brand B. **23.** Ship 35 crates of oranges, 35 crates of grapefruits, and 30 crates of avocados. **25.** In Detroit make 100 cars and 300 trucks. In Cleveland make 500 cars and 0 trucks. **27.** Produce 90,000 gallons of gasoline, 5000 gallons of jet fuel, and 5000 gallons of diesel fuel. **29.** Buy 9 high-capacity trucks and 21 low-capacity trucks.

Supplementary Exercises

1. Use 10 type A planes and 3 type B planes. **2.** Use 2 ounces of wheat germ and 1 ounce of enriched oat flour **3.** Produce 9 hardtops and 16 sports cars. **4.** Make 500 boxes of mixture A and 200 boxes of mixture B. **5.** Publish 60 elementary books, 8 intermediate books, and 4 advanced books. **6.** Transport 80 computers from Rochester and 45 computers from Queens. **7.** Transport no computers from warehouse A to outlet I, 200 computers from warehouse A to outlet II, 200 computers from warehouse B to outlet I, and 100 computers from warehouse B to outlet II.

Exercises 3.2

19.

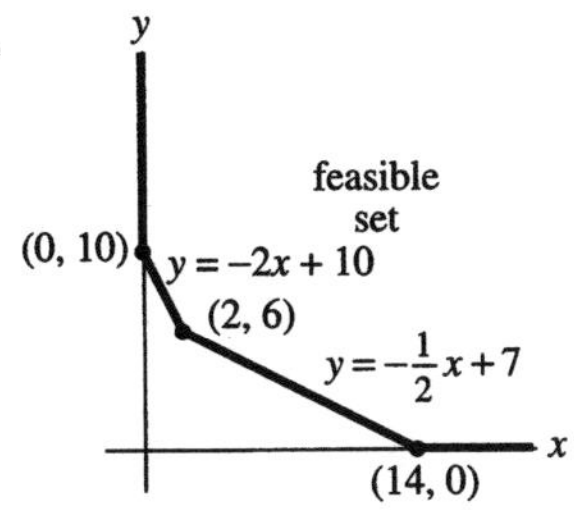

21.

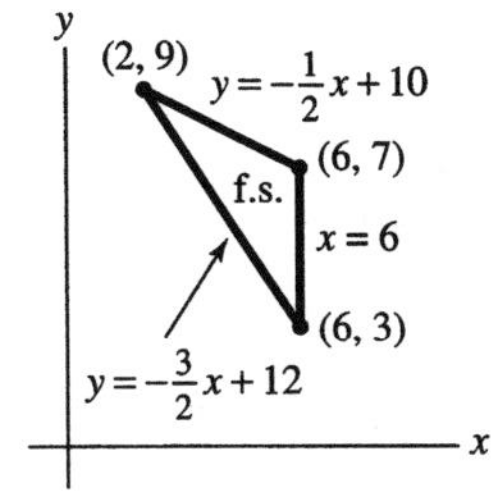

23.

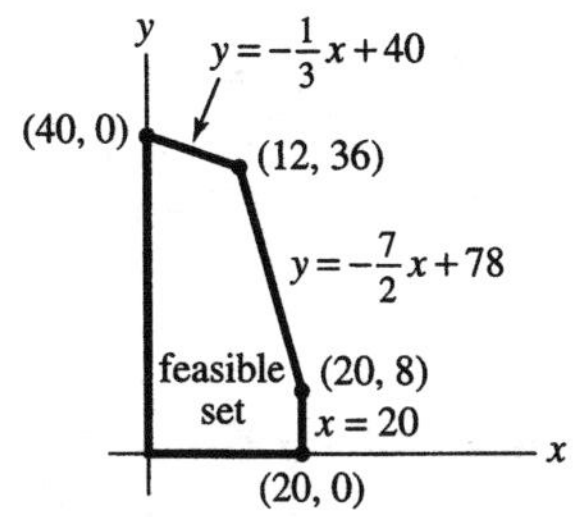

25.

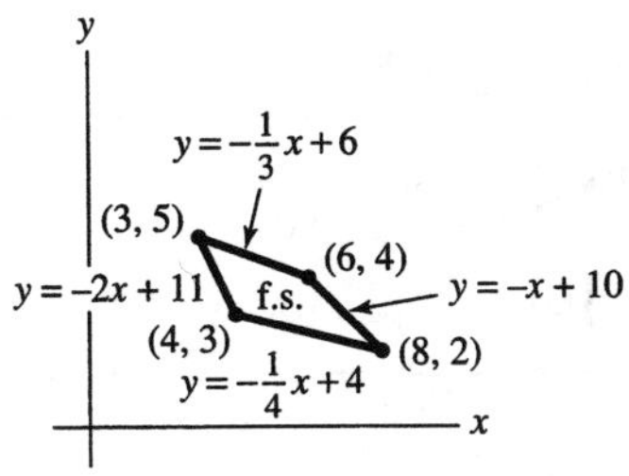

27.

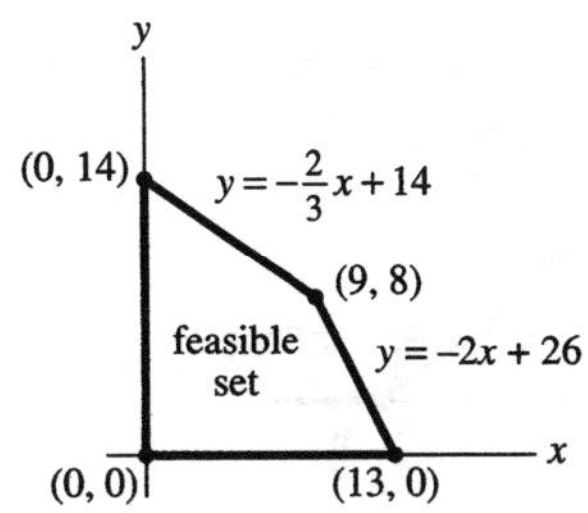

29.

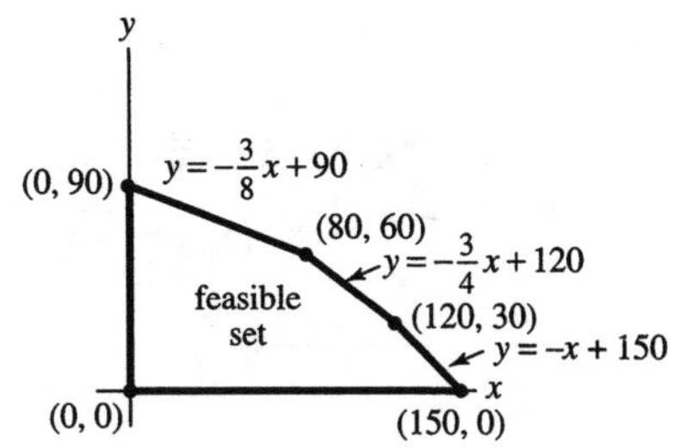

31.

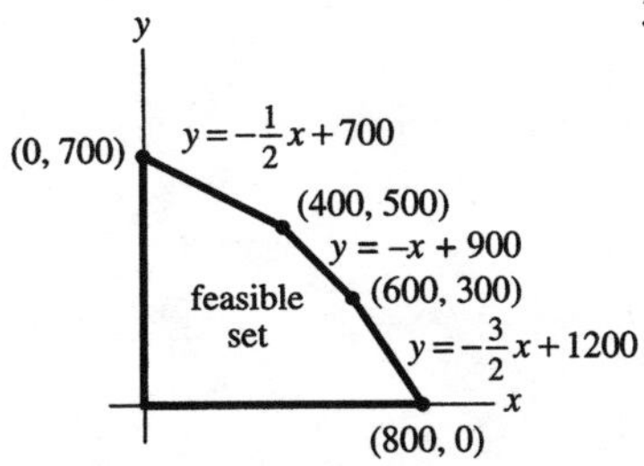

33.

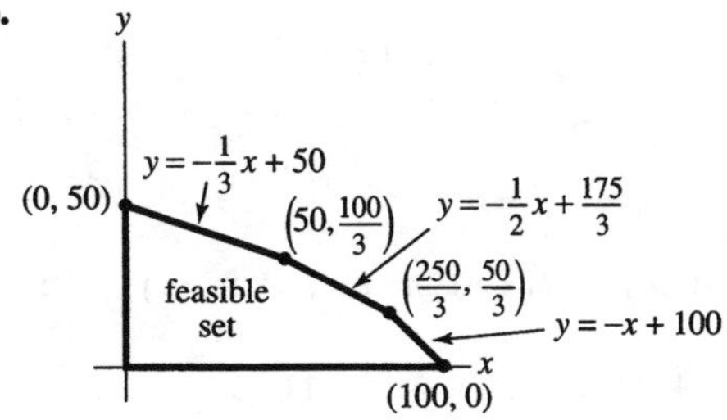

35.

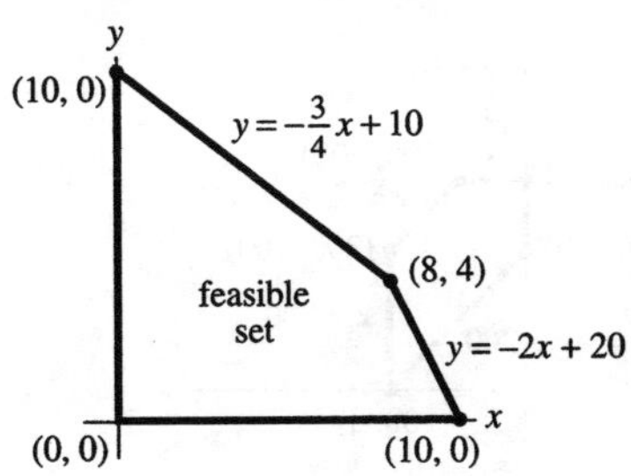

Exercises 3.3

21.

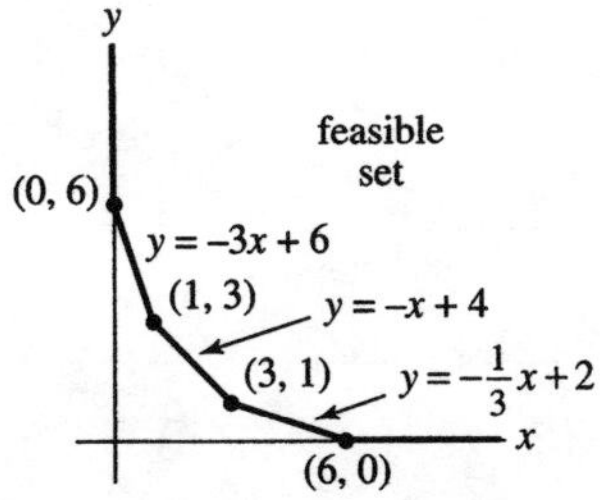

23.

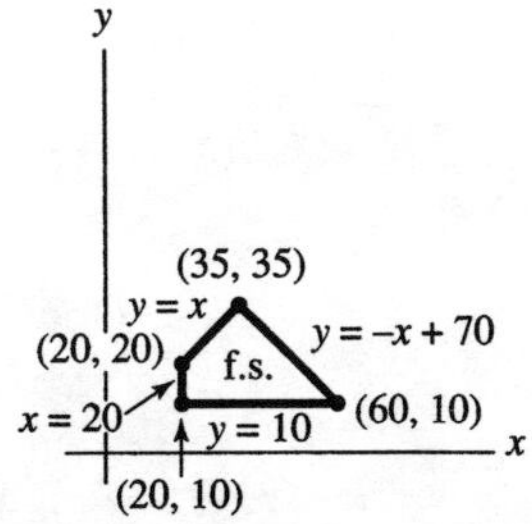

25.

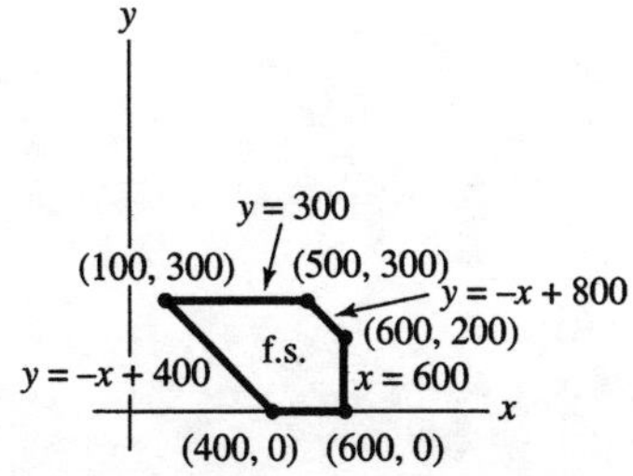

27.

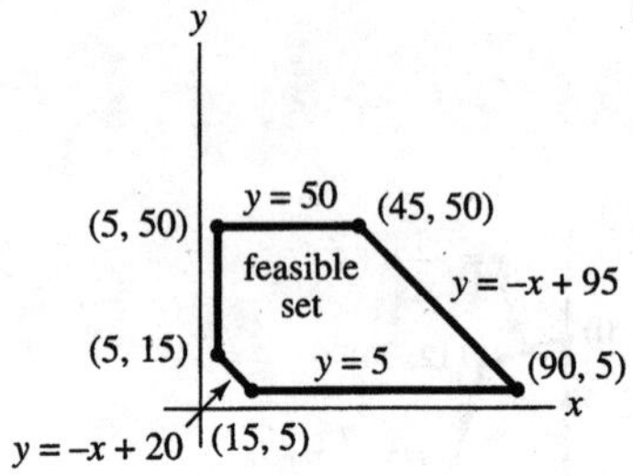

29.

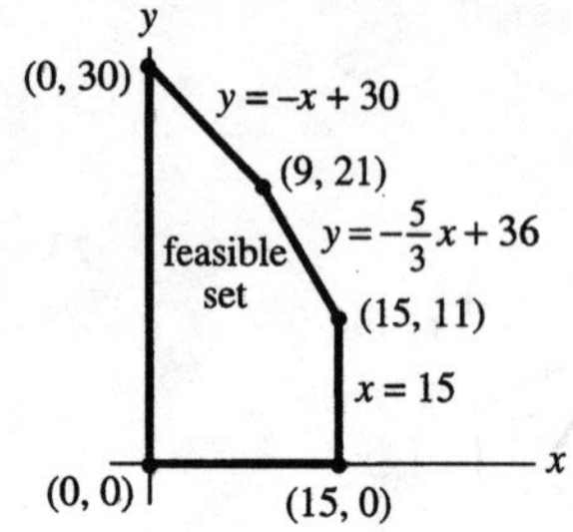

Supplementary Exercises

1.

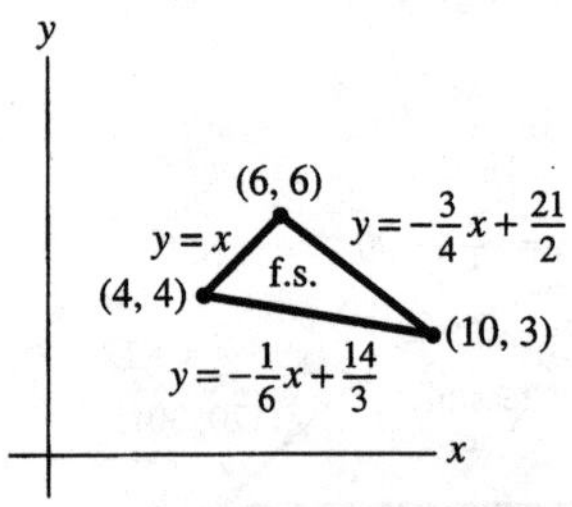

2.

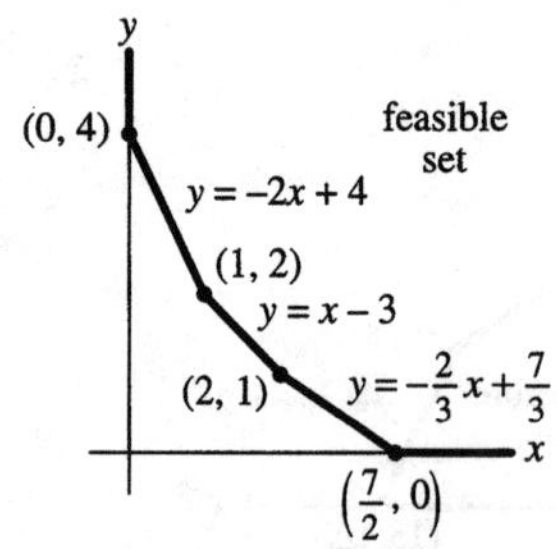

3.

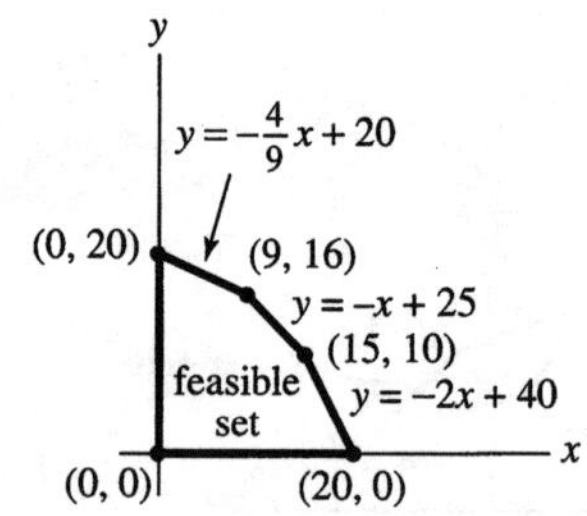

4.

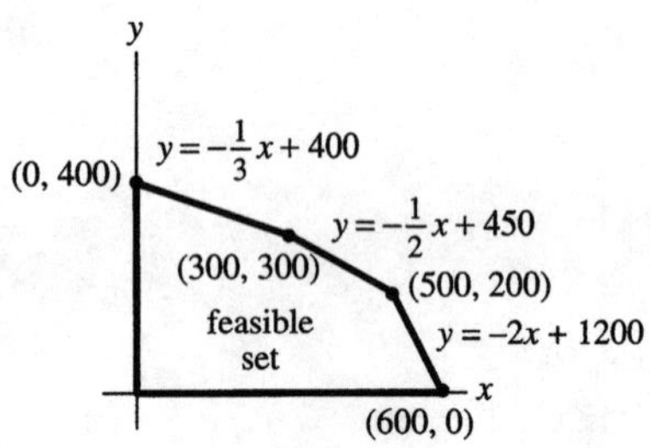

5.

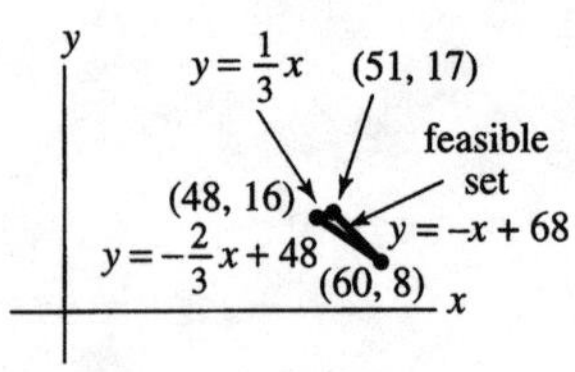

6.

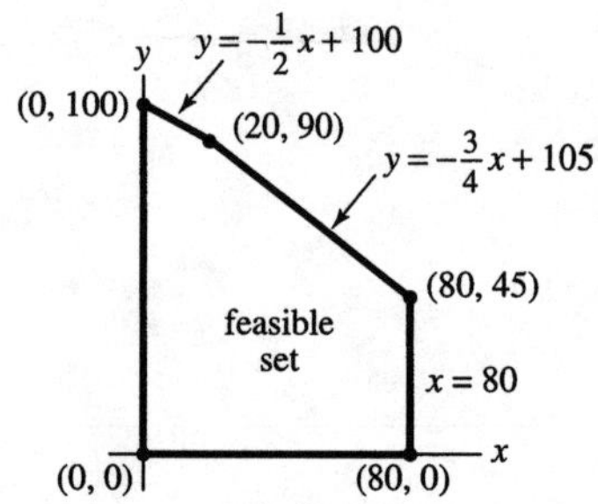

7.

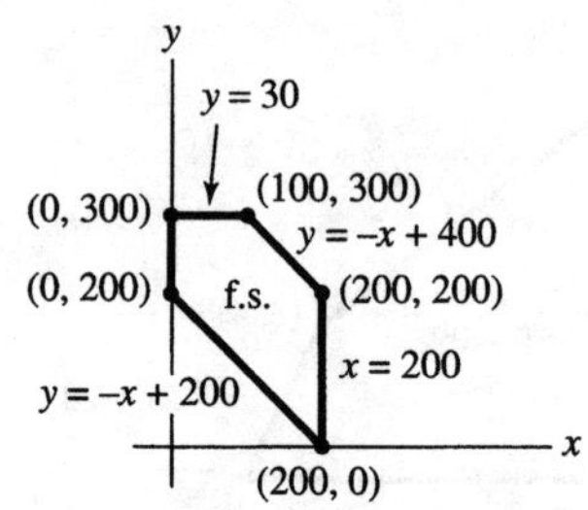

CHAPTER 9

Exercises 10.1

1. (a) $i = .01, n = 24$ (b) $i = .02, n = 20$ (c) $i = .05, n = 40$ **3.** (a) $i = .06, n = 4, P = \$500, F = \$631.24$ (b) $i = .005, n = 120, P = \$800, F = \$1455.52$ (c) $i = .02, n = 19, P = \$6177.88, F = \$9000$ **5.** \$1127.16 **7.** \$22,396.57 **9.** \$7180.08; 1180.08 **11.** (a) \$3548.74 (b) \$451.26 (c)

Month	Interest	Balance
0		\$3548.74
1	\$17.74	\$3566.48
2	\$17.84	\$3584.32
3	\$17.92	\$3602.24

13. \$12,824.32 **15.** \$8874.49 **17.** (a) 6% (b) \$5.05; \$1015.08 (c) \$5.61; \$1127.16 **19.** \$1700 in 9 years **21.** Worse **23.** \$8874.49 **25.** (a) $r = .04, n = \frac{1}{2}, P = \$500, A = \$510$ (b) $r = .05, n = 2, P = \$500, A = \$550$ **27.** \$1150 **29.** \$2500 **31.** 4.08% **33.** 20 years **35.** $P = \dfrac{A}{1 + nr}$ **37.** (a) **39.** \$104.06; 4.06% **41.** 4.04% **43.** 6.17% **45.** $r + \dfrac{r^2}{4}$ **47.** $\left(1 + \dfrac{r}{n}\right)^n - 1$ **49.** \$10,252.51, \$10,511.40, \$10,776.83; 30; 53 **51.** 11.75

Exercises 10.2

1. (a) $i = .005, n = 120, R = \$50, F = \$8193.97$ (b) $i = .02, n = 20, R = \$2675.19, F = \$65,000$ **3.** \$6977 **5.** \$4698.97 **7.** (a) \$27,048.92 (b) \$3048.92 (c)

Month	Interest	Balance
1		500
2	2.50	1002.50
3	5.01	1507.51

9. \$305.06; \$10982.16, \$1017.84 **11.** \$200 each month; \$343.75 **13.** \$24,649.92 **15.** 877.91 **17.** \$1000 at the end of each month **19.** \$7239.32 **21.** \$17,584.62 **23.** (a) **25.** \$24,000 **27.** \$4339.35 **29.** \$59,189.85 **31.** (a) \$1 (b) $s_{\overline{n}|i} \cdot i$ **35.** \$3623.52 **37.** (a) \$1269.67 (b) \$65,134.88 **39.** \$1000, \$2050, \$3152.50; 19; 26 **41.** After 33 weeks

Exercises 10.3

1. \$193.33 **3.** \$11,469.92 **5.** (a) \$583.31 (b) \$16.69 (c) \$58,314.31 (d) \$26,973.02 (e) \$4188.65 (f) \$269.73 **7.** (a) \$265.71 (b) \$9565.56 (c) \$1565.56 (d) \$5644.58 (e) \$2990.59 (f) \$534.53

(g)

Payment number	Amount	Interest	Applied to principal	Unpaid balance
1	\$265.71	\$80.00	\$185.71	\$7814.29
2	265.71	78.14	187.57	7626.72
3	265.71	76.27	189.44	7437.28
4	265.71	74.37	191.34	7245.94

9. \$8950 **11.** \$14,042.36

13.

Payment number	Amount	Interest	Applied to principal	Unpaid balance
1	\$256.28	\$10.00	\$246.28	\$753.72
2	256.28	7.54	248.74	504.98
3	256.28	5.05	251.23	253.74
4	256.28	2.54	253.74	0.00

15. \$804.62 **17.** \$119,161.62 **19.** \$1196.68 **21.** (a) **23.** (a) \$183,927.96 (b) \$24,989.92 (c) \$105,266.63 **25.** \$2.5 billion **27.** Yes **29.** \$2105.33, \$2021.12, \$1936.28, 24 months. **31.** \$4258.21; after 46 months **33.** 260

Supplementary Exercises

1. (b) **2.** \$488.16 **3.** \$49,712.75 **4.** \$53.79 **5.** 10% compounded annually **6.** \$13,954.01 **7.** (a) \$2400.34 (b) \$167,304.68 **8.** \$43,665.52 **9.** \$27,481.64 **10.** \$13,050.08 **11.** \$211.37 **12.** \$12,050.34 **13.** \$782.92 **14.** \$6872.11 **15.** \$100,451.50 **16.** 139,401.04 **17.** Investment A **18.** Yes, it is a bargain since the present value is \$879.57. **19.** 10.25% **20.** 19.56% **21.** \$146,861.85

22.

Payment number	Amount	Interest	Applied to principal	Unpaid balance
1	\$304.22	\$50.00	\$254.22	\$9745.78
2	304.22	48.73	255.49	9490.29
3	304.22	47.45	256.77	9233.52
4	304.22	46.17	258.05	8975.47
5	304.22	44.88	259.34	8716.13
6	304.22	43.58	260.64	8455.49

23. \$151,843.34 **24.** \$5799.84 **25.** \$1206.93

CHAPTER 11

Exercises 11.1